Lecture Notes in Chemistry

Volume 111

The series Lecture Notes in Chemistry (LNC), reports new developments in chemistry and molecular science - quickly and informally, but with a high quality and the explicit aim to summarize and communicate current knowledge for teaching and training purposes. Books published in this series are conceived as bridging material between advanced graduate textbooks and the forefront of research. They will serve the following purposes:

- provide an accessible introduction to the field to postgraduate students and nonspecialist researchers from related areas,
- provide a source of advanced teaching material for specialized seminars, courses and schools, and
- be readily accessible in print and online.

The series covers all established fields of chemistry such as analytical chemistry, organic chemistry, inorganic chemistry, physical chemistry including electrochemistry, theoretical and computational chemistry, industrial chemistry, and catalysis. It is also a particularly suitable forum for volumes addressing the interfaces of chemistry with other disciplines, such as biology, medicine, physics, engineering, materials science including polymer and nanoscience, or earth and environmental science.

Both authored and edited volumes will be considered for publication. Edited volumes should however consist of a very limited number of contributions only. Proceedings will not be considered for LNC.

The year 2010 marked the relaunch of LNC.

Natalja Vogt · Jean Demaison

Equilibrium Structure of Free Molecules

Theory, Experiment, and Data Analysis

Springer

Natalja Vogt
Faculty of Sciences
University of Ulm
Ulm, Germany

Jean Demaison
Physique des Lasers Atomes et Molécules
University of Lille
Villeneuve d'Ascq, France

ISSN 0342-4901 ISSN 2192-6603 (electronic)
Lecture Notes in Chemistry
ISBN 978-3-031-36044-2 ISBN 978-3-031-36045-9 (eBook)
https://doi.org/10.1007/978-3-031-36045-9

This Springer imprint is published by the registered company Springer Nature Switzerland AG
The registered company address is: Gewerbestrasse 11, 6330 Cham, Switzerland

Foreword

The atoms ... are driven abroad and vexed
by blow on blow, even from all time of old.
They thus at last, after attempting all
the kinds of motion and conjoining, come
into those great arrangements out of which
this sum of things established is created.
Lucretius, De Rerum Natura, Book I.

The chemical bond is the core concept of chemistry. Men have explored for centuries an astonishing number of natural and artificial compounds in the quest to understand how elements are combined and how to produce food, materials, and tools. However, no progress on the chemical constitution of matter can be achieved on purely empirical methods. The mathematization of chemistry and the construction of theoretical models were thus a necessary step to jump from observation to prediction. The construction of chemical models was slow and sometimes deemed irresoluble, but it eventually resulted in the interpretation of macroscopic phenomena like reaction energetics, chemical kinetics, or electrochemistry. On the other hand, the comprehension of the microscopic world was a stumbling block which required the new quantum paradigm. Following ninety years of experimental verification, quantum mechanics is now a solid theoretical description of the natural phenomena at the chemical scale, i.e., the electromagnetism of electrons and nuclei at moderate energies. The consequences of this quantum mechanics revolution cannot be overemphasized: Following the understanding of chemistry scientists could finally explain life.

Quantum mechanics is naturally intermingled with spectroscopy since its beginnings. The interaction of light and matter quickly confirmed the quantum hypotheses, and atoms and molecules could be described in terms of dynamic electronic clouds around nearly static nuclei. Spectroscopy has revealed also the internal motions of nuclei and provided structural data with astounding accuracy. As a result, macroscopic phenomena can now be predicted from an atomic point of view, advancing the promise of a totally predictive chemistry. Certainly, quantum mechanical equations are not easy. As a reflection of the complexities of nature, the Schrödinger equation can only be solved analytically for the hydrogen atom, all other chemical species relying on numerical methods. However, in an ironical twist of times, the invention of electronic devices made possible by quantum mechanics has now permitted to solve the Schrödinger equation using computers.

The present book examines different topics related to the molecular constitution of matter, with focus in small- and middle-size free molecules that can be described in high detail through computations and high-resolution spectroscopic and electron diffraction experiments. Starting with a description of computational chemistry methods, it later explains the influence of vibrational and rovibrational effects on experimental molecular structure and the determination of experimental and semiexperimental equilibrium structures. Fluxional molecules, characterized by large amplitude motions, are discussed separately. The volume covers different gas-phase techniques, both high-resolution spectroscopic and electron diffraction, exploiting the synergy of theory and experiment. It also presents comprehensive databases containing equilibrium structures of free molecules.

The present volume puts the large research experience of the authors to the service of readers interested in molecular structure, high-resolution spectroscopy, and electron diffraction. It will thus be a very valuable addition to the literature and inspiration for future endeavors.

Alberto Lesarri
Universidad de Valladolid
Valladolid, Spain

Preface

A considerable amount of work in chemistry has been devoted to the determination and explanation of molecular structures and its resulting application to the dynamics of molecules and their chemical reactions under laboratory conditions, in planetary atmospheres and in interstellar clouds.

Many reviews and several books have already been published on the subject of structure determinations (including by us). However, the field of molecular structure has developed so extensively that it is not possible to give a comprehensive coverage of all its aspects in a single volume.

The main objective of the present book is to discuss some aspects that were not (or only briefly) reviewed previously. Nevertheless, this book is written as a stand-alone book covering all fields of structure determinations, namely theory, experiment, structural analysis, and systematization of new structural data. Particular emphasis is put on the more recent developments, especially on the quantum-chemical methods, the determination of equilibrium molecular structure by experimental methods, and the structure of floppy molecules. The different ways to estimate a reliable accuracy are also discussed in great details.

This book is intended for advanced undergraduate and graduate students, as well as for research workers interested in molecular structure.

Ulm, Germany Natalja Vogt
Lille, France Jean Demaison

Acknowledgements We are very grateful to Dr. Zachary Evenson and Dr. Antje Endemann from the Springer Nature editorial office as well as to Ravi Vengadachalam and Jegadeeswari Diravidamani from the book production service for their support and efforts. Our special thanks go to Dr. Jürgen Vogt and Marie Demaison for their assistance.

Contents

Abbreviations

BOB	Breakdown of the Born–Oppenheimer approximation
CCSD(T)	Coupled-cluster methods including single and double excitations with a perturbational estimate of the effects of connected triple excitations
CEI	Coulomb explosion imaging
CI	Configuration interaction
DFT	Density functional theory
DPF	Direct potential fit
EMO	Expanded Morse Oscillator
fs-DFWM	Time-resolved femtosecond degenerate four-wave mixing
FTIR	Fourier transform infrared spectroscopy
FTMW	Fourier transform microwave spectroscopy
GED	Gas-phase electron diffraction
HF	Hartree-Fock
HR	High resolution (spectroscopy)
IR	Infrared (spectroscopy)
LAM	Large-amplitude motion
LCNMR	Liquid crystal nuclear magnetic resonance spectroscopy
MB	Molecular beam
MLR	Morse long-range potential
MP2	Second-order Møller Plesset perturbation theory
MW	Microwave, millimeter-wave, and/or submillimeter-wave (spectroscopy)
NMR	Nuclear magnetic resonance spectroscopy
NOE	Nuclear Overhauser effect
PEF	Potential energy function
PES	Photoelectron spectroscopy
PES	Potential energy surface
QC	Quantum chemical (*ab initio* and DFT calculations)
Ra	Raman spectroscopy
RCS	Rotational coherence spectroscopy
SMM	Submillimeter-wave spectroscopy
TRFD	Time-resolved fluorescence depletion

Different Structures

r_e^{BO} Born–Oppenheimer equilibrium structure: corresponds to a minimum of the potential energy hypersurface defined within the Born–Oppenheimer approximation

r_e^{exp} Experimental equilibrium structure: obtained from a fit of the structural parameters to the experimental moments of inertia
or
Experimental equilibrium structure: obtained from a fit of the electron scattering intensities expressed in the terms of anharmonic potential energy function to the experimental intensities $sM(s)$

r_g Thermal average internuclear distance corresponding to the center of gravity of the $P(r)$ distribution, $P(r)$ being the probability distribution function for the internuclear distance r

r_a Effective internuclear distance corresponding to the center of gravity of the $P(r)/r$ distribution, $r_a = r_g - u^2/r_e$, u being the root-mean-square amplitude of vibration

r_e^{se} Semiexperimental equilibrium structure determined from a fit of experimental electron scattering intensities $sM(s)$ to their theoretical counterparts taking into account vibrational corrections to experimental internuclear distances, $\Delta r = r_g - r_e$, calculated with quantum-chemical harmonic and anharmonic (cubic) force fields
and/or
Semiexperimental equilibrium structure: determined from a fit of the structural parameters to the equilibrium moments of inertia, obtained from the experimental ground state rotational constants corrected by the rovibrational contribution calculated using an *ab initio* or DFT cubic force field

$r_z = r_{\alpha,0}$ Zero-point average structure: a temperature independent average structure corresponding to the average nuclear positions in the ground vibrational state

r_m Mass-dependent structure: obtained from a fit of the structural parameters to the mass-dependent moments of inertia $I_m = 2I_s - I_0$ where I_s is the moments of inertia calculated from the substitution coordinates r_s, and I_0 the experimental ground state moments of inertia

r_c Improvement of the r_m structure by using complementary sets of isotopologues

r_m^ρ Fit of the structural parameters to the moments of inertia $I_{m,g}^\rho = (2\rho_g - 1)I_g^0(i)$ with $\rho_g = I_g^s(1)/I_g^0(1)$ where 1 is for the parent isotopologue and $g = a, b, c$

$r_m^{(1)}$ Mass-dependent structure: fit of the structural parameters to the moments of inertia $I_0^g = I_m^g + c_g\sqrt{I_m^g}$ where I_m^g is an approximation of I_e^g and $g = a, b, c$

$r_m^{(2)}$	To the equation for the $r_m^{(1)}$ method, $d_g\left(\frac{m_1 m_2 \cdots m_n}{M}\right)^{1/(2n-2)}$ is added where n is the number of atoms, m_i their respective masses, and M the total mass of the molecule
r_0	Effective structure: obtained from a fit of the structural parameters to the ground state moments of inertia I_0 of a set of isotopologues
r_s	Substitution structures: derived from the Cartesian coordinates of the individual atoms that are calculated with the Kraitchman's equations from the difference of moments of inertia $\Delta I = I_0$ (isotopologue) $- I_0$ (parent)

Chapter 1
Introduction

Abstract The goal of the book is explained, and the content of different chapters is briefly described.

There are many ways to characterize a molecule. For instance, it is common to use its infrared spectrum and, in some cases, its rotational spectrum (in particular for radioastronomical applications). It is also possible to use many molecular characteristics such as the electric dipole moment, the melting and boiling points, the viscosity, etc.

However, one of the most useful piece of information is the three-dimensional geometrical structure because it is well established that the properties and behaviors of molecules are essentially defined by their structure. Therefore, for example, pharmaceutical and chemical companies generate computed structures before doing laboratory work.

Ideally, the structure of a molecule should be determined when it does not suffer external influences. It is indeed well verified that structures in the gas and solid states can be extremely different [1].

The equilibrium structure corresponds to the minimum of the potential energy hypersurface. Although the concept of molecule was rather well established at the beginning of the twentieth century, the determination of the molecular structure was permitted by three breakthroughs:

- The discovery of the diffraction of X-rays by crystals around 1912 [2] and of electrons by a gas in 1930 [3]. Debye et al. were the first to determine the structure of an isolated molecule (CCl_4) by diffraction of X-rays [4]. Nowadays, electron diffraction is one of the best methods to determine the structure of a molecule in gas phase. After their discovery, many molecules were studied in a short time and permitted to develop the concept of chemical bond [5, 6].
- The advent of rotational spectroscopy after the Second World War. It is generally considered that the determination of the rotational constants is the most precise method for obtaining the molecular geometry in gas phase because the rotational constants can be determined with a very high accuracy [7].

N. Vogt and J. Demaison, *Equilibrium Structure of Free Molecules*, Lecture Notes in Chemistry 111, https://doi.org/10.1007/978-3-031-36045-9_1

- Progress in quantum chemistry. The structure of simple molecules such as H_2^+ [8] and H_2 [9] was computed quite early (1927). However, the accurate computation of molecular structures was only possible at the end of the twentieth century thanks to the apparition of powerful computers and the development of efficient ab initio methods.

In the earlier studies, it was assumed that the molecule was rigid which limited the accuracy of the results because the vibrations may significantly affect the accuracy. The problem is that it was arduous to take into account the contribution of the vibrations because it required a great lot of difficult experimental studies. For this reason, up to the seventies, accurate structures were only determined for rather small molecules (three to four degrees of freedom). Then, the situation considerably changed with the advent of sophisticated ab initio methods and powerful computers. It is now relatively easy to determine the equilibrium structure of a molecule of about 30 atoms (up to 66 independent parameters) [10, 11] by using the experimental ground state rotational constants and the rovibrational corrections computed from an ab initio anharmonic force field (the so-called semiexperimental method). Tables 1.1 and 1.2 illustrate the improvement of accuracy with time.

The determination of equilibrium structure by electron diffraction is presently possible even for larger molecules (with about hundreds of atoms), particularly if they have high symmetry, and if an accurate anharmonic force field is available.

Chapter 2 presents the molecular Hamiltonian. It also describes the different quantum-chemical computation methods to calculate a structure. The accuracy of the different methods is discussed.

Chapter 3 is devoted to the spectroscopy of diatomic molecules. Because the structure is defined by only one parameter (the interatomic distance), the Schrödinger equation is effectively one-dimensional and can be solved using standard numerical methods. Therefore, the bond length can be determined with a high accuracy. Furthermore, it is possible to point out the breakdown of the Born–Oppenheimer approximation, and when a heavy atom is present, the effect of the size of its nucleus can be seen. For these reasons, diatomic molecules are used as probes of time-reversal-symmetry violation, constancy of the proton-to-electron mass ratio [12], etc.

Chapter 4 explains how to determine rotational constants by rotational spectroscopy, high-resolution infrared spectroscopy, and Raman spectroscopy. The rotational coherence spectroscopy method and the photoelectron spectroscopy are also briefly presented.

Chapter 5 explains the different ways to determine the equilibrium structure of a semirigid molecule from the rotational constants. In addition to the experimental and semiexperimental methods, several empirical methods permitting to determine an accurate equilibrium structure are also discussed and compared.

Chapter 6 is dedicated to the structure determination of molecules affected by a large-amplitude motion, namely by inversion, ring-puckering, internal rotation, as well as quasi-linear molecules, and weakly bound complexes, using spectroscopic methods.

Table 1.1 Improvement of accuracy with time (distances in Å and angles in degrees)

Molecule	Bond/angle	First determination			Most recent determination		
		Value	Year	Ref	Value	Year	Ref
HOF	OF	1.413(19)	1953	1	1.43447(11)	2002	2
	HOF	103.8(15)			97.86(2)		
NF_3	NF	1.37(2)	1950	3	1.3675(10)	2003	4
	FNF	102.5(15)			101.86(10)		
CH_3F	CF	1.43	1932	5	1.383(1)	1999	6
CHF_3	CH	1.098	1952	7	1.0850	2004	8
c-C_6H_6	CH	1.04	1947	9	1.0802(20)	2000	10
c-C_6H_5F	CF	1.31	1947	9	1.3435(10)	2013	11
c-C_6H_{12}	CC	1.54	1947	9	1.5258(6)	2015	12
	CCC	109.47			111.114(9)		

[1] Ibers JA, Schomaker V (1953) J Phys Chem 57: 699–701
[2] Pawlowski F, Jørgensen P, Olsen J, Hegelund F, Helgaker T, Gauss J, Bak KL, Stanton JF (2002) J Chem Phys 116: 6482–6496
[3] Schomaker V, Lu C-S (1950) J Am Chem Soc 72: 1182–1185
[4] Breidung J, Constantin L, Demaison J, Margulès L, Thiel W (2003) Mol Phys 101: 1113–1122
[5] Pauling L (1932) Proc Nat Acad Sci 18: 293–297
[6] Demaison J, Breidung J, Thiel W, Papousek D (1999) Struct Chem 10: 129–133
[7] Ghosh SN, Trambarulo R, Gordy W (1952) J Chem Phys 20: 605–607
[8] Breidung J, Cosléou J, Demaison J, Sarka K, Thiel W (2004) Mol Phys 102: 1827–1841
[9] Hassel O, Viervoll H (1947) Acta Chem Scand 1: 149–168
[10] Gauss J, Stanton JF (2000) J Phys Chem A 104: 2865–2868
[11] Demaison J, Rudolph HD, Császár AG (2013) Mol Phys 111: 1539–1562
[12] Demaison J, Craig NC, Groner P, Écija P, Cocinero EJ, Lesarri A, Rudolph HD (2015) J Phys Chem A 119: 1486–1493

Chapter 7 is devoted to the determination of equilibrium structure by gas-phase electron diffraction. Being used in combination with rotational spectroscopy and/or ab initio calculations, it is a powerful technique for the structure determination of both small size and relatively large molecules.

Chapter 8 is mainly dedicated to the determination of the X–H bond length (X = C, N, O, ...). Indeed, it is difficult to accurately determine the coordinates of a hydrogen atom using the standard methods. The scattering cross-section of the hydrogen atom is quite small, and, thus, gas-phase electron diffraction is not a good technique for this purpose. Likewise, because of its small mass, the hydrogen atom does not contribute much to the moments of inertia. Furthermore, when a hydrogen atom is isotopically substituted by deuterium, the rovibrational correction changes a great deal and, therefore, also its error.

Fortunately, there are other ways to obtain the X–H bond length which are discussed in this chapter.

Chapter 9 is devoted to the systematization of equilibrium molecular structures determined by experimental methods and published after 2017. The presentation of

Table 1.2 Improvement of the structure of NH_3 with time (distances in Å and angles in degrees)

Year	NH	HNH	Type	Ref
1940	1.014	107.3	r_e	1
1945	1.014	106.78	r_e	2
1951	1.016	107	r_0	3
1957	1.0144	107	r_0	4
1957	1.0124	106.67	r_e	5
1964	1.0116	106.7	r_e	6
1968	1.011	108.2	r_e	7
1969	1.0136	107.05	r_s	8
1971	1.0156	107.28	r_0	9
1971	1.0156	107.28	r_0	10
1972	1.025	107	r_e	11
1974	1.0128	107.03	r_s	12
1981	1.01718	106.75	r_0	13
1982	1.0162	107.47	r_0	14
1989	1.018	107.28	r_e	15
1989	1.0141	107.49	r_e	15
1989	1.0149	107.4	r_e	15
1992	1.0124	106.18	r_e	16
1998	1.0126	106.61	r_e	17
2001	1.0112	106.36	r_e	18
2002	1.01139	107.17	r_e	19
2008	1.010936	106.8147	r_e	20
2021	1.01099	106.7474	r_e	21
Range	0.0141	2.02		

(continued)

Table 1.2 (continued)

Year	NH	HNH	Type	Ref
Range r_e	0.0141	2.02		
Median	1.0138	107.0		

[1] Dennison DM (1940) Rev Mod Phys 12: 175–214
[2] Herzberg G in: D. van Nostrand (Ed.), Infrared and Raman Spectra of Polyatomic Molecules, New York, 1945
[3] Weiss MT, Strandberg MPW (1951) Phys Rev 83: 567–575
[4] Erlandsson G, Gordy W (1957) Phys Rev 106: 513–515
[5] Benedict WS, Plyler K (1957) Can J Phys 35: 1235–1241
[6] Duncan JL, Mill IM (1964) Spectrochim Acta 20: 523–546
[7] Kuchitsu K, Guillory JP, Bartell LS (1968) J Chem Phys 49: 2488–2493
[8] Helminger P, Gordy W (1969) Phys Rev 188: 100–108
[9] Helminger P, DeLucia FC, Gordy W (1971) J Mol Spectrosc 39: 94–97
[10] Krupnov AF, Gershtein LI, Shustrov VG (1971) Opt Spektrosk 30: 790/427
[11] Hoy AR, Mills IM, Strey G (1972) Mol Phys 24: 1265–1290
[12] Helminger P, DeLucia FC, Gordy W, Morgan HW, Staats PA (1974) Phys Rev A 9: 12–16
[13] Marshall MD, Muenter JS (1981) J Mol Spectrosc 85: 322–326
[14] Cohen EA, Pickett HM (1982) J Mol Spectrosc 93: 83–100
[15] Spirko V, Kraemer WP (1989) J Mol Spectrosc 133: 331–344
[16] Martin JML, Lee TJ, Taylor PR (1992) J Chem Phys 97: 8361–8371
[17] Peterson KA, Xantheas SS, Dixon DA, Dunning Jr TH (1998) J Phys Chem A 102: 2449–2454
[18] Bak KL, Gauss J, Jørgensen P, Olsen J, Helgaker T, Stanton JF (2001) J Chem Phys 114: 6548–655
[19] Pawlowski F, Jørgensen P, Olsen J, Hegelund F, Helgaker T, Gauss J, Bak KL, Stanton JF (2002) J Chem Phys 116: 6482–6496
[20] Puzzarini C, Heckert M, Gauss J (2008) J Chem Phys 128: 194108/1–9
[21] Egorov O, Rey MM, Nikitin AV, Viglaska D (2021) J Phys Chem A 125: 10568–10579

structural datasets is similar to that in the handbook "Structure Data of Free Polyatomic Molecules" [13]. Therefore, this chapter can be used as a current supplement to this handbook.

References

1. Sim GA, Sutton LE, Beagley B (1975) Gases and crystals: a comparative survey. Mol Struct Diffr Methods 3:52–71
2. Friedrich W, Knipping P, von Laue M (1912) Interferenz-Erscheinungen bei Röntgenstrahlen. Sitzungsberichte der Mathematisch-Physikalischen Classe der Königlich-Bayerischen Akademie der Wissenschaften zu München 1912:303
3. Mark H, Wierl R (1930) Über Elektronenbeugung am einzelnen Molekül. Naturwissenschaften 18:205–205
4. Debye P, Bewiloga L, Ehrhardt P (1929) Zerstreuung von Röntgenstrahlen an einzelen Molekeln. Phys Z 30:84–87
5. Pauling L (1931) The nature of the chemical bond. Application of results obtained from the quantum mechanics and from a theory of paramagnetic susceptibility to the structure of molecules. J Am Chem Soc 53:1367–1400
6. Pauling L (1960) The nature of the chemical bond and the structure of molecules and crystals: an introduction to modern structural chemistry. Cornell University Press, Ithaca, NY

7. Townes CH, Schawlow AL (1955) Microwave spectroscopy. McGraw-Hill, New York
8. Burrau Ø (1927) Berechnung des Energiewertes des Wasserstoffmolekül-ions (H_2^+) im Normalzustand. Kgl Danske Vid Selsk Math-Fys Medd 7(14)
9. Heitler W, London F (1927) Wechselwirkung neutraler Atome und homöopolare Bindung nach der Quantenmechanik. Z Physik 44:455–472
10. Vogt N, Demaison J, Krasnoshchekov SV, Stepanov NF, Rudolph HD (2017) Determination of accurate semiexperimental equilibrium structure of proline using efficient transformations of anharmonic force fields among the series of isotopologues. Mol Phys 115:942–951
11. Vogt N, Demaison J, Cocinero EJ, Écija P, Lesarri A, Rudolph HD, Vogt J (2016) The equilibrium molecular structures of 2-deoxyribose and fructose by the semiexperimental mixed estimation method and coupled-cluster computations. Phys Chem Chem Phys 18:15555–15563
12. DeMille D (2015) Diatomic molecules, a window onto fundamental physics. Phys Today 68:34–40
13. Vogt N, Vogt J (2019) Structure data of free polyatomic molecules. Switzerland, Springer Nature, p 926

Chapter 2
Quantum-Chemical Methods

Abstract This chapter describes the different methods used in computational chemistry to determine the equilibrium structure of an isolated molecule. It starts with the presentation of the Hamiltonian. Then, the different methods (ab initio and density functional theory) are reviewed. The approximations and their effect on the accuracy are discussed.

2.1 Introduction

Ab initio electronic structure methods permit to compute equilibrium geometries of small- and middle-size molecules (a few tens of atoms) with a high accuracy: a few thousandths of an Ångström for bond lengths and a few tenths of a degree for bond angles. These excellent results are due to the development of efficient methods and to the gigantic advances in computer technology. Electronic structure calculations have significant advantages over experimental methods: they are easier to use, faster, and cheaper. Furthermore, their accuracy decreases less rapidly with the increasing number of atoms.

The methods of quantum chemistry that are capable to compute a structure with a high accuracy are reviewed in this chapter. A recent book on this subject is Helgaker et al. [1]. There are also many reviews, for instance [2–7].

2.2 Ab Initio Methods

A molecule is made up of positive nuclei and negative electrons held together by electrostatic forces. If we assume that the electrons and nuclei are point masses (an excellent approximation, however see Sect. 3.4) and that relativistic effects are negligible, it is easy to write the molecular Hamiltonian. When there are heavy atoms, relativistic effects are no more negligible, but it is possible to take them into account, see Sect. 2.7.

N. Vogt and J. Demaison, *Equilibrium Structure of Free Molecules*, Lecture Notes in Chemistry 111, https://doi.org/10.1007/978-3-031-36045-9_2

This molecular Hamiltonian is still much too complicated to calculate its eigenvalues, but it may be significantly simplified thanks to the Born–Oppenheimer approximation [8]. This approximation is based on the fact that the electrons are much lighter than the nuclei (proton mass/electron mass $\approx$ 1836). Thus, the motion of the electrons is much faster, and it is possible to study it assuming that the nuclei are stationary. The coordinates of the nuclei are parameters of the electronic energy. Varying the coordinates of the nuclei permits to define the potential energy surface of the nuclei. The minimum of this surface defines the equilibrium geometry of the molecule. In the frame of the Born–Oppenheimer approximation, the total wave function of the molecular Hamiltonian is the product of two wave functions, one for the nuclei and another one for the electrons. It permits to separate the Hamiltonian into electronic and nuclear terms, the cross-terms between electrons and nuclei being neglected. This Born–Oppenheimer approximation is an excellent one, but it may break down in some cases (see Sect. 3.3).

The electronic Hamiltonian is still too complicated to be solved except in the case of monoelectronic molecules such as H_2^+. However, it can be considerably simplified by using the Hartree–Fock (HF) method [9] whose basic idea is that each electron interacts with the average field of every other electron. Thus, instead of an n-electrons problem, we have n monoelectronic Hamiltonians which are easy to be solved.

To calculate the eigenvalues of these monoelectronic Hamiltonians, we need a basis set that is used to represent the electronic wave function. This basis set is formed by molecular orbitals φ_i which are a linear combination of atomic orbitals, χ_k. These atomic orbitals are usually a linear combination of Gaussian-type orbitals GTOs:

$$\varphi_i = \sum_k c_{ik}\chi_k \tag{2.1}$$

The main advantage of the Gaussian orbitals is that they render the computations much faster. The electrons being Fermions, the many-electron wave function Ψ must be antisymmetric upon exchange of two electrons. The simplest form of Ψ is a determinant called Slater determinant:

$$\psi = \frac{1}{\sqrt{n!}}|\varphi_1\varphi_2\ldots\varphi_n| \tag{2.2}$$

For a molecule with n electrons, the $n/2$ orbitals of smallest energy are the occupied orbitals and the remaining orbitals of higher-energy are the virtual orbitals.

The Hartree–Fock method permits to handle large molecules (up to about 1000 atoms), and the error in bond distances is about 1%.

To improve the accuracy, it is necessary to take into account the electron correlation, i.e., the instantaneous interactions between the electrons. Different methods are available. The simplest one is the second-order perturbation theory, usually called Møller-Plesset perturbation theory, abbreviated as MP2 [10]. It greatly improves the accuracy, but it is not yet completely satisfactory.

A better approximation is obtained by using further determinants obtained by replacing one or several occupied orbitals φ_i, φ_j, ... with virtual orbitals φ_a, φ_b, ... A better approximation is thus obtained

$$\psi = c_0\psi_0 + \sum_{ia} c_i^a \psi_i^a + \sum_{ijab} c_{ij}^{ab} \psi_{ij}^{ab} + \cdots \tag{2.3}$$

where ψ_0 is the HF solution, the first sum refers to single excitations applied to ψ_0, and the second sum to double excitations. When all virtual orbitals and all degrees of excitation are included, it gives the full configuration interaction (CI).

One of the simplest practical methods to estimate the correlation energy is to use the coupled-cluster method [3]. The double-excitation terms are the most important. When they are taken into account together with the single excitation terms, it gives the CCSD method [11]. If the triple excitation terms are also added, it gives the CCSDT method. However, to reduce the computation cost, the triple excitations can be taken into account by a perturbation calculation giving the CCSD(T) method [12]. In conclusion, there is a hierarchy of methods of increasing accuracy:

$$\begin{aligned} \text{HF} < \text{MP2} < \text{CCSD} < \text{CCSD(T)} \approx \text{CCSDT} \\ < \text{CCSDT(Q)} \approx \text{CCSDTQ} < \ldots \end{aligned} \tag{2.4}$$

The CCSDTQ method takes into account the quadruple excitations [13], whereas the CCSDT(Q) method is an approximation, the quadruple excitations being treated perturbatively [14, 15]. This CCSDT(Q) approximate method has a tendency to overestimate the quadruple energy. For instance, the CCSDT(Q) method overestimates the NN bond lengths in N_2 by 0.00032 Å and in N_2O by 0.00077 Å [16] and the CO bond length in H_2CO by 0.00027 Å [17]. Although this overestimation is small, it is not negligible when compared to the correction CCSDTQ–CCSD(T) which is 0.00075 Å for N_2. However, what is interesting is that, in some cases, the CCSDT(Q)–CCSDTQ correction seems to be of the same order of magnitude as the CCSDTQP–CCSDTQ correction, P meaning that the pentuple excitations are taken into account. For instance, for N_2, CCSDTQP–CCSDTQ = 0.00022 Å.

The coupled-cluster method gives excellent results, except when the system has a multiconfigurational character, i.e., when one of several c_i coefficients of Eq. (2.3) is large. In this case, multireference methods (CASSCF, CASPT2, MRCI, ...) have to be used [1].

The improvement with method, Eq. (2.4) has been checked by many authors. In particular, Bak et al. [18] have studied a selection of 19 molecules and found the following decrease of the mean absolute error for the bond lengths (in Å): HF (0.260); MP2 (0.0046); CCSD (0.0067); and CCSD(T) (0.0009), see Table 2.1. In this selection, there were only first-row atoms. Coriani et al. [19] made a similar study with 31 molecules containing second-row elements, and they found similar results: HF (0.213); MP2 (0.0071); CCSD (0.0080); and CCSD(T) (0.0013). The improvement by using methods beyond CCSD(T) has also been analyzed by many

Table 2.1 Mean absolute error (in Å) for a selection of 19 molecules [a]

Basis set	cc-pCVDZ	cc-pCVTZ	cc-pCVQZ
HF	**0.0182**	0.0252	0.0260
MP2	0.0128	**0.0046**	0.0046
CCSD	0.0108	**0.0045**	0.0067
CCSD(T)	0.0161	0.0022	**0.0009**

[a] Refs. [1, 4, 5, 13]

authors [20–23]. Specially, Halkier et al. [20] have found that the CCSD(T) method is more accurate than the CCSDT one (this is an important point which will be discussed in more detail in Sect. 2.6) and that the CCSDTQ method decreases the mean absolute error by a factor about two compared to the CCSD(T) results [21]. Of course, there are large variations depending on the molecule studied. This point is further discussed in Sect. 2.7.

Finally, it is worth noting that the bond length increases when a higher-level method is used. For instance, Peterson and Dunning [24] using the cc-pVQZ basis set have found the following values for the C–H bond length in methane (in Å): HF (1.0815); MP2 (1.0841); CCSD (1.0865); and CCSD(T) (1.0879).

2.3 Density Functional Theory

Coupled-cluster approaches permit to take into account the electron correlation. However, they are computer intensive. A useful alternative is the density functional theory (DFT) because it also accounts for the electron correlation but is not more expensive than the HF method [25, 26]. The idea is to use the electronic density, $\rho(\mathbf{r})$, which only depends on the coordinates x, y, and z instead of the many-electron wave function which depends on many variables.

Hohenberg and Kohn [27] demonstrated that the DFT method is exact, provided that the true functional is known, which is unfortunately not the case. To find this density functional, Kohn and Sham [28] modified the HF equations by introducing the local exchange–correlation term, E^{XC}, that accounts for the exchange phenomenon (originating from the antisymmetry of the wave function) and the dynamic correlation in the motion of the electrons. In other words, this E^{XC} term is the sum of all electronic energy contributions that we do not know how to express exactly. The energy is partitioned in several terms:

$$E = E^{\mathrm{T}} + E^{\mathrm{V}} + E^{\mathrm{XC}} \tag{2.5}$$

where E^{T} is the kinetic energy of the non-interacting electrons and E^{V} is the electrostatic Coulomb interaction between two charges (nucleus/electron, nucleus/nucleus, and electron/electron). $E^{\mathrm{T}} + E^{\mathrm{V}}$ corresponds to the classical energy of the charge distribution ρ. The problem is to find and approximate E^{XC} which is usually divided into two parts: exchange (X) and correlation (C)

$$E^{XC} = E^{X} + E^{C}. \tag{2.6}$$

The simplest approximation is the local density approximation (LDA), where the XC (exchange–correlation) functional only depends on the density ρ. A better approximation is to introduce the reduced density gradient, leading to the various GGA (generalized gradient approximation) functionals. The introduction of the Laplacian (or the kinetic energy density, which contains similar information) is still a better approximation, the meta-GGAs. A further improvement is the hybrid functional where a portion of exact exchange from HF theory is mixed with the rest of the exchange–correlation energy from other DFT E^{XC}. Finally, the next improvement is the double-hybrid functional where linear combination of GGA correlation and MP2 correlation from HF orbitals is used.

However, contrary to the electron correlation treatment in ab initio calculations, there is no systematic path toward improving the exchange–correlation functional, and calibration studies have to be made.

Quantum chemistry programs offer a vast number of choices of functionals. It is beyond the scope of this chapter to describe all of them. Only those which are frequently used in structure determination will be mentioned:

- A popular one is the hybrid functional B3LYP [29] which mixes for Becke three-parameter, exchange functional [30] with the Lee–Yang–Parr correlation functional [31]. This functional was shown to permit the calculation of accurate anharmonic force fields with double zeta basis sets [32].
- The double-hybrid functional, B2PLYP functional [33], was shown to be an improvement over B3LYP [34]. With a triple zeta basis set, it gives accurate results for both equilibrium structures and rovibrational corrections.
- Finally, the recent double-hybrid rev-DSD-PBEP86 functional [35] was found still better [36, 37].

One problem with these functionals is the treatment of the dispersion interactions. They are not able to properly account for long-range interactions. A vast amount of work is dedicated to improving the description of non-covalent interactions. One popular solution is to use the dispersion tails developed by Grimme et al. [38]. Four generations of dispersion corrections are available: D1, D2, D3 [39–41], and D4. The recent D4 dispersion correction [42, 43] being clearly superior.

The performance of the B2PLYP functional was assessed by Penocchio et al. [44] and was also compared to the B3LYP one.

In conclusion, this is a very dynamic field, and future improvements are probable.

2.4 Basis Sets

Several reviews on basis sets are available [45–47].

Many different Gaussian-type orbitals are available. The most widely used for post-Hartree–Fock calculations (which will be described below) are due to Dunning et al. because extrapolations methods permit to converge to the infinite basis set limit. The simplest ones called correlation-consistent polarized valence basis sets are cc-pVnZ with $n =$ D(ouble), T(riple), Q(uadruple), 5, ... [48]. For the second-row elements, Al-Ar, the original d polarization functions in cc-pVnZ are not tight enough, and the revised cc-pV($n + d$)Z basis sets should be used [49]. When heavier atoms are present, pseudopotential basis sets are used: cc-pVnZ-PP [50, 51]. These basis sets are for valence electrons. To take into account the inner electrons, they are augmented becoming weighted core-valence correlation-consistent polarized valence basis sets, cc-pCVnZ [52] or cc-pwCVnZ [53].

Diffuse functions can also be added for describing the outer electrons. They are important for electronegative atoms and to describe long-range interactions. This gives aug(mented) basis sets [54].

cc-pVnZ-F12 are correlation-consistent basis sets optimized for use with F12 methods (see Sect. 2.5.2) [55, 56].

Whereas the bond length increases when the level of theory increases, it decreases when the basis set increases. For instance, for the CO bond length of CO at the CCSD(T)_fc level of theory in the frozen core approximation (fc), the results are the following (in Å): cc-pVTZ: 1.135711; cc-pVQZ: 1.131392; cc-pV5Z: 1.130730; cc-pV6Z: 1.130496; and cc-pV∞Z: 1.130210 [57]. This is an interesting observation because for each level of theory there is a basis set for which the error is minimal due to its partial compensation, see Table 2.1.

The atomic natural orbital (ANO) basis sets of Almöf and Taylor [58, 59] are of similar quality, and they are often used. These ANO basis sets benefit from fortuitous cancelation of errors and, therefore, perform exceptionally well, in particular for the calculation of harmonic frequencies with the CCSD(T) method [60].

Although the Dunning's basis sets have been optimized for post-Hartree–Fock calculations, they are often used for DFT calculations. However, it would be better to use Jensen's polarization-consistent (pc-n) basis which converges more quickly [61–64].

The small polarized double-ζ SNSD basis set has also to be noted [65]. It is obtained by adding to the 6-31G** basis set one contracted and one diffuse function on each atom, together with diffuse p and d functions on non-hydrogen atoms, all with optimized orbital exponents. Several papers have shown that the SNSD basis set used with the B3LYP method represents an excellent compromise between accuracy and computational cost for several spectroscopic properties and in particular for the computation of anharmonic force fields [44].

Most basis sets may be found in the basis set library of program packages. They are also available on the basis set exchange web page: https://www.basissetexchange.org [66–68].

2.5 Convergence of the Basis Sets

The correlation energy converges toward the basis set limit at a rate of about $\ell_{\max}^{-3}$, where $\ell_{\max}$ is the largest angular momentum in the basis set. This is due to a poor description of the Coulomb cusp when Gaussian basis sets are used. The consequence is that very large basis sets are required. For instance, the bond length of HF at the CCSD(T)_fc/aug-cc-pVTZ level of theory is 0.9210 Å. It decreases by 0.0033 Å when going from aug-cc-pVTZ to aug-cc-pVQZ and by 0.0004 Å when going from aug-cc-pVQZ to aug-cc-pV5Z [21].

This is one of the most problematic issues. Instead of using larger and larger basis sets until acceptable convergence is achieved, there are alternative solutions. The first one is to simply extrapolate the results given by a well-defined hierarchy of basis sets, and the other one is to use basis sets that contain the interelectronic distance (such methods are known as "explicitly correlated" approaches). There is also a third way: the extrapolation (or interpolation) method of the computed rotational constants which makes use of the semiexperimental equilibrium rotational constants. It is described in Sect. 5.6.

2.5.1 Extrapolation Methods

The Hartree–Fock energy converges much faster than the correlation correction. For this reason, the basis set limit is estimated by extrapolating separately the HF energy and the correlation correction. Using correlation-consistent basis sets, the following empirical formula may be used for the HF energy [69]:

$$E_{\mathrm{HF}}^{\infty} = E_{\mathrm{HF}}^{n} + ae^{-bn} \tag{2.7}$$

where n is the cardinal number of the basis set. Three calculations with different n are required. For the correlation correction, the following formula may be used [70]:

$$\Delta E_{\mathrm{corr}}^{\infty} = \Delta E_{\mathrm{corr}}^{n} - \frac{c}{n^3} \tag{2.8}$$

This extrapolation only requires two calculations. Other empirical equations are compared in Ref. [71]. The conclusion was that the empirical "geometry" is a cost-effective approach and provides reliable results. A mixed exponential/Gaussian function may be used [72]

$$r(n) = r_{\infty} + B\mathrm{e}^{-(n-1)} + C\mathrm{e}^{-(n-1)^2} \tag{2.9}$$

Another possibility [70] is to calculate the correlation contribution by an extrapolation formula with $n = \mathrm{Q}$ and 5 (or better $n = 5$ and 6):

$$\Delta r^{\text{corr}}(n) = \Delta r_{\infty}^{\text{corr}} + An^{-3} \tag{2.10}$$

and to add this correlation contribution to the HF-SCF CBS limit, assumed to be reached at the HF/cc-pV6Z level:

$$r(\text{CBS}) = r_{\infty}^{\text{SCF}} + \Delta r_{\infty}^{\text{corr}} \tag{2.11}$$

In all cases, the errors were found smaller than 0.0005 Å.

2.5.2 *Explicitly Correlated Methods*

The convergence can be improved by including terms into the wave function that depend explicitly on the interelectronic distances r_{ij} [73, 74]. One of the best solutions is to use a correlation factor in terms of Slater functions $\exp(\gamma\ r_{12})$ which are approximated by a linear combination of Gaussians [75]. With this method, a triple zeta basis set is enough to reach quintuple zeta quality, see Table 2.2.

2.6 CCSD(T)_ae/cc-pwCVQZ Level of Theory

The CCSD(T)_ae/cc-pwCVQZ level of theory is known to give satisfactory results because of a partial compensation of errors: the introduction of connected quadruples (CCSDTQ) increases the bond length by 0.001–0.002 Å (see Sect. 2.7), whereas the basis set extension from quadruple zeta to sextuple zetas shortens it by about 0.001 Å.

In Tables 2.3, 2.4 and 2.5, the CCSD(T)_ae/cc-pwCVQZ bond length is compared to the equilibrium one for several molecules. For X–H bonds (X = B, C, N, O), the median of deviations is −0.0001 Å, i.e., negligible. For bonds between two first-row atoms (B, C, N, O, F), the median of deviations is −0.00035 Å, and for bonds between a first-row atom and a second-row atom (P, Si, Cl), the median of deviations is −0.0006 Å. It may be concluded that the CCSD(T)_ae/cc-pwCVQZ level of theory gives bond lengths which are slightly too long but with an accuracy not worse than about 0.001 Å. Of course, there are a few outliers. In particular, for N_2O, the computed bond lengths are too short, but it is known that the non-dynamical correlation is not small for this molecule, see Sect. 5.6 and Ref. [76].

As a further check, Table 2.6 confirms that the extension of the basis set from quadruple zeta to quintuple zeta does not improve much the accuracy.

For large molecules, the cc-pwCVQZ basis set may be too large. In this case, it is possible to assume that small corrections are additive. The structure is first optimized at the CCSD(T)_ae/cc-pwCVTZ level of theory, and the small effect of basis set enlargement (cc-pwCVTZ → cc-pwCVQZ) is then estimated at the MP2 level. In other words

Table 2.2 Convergence of the CCSD(T)-F12 method. Bond lengths in Å and angles in degrees

Molecule, parameter	r_e [a]	CCSD(T)_ae	CCSD(T)-F12_ae		Ref
		aug-cc-pwCVQZ	cc-pCVTZ-F12	cc-pCVQZ-F12	
N_2, r(N≡N)	1.0976	1.0982	1.0970	1.0968	1
HC≡N, r(C–H)	1.0651	1.0657	1.0652	1.0652	1
HC≡N, r(C≡N)	1.1533	1.1539	1.1527	1.1525	1
HNC, r(N–H)	0.9954	0.9956	0.9954	0.9953	1
HNC, r(N≡C)	1.1685	1.1693	1.1682	1.1681	1
N≡C–C≡N, r(C–C)	1.3833	1.3849	1.3843	1.3842	1
N≡C–C≡N, r(C≡N)	1.1584	1.1586	1.1575	1.1573	1
SO_2, r(S=O)	1.4308	1.4311		1.4298	2
SO_2, ∠(OSO)	119.329	119.389		119.362	2
H_2O_2, r(O–H)	0.9617	0.9619	0.963	0.963	3
H_2O_2, r(O–O)	1.4524	1.4497	1.451	1.450	3
H_2O_2, ∠(OOH)	99.76	99.99	100.1	100.1	3
H_2O_2, τ(HOOH)	113.6		112.7	112.7	3
NH_3, r(N–H)	1.0110	1.0115		1.0107	4
NH_3,∠(HNH)	106.75	106.71		106.81	4
H_2CS, r(C=S)	1.6091	1.6100	1.6125	1.6121	5
H_2CS, r(C–H)	1.0853	1.0854	1.0869	1.0867	5
H_2CS, ∠(HCH)	116.484	116.294	116.42	116.42	5

[a] Experimental or semiexperimental equilibrium structures
[1] Breidung J, Thiel W (2019) J Phys Chem C 123:7940–7951
[2] Demaison J, Liévin J (2021) Mol Phys https://doi.org/10.1080/00268976.2021.1950857
[3] Hollman D S, Schaefer HF (2012) J Chem Phys 136: 084302
[4] Egorov O, Rey MM, Nikitin AV, Viglaska D (2021) J Phys Chem A 125: 10568–10579
[5] Yachmenev A, Yurchenko SN, Ribeyre T, Thiel W (2011) J Chem Phys 135: 074302

$$\begin{aligned} &r_e[\text{CCSD(T)\_ae/cc-pwCVQZ}] \\ &\quad \approx r_e[\text{CCSD(T)\_ae/cc-pwCVTZ}] + r_e[\text{MP2\_ae/cc-pwCVQZ}] \\ &\quad - r_e[\text{MP2\_ae/cc-pwCVTZ}]. \end{aligned} \tag{2.12}$$

There is a lot of documented evidence confirming the accuracy of this method, and Table 2.7 shows that the loss of accuracy is quite small, the median absolute deviation being 0.0002 Å for bond length and 0.02° for bond angles.

There are other approximate methods to obtain a good approximation of the CCSD(T)_ae/cc-pwCVQZ structure. Frequently, the frozen core (fc) approximation is used. For instance, the structure is first optimized at the CCSD(T)_fc/cc-pVTZ

Table 2.3 Comparison of the CCSD(T)_ae/cc-pCVQZ X–H bond lengths with the experimental or semiexperimental equilibrium values r_e (in Å), see also Tables 2.4 and 2.5

Molecule	Bond	r_e	cc-pCVQZ	r_e − cc-pCVQZ	Ref.
HBS	B–H	1.1696	1.1699	−0.0003	a
BeH_2	Be–H	1.3264	1.3265	−0.0001	b
CH_2	C–H	1.0757	1.0756	0.0001	c
CH_3	C–H	1.0767	1.0762	0.0005	c
CH_4	C–H	1.0862	1.0864	−0.0003	a
CH_2F_2	C–H	1.0870	1.0871	−0.0001	d
CH_2Cl_2	C–H	1.0816	1.0817	−0.0001	d
CH_3CN	C–H	1.0865	1.0867	−0.0002	e
C_2H_6	C–H	1.0887	1.0892	−0.0005	f
c-C_3H_6	C–H	1.0790	1.0781	0.0009	g
c-C_6H_6	C–H	1.0802	1.0818	−0.0016	h
CH_2O	C–H	1.1005	1.1008	−0.0003	i
CH_2=CH_2	C–H	1.0809	1.0809	0.0000	i
H_2C=S	C–H	1.0853	1.0854	−0.0001	a
CH_2=C=CH_2	C–H	1.0808	1.0810	−0.0002	j
CH_2=S=O	C–H *cis*	1.0796	1.0803	−0.0007	k
CH_2=S=O	C–H *trans*	1.0817	1.0805	0.0012	k
c-CH_2N_2	C–H	1.0769	1.0770	−0.0001	l
c-C_2H_2Si	C–H	1.0802	1.0803	−0.0001	m
HCCH	C–H	1.0618	1.0620	−0.0003	n
HCN	C–H	1.0652	1.0655	−0.0003	i
HCP	C–H	1.0702	1.0707	−0.0005	a
HF	F–H	0.9171	0.9158	0.0013	i
HNNH	N–H	1.0283	1.0284	−0.0001	i
HN_3	N–H	1.0158	1.0157	0.0001	o
$HNCCN^+$	N–H	1.0132	1.0136	−0.0004	p
NH_3	N–H	1.0109	1.0112	−0.0003	i
HNC	N–H	0.9954	0.9953	0.0001	i
H_2O	O–H	0.9759	0.9757	0.0002	i
HOF	O–H	0.9666	0.9657	0.0009	i
HOCl	O–H	0.9644	0.9631	0.0012	a
H_2O_2	O–H	0.9617	0.9619	−0.0002	i
HSOH	O–H	0.9606	0.9601	0.0005	q
PH_3	P–H	1.4117	1.4115	0.0002	a
SH_3^+	S–H	1.3499	1.3492	0.0007	r

(continued)

Table 2.3 (continued)

Molecule	Bond	r_e	cc-pCVQZ	r_e − cc-pCVQZ	Ref.
HSOH	S–H	1.3420	1.3414	0.0006	q
H_2S	S–H	1.3362	1.3353	0.0009	a
HSiCl	Si–H	1.5140	1.5147	−0.0007	a
SiH_2	Si–H	1.514	1.514	0.0000	a
SiH_4	Si–H	1.4741	1.4742	−0.0001	a
SiH_3Cl	Si–H	1.4688	1.4691	−0.0003	s
Median				−0.0001	
Max (abs)				0.0016	

[a] Coriani S, Marchesan D, Gauss J, Hätting C, Helgaker T, Jørgensen P (2005) J Chem Phys 123: 184107
[b] Koput J, Peterson KA (2006) J Chem Phys 125: 044306/1–7
[c] Peterson KA, Dunning TH Jr (1997) J Chem Phys 106: 4119–41140
[d] Vogt N, Demaison J, Rudolph HD (2014) Mol Phys 112: 2873–2883
[e] Puzzarini C, Cazzoli G (2006) J Mol Spectrosc 240: 260–264
[f] Puzzarini C, Taylor PR (2005) J Chem Phys 122: 054315/1–10
[g] Gauss J, Cremer D, Stanton JF (2000) J Phys Chem A 104: 1319–1324
[h] Demaison J, Rudolph HD, Császár AG (2013) Mol Phys 111: 1539–1562
[i] Bak KL, Gauss J, Jørgensen P, Olsen J, Helgaker T, Stanton JF (2001). J Chem Phys 114: 6548–6556
[j] Auer AA, Gauss J (2001) Phys Chem Chem Phys 3: 3001–3005
[k] Demaison J, Vogt N, Ksenafontov DN (2020) J Mol Struct 1206: 127676
[l] Puzzarini C, Gambi A, Cazzoli G (2004) J Mol Struct 695–696: 203–210
[m] Thorwirth S, Harding ME (2009) J Chem Phys 130: 214303
[n] Liévin J, Demaison J, Herman M, Fayt A, Puzzarini C (2011) J Chem Phys 124: 064119/1–8
[o] Owen AN, Sahoo NP, Esselman BJ, Stanton JF, Woods RC, McMahon RJ (2022) J Chem Phys in press
[p] Puzzarini C, Cazzoli G (2009) J Mol Spectrosc 256: 53–56
[q] Baum O, Esser S, Gierse N, Brünken S, Lewen F, Hahn J, Gauss J, Schlemmer S, Giesen TF (2006) J Mol Struct 795: 256–262
[r] Puzzarini C (2007) J Mol Spectrosc 242: 70–75
[s] Demaison J, Sormova H, Bürger H, Margulès L, Constantin FL, Ceausu-Velcescu A (2005) J Mol Spectrosc 232: 323–330

level of theory. It is relatively cheap because only the valence electrons are correlated. Then, the small correction VTZ → VQZ is estimated with the MP2_fc method. This approximation is widely used and is known to be accurate [22]. Finally, the core-correlation correction fc → ae has to be taken into account. It is non-negligible only for the bond lengths. It can also be calculated with the MP2 method. However, it is known that MP2 overestimates the correction. Fortunately, the cc-pwCVTZ basis set underestimates it, see Table 2.8. Thus, for large molecules, an acceptable compromise is to estimate the core-correlation correction in the following way:

$$\Delta r(\text{fc} \rightarrow \text{ae}) = r_e(\text{MP2\_ae/cc-pwCVTZ}) - r_e(\text{MP2\_fc/cc-pwCVTZ}). \quad (2.13)$$

Table 2.4 Comparison of the CCSD(T)_ae/cc-pCVQZ bond lengths with the experimental or semiexperimental equilibrium values r_e (in Å) for bonds between first-row atoms, see also Tables 2.3 and 2.5

Molecule	Bond	r_e	cc-pCVQZ	r_e − cc-pCVQZ	Ref
FBS	B–F	1.2762	1.2775	−0.0013	a
CH_2F_2	C–F	1.3532	1.3518	0.0014	b
CF_3Cl	C–F	1.3200	1.3208	−0.0008	b
$F_2C{=}S$	C–F	1.3084	1.3083	0.0001	a
CF_4	C–F	1.3153	1.3152	0.0001	b
C_2H_6	C–C	1.5521	1.5229	0.0292	c
c-C_3H_6	C–C	1.5030	1.5019	0.0011	d
CH_3CN	C–C	1.4586	1.4596	−0.0010	e
c-C_6H_6	C–C	1.3914	1.3916	−0.0002	f
$HNCCN^+$	C–C	1.3723	1.3745	−0.0022	g
c-C_2H_2Si	C=C	1.3439	1.3445	−0.0006	h
$CH_2{=}CH_2$	C=C	1.3307	1.3312	−0.0005	i
$CH_2{=}C{=}CH_2$	C=C	1.3069	1.3074	−0.0005	j
HCCH	C≡C	1.2029	1.2034	−0.0005	k
c-CH_2N_2	C–N	1.4756	1.4760	−0.0004	l
$HNCCN^+$	C≡N	1.1634	1.1629	0.0005	g
CH_3CN	C≡N	1.1554	1.1557	−0.0003	e
HCN	C≡N	1.1532	1.1538	−0.0006	i
CH_2O	C=O	1.2046	1.2043	0.0003	i
CO_2	C=O	1.1600	1.1604	−0.0005	i
OCS	C=O	1.1562	1.1562	0.0000	a
CO	C≡O	1.1284	1.1289	−0.0005	i
F_2	F–F	1.4108	1.4113	−0.0005	i
HOF	F–O	1.4328	1.4326	0.0002	i
HNC	N≡C	1.1687	1.1693	−0.0006	i
$HNCCN^+$	N≡C	1.1407	1.1406	0.0001	g
NF_3	N–F	1.3678	1.3662	0.0016	m
HNNH	N=N	1.2452	1.2467	−0.0015	i
HN_3	N1≡N2	1.2418	1.2415	0.0003	n
c-CH_2N_2	N=N	1.2262	1.2296	−0.0034	l
HN_3	N2≡N3	1.1307	1.1304	0.0003	n
N_2	N≡N	1.0975	1.0981	−0.0006	i
HNO	N=O	1.2087	1.2085	0.0002	i
H_2O_2	O–O	1.4515	1.4497	0.0018	i

(continued)

Table 2.4 (continued)

Molecule	Bond	r_e	cc-pCVQZ	r_e − cc-pCVQZ	Ref
Median				−0.00035	
MAD				0.00045	

[a] Coriani S, Marchesan D, Gauss J, Hätting C, Helgaker T, Jørgensen P (2005) J Chem Phys 123: 184107
[b] Vogt N, Demaison J, Rudolph HD (2014) Mol Phys 112: 2873–2883
[c] Puzzarini C, Taylor PR (2004) J Chem Phys 122: 054315/1–10
[d] Gauss J, Cremer D, Stanton JF (2000) J Phys Chem A 104: 1319–1324
[e] Puzzarini C, Cazzoli G (2006) J Mol Spectrosc 240: 260–264
[f] Demaison J, Rudolph HD, Császár AG (2013) Mol Phys 111: 1539–1562
[g] Puzzarini C, Cazzoli G (2009) J Mol Spectrosc 256: 53–56
[h] Thorwirth S, Harding ME (2009) J Chem Phys 130: 21430
[i] Bak KL, Gauss J, Jørgensen P, Olsen J, Helgaker T, Stanton JF (2001) J Chem Phys 114: 6548–6556
[j] Auer AA, Gauss J (2001) Phys Chem Chem Phys 3: 3001–3005
[k] Liévin J, Demaison J, Herman M, Fayt A, Puzzarini C (2011) J Chem Phys 124: 064119/1–8
[l] Puzzarini C, Gambi A, Cazzoli G (2004) J Mol Struct 695–696: 203–210
[m] Breidung J, Constantin L, Demaison J, Margulès L, Thiel W (2003) Mol Phys 101: 1113–1122
[n] Owen AN, Sahoo NP, Esselman BJ, Stanton JF, Woods RC, McMahon RJ (2022) J Chem Phys in press

The loss of accuracy due to this approximation is not much larger than 0.001 Å.

Finally, when an electronegative atom such as fluorine is present, it may be necessary to add diffuse functions. Again, the small correction can be calculated at the MP2 level of theory

$$r_e(\text{MP2\_fc/aug-cc-pVQZ}) - r_e(\text{MP2\_fc/cc-pVQZ}), \tag{2.14}$$

see Table 2.9.

The diffuse functions are also necessary for the weakly bound cluster molecules. More approximate methods will be discussed in Sect. 2.8.

2.7 Higher-Level Methods

The accuracy of the CCSD(T) method is excellent, about 0.001–0.002 Å, at least when the non-dynamical correlation is negligible. However, a better accuracy is sometimes desired. When the non-dynamical correlation is large, it may be necessary to use multireference approaches such as the full configuration interaction (full-CI method, Eq. (2.3)). If all possible determinants are retained in the expansion, it gives the exact solution. Actually, the full-CI method can be used only for small molecules and for relatively small basis sets. Approximations have to be made. For instance, in the multireference configuration interaction methods (MRCI), all important determinants are first identified and treated equivalently to generate the MRCI expansion space [77]. For more details, see [1].

Table 2.5 Comparison of the CCSD(T)_ae/cc-pCVQZ bond lengths with the experimental or semiexperimental equilibrium values r_e (in Å) for bonds between a first-row and a second-row atoms, see also Table 2.4

Molecule	Bond	r_e	cc-pCVQZ	r_e − cc-pCVQZ	Ref.
ClBS	B-Cl	1.6806	1.6821	−0.0015	a
ClBS	B=S	1.6049	1.6062	−0.0012	a
FBS	B=S	1.6091	1.6099	−0.0008	a
HBS	B=S	1.5978	1.5995	−0.0017	a
CF_3Cl	C–Cl	1.7520	1.7495	0.0025	b
CCl_4	C–Cl	1.7610	1.7630	−0.0020	b
CH_2Cl_2	C–Cl	1.7643	1.7643	0.0000	b
CCl_2	C–Cl	1.7113	1.7125	−0.0012	a
HCP	C≡P	1.5399	1.5405	−0.0006	a
$H_2C=S$	C=S	1.6091	1.6097	−0.0006	a
$CH_2=S=O$	C=S	1.6084	1.6067	0.0017	c
$F_2C=S$	C=S	1.5913	1.5910	0.0003	a
OCS	C=S	1.5614	1.5625	−0.0011	a
CS_2	C=S	1.5554	1.5533	0.0021	a
c-C_2H_2Si	C–Si	1.8172	1.8176	−0.0004	d
HOCl	O–Cl	1.6891	1.6902	−0.0011	a
Cl_2O	O–Cl	1.6959	1.6964	−0.0005	a
$SOCl_2$	S–Cl	2.0700	2.0710	−0.0010	e
SCl_2	S–Cl	2.0111	2.0131	−0.0020	f
HSOH	S–O	1.6616	1.6619	−0.0003	g
$CH_2=S=O$	S=O	1.4650	1.4639	0.0011	c
SO_2	S=O	1.4308	1.4311	−0.0004	h
$SOCl_2$	S=O	1.4330	1.4320	0.0010	e
SO_3	S=O	1.4173	1.4179	−0.0006	a
S_2O	S=O	1.4570	1.4555	0.0015	a
S_2O	S=S	1.8870	1.8853	0.0017	a
HSiCl	Si–Cl	2.0724	2.0712	0.0012	a
$SiCl_2$	Si–Cl	2.0653	2.0680	−0.0026	a
SiH_3Cl	Si–Cl	2.0458	2.0479	−0.0021	i
SiF_2	Si–F	1.5901	1.5919	−0.0018	a

(continued)

Table 2.5 (continued)

Molecule	Bond	r_e	cc-pCVQZ	r_e − cc-pCVQZ	Ref.
Median				−0.0006	
Max				0.0026	

[a] Coriani S, Marchesan D, Gauss J, Hätting C, Helgaker T, Jørgensen P (2005) J Chem Phys 123: 184107
[b] Vogt N, Demaison J, Rudolph HD (2014) Mol Phys 112: 2873–2883
[c] Demaison J, Vogt N, Ksenafontov DN (2020) J Mol Struct 1206: 127676
[d] Thorwirth S, Harding ME (2009) J Chem Phys 130: 214303
[e] Martin-Drumel MA, Roucou A, Brown GG, Thorwirth S, Pirali O, Mouret, G, Hindle F, McCarthy MC, Cuisset A (2016) J. Chem. Phys. 144: 084305
[f] Demaison J, Vogt N (2022) J Mol Spectrosc 387: 111661
[g] Baum O, Esser S, Gierse N, Brünken S, Lewen F, Hahn J, Gauss J, Schlemmer S, Giesen TF (2006) J Mol Struct 795: 256–262
[h] Demaison J, Liévin J (2021) Mol Phys 120: e1959857
[i] Demaison J, Sormova H, Bürger H, Margulès L, Constantin FL, Ceausu-Velcescu A (2005) J Mol Spectrosc 232: 323–330

More generally, to improve the accuracy, the usual way is to go beyond CCSD(T), i.e., to use the CCSDTQ method. A widely used approach is based on the assumption of the additivity of the corrections which is well established for the energy. It may be written in the following way:

$$E_{\text{total}} = E_{\text{CCSD(T)}}(\text{A}) + \Delta E_{\text{CCSDT}}(\text{B}) + \Delta E_{\text{CCSDTQ}}(\text{C}) + \Delta E_{\text{core}}(\text{D}) \tag{2.15}$$

with

$$\Delta E_{\text{CCSDT}}(\text{B}) = E_{\text{CCSDT}}(\text{B}) - E_{\text{CCSD(T)}}(\text{B}), \tag{2.16}$$

$$\Delta E_{\text{CCSDTQ}}(\text{C}) = E_{\text{CCSDTQ}}(\text{C}) - E_{\text{CCSDT}}(\text{C}), \tag{2.17}$$

and

$$\Delta E_{\text{core}}(\text{D}) = E_{\text{ae}}[\text{CCSD(T)/D}] - E_{\text{fc}}[\text{CCSD(T)/D}] \tag{2.18}$$

The basis set A is as large as possible, the basis set B is smaller, and the basis set C is still smaller, all calculations being performed in the frozen core (fc) approximation. ΔE_{core}(D) is calculated, if possible, with a basis set of quadruple zeta quality (e.g., cc-pwCVQZ). This correction is the difference between the energy computed with all electrons correlated (ae) and the energy obtained in the frozen core approximation. The basis set C may be smaller because the connected quadruple excitations to bond distances converge faster with basis set size than those from connected triples [21].

The contribution of the quadruple excitations has been studied by many authors [13–17, 20–23, 57]. For a X–H bond length (X = C, N, O, F), the median absolute deviation (MAD) of the quadruple excitations is 0.00013 Å, the largest value

Table 2.6 Comparison of the CCSD(T)_ae/cc-pCVQZ bond lengths between heavy atoms with the CCSD(T)_ae/cc-pCV5Z ones and with the most accurate experimental or semiexperimental equilibrium structures r_e (all values in Å)

Molecule	Bond	r_e	Ref	cc-pCVQZ	cc-pCV5Z	r_e − cc-pCVQZ	r_e − cc-pCV5Z
CO	C≡O	1.1284	a	1.1289	1.1280	−0.0005	0.0004
CO_2	C=O	1.1600	b	1.1604	1.1598	−0.0004	0.0002
HCCH	C≡C	1.2028	c	1.2034	1.2028	−0.0006	0.0000
N_2	N≡N	1.0975	d	1.0981	1.0971	−0.0006	0.0004
F_2	F–F	1.4108	d	1.4113	1.4092	−0.0005	0.0016
HF	H–F	0.9171	d	0.9158	0.9154	0.0013	0.0017
HCN	C≡N	1.1532	e	1.1538	1.1530	−0.0006	0.0002
CH_2O	C=O	1.2045	e	1.2043	1.2041	0.0002	0.0005
CH_2=CH_2	C=C	1.3308	e	1.3312	1.3306	−0.0004	0.0002
HNC	C≡N	1.1687	e	1.1693	1.1686	−0.0006	0.0001
HNO	N=O	1.2081	e	1.2085	1.2077	−0.0004	0.0003
HN_3	N1=N2	1.2418	f	1.2415	1.2410	0.0003	0.0008
HN_3	N2≡N3	1.1307	f	1.1304	1.1295	0.0003	0.0012
CH_3CCH	C≡C	1.2046	g	1.2050	1.2045	−0.0004	0.0001
CH_3CCH	C–C	1.4589	g	1.4600	1.4595	−0.0011	−0.0006
					Median	−0.0004	0.0004
					MAD	0.0005	0.0004

[a] Experimental: Authier N, Bagland N, Le Floch A (1993) J Mol Spectrosc 160: 590–592
[b] Experimental: Teffo J-L, Ogilvie JF (1993) Mol Phys 80: 1507–1524
[c] Experimental: Tamassia F, Cané E, Fusina L, Di Lonardo G (2016) Phys Chem Chem Phys 18: 1937–1944
[d] CCSD(T)/V∞Z + core + ΔT + ΔQ: Ruden TA, Helgaker T, Jørgensen P, Olsen J (2004) J Chem Phys 121: 5874–5884
[e] CCSD(T)/V∞Z + core + ΔT + ΔQ: Puzzarini C, Heckert M, Gauss J (2008) J Chem Phys 128: 194108
[f] Semiexperimental: Owen AN, Sahoo NP, Esselman BJ, Stanton JF, Woods RC, McMahon RJ (2022) J Chem Phys in press
[g] Semiexperimental: Müller HSP, Thorwirth S, Lewen F (2020) J Mol Struct 1207: 127769

being 0.00028 for HF. For a bond length between two first-row atoms, the MAD is 0.00076 Å, the largest value being 0.00259 Å for F_2. For the bond angles, the MAD is only 0.0011°, see Table 2.10.

It is possible to improve the accuracy of the method by extrapolating. For details, see [23] and Sect. 2.5.1. Another extrapolation (or interpolation) method, close to the semiexperimental method, can also be used, see Sect. 5.6.

For small molecules, it is possible to take into account the pentuple excitations (CCSDTQP) [21, 23] and even an approximation of the hextuple excitations [16]. Typical examples include: N_2 and N_2O [16], C_3 [78], NH_3 [79], and CO_2 and N_3^- [80]. For instance, for N_2, the contribution of the pentuple excitations is P–Q =

Table 2.7 Comparison of the CCSD(T)_ae/cc-pwCVQZ structural parameters with those of Eq. (2.12) (distances in Å and angles in degrees)

Molecule	Bond	cc-pwCVQZ	Equation (2.12)[d]	Res	Angle	cc-pwCVQZ	Equation (2.12)[d]	Res
Azirine [a]	N1=C2	1.2542	1.2546	−0.0004	N1C2H	139.062	139.071	−0.009
	N1–C3	1.5455	1.5464	−0.0009	C3C2H	151.577	151.546	0.031
	C2–C3	1.4476	1.4479	−0.0003	N1C3H	115.868	115.849	0.019
	C2–H	1.0776	1.0775	−0.0001	C2C3H	120.419	120.412	0.007
	C3–H	1.0809	1.0807	−0.0002	HC3H	116.594	116.621	−0.027
					H2C2C3H3	−80.606	−80.628	0.022
Azete [a]	N1=C2	1.2882	1.2890	−0.0008	N1C2C3	96.013	96.017	−0.004
	C2–C3	1.5419	1.5417	0.0002	C2C3C4	84.559	84.584	−0.025
	C3=C4	1.3362	1.3367	−0.0005	C3C2H2	136.633	136.638	−0.005
	N1–C4	1.5509	1.5515	−0.0006	C2C3H3	138.596	138.572	0.024
	C2–H2	1.0831	1.0830	0.0001	H4C4C3	139.106	139.111	−0.005
	C3–H3	1.0771	1.0770	0.0001				
	C4–H4	1.0776	1.0776	0.0000				
BF_2OH[b]	B–F_{syn}	1.3231	1.3230	0.0001	OBF_{syn}	122.24	122.28	−0.04
	B–F_{anti}	1.3131	1.3131	0.0000	BOH	113.14	112.98	0.16
	B–O	1.3449	1.3453	−0.0004	OBF_{anti}	119.39	119.34	0.05
	O–H	0.9575	0.9576	−0.0001	FBF	118.37	118.38	−0.01
MAD[c]				0.0002				0.019

[a] Császár AG, Demaison J, Rudolph HD (2015) J Phys Chem A 119: 1731–1746
[b] Vogt N, Demaison J, Rudolph HD, Perrin A (2015) Phys Chem Chem Phys 17: 30440–30449
[c] Median absolute deviation
[d] $r_e^{BO} = r_e$(CCSD(T)_ae/cc-pwCVTZ) + r_e(MP2_ae/cc-pwCVQZ)–r_e(MP2_ae/cc-pwCVTZ)

Table 2.8 Calculation of the core correlation (fc–ae, in Å) at various levels of theory for a few bond lengths

Molecule	Bond	CCSD(T)			MP2	MP2	[a]	Ref.
		cc-pwCVTZ	cc-pwCVQZ	cc-pwCV5Z	cc-pwCVQZ	cc-wCVTZ		
HCCH	CC	0.00221	0.00262	0.00274	0.00290	0.00255	0.00256	[b]
	CH	0.00107	0.00126	0.00132	0.00130	0.00113	0.00124	[b]
BF_2OH	BO	0.00268	0.00318	0.00332	0.00339	0.00289	0.00318	[b]
	OH	0.00074	0.00085	0.00088	0.00090	0.00079	0.00085	[b]
	BF	0.00220	0.00270	0.00284	0.00283	0.00233	0.00270	[b]
BF_3	BF	0.00223	0.00274	0.00288	0.00286	0.00236	0.00273	[b]
CS	CS	0.00364	0.0045	0.00473	0.0049	0.00409	0.00445	[b]
CCl_2	CCl		0.0049		0.0056	0.00494		[b]
CH_3Cl	CCl	0.00370	0.0038		0.0046	0.00416	0.00414	[b]
$(CH_3S)_2$	SS	0.00462	0.00540		0.00662	0.00558	0.00566	[b]
$(CH_3S)_2$	CS	0.00405	0.00464		0.00529	0.00466	0.00468	[b]
$N_2O{\cdot}CO$	CO	0.0020	0.0023	0.0025		0.0021		[c]
$N_2O{\cdot}CO$	NN	0.0018	0.0021	0.0023		0.0025		[c]
$N_2O{\cdot}CO$	C···O	0.0024	0.0020	0.0017		0.0055		[c]

[a] $r_e^{BO} = r_e(\text{CCSD(T)\_ae/cc-pwCVTZ}) + r_e(\text{MP2\_ae/cc-pwCVQZ}) - r_e(\text{MP2\_ae/cc-pwCVTZ})$

[b] Demaison J, Vogt N (2020) Accurate Structure Determination of Free Molecules. Springer-Nature, Switzerland, 277 p

[c] Demaison J, Vogt N, Jin Y, Saragi RT, Juanes M, Lesarri A (2021) J Chem Phys 154: 194302

Table 2.9 Effect of diffuse functions on the geometrical parameters: aug-cc-pV*n*Z—cc-pV*n*Z (bond lengths in Å and angles in degrees)

Molecule	Bond	VTZ	VQZ	V5Z	V6Z	Ref.
HF	H–F	0.0038	0.0015	0.0006	−0.0004	a
N_2	N≡N	0.0002	0.0002	0.0001	0.0000	a
F_2	F–F	0.0023	0.0001	−0.0001	−0.0001	a
CO	C≡O	0.0003	0.0004	0.0002	0.0001	a
HC≡CH	C≡C	0.0005	0.0004	0.0001	0.0001	a
	C–H	0.0003	0.0002	0.0001	0.0001	a
NH_3	N–H	0.0006	0.0004	0.0003	0.0002	a
	HNH	0.72	0.34			a
CH_3F	C–F	0.0078	0.0029	0.0008		a
	C–H	0.0003	0.0002	0.0002		a
	HCF	−0.46	−0.17	−0.04		a
H_2O	O–H	0.0022	0.0010	0.0004	0.0002	a
	HOH		0.24			a
H_2O_2	O–H	0.0025	0.0008	0.0004		a
	O–O	0.0031	0.0010	0.0003		a
$N_2O{\cdot}CO$	C≡O	0.0003	0.0005	0.0002	0.0002	b
	N≡N	0.0003	0.0003	0.0002	0.0001	b
	C···O	−0.0423	−0.0307	−0.0097	−0.0084	b
	OCO	1.73	−0.04	−0.12	0.01	b

[a] Demaison J, Császár AG (2012) Equilibrium CO bond lengths. J Mol Struct 1023: 7–14
[b] Demaison J, Vogt N, Jin Y, Saragi RT, Juanes M, Lesarri A (2021) J Chem Phys 154: 194302
Most calculations were performed at the CCSD(T) level of theory, with the exception of $N_2O \cdot CO$ for which the MP2 method was used

0.00022 Å and that of the hextuple ones is H–P = 0.00003 Å [16]. For the sake of comparison, the Q–(T) correction is 0.00075 Å. For N_2O, the contributions for the NN bond length are: P–Q = 0.00043 Å and H–P = 0.00007 Å; for the NO bond length, the contributions are: P–Q = −0.00002 Å and H–P = −0.00001 Å [16]. The Q–(T) correction is 0.00010 Å for the NN bond and 0.00011 Å for the NO bond. Obviously, when the quintuple excitations are taken into account, the other contributions of the same order of magnitude such as the scalar relativistic correction have to be considered. For instance, for N_2, the scalar relativistic correction is −0.00018 Å, i.e., of the same order of magnitude as the pentuple excitations correction [16].

The relativistic effects are non-negligible when heavy atoms are present. These effect as they scale up as Z^4 (see for instance Table 2.12 in Ref. [4]) and, therefore, small for the majority of atoms. For this reason, perturbation theory may be used. A number of methods have been developed:

Table 2.10 Contribution of the quadruple excitations CCSDTQ–CCSD(T) (distances in Å and angles in degrees)

Molecule	Bond	Q–(T)	Ref.	Molecule	Bond	Q–(T)	Ref.
HF	HF	**0.00028**	1	N_2	NN	−0.00106	1
H_2O	OH	0.00026	1	F_2	FF	**0.00259**	1
NH_3	NH	0.00018	1	CO	CO	0.00045	1
CH_4	CH	0.00012	1	HCN	CN	0.00053	1
CH_2	CH	−0.00014	2	HNC	NC	0.00041	1
HCN	CH	−0.00010	1	HCCH	CC	0.00042	1
HNC	NH	−0.00002	1	CO_2	CO	0.00054	1
HCCH	CH	−0.00013	1	N_2H_2	NN	0.00078	1
N_2H_2	NH	−0.00001	1	C_2H_4	CC	0.00051	1
C_2H_4	CH	0.00000	1	H_2CO	CO	0.00062	1
H_2CO	CH	0.00006	1	HOF	OF	0.00210	1
HOF	OH	0.00021	1	HNO	NO	0.00106	1
HNO	HN	−0.00015	1	N_2O	NN	0.00100	3
MAD		0.00013		N_2O	NO	0.00105	3
Max		0.00028		C_3	CC	0.00074	4
				SO_2	SO	0.0009	5
Molecule	Angle	Q–(T)	Ref.	MAD		0.00076	
H = O	HOH	0.0139	1	Max		0.00259	
NH_3	HNH	0.006	2				
CH_2	HCH	−0.210	2				
N_2H_2	HNN	−0.0084	1				
C_2H_4	HCC	0.0055	1				
H_2CO	HCO	0.0006	1				
HOF	HOF	−0.0512	1				
HNO	HNO	0.023	1				
SO_2	OSO	0.04	5				
MAD		0.0112					
Max		0.210					

[1] Puzzarini C. Heckert M. Gauss J (2008) J Chem Phys 128: 194108
[2] Heckert M, Kállay M, Gauss J (2005) Mol Phys 103: 2109–2115
[3] Schröder B. Sebald P. Stein C. Weser O. Botschwina P (2015) Z Phys Chem 229: 1663–1690
[4] Schröder B. Sebald P (2016) J Chem Phys 144: 044307
[5] Demaison J, Liévin J (2022) Mol Phys 120: 1950857

- the second-order Douglas-Kroll-Hess no-pair Hamiltonian [81–83],
- the exact two component relativistic Fock operator [84, 85],
- the Cowan-Griffin operator [86].

However, the simplest method is the pseudopotential approximation [50, 51].

2.8 Lower-Level Methods

For a large molecule, the CCSD(T) method may be too expensive. In such a frequent case, the solution is to use lower-level methods. It was already pointed out in Sect. 2.3 that the double-hybrid DFT methods can give results close to the equilibrium structure. It is also well established that the MP2_fc/cc-pVTZ level of theory can also give satisfactory results, see Table 2.1. In particular, it furnishes accurate values for the C–H bond length [87]. This level of theory still performs well for C–C and C–N bond lengths [88] provided that the π-character is negligible. See also Sect. 8.8.

For the bond angles, the MP2 method is able to predict them with an accuracy better than 0.3–0.4° in most cases [89] with the cc-pVQZ basis set being slightly more accurate than the cc-pVTZ one and the relatively small 6–311 + G(3df,2pd) basis set being a good compromise.

However, it gives poor results for dihedral angles, the error being sometimes as large as several degrees [90]. This can be explained by the fact that it requires much less energy to modify a dihedral angle than a bond angle. The accuracy was found to be much less sensitive to the basis set than to the method, and the CCSD/cc-pVTZ level of theory was found to be a significant improvement over the MP2 method.

It is also possible to estimate the dihedral angles using an empirical correlation. The residuals of 103 dihedral angles of organic molecules have been analyzed [90], and a nice correlation was observed between τ_e–τ[MP2_fc/cc-pVTZ] and τ[MP2_fc/cc-pVTZ]–τ[B3LYP/6-311 + G(3df,2pd)]. This fact allows us to predict the torsional angles with a standard deviation of 0.37°, which is acceptable for predicates. The resulting equation is given by

$$\tau_e - \tau\left[\text{MP2\_fc/cc-pVTZ}\right] = -0.2916(15) \times \tau\left[\text{MP2\_fc/cc-pVTZ}\right] - \tau[\text{B3LYP/6-311+G(3df, 2pd)}]. \tag{2.19}$$

With some exceptions mentioned above, the situation is more complicated for bond lengths because some systematic errors remain. These errors, assumed to be systematic and called offsets, may be estimated with the help of molecules whose structure is accurately known and can then be used to correct the ab initio structure of a new molecule. This simple approach is expected to be reliable only for molecules that have a similar bonding as those used to derive the offset.

One way to improve this method is to use a linear regression approach:

$$r_e = a \times r_x + b \tag{2.20}$$

where x = MP2, DFT, etc.

This equation has already been used with success for many bonds, for instance, for the C–F bond [90], the CO bond [91]. Several examples using the B2PLYP functional are given by Penocchio et al. [34].

2.9 Calculation of the Force Field

This topic was reviewed by Császár [92].

In the framework of the Born–Oppenheimer approximation, the electronic energy at a given set of nuclear positions is the potential energy whose minimum defines the equilibrium structure. The potential energy surface governs the vibrations of the molecules, and its knowledge is important for a better understanding of reaction kinetics.

When one is interested in equilibrium structure and vibrations in the ground electronic state, it is enough to know the potential hypersurface near its minimum. In other words, the potential energy may be developed in Taylor series as a function of the nuclear displacement coordinates around the equilibrium. The coefficients of this expansion are the force field. The first and most important term of the expansion is the harmonic (quadratic) force field, the first-order force constants being zero because the expansion is made around the minimum. The following terms in the expansion are the anharmonic force field (cubic, quartic, ...).

Reliable experimental anharmonic force fields are available for only a tiny number of extremely small molecules (mainly triatomic molecules). Ab initio computations are the simplest way to obtain a complete force field up to semidiagonal quartic terms that have been computed for a molecule as large as proline [93]. The potential of this 17-atom molecule is defined by 62,835 force constants.

Expansion of the potential energy surface around equilibrium is written as:

$$V = \frac{1}{2}\sum_{ij} f_{ij} R_i R_j + \frac{1}{6}\sum_{ijk} R_i R_j R_k + \frac{1}{24}\sum_{ijkl} R_i R_j R_k R_l + \cdots, \qquad (2.21)$$

where $\boldsymbol{R}$ denotes an arbitrary set of complete and non-redundant nuclear displacement coordinates. Different coordinates may be used: normal coordinates (Q), dimensionless normal coordinates (q), Cartesian coordinates (x), and internal valence coordinates (R, displacements in internuclear distances r, bond angles θ, and torsional angles τ from their equilibrium values). These sets of coordinates are rectilinear, but the internal coordinates can be curvilinear which give a more accurate representation because atoms move around arc of circles in bending and torsion vibrations (the off-diagonal terms are smaller). The rovibrational Hamiltonian is formulated in normal coordinates, but these coordinates are not isotope independent. On the other hand, Cartesian coordinates are useful in some calculations but do not have any physical meaning. For comparisons (or transfer), internal coordinates have to be chosen. One additional difficulty is that the transformation of one set of coordinates to another one is a nonlinear transformation [94].

Several computer programs compute analytically second (f_{ij}) derivatives for most methods. There are two ways to obtain all force constants: (i) a least-squares fitting of the energy [95] and (ii) numerical differentiation which is implemented in many programs. The numerical differentiation method is obviously easier. The calculation of the potential hypersurface was made for many small molecules, for instance: N_2 and N_2O [16], N_3^- and CO_2 [80], C_3H^- [96], HCN/HNC [97], H_2O [98], and H_3^+ [99].

References

1. Helgaker T, Jørgensen P, Olsen J (2000) Molecular electronic-structure theory. Wiley, Chichester
2. Allen WD, Császár AG (2011) Quantum theory of equilibrium molecular structures. In: Demaison J, Boggs JE, Császár AG (eds) Equilibrium molecular structures. CRC Press, Boca Raton, pp 1–28
3. Bartlett RJ, Stanton JF (1994) Applications of post-Hartree-Fock methods: a tutorial. In: Lipkowitz KB, Boyd DB (eds) Reviews in computational chemistry, vol V. VCH, New York, pp 65–117
4. Demaison J, Vogt N (2020) Accurate structure determination of free molecules. Springer-Nature, Switzerland, p 7-52
5. Helgaker T, Ruden TA, Jørgensen P, Olsen J, Klopper W (2004) A priori calculations of molecular properties to chemical accuracy. J Phys Org Chem 17:913–933
6. Puzzarini C, Barone V (2009) Benchmark calculations for molecules in the gas phase: state-of-the-art coupled-cluster computations. Int J Quantum Chem 110:637–655
7. Puzzarini C, Stanton JF, Gauss J (2010) Quantum-chemical calculation of spectroscopic parameters for rotational spectroscopy. Int Rev Phys Chem 29:273–367
8. Bunker PR, Jensen P (2000) The Born-Oppenheimer approximation. In: Jensen P, Bunker PR (eds) Computational molecular spectroscopy. Wiley, Chichester, pp 3–11
9. Roothaan CCJ (1951) New developments in molecular orbital theory. Rev Mod Phys 23:69–89
10. Møller C, Plesset MS (1934) Note on an approximation treatment for many-electron systems. Phys Rev 46:618–622
11. Purvis GD III, Bartlett RJ (1982) A full coupled-cluster singles and doubles model: the inclusion of disconnected triples. J Chem Phys 76:1910–1918
12. Raghavachari K, Trucks GW, Pople JA, Head-Gordon M (1989) A fifth-order perturbation comparison of electron correlation theories. Chem Phys Lett 157:479–483
13. Kállay M, Surján PR (2001) Higher excitations in coupled-cluster theory. J Chem Phys 115:2945–2954
14. Bomble YJ, Stanton JF, Kállay M, Gauss J (2005) Coupled-cluster methods including noniterative corrections for quadruple excitations. J Chem Phys 123:054101
15. Kállay M, Gauss J (2005) Approximate treatment of higher excitations in coupled cluster theory. J Chem Phys 123:214105
16. Schröder B, Sebald P, Stein C, Weser O, Botschwina P (2015) Challenging high-level ab initio rovibrational spectroscopy: the nitrous oxide molecule. Z Phys Chem 229:1663–1690
17. Morgan WJ, Matthews DA, Ringholm M, Agarwal J, Gong JZ, Ruud K, Allen WD, Stanton JF, Schaefer HF (2018) Geometric energy derivatives at the complete basis set limit: application to the equilibrium structure and molecular force field of formaldehyde. J Chem Theory Comput 14:1333–1350

18. Bak KL, Gauss J, Jørgensen P, Olsen J, Helgaker T, Stanton JF (2001) The accurate determination of molecular equilibrium structures. J Chem Phys 114:6548–6556
19. Coriani S, Marchesan D, Gauss J, Hättig C, Helgaker T, Jørgensen P (2005) The accuracy of ab initio molecular geometries for systems containing second-row atoms. J Chem Phys 123:184107
20. Halkier A, Jørgensen P, Gauss J, Helgaker T (1997) CCSDT calculations of molecular equilibrium geometries. Chem Phys Lett 274:235–241
21. Ruden TA, Helgaker T, Jørgensen P, Olsen J (2004) Coupled-cluster connected quadruples and quintuples corrections to the harmonic vibrational frequencies and equilibrium bond distances of HF, N_2, F_2, and CO. J Chem Phys 121:5874–5884
22. Feller D, Peterson KA, Dixon DA (2008) A survey of factors contributing to accurate theoretical predictions of atomization energies and molecular structures. J Chem Phys 129:204105
23. Heckert M, Kállay M, Tew DP, Klopper W, Gauss J (2006) Basis-set extrapolation techniques for the accurate calculation of molecular equilibrium geometries using coupled-cluster theory. J Chem Phys 125:044108
24. Peterson KA, Dunning TH Jr (1997) Benchmark calculations with correlated molecular wave functions. VIII. Bond energies and equilibrium geometries of the CH_n and C_2H_n ($n = 1$–4) series. J Chem Phys 106:4110–4140
25. Bauschlicher CW, Ricca A, Partridge H (1997) Chemistry by density functional theory. In: Chong DP (ed) Recent advances in density functional methods. World Scientific, Singapore, pp 165–227
26. Mardirossian N, Head-Gordon M (2017) Thirty years of density functional theory in computational chemistry: an overview and extensive assessment of 200 density functionals. Mol Phys 115:2315–2372
27. Hohenberg P, Kohn W (1964) Inhomogeneous electron gas. Phys Rev B 136:864–871
28. Kohn W, Sham LJ (1965) Self-consistent equations including exchange and correlation effects. Phys Rev A 140:1133–1138
29. Becke AD (1993) A new mixing of Hartree-Fock and local density-functional theories. J Chem Phys 98:1372–1377
30. Becke AD (1988) Density-functional exchange-energy approximation with correct asymptotic behavior. Phys Rev A 38:3098–3100
31. Lee CT, Yang WT, Parr RG (1988) Development of the Colle-Salvetti correlation-energy formula into a functional of the electron density. Phys Rev B 37:785–789
32. Riccardo M, Penocchio E, Puzzarini C, Biczysko M, Barone V (2015) Semi-experimental equilibrium structure determinations by employing B3LYP/SNSD anharmonic force fields: validation and application to semirigid organic molecules. J Phys Chem A 119:2058–2082
33. Grimme S (2006) Semiempirical hybrid density functional with perturbative second-order correlation. J Chem Phys 124:034108–034116
34. Penocchio E, Riccardo M, Barone V (2015) Semiexperimental equilibrium structures for building blocks of organic and biological molecules: the B2PLYP route. J Phys Chem A 11:4689–4707
35. Santra G, Sylvetsky N, Martin JML (2019) Minimally empirical double-hybrid functionals trained against the GMTKN55 database: revDSD-PBEP86-D4, revDOD-PBE-D4, and DOD-SCAN-D4. J Phys Chem A 123:5129–5143
36. Melli A, Tonolo F, Barone V, Puzzarini C (2021) Extending the applicability of the semi-experimental approach by means of template molecule and linear regression models on top of DFT computations. J Phys Chem A 125:9904–9916
37. Barone V, Ceselin G, Fuse M, Tasinato N (2020) Accuracy meets interpretability for computational spectroscopy by means of hybrid and double-hybrid functionals. Front Chem 8:584203
38. Grimme S (2006) Semiempirical GGA-type density functional constructed with a long-range dispersion correction. J Comput Chem 27:1787–1799
39. Grimme S, Ehrlich S, Goerigk L (2011) Effect of the damping function in dispersion corrected density functional theory. J Comp Chem 32:1456–1465

40. Schröder H, Creon A, Schwabe T (2015) Reformulation of the D3 (Becke-Johnson) dispersion correction without resorting to higher than C6 dispersion coefficients. J Chem Theory Comput 11:3163–3170
41. Smith DGA, Burns LA, Patkowski K, Sherrill CD (2016) Revised damping parameters for the D3 dispersion correction to density functional theory. J Phys Chem Lett 7:2197–2203
42. Grimme S, Bannwarth C, Caldeweyher E, Pisarek J, Hansen AA (2017) General intermolecular force field based on tight-binding quantum chemical calculations. J Chem Phys 147:161708
43. Caldeweyher E, Ehlert S, Hansen A, Neugebauer H, Spicher S, Bannwarth C, Grimme S (2019) Generally applicable atomic-charge dependent London dispersion correction. J Chem Phys 150:154122
44. Penocchio E, Riccardo M, Barone V (2015) Semiexperimental equilibrium structures for building blocks of organic and biological molecules: the B2PLYP route. J Chem Theory Comput 11:4689–4707
45. Ahlrichs R, Taylor PR (1981) The choice of Gaussian basis sets for molecular electronic structure calculations. J Chem Phys 78:315–324
46. Feller D, Davidson ER (1986) Basis set selection for molecular calculations. Chem Rev 86:681–696
47. Jensen F (2012) Atomic orbital basis sets. WIREs Comput Mol Sci 3:273–295
48. Dunning TH Jr (1989) Gaussian basis sets for use in correlated molecular calculations. I. the atoms boron through neon and hydrogen. J Chem Phys 90:1007–1023
49. Dunning TH Jr, Peterson KA, Wilson AK (2001) Gaussian basis sets for use in correlated molecular calculations. X. The atoms aluminum through argon revisited. J Chem Phys 114:9244–9253
50. Peterson KA (2003) Systematically convergent basis sets with relativistic pseudopotentials. I. Correlation consistent basis sets for the post-d group 13–15 elements. J Chem Phys 119:11099–11112
51. Peterson KA, Figgen D, Goll E, Stoll H, Dolg M (2003) Systematically convergent basis sets with relativistic pseudopotentials. II. Small-core pseudopotentials and correlation consistent basis sets for the post-d group 16–18 elements. J Chem Phys 119:11113–11123
52. Woon DE, Dunning TH Jr (1995) Gaussian basis sets for use in correlated molecular calculations. V. Core-valence basis sets for boron through neon. J Chem Phys 103:4572–4585
53. Peterson KA, Dunning TH Jr (2002) Accurate correlation consistent basis sets for molecular core-valence correlation effects: the second-row atoms Al-Ar, and the first-row atoms B–Ne revisited. J Chem Phys 117:10548–10560
54. Kendall RA, Dunning TH Jr, Harrison RJ (1992) Electron affinities of the first-row atoms revisited. Systematic basis sets and wave functions. J Chem Phys 96:6796–6806
55. Peterson KA, Adler TB, Werner H-J (2008) Systematically convergent basis sets for explicitly correlated wavefunctions: the atoms H, He, B-Ne, and Al–Ar. J Chem Phys 128:084102
56. Mazumder S, Peterson KA (2010) Correlation consistent basis sets for molecular core-valence effects with explicitly correlated wave functions: the atoms B-Ne and Al–Ar. J Chem Phys 132:05418
57. Puzzarini C, Hecker M, Gauss J (2008) The accuracy of rotational constants predicted by high-level quantum-chemical calculations. I. molecules containing first-row atoms. J Chem Phys 128:194108
58. Almlöf J, Taylor PR (1987) General contraction of Gaussian basis sets. I. Atomic natural orbitals for first- and second-row atoms. J Chem Phys 86:4070–4077
59. Almlöf J, Taylor PR (1990) General contraction of Gaussian basis sets. II. Atomic natural orbitals and the calculation of atomic and molecular properties. J Chem Phys 90:551–560
60. McCaslin L, Stanton JF (2013) Calculation of fundamental frequencies for small polyatomic molecules: a comparison between correlation consistent and atomic natural orbital basis sets. Mol Phys 111:1492–1496
61. Jensen F (2001) Polarization consistent basis sets: principles. J Chem Phys 115:9113–9125; Erratum: J Chem Phys: 116:3502–3502

62. Jensen F (2002) Polarization consistent basis sets. II. Estimating the Kohn-Sham basis set limit. J Chem Phys 116:7372–7379
63. Jensen F (2007) Polarization consistent basis sets. 4. The elements He, Li, Be, B, Ne, Na, Mg, Al, and Ar. J Phys Chem A 111:11198–11204
64. Jensen F (2012) Polarization consistent basis sets. VII. The elements K, Ca, Ga, Ge, As, Se, Br, and Kr. J Chem Phys 136:114107
65. Carnimeo I, Puzzarini C, Tasinato N, Stoppa P, Pietropolli- Charmet A, Biczysko M, Cappelli C, Barone V (2013) Anharmonic theoretical simulations of infrared spectra of halogenated organic compounds. J Chem Phys 139:074310
66. Pritchard BP, Altarawy D, Didier B, Gibsom TD, Windus TL (2019) A new basis set exchange: an open, up-to-date resource for the molecular sciences community. J Chem Inf Model 59:4814–4820
67. Feller D (1996) The role of databases in support of computational chemistry calculations. J Comput Chem 17:1571–1586
68. Schuchardt KL, Didier BT, Elsethagen T, Sun L, Gurumoorthi V, Chase J, Li J, Windus TL (2007) Basis set exchange: a community database for computational sciences. J Chem Inf Model 47:1045–1052
69. Feller D (1993) The use of systematic sequences of wave functions for estimating the complete basis set, full configuration interaction limit in water. J Chem Phys 98:7059–7071
70. Helgaker T, Klopper W, Koch H, Noga J (1997) Basis-set convergence of correlated calculations on water. J Chem Phys 106:9639–9646
71. Puzzarini C (2009) Extrapolation to the complete basis set limit of structural parameters: comparison of different approaches. J Phys Chem A 113:14530–14535
72. Peterson KA, Woon DE, Dunning TH (1994) Benchmark calculations with correlated molecular wave functions. IV. The classical barrier height of the $H+H_2 \rightarrow H_2+H$ reaction. J Chem Phys 100:7410–7415
73. Hättig C, Klopper W, Köhn A, Tew DP (2012) Explicitly correlated electrons in molecules. Chem Rev 112:4–74
74. Kong L, Bischoff FA, Valeev EF (2012) Explicitly correlated R12/F12 methods for electronic structure. Chem Rev 112:75–107
75. Adler TB, Knizia G, Werner H-J (2007) A simple and efficient CCSD(T)-F12 approximation. J Chem Phys 127:221106
76. Demaison J, Liévin J, Vogt N (2023) Accurate equilibrium structures of some challenging molecules: FNO, ClNO, HONO, and FNO_2. J Mol Spectrosc 394:111788
77. Szalay PG, Müller T, Gidofalvi G, Lischka H, Shepard R (2012) Multiconfiguration self-consistent field and multireference configuration interaction methods and applications. Chem Rev 112:108–181
78. Schröder B, Sebald P (2016) High-level theoretical rovibrational spectroscopy beyond fc-CCSD(T): the C_3 molecule. J Chem Phys 144:044307
79. Egorov O, Rey MM, Nikitin AV, Viglaska D (2021) New ab initio potential energy surfaces for NH_3 constructed from explicitly correlated coupled cluster methods. J Phys Chem A 125:10568–10579
80. Sebald P, Stein C, Oswald R, Botschwina P (2013) Rovibrational states of N_3^- and CO_2 up to high J: a theoretical study beyond fc-CCSD(T). J Phys Chem A 117:13806–13814
81. Douglas M, Kroll NM (1974) Quantum electrodynamical corrections to the fine structure of helium. Ann Phys 82:89–155
82. Hess BA (1985) Applicability of the no-pair equation with free-particle projection operators to atomic and molecular structure calculations. Phys Rev 32:756–763
83. de Jong WA, Harrison RJ, Dixon DA (2001) Parallel Douglas-Kroll energy and gradients in NWChem: estimating scalar relativistic effects using Douglas-Kroll contracted basis sets. J Chem Phys 114:48–53
84. Kutzelnigg W, Liu W (2005) Quasirelativistic theory equivalent to fully relativistic theory. J Chem Phys 123:241102

85. Peng D, Reiher M (2012) Exact decoupling of the relativistic Fock operator. Theor Chem Acc 131:1081
86. Cowan RD, Griffin DC (1976) Approximate relativistic corrections to atomic radial wave functions. J Opt Soc Am 66:1010–1014
87. Demaison J, Craig NC (2011) Semiexperimental equilibrium structure for cis, trans-1,4-difluorobutadiene by the mixed estimation method. J Phys Chem A 115:8049–8054
88. Demaison J, Craig NC, Cocinero EJ, Grabow J-U, Lesarri A, Rudolph HD (2012) Semiexperimental equilibrium structures for the equatorial conformers of *N*-methylpiperidone and tropinone by the mixed estimation method. J Phys Chem A 116:8684–8692
89. Margulès L, Demaison J, Boggs JE (2010) Ab initio and equilibrium bond angles. Structures of HNO and H_2O_2. J Mol Struct (Theochem) 500:245–258
90. Juanes M, Vogt N, Demaison J, León I, Lesarri A, Rudolph HD (2017) Axial–equatorial isomerism and semiexperimental equilibrium structures of fluorocyclohexane. Phys Chem Chem Phys 19:29162–29169
91. Demaison J, Császár AG (2012) Equilibrium CO bond lengths. J Mol Struct 1023:7–14
92. Császár AG (2012) Anharmonic molecular force fields. WIREs Comput Mol Sci 2:273–289
93. Allen WD, Czinki E, Császár AG (2004) Molecular structure of proline. Chem Eur J 10:4512–4517
94. Hoy AR, Mills IM, Strey G (1972) Anharmonic force constants calculations. Mol Phys 24:1265–1290
95. Császár AG, Allen WD, Yamaguchi Y, Schaefer HF (2000) Ab initio determination of accurate potential energy hypersurfaces for the ground electronic states of molecules. In: Jensen P, Bunker PR (eds) Computational molecular spectroscopy. Wiley, New York, pp 15–16
96. Lakin NM, Hochlaf M, Chambaud G, Rosmus P (2001) The potential energy surface and vibrational structure of C_3H^-. J Chem Phys 115:3664–4367
97. van Mourik T, Harris GJ, Polyansky OL, Tennyson J, Császár AG, Knowles PJ (2001) Ab initio potential, dipole, adiabatic and relativistic correction surfaces for the HCN/HNC system. J Chem Phys 115:3706–3718
98. Polyansky OL, Császár AG, Shirin SV, Zobov NF, Barletta P, Tennyson J, Schwenke D, Knowles PJ (2003) High-accuracy rotation–vibration transitions for water. Science 299:539–542
99. Munro JJ, Ramanlal J, Tennyson J, Mussa HY (2006) Properties of high-lying vibrational states of the $H_3{}^+$ molecular ion. Mol Phys 104:115–125

Chapter 3
Rovibrational Spectroscopy and Structure of Diatomic Molecules

Abstract The results of the rovibrational spectroscopy of the diatomic molecule are presented. The determination of the equilibrium structure is discussed as well as the influence of the breakdown of the Born–Oppenheimer approximation and the effect of the size of the nuclei.

3.1 Introduction

The theory available to calculate the rovibrational energies of a diatomic molecule is much more sophisticated than for polyatomic molecules. In particular, the equilibrium bond length can be determined with a much higher accuracy, and the breakdown of the Born–Oppenheimer (BO) approximation can be pointed out.

This chapter mainly describes the rovibrational spectroscopy of diatomic molecules in the electronic ground state $^1\sum^+$ or $0^{\pm}$. The other states will be briefly discussed at the end of this chapter.

References for this section are: Gordy and Cook [1], Herzberg [2], Le Roy [3], Tiemann [4], and chapter 3 of Ref.[5].

3.2 Dunham's Theory [6]

One of the most fundamental approximations in molecular physics is the Born–Oppenheimer approximation, see Chap. 2, Sect. 2.2. It assumes that the motions of the nuclei and the electrons can be treated separately because the nuclei, being much heavier, move much more slowly. In the frame of this approximation, it is possible to define a potential for the molecule whose minimum is the equilibrium structure. A consequence is that this structure is invariant upon isotopic substitution.

It will be assumed that atoms are point masses (more exactly, the atoms are small spheres with all the mass concentrated at their center). It ensues that the center of mass of the electrons coincides with the nucleus. Although it is a rather good

N. Vogt and J. Demaison, *Equilibrium Structure of Free Molecules*, Lecture Notes in Chemistry 111, https://doi.org/10.1007/978-3-031-36045-9_3

approximation, we will see that it is not a perfect one and that it is possible to make a correction.

The electronic energies U as a function of the instantaneous internuclear distance r may be approximated by a power series using the following expansion parameter:

$$\xi = \frac{r - r_e}{r_e} \tag{3.1}$$

where r_e is the equilibrium bond length. One gets

$$U(r) = a_0 \xi^2 \left(1 + \sum_{i=1}^{\infty} a_i \xi^i\right) \tag{3.2}$$

where the a_i are the Dunham potential coefficients. The equilibrium rotational constant B_e is a function of r_e

$$B_e = \frac{h}{8\pi^2 \mu r_e^2} = \frac{h}{8\pi^2 I_e} \tag{3.3}$$

h is the Planck constant and μ the reduced mass

$$\mu = \frac{m_A m_B}{m_A + m_B - C m_e} \tag{3.4}$$

where m_A and m_B are the masses of atoms A and B of the molecule AB, m_e the electron mass, C the charge number of the molecule (for an ion), and $I_e = \mu r^2$ is the equilibrium moment of inertia.

Using the series expansion of the potential, Eq. (3.2), Dunham has derived the rovibrational energy of a molecule in a vibrational state v and a rotational state J

$$\frac{1}{h} E(v, J) = \sum_{lk} Y_{lk} \left(v + \frac{1}{2}\right)^l (J)^k (J + 1)^k \tag{3.5}$$

$$\frac{1}{h} E(v, J) = B_v J(J + 1) - D_v J^2 (J + 1)^2 + H_v J^3 (J + 1)^3 + \cdots + \sum_l Y_{l0} \left(v + \frac{1}{2}\right)^l \tag{3.5'}$$

The rotational constant in a vibrational state v is

$$B_v = \sum_l Y_{l1} \left(v + \frac{1}{2}\right)^l \tag{3.6}$$

The first Dunham coefficient Y_{01} is close to the equilibrium rotational constant, but the exact relationship is

$$Y_{01} = B_e + \Delta Y_{01}^{(D)} \tag{3.7}$$

where $\Delta Y_{01}^{(D)}$ is a small correction called Dunham's correction. It may be calculated using the potential coefficients

$$\Delta Y_{01}^{(D)} = \frac{B_e^3}{2\omega_e^2}\left(15 + 14a_1 - 9a_2 + 15a_3 - 23a_1a_2 + \frac{21}{2}\left[a_1^2 + a_1^3\right]\right) \tag{3.8}$$

where $\omega_e = 2\sqrt{a_0 B_e}$ is the harmonic vibrational frequency. For most molecules, the ratio B_e^3/ω_e^2 is the order of 10^{-6}. This correction is small except for light molecules. For $^{12}C^{16}O$, $\Delta Y_{01}^{(D)} = -0.071$ MHz, and for $H^{35}Cl$, $\Delta Y_{01}^{(D)} = -5.3$ MHz.

The vibration–rotation interaction constant is $\alpha_e = -Y_{11}$.

For an isotopologue α, a Dunham isotopic-independent parameter U_{lk} is defined

$$Y_{lk}^{(\alpha)} = U_{lk}\mu_\alpha^{-(l+2k)/2} \tag{3.9}$$

where μ_α is the reduced mass of isotopologue α, see Eq. (3.4).

3.3 Breakdown of the Born–Oppenheimer Approximation (BOB)

Actually, U_{lk} is not exactly isotope-independent. It is due to

- the breakdown of the Born–Oppenheimer approximation,
- the neglect of $\Delta Y_{01}^{(D)}$ and the assumption that the center of mass of the electrons coincides with the nucleus.

Watson [7] has shown that

$$Y_{lk} = U_{lk}\mu_\alpha^{-(l+2k)/2}\left[1 + \frac{m_e}{m_A}\Delta_{lk}^{A} + \frac{m_e}{m_B}\Delta_{lk}^{B}\right] \tag{3.10}$$

where m_e is the electron mass, m_A and m_B are the masses of atoms A and B, and Δ_{lk}^{i} the Born–Oppenheimer breakdown parameters. For a structure determination, only the Δ_{01}^{i} are significant. They may be expanded as

$$\Delta_{01}^{A} = \left(\Delta_{01}^{A}\right)^{ad} + \frac{(\mu g_J)_B}{M_p} + \frac{\mu\Delta Y_{01}^{(D)}}{m_e B_e} \tag{3.11}$$

$\left(\Delta_{01}^{A}\right)^{ad}$ is the pure adiabatic part, M_p is the mass of the proton, $(\mu g_J)_B$ is the non-adiabatic part, it is the isotopically independent value of μg_J referred to the nucleus B as origin, and g_J is the molecular rotational g-factor. It can be obtained experimentally from the analysis of the Zeeman effect on the rotational spectrum or computed ab initio, see Sect. 4.7. This non-adiabatic correction originates from coupling of the electronic state with other electronic states. The value of g is large for light molecules and decreases rapidly with the value of the rotational constant.

If the pure adiabatic part, $\left(\Delta_{01}^{\mathrm{A}}\right)^{\mathrm{ad}}$, is neglected

$$Y_{01} = B_{\mathrm{e}}^{\mathrm{ad}}\left[1 + \frac{\Delta Y_{01}^{(\mathrm{D})}}{B_{\mathrm{e}}} + \frac{m_{\mathrm{e}}}{M_{\mathrm{p}}} g_J\right] \tag{3.12}$$

When the structure is determined from $B_{\mathrm{e}}^{\mathrm{ad}}$, it gives the adiabatic bond length, $r_{\mathrm{e}}^{\mathrm{ad}}$

$$r_{\mathrm{e}}^{\mathrm{ad}} = r_{\mathrm{e}}^{\exp}\left[1 + \frac{\Delta Y_{01}^{(\mathrm{D})}}{B_{\mathrm{e}}} + \frac{m_e g_J}{M_{\mathrm{p}}}\right]^{1/2} \tag{3.13}$$

$r_{\mathrm{e}}^{\exp}$ is the bond length obtained from Y_{01}. When the three terms of Eq. (3.11) are taken into account, it gives the Born–Oppenheimer bond length, $r_{\mathrm{e}}^{\mathrm{BO}}$.

The different structures of CO are given in Table 3.1 using the same isotopologues as in the work of Watson [7], except for the isotopologue $^{14}C^{16}O$ whose Y_{01} value is inaccurate.

Table 3.1 shows that the experimental equilibrium bond length, $r_{\mathrm{e}}^{\exp}$, decreases as the moment of inertia increases (i.e., the mass of the molecule) leading to the conclusion that r_{e} is not invariant upon isotopic substitution. Taking into account the non-adiabatic correction, it greatly reduces the variation, but there is still a small effect due to the adiabatic correction. When this adiabatic correction is taken into account

Table 3.1 Equilibrium structure (in Å) of CO[a]

	Y_{01}/MHz	$r_{\mathrm{e}}^{\exp}$	Corrections/MHz			
			ΔB[b]	$\Delta Y_{01}^{(D)}$	$r_{\mathrm{e}}^{\mathrm{ad}}$	Ref.
$^{12}C^{16}O$	57,898.342	1.1283233	−8.485	−0.071	1.1282399	1
$^{12}C^{17}O$	56,432.467	1.1283209	−8.061	−0.067	1.1282396	2
$^{13}C^{16}O$	55,346.235	1.1283191	−7.748	−0.065	1.1282395	3
$^{12}C^{18}O$	55,135.244	1.1283187	−7.694	−0.064	1.1282393	3
$^{13}C^{18}O$	52,583.094	1.1283146	−6.993	−0.058	1.1282389	3
Range		0.0000087			0.0000010	

[a] See also: Watson JKG (1973) J Mol Spectrosc 45: 99–113
[b] Electronic correction: $m_e g_J B_e / M_p$; g-factor from: Meerts WL, De Leeuw FH, Dymanus A (1977). Chem Phys 22: 319–324
[1] Le Floch A (1991) Mol Phys 72: 133–144
[2] Lovas FJ, Tiemann E (1974) J Phys Chem Ref Data 3: 609–770
[3] Dale RM, Herman M, Johns JWC, McKellar ARW, Nagler S, Strathy JKM (1979) Can J Phys 57: 677–686

Table 3.2 Derived parameters for the breakdown of the Born–Oppenheimer approximation

	Δ_{01}	$\frac{\mu \Delta Y_{01}^{(D)}}{m_e B_e}$	$\frac{(\mu g_J)_B}{M_p}$	$(\Delta_{01})^{ad}$	r_e/Å	Ref.
CO/C	2.0545(12)	−0.01526(6)	−1.8484(36)	−0.197(35)	1.128230(1)	1,2
CO/O	−2.0982(13)	−0.01526(6)	−1.8072(36)	−0.296(50)		
CS/C	−2.5518(70)	−0.0148	−2.4758	−0.105(55)	1.53482211(41)	2,3
CS/S	−2.358(20)	−0.0148	−1.9446	−0.264(110)		
SiS/Si	−1.3935(42)	−0.0106	−1.176	−0.205(69)	1.92926351(26)	2,4
SiS/S	−1.8728(55)	−0.0106	−1.5494	−0.310(75)		
GeS/Ge	−1.452(111)	0.0008	−1.2244	−0.239(70)	2.01204279(34)	2,5
GeS/S	−1.884(50)	0.0008	−1.6384	−0.233(45)		
SnS/Sn	−1.76(19)	0.0058	−1.0608	−0.70(19)	2.2089829(22)	2
SnS/S	−1.821(65)	0.0058	−1.6602	−0.167(25)		

[1] Authier N, Bagland N, Le Floch A (1993) J Mol Spectrosc 160: 590–592
[2] Tiemann E, Arnst H, Stieda WU, Törring T, Hoeft J (1982) Chem Phys 67:133–138
[3] Uehara H, Horiai K, Sakamoto Y (2015) J Mol Spectrosc 313: 19–39
[4] Müller HSP, McCarthy MC, Bizzocchi L, Gupta H, Esser S, Lichau H, Caris M, Lewen F, Hahn J, Degli Esposti C, Schlemmer S, Thaddeus P (2007) Phys Chem Chem Phys 9: 1579–1586
[5] Uehara H, Horiai K, Ozaki Y, Konno T (1995) J Mol Struct 352/353: 395–405

$\left((\Delta_{01}^{C})^{ad} = -2.05455;\ (\Delta_{01}^{O})^{ad} = -2.09818\right)$, the bond length is 1.128230(1) Å and the isotopic dependence completely vanishes. It is worth noting that the Dunham correction, $\Delta Y_{01}^{(D)}$, decreases the bond length by only 0.0000007 Å. Its effect is quite small for CO.

Table 3.2 reports the results for some molecules. It shows that the largest contribution to the BOB comes from the non-adiabatic term (i.e., the *g*-factor) and that the Dunham correction $\Delta Y_{01}^{(D)}$ is negligible for heavy molecules.

3.4 Effect of the Size of the Nuclei

$(\Delta_{01}^{A})^{ad}$ is expected to be of the same order of magnitude and close to unity, for both atoms in agreement with the prediction from Watson [8], see also Table 3.2. However, for molecules with a heavy nucleus, large differences are observed, see Table 3.3.

This discrepancy was explained by Tiemann et al. [9] as due to the finite size of the nucleus. When at least one of the atoms is heavy, its nucleus can no longer be considered as a point charge, and the charge is distributed over a volume to produce the so-called field-shift contribution. The nuclear size will vary slightly

Table 3.3 Δ_{01} parameters for some molecules AB, see also Tables 3.2 and 3.4

Molecule	U_{01}/MHz	Δ_{01}^{A}	Δ_{01}^{B}	Ref.
HeH^+	842,866(12)	0.878(35)	−0.200(22)	1
ArH^+	250,214.36(30)	0.91(10)	0.1244(37)	2
KrH^+	250,232.03(30)	0.682(70)	0.1215(40)	3
XeH^+	196,721.2(15)	0.5414(20)	−0.0150(30)	4
CO	397,029.003(24)	−2.0545(12)	−2.0982(13)	5
ClH	311,077.90(96)	−0.26(20)	0.1262(8)	6
BrH	252,592.768(65)	−0.132	0.107	7
ZrO	172,480.086(98)	−4.872(39)	−6.1888(25)	8
ZrS	108,670.07(19)	−5.325(82)	−6.523(39)	8
SbF	137,438.3(40)	−36.17(81)	−36.2(10)	9
SbCl	92,654.5(11)	−26.1(23)	−26.19(30)	9
HfS	108,708.38(27)	−4.18(53)	−5.820(49)	10

[1] Crofton MW, Altman RS, Haese NN, Oka T (1989) J Chem Phys 91: 5882–5886
[2] Johns JWC (1984) J Mol Spectrosc 106: 124–133
[3] Warner HE, Conner WT, Woods RC (1984) J Chem Phys 81: 5413–5416
[4] Peterson KA, Petrmichl RH, McClain RL, Woods RC (1991) J Chem Phys 95: 2352–2360
[5] Authier N, Bagland N, Le Floch A (1993) J Mol Spectrosc 160: 590–592
[6] Guelachvili G, Niay P, Bernage P (1981) J Mol Spectrosc 85: 271–281
[7] Odashima H (2006) J Mol Spectrosc 240: 69–74
[8] Beaton SA, Gerry MCL (1999) J Chem Phys 110: 10715–10724
[9] Cooke SA, Gerry MCL (2005) J Mol Spectrosc 234: 195–203
[10] Cooke SA, Gerry MCL (2002) J Mol Spectrosc 216: 122–130

upon isotopic substitution giving rise to small changes in the Coulomb interaction between the electrons and the nucleus. Equation (3.10) is slightly modified where the mean square nuclear charge radius $\langle r^2\rangle_{\mathrm{A,B}}$ is used as expansion parameter and the new molecular parameter $V_{01}^{\mathrm{A,B}}$ is introduced

$$Y_{01} = \frac{\overline{U}_{01}}{\mu^{(l+2k)/2}}\left[1 + m_{\mathrm{e}}\left(\frac{\Delta_{01}^{\mathrm{A}}}{m_{\mathrm{A}}} + \frac{\Delta_{01}^{\mathrm{B}}}{m_{\mathrm{B}}}\right) + V_{01}^{\mathrm{A}}\delta\langle r^2\rangle_{\mathrm{AA'}} + V_{01}^{\mathrm{B}}\delta\langle r^2\rangle_{\mathrm{BB'}}\right] \tag{3.14}$$

with

$$\overline{U}_{01} = U_{01}\left[1 + V_{01}^{\mathrm{A}}\delta\langle r^2\rangle_{\mathrm{A}} + V_{01}^{\mathrm{B}}\delta\langle r^2\rangle_{\mathrm{B}}\right] \tag{3.15}$$

The field-shift parameter $V_{01}^{\mathrm{A,B}}$ depends mainly on the electron density and its derivatives with respect to the internuclear distance at the nucleus A or B. It is isotopically independent. The difference $\delta\langle r^2\rangle_{\mathrm{A,B}}$ is the mean square nuclear charge radius on isotopic substitution A → A′. This finite nuclear size correction is only significant when the accuracy of the measurements is extremely high and when the mass number of the atom is not too small (>40). The field-shift parameter can be

Table 3.4 Derived correction parameters of the rotational constant Y_{01} for the breakdown of the Born–Oppenheimer approximation and for the field shift[a]

Molecule	Δ_{01}^{A}	Δ_{01}^{B}	V_A (10^4 Å^{-2})[a]	Ref.
$^{208}Pb^{32}S$	−12.94(141)	−1.997(71)	0	1
	−1.333	−1.988(70)	2.45(19)	
$^{208}Pb^{80}Se$	−11.86(92)	−2.120(76)	0	1
	−1.520	−2.094(72)	2.21(19)	
$^{208}Pb^{130}Te$	−11.98(21)	−1.794(110)	0	1
	−1.405	−1.84(11) ,	2.12(16)	
$^{205}Tl^{35}Cl$	−18.96(200)	−1.243(49)	0	1
	−0.500	−1.257(73)	4.09(55)	
$^{195}Pt^{32}S$	−42.60(74)	−62.466(49)		2
	−62.5(10)[b]	−62.46(5)	−10.4(9)	
$^{195}Pt^{28}Si$	10.75(68)	−2.99(4)		3
	−3(1)[b]	−2.99(4)	−7.2(12)	

[a] V_B is fixed at zero

[b] Estimated value based on the assumption that $\Delta_{01}^{A} = \Delta_{01}^{A} \pm 1$

[1] Schlembach J, Tiemann E (1982) Chem Phys 68: 21–28

[2] Cooke SA, Gerry MCL (2004) J Chem Phys 121: 3486–3494

[3] Cooke SA, Gerry MCL, Brugh DJ, Suenram RD (2004) J Mol Spectrosc 223: 185–194

determined experimentally or calculated ab initio. The effect of taking into account the field-shift shift is shown in Table 3.4.

3.5 Morse Potential

Dunham's expansion, Eq. (3.2), is widely used although the expansion parameter, ξ, Eq. (3.1), has limitations. Therefore, better expansion parameters have been proposed. For a discussion, see [3]. However, there is another approximation of the potential function, the Morse potential [10, 11]

$$V_M(r) = \tilde{D}_e\left(1 - e^{-a(r-r_e)}\right)^2 \tag{3.16}$$

where $\tilde{D}_e$ is the dissociation energy measured from the bottom of the potential well and a is the Morse force constant. The main inconvenient of this equation is that it was not convenient for calculations, at least in the past. However, it is now well established that the Morse equation is a good approximation for the diagonal stretching force constants. Therefore, this equation is now often used in direct potential fits (see next Sect. 3.6) for diatomic molecules and, for instance, in the method Morse Oscillator Rigid Bender Internal Dynamics (MORBID) for polyatomic molecules, see Sect. 6.5.

Another advantage of the Morse equation is that, by differentiating it, it is possible to estimate the cubic and quartic diagonal force constants from the quadratic one. There is also a simple relationship between the Dunham constants a_i and Morse parameter a

$$a_1 = -ar_e \tag{3.17}$$

and

$$a_2 = \frac{7}{12}(ar_e)^2 \tag{3.18}$$

The value of the a-parameter is about 2 Å^{-1} for many bonds (theoretically, it depends on the bond length r_e, but the range of r_e is small) [12].

3.6 Direct Potential Fit (DPF) [3]

The Hamiltonian of a diatomic molecule is one-dimensional and can be solved efficiently by standard numerical methods. Thus, it is not necessary to calculate the traditional molecular spectroscopic constants, and it has the further advantage to require fewer parameters.

The vibration–rotation energies of a diatomic molecule are the eigenvalues of the radial Schrödinger equation

$$-\frac{\hbar^2}{2\mu}\frac{d^2\psi(r)}{dr^2} + \left[V(r) + \frac{\hbar^2}{2\mu r^2}J(J+1)\right]\psi(r) = E\psi(r) \tag{3.19}$$

$V(r)$ is the electronic potential, and the next term is the centrifugal potential due to the molecular rotation. One starts with an analytic model of the potential energy and solves the radial Schrödinger equation for the upper and lower levels of every transition and uses a nonlinear least-squares method to fit the experimental data in order to optimize the potential.

It is essential to use an analytic potential function which represents well the true potential and which is easy to use. The simplest form is the Dunham expansion, Eq. (3.2). The problem is that $V_{\mathrm{Dunham}}(r) \to \infty$ when $r \to \infty$. Several better expansion variables have been proposed, but more sophisticated expansions are in common use: the Expanded Morse Oscillator (EMO) function and the Morse Long-Range (MLR) potential.

3.6.1 *Expanded Morse Oscillator (EMO)*

$$V(r) = D_e\left[1 - e^{\beta(r)(r-r_e)}\right]^2 \tag{3.20}$$

in which D_e is the well depth, r_e the equilibrium internuclear distance, and

$$\beta(r) = \sum_{i=0}^{N_A} \beta_i y_q(r)^i \tag{3.21}$$

$y_q(r)$ is a dimensionless variable

$$y_q(r) = \frac{r^q - r_e^q}{r^q + r_e^q} \tag{3.22}$$

where $q \geq 3$. Instead of using r_e in $y_q(r)$, it is advantageous to use $r_{ref} \approx 1.5r_e$.

The EMO potential form is widely used and is able to give accurate equilibrium bond lengths. However, due to its exponential nature, it is not able to incorporate the inverse-power-sum characteristic of all long-range intermolecular potentials.

3.6.2 Morse Long-Range (MLR) Potential

$$V(r) = D_e\left[1 - \frac{u_{LR}(r)}{u_{LR}(r_e)} e^{-\beta(r) y_q(r)}\right]^2 \tag{3.23}$$

u_{LR} is a function which defines the attractive long-range behavior of the potential energy function, and it is defined as a sum of inverse-power terms

$$u_{LR} = \frac{C_{m_1}}{r^{m_1}} + \frac{C_{m_2}}{r^{m_2}} + \cdots \tag{3.24}$$

The integer powers, $m_1 < m_2 < \ldots$, characterizing the terms contributing to the sum are determined by the nature of the atoms to which the molecular state dissociates, and the coefficients C_i may be computed from theory.

A comparison of a fit using either the EMO potential or the MLR potential was made for MgH [13]. A detailed study of the use of the LMR potential for Li_2 may be found in [14].

Computer programs to determine these potentials may be found at: leroy.uwaterloo.ca. However, it has to be noted that the use of these programs requires some expertise.

3.6.3 Born–Oppenheimer Breakdown

There is no difficulty to take into account the BOB. It is enough to slightly modify Eq. (3.19). For an isotopologue α, it may be written as

$$\left\{-\frac{\hbar^2}{2\mu_\alpha}\frac{d^2}{dr^2}+\left[V_{\text{ad}}^{(1)}(r)+\Delta V_{\text{ad}}^{(\alpha)}(r)\right]+\frac{J(J+1)\hbar^2}{2\mu_\alpha r^2}\left[1+g^{(\alpha)}(r)\right]\right\}$$
$$\psi_{v,J}(r)=E_{v,J}\psi_{v,J}(r) \tag{3.25}$$

$V_{\text{ad}}^{(1)}(r)$ is the total electronic internuclear potential for a selected reference isotopologue ($\alpha = 1$), $\Delta V_{\text{ad}}^{(\alpha)}(r)$ is the difference between the effective adiabatic potentials for isotopologue α and for reference isotopologue, and $g^{(\alpha)}(r)$ is the non-adiabatic centrifugal potential correction function for isotopologue α. Both $\Delta V_{\text{ad}}^{(\alpha)}(r)$ and $g^{(\alpha)}(r)$ are written as the sum of two terms

$$\Delta V_{\text{ad}}^{(\alpha)}(r)=\frac{\Delta m_{\text{A}}^{(\alpha)}}{m_{\text{A}}^{(\alpha)}}\tilde{S}_{\text{ad}}^{\text{A}}(r)+\frac{\Delta m_{\text{B}}^{(\alpha)}}{m_{\text{B}}^{(\alpha)}}\tilde{S}_{\text{ad}}^{\text{B}}(r) \tag{3.26}$$

$$g^{(\alpha)}(r)=\frac{m_{\text{A}}^{(1)}}{m_{\text{A}}^{(\alpha)}}\tilde{R}_{\text{na}}^{\text{A}}(r)+\frac{m_{\text{B}}^{(1)}}{m_{\text{B}}^{(\alpha)}}\tilde{R}_{\text{na}}^{\text{B}}(r) \tag{3.27}$$

In Eq. (3.26), $\Delta m_{\text{A}}^{(\alpha)} = m_{\text{A}}^{(\alpha)} - m_{\text{A}}^{(1)}$.

The only difference is that the parameters $\tilde{S}_{\text{ad}}^{\text{A/B}}(r)$ and $\tilde{R}_{\text{na}}^{\text{A/B}}(r)$ have to be also taken into account.

3.7 Equilibrium Structure from the Ground State Rotational Constants

In some difficult cases, the rotational spectrum can only be measured in the vibrational ground state $v = 0$, i.e., only the ground state rotational constant, $B_0 = B_e - \alpha_e/2$ can be determined. However, from B_0, it is possible to calculate the effective structure r_0

$$r_0 = r_e\left(1+\frac{\alpha_e}{4B_e}\right) \tag{3.28}$$

As α_e/B_e varies as $\mu^{-1/2}$

$$r_0 = r_e + \frac{C}{\sqrt{\mu}} \tag{3.29}$$

If it is possible to obtain r_0 for several isotopologues, Eq. (3.29) permits to obtain r_e. The value of r_e may be less precise than r_0 by an order of magnitude, but as r_0 is usually extremely accurately known, it does not matter much. However, the higher-order neglected terms may have a non-negligible influence on the accuracy.

Another method is to use the substitution structure r_s which is obtained from the difference of the moments of inertia of two isotopologues

$$r_s = \sqrt{\frac{\Delta I_0}{\Delta \mu}} = r_e\left(1 + \frac{\alpha_e}{8B_e}\right) \tag{3.30}$$

giving

$$r_e = 2r_s - r_0 \tag{3.31}$$

As an example, the equilibrium structure of CSe is calculated using Eqs. (3.29) and (3.31), see Table 3.5 for the ground state rotational constants and the r_0 values. The non-adiabatic correction was taken into account using $g(^{2}C^{80}Se) = -0.2431$. The rotational constants and the g-factor are from Refs. [15, 16].

Equation (3.29) gives $r_e = 1.676052(32)$ Å, whereas Eq. (3.31) gives 1.676015 Å to be compared to $r_e^{BO} = 1.676086(64)$ Å. In other words, the agreement is satisfactory. Unfortunately, it does not work as well for weakly bound complexes. For instance, using the ground state rotational constants of seven isotopologues of ArXe [17], Eq. (3.29) gives 4.09407(5) Å and Eq. (3.31) 4.0948 Å, whereas the value obtained by adjusting the potential is 4.09225 Å. In theory, it is possible to determine the Y_{21} term only using ground state rotational constants. From Eq. (3.6)

$$B_0 = \sum_k \frac{Y_{k1}}{2^k} = Y_{01} + \frac{Y_{11}}{2} + \frac{Y_{21}}{4} \tag{3.32}$$

or using Dunham's expressions for Y_{k1}

$$B_0 = \frac{\beta_0}{\mu} + \frac{\beta_1}{\mu^{3/2}} + \frac{\beta_2}{\mu^2} \tag{3.33}$$

where the β_i are mass-independent and $\beta_0 = h/8\pi^2 r_e^2$. Unfortunately, a least-squares fit to Eq. (3.33) is ill-conditioned [17].

Table 3.5 Effective structure of CSe

Isotopologue	B_0 /MHz[a]	r_0/Å
$^{12}C^{80}Se$	17,182.77	1.6788859
$^{12}C^{78}Se$	17,240.20	1.6788917
$^{12}C^{77}Se$	17,270.00	1.6788932
$^{12}C^{76}Se$	17,300.62	1.6788978
$^{12}C^{74}Se$	17,364.35	1.6789000
$^{12}C^{82}Se$	17,128.09	1.6788816
$^{13}C^{80}Se$	16,032.00	1.6787896

[a] Uehara H, Horiai K, Sakamoto T (2015) J Mol Spectrosc 313: 19–39

3.8 Radicals [1, 2, 15, 18]

All molecules with a non-$^1\sum$ electronic state are called radicals. As a diatomic molecule has a cylindrical symmetry, the projection Λ of the angular momentum along the symmetry axis is a conserved quantity. The states belonging to different Λ values are identified by Greek capital letters.

Λ	1	2	3	4
Symbol	$\sum$	$\prod$	Δ	Φ

There are two kinds of $\sum$ states, those that change sign on reflection in a plane containing the symmetry axis and those that do not. They are labeled respectively as $\sum^-$ and $\sum^+$.

The state is also identified by its spin multiplicity. When $\Lambda \neq 0$, $\sum$ (not to be confused with the state) is the projection of the total electronic angular momentum (the sum of electron spins in unfilled shells), it has 2S + 1 values, $\sum = S, S - 1, \ldots, -S$. Finally, $\Omega = \sum + \Lambda$ is the sum of projections along internuclear axis of electron spin and orbital angular momenta. The complete notation of a state is: $^{2S+1}\Lambda^{(+/-)}_{\Omega,(g/u)}$. *g*/*u* is the effect of the point group operation *I* (reflection through an inversion center), and $\pm$ is the reflection symmetry along an arbitrary plane containing the internuclear axis.

In the case of large spin-orbit interaction, Λ and S are not defined, and the lowest fine structure component of the ground state may be designated by $\Omega = 0$ and well separated from the other components. In this case, the rotational fine structure is described in the same manner as for $^1\sum$ states. Note that, when heavy atoms are presents, it is not always obvious to deduce the correct symmetry. For instance, the symmetry of PtO was first believed to be $^1\sum^+$ and later corrected to $^3\sum^-$. In such a case, only Ω is given, see [19].

For radicals, an effective Hamiltonian may be written

$$\mathbf{H} = \mathbf{H}_{\text{vib}} + \mathbf{H}_{\text{rot}} + \mathbf{H}_{\text{SO}} + \mathbf{H}_{\text{SS}} + \mathbf{H}_{\text{SR}} + \mathbf{H}_{\text{HFS}} \tag{3.34}$$

where

- $\mathbf{H}_{\text{vib}}$ is the vibrational Hamiltonian.
- $\mathbf{H}_{\text{rot}}$ is the rotational Hamiltonian.
- $\mathbf{H}_{\text{SO}}$ is the Hamiltonian for spin-orbit interaction.
- $\mathbf{H}_{\text{SS}}$ is the Hamiltonian for spin-spin interaction.
- $\mathbf{H}_{\text{SR}}$ is the Hamiltonian for spin-rotation interaction.
- $\mathbf{H}_{\text{HFS}}$ is the Hamiltonian for hyperfine interaction.

Effective Hamiltonians are used to fit the experimental data, but different forms of the effective Hamiltonian are in use, yielding different values of the molecular parameters.

The most frequent states are described below. For the rarer ones, see the original literature and refs. [2, 18].

3.8.1 $^2\Sigma$ Electronic State

The approximate electronic quantum numbers of the state are the spin $S = ½$, the projection of the electronic angular momentum onto the molecular axis, $|\Lambda| = 0$, and the reflection symmetry $\pm$ in a plane containing the molecular axis. The effective Hamiltonian is written

$$\mathbf{H} = B\mathbf{N}^2 + \gamma \mathbf{N} \cdot \mathbf{S} \tag{3.35}$$

N is the sum of the rotational angular momentum **R** and the electronic orbital angular momentum **L**, B is the rotational constant of the state, and γ is the spin-rotation interaction parameter. Both parameters are dependent on the vibrational quantum number, and generally, a Dunham expansion is used as for the $^1\Sigma$ state

$$B_{vl} = \sum_l Y_{l1}\left(v + \frac{1}{2}\right)^l \tag{3.36}$$

The rotational levels are split into doublets.

3.8.2 $^3\Sigma$ Electronic State

The difference with the $^2\Sigma$ state is that $S = 1$. The rotational Hamiltonian contains additionally the spin–spin interaction

$$\mathbf{H} = B\mathbf{N}^2 + \gamma \mathbf{N} \cdot \mathbf{S} + \frac{2}{3}\lambda\left(3S_z^2 - \mathbf{S}^2\right) \tag{3.37}$$

S_z is the total electronic spin projected onto the molecular axis z, γ is the spin-rotation interaction parameter, and λ is the spin–spin interaction parameter.

The rotational levels are split into triplets.

The selection rules are: $\Delta N = \pm 1$, $\Delta J = 0, \pm 1$, and $\Delta F = 0, \pm 1$ (if there are nuclei with nonzero nuclear spin.

3.8.3 $\sum$ States with Higher Multiplicity

The effective Hamiltonian is similar to that of $^3\sum$ states. The Hamiltonian and matrix elements are given by Nelis et al. [20].

3.8.4 $^2\prod$ Electronic State

There are additional splittings due to the electronic spin and orbital angular momentum interactions. The rotational levels are defined by the quantum numbers Ω, J, and F (if there are nuclei with nonzero nuclear spin) and the parity. The selection rules are: $\Delta\,\Omega = 0$, $\Delta J = \pm 1$, and $+ \nleftrightarrow -$. The Hamiltonian may be written

$$\mathbf{H} = B(\mathbf{J}^2 - L_z^2 + \mathbf{S}^2) + AL_zS_z - 2B\mathbf{J}\cdot\mathbf{S} + \left(B + \frac{A}{2}\right)(L_+S_- + L_-S_+) + \\ - B(J_+L_- + J_-L_+) + \gamma(\mathbf{J} - \mathbf{S})\cdot\mathbf{S} \tag{3.38}$$

z is the axis along the molecular axis, L_z, L_+, and L_- are the spherical components of the electronic orbital angular momentum, S_z, S_+, and S_- spherical components of the spin, and J_z, J_+, and J_- are the spherical components of the total angular momentum $\mathbf{J}$ (including electron spin, i.e., $\mathbf{J} = \mathbf{N} + \mathbf{S} = \mathbf{R} + \mathbf{L} + \mathbf{S}$). A is the spin-orbit constant, B is the rotational constant, and γ is the spin-rotation interaction parameter. B_{v1} and B_{v2} are the effective rotational constants for the $^2\prod_{1/2}$ and $^2\prod_{3/2}$ states, respectively. The rotational constant B_v in vibrational state v can be evaluated from B_{v1} and B_{v2}, if assumptions are made.

3.8.5 $^3\prod$ Electronic State

The Hamiltonian has the same structure as for $^2\prod$ states but with two additional terms: the spin–spin coupling molecular (parameter λ) and the Λ-doubling (parameter o). The Hamiltonian is given by Brown et al. [21, 22].

3.8.6 Electronic States with Orbital Angular Momentum $\Lambda \geq 2$ and Spin $S \geq 1/2$

The Hamiltonian was given by Brown et al. [23]. The effective Hamiltonian can be written

$$\mathbf{H} = B\mathbf{N}^2 - D\mathbf{N}^4 + \gamma\mathbf{N}\cdot\mathbf{S} + AL_zS_z + \frac{2}{3}\lambda\left(3S_z^2 - \mathbf{S}^2\right) + \eta L_zS_z\frac{1}{5}\left(S_z^2 - \left(3\mathbf{S}^2 - 1\right)\right)$$
$$+ \frac{1}{12}\theta\left(35S_z^4 - 30\mathbf{S}^2S_z^2 + 25S_z^2 - 6\mathbf{S}^2 + 3\mathbf{S}^4\right) \quad (3.39)$$

Besides the usual parameters B for the rotation, γ for the spin-rotation, A for the spin–orbit, and λ for the effective spin–spin interaction, two new parameters are introduced: η, the effective cross-term between the spin–orbit and spin–spin interactions, and θ, the effective higher-order spin–spin interaction.

3.9 Experimental Data

Experimental equilibrium structures of diatomic molecules in their $^1\sum$ ground electronic state and radicals are given in Tables 3.6 and 3.7, respectively. The experimental structures derived by spectroscopy are also listed in the Landolt-Börnstein tables: molecular constants, mostly from microwave, molecular beam, and sub-Doppler laser spectroscopy, see for instance Refs. [24, 25].

There is also Huber and Herzberg's classic book on the constants of diatomic molecules published in 1979 [26]. The National Institute of Standards and Technology (NIST) has placed this book online (http://webbook.nist.gov/chemistry) [27]. For additional references, see Sect. 5.14.

There is a database of references, called Diatomic Reference Database (DiRef), compiled by Bernath and McLeod (University of Waterloo): diref.uwaterloo.ca.

Table 3.6 Equilibrium structure of diatomic molecules in their $X^1\Sigma$ ground state

Molecule		r_e(Å)	Method[a]	Reference
AgBr	$^{107}Ag^{79}Br$	2.3931074(83)	Y_{01}	Nair KPR, Hoeft J (1987) Phys Rev A 35: 668
AgCl		2.2807867(11)	U_{01}	Le Floch AC, Rostas J (1982) J Mol Spectrosc 92: 276
AgCu		2.438(13)	r_0	Bishea GA, Marak N, Morse MD (1991) J Chem Phys 95: 5618
AgF	$^{107}Ag^{19}F$	1.983171(1)	Y_{01}	Hoeft J, Lovas FJ, Tiemann E, Törring T (1970) Z Naturforsch A 25: 35
AgH		1.6179162(1)	U_{01}	Le Roy RJ, Appadoo DRT, Anderson K, Shayesteh A, Gordon IE, Bernath PF (2005) J Chem Phys 123: 204304–1 to 12
AgI	$^{107}Ag^{127}I$	2.544626(9)	Y_{01}	Hoeft J, Nair KPR (1986) Chem Phys Lett 129: 538
Ag_2		2.5303(2)	B_e	Beutel V, Krämer H-G, Bhale GL, Kuhn M, Weyers K, Demtröder W (1993) J Chem Phys 98: 2699–2708
AlBr	$^{27}Al^{79}Br$	2.29485976(50)	B_e	Fleming PE, Mathews CW (1996) J Mol Spectrosc 175: 31–36
AlCl		2.1301672(50)	DPF	Yousefi M, Bernath PF (2018) Astrophys J Supp Series 237: 8
AlCu		2.3389(4)	r_0	Behm JM, Arrington CA, Langenberg JD, Morse MD (1993) J Chem Phys 99: 6394
AlF	$^{27}Al^{19}F$	1.654368(10)	DPF	Yousefi M, Bernath PF (2018) Astrophys J Supp Series 237: 8
AlH		1.6453669(14)	DPF	Sauer SPA, Ogilvie JF (1994) J Phys Chem 98: 8617–8621
AlI	$^{27}Al^{127}I$	2.53709241(29)	B_e	Wyse FC, Gordy W (1972) J Chem Phys 56: 2130
ArCd	CdAr	4.33(4)	B_e	Kvaran A, Funk DJ, Kowalski A, Breckenridge WH (1989) J Chem Phys 89: 6069–6080
ArD^+		1.280375(7)	U_{01}	Laughin KB, Blake GA, Cohen RC, Hovde DC, Saykally RJ (1987) Phys Rev Lett 58: 996–999
ArF^+	$^{40}Ar^{19}F^+$	1.6200	B_e	Bogey M, Cordonnier M, Demuynck C, Destombes JL (1992) J Mol Spectrosc 155: 217
ArH^+		1.28031178(6)	DPF	Coxon JA, Hajigeorgiou PG (2016) J Mol Spectrosc 330: 63–71
ArHg		4.01419(26)	r_0	Ohshima Y, Iida M, Endo Y (1990) J Chem Phys 92: 3990
ArKr		3.89381775	DPF	Xu Y, Jäger W, Djauhari J, Gerry MCL (1995) J Chem Phys 103: 2827

(continued)

Table 3.6 (continued)

Molecule		r_e(Å)	Method[a]	Reference
ArMg	$^{40}Ar^{24}Mg$	4.49(10)	B_e	Bennett RR, McCaffrey JG, Wallace I, Funk DJ, Kowalski A, Breckenridge WH (1989) J Chem Phys 90: 2139
ArNe	NeAr	3.48118195	DPF	Grabow J-U, Pine AS, Fraser GT, Lovas FJ, Suenram RD, Emilsson T, Arunan E, Gutowsky HS (1995) J Chem Phys 102: 1181
ArZn	$^{40}Ar^{64}Zn$	4.18(7)	B_e	Wallace I, Bennett RR, Breckenridge WH (1988) Chem Phys Lett 153: 127
Ar_2		3.766(2)	B_e	Myatt PT, Dham AK, Chandrasekhar P, McCourt FRW, Le Roy (2018) Mol Phys 116: 1598–1623
AsBi	$^{209}Bi^{75}As$	2.39075(5)	B_e	Breidohr R, Shestakov O, Setzer KD, Fink EH (1995) J Mol Spectrosc 172: 369
AsP		1.9995440(2)	B_e	Leung F, Cooke SA, Gerry MCL (2006) J Mol Spectrosc 238: 36–41
AuBr		2.318410(1)	B_e	Evans CL, Gerry MCL (2000) J Mol Spectrosc 203: 105–117
AuCl	$Au^{35}Cl$	2.1990287(9)	B_e	Evans CL, Gerry MCL (2000) J Mol Spectrosc 203: 105–117
AuF		1.918449(5)	B_e	Evans CJ, Gerry MCL (2000) J Am Chem Soc 122: 1560–1561
AuH		1.5236752(13)	DPF	Seto JY, Morbi Z, Charron F, Lee SK, Bernath PF (1999) J Chem Phys 110: 11,756–11,767
AuI		2.4711022(4)	B_e	Reynard LM, Evans CJ, Gerry MCL (2001) J Mol Spectrosc 205: 344–346
AuO^-		1.899(6)	PES	Ichino T, Gianola AJ, Andrew DH, Lineberger WC (2004) J Phys Chem A 108: 11,307–11,313
Au_2		2.4715	B_e	Simard B, Hackett PA (1990) J Mol Spectrosc 142: 310–318
BBr	$^{11}B^{79}Br$	1.88658632(79)	B_e	Nomoto M, Okabayashi T, Klaus T, Tanimoto M (1997) J Mol Struct 413–414: 471–476
BCl	$^{11}B^{35}Cl$	1.715283(31)	Y_{01}	Maki AG, Lovas FJ, Suenram RD (1982) J Mol Spectrosc 91:424–429
BF		1.26271167(42)	U_{01}	Zhang K-Q, Guo B, Braun V, Dulick M, Bernath PF (1995) J Mol Spectrosc 170: 82–93
BH	^{11}BH	1.23218(1)	Y_{01}	Shayesteh A, Ghazizadeh E (2015) J Mol Spectrosc 312: 110–112
BH	BD	1.23087(2)	Y_{01}	Shayesteh A, Ghazizadeh E (2015) J Mol Spectrosc 312: 110–112
BaO	$^{138}Ba^{16}O$	1.93968764(19)	B_e	Li H, Focsa C, Pinchemel B, Le Roy RJ, Bernath PF (2000) J Chem Phys 113: 3026–3033
BaS	$^{138}Ba^{32}S$	2.50731595(22)	B_e	Morbi Z, Bernath PF (1995) J Mol Spectrosc 171: 210–222

(continued)

Table 3.6 (continued)

Molecule		r_e(Å)	Method[a]	Reference
BeO	$^{9}Be^{18}O$	1.32351(22)	B_e	Antic-Jovanovic A, Pesic DS, Bojovic V, Vukelic N (1991) J Mol Spectrosc 145: 403
BeS		1.7415	B_e	Cheetham, CJ, Gissane WJM, Barrow RF (1965) Trans Farad Soc 61: 1308–1316
Be_2		2.445(5)	DPF	Meshkov VV, Stolyarov AV, Heaven MC, Haugen C, Le Roy RJ (2014) J Chem Phys 140: 064,315
BiN	BiN	1.9348880(33)	U_{01}	Cooke SA, Michaud JM, Gerry MCL (2004) J Mol Struct 695–696: 13–22
BiP	$^{208}Bi^{31}P$	2.29345(8)	B_e	Breidohr R, Shestakov O, Etzer KD, Fink EH (1995) J Mol Spectrosc 172: 369
Bi_2	$^{208}Bi_2$	2.65964(9)	B_e	Breidohr R, Setzer KD, Shestakov O, Fink EH, Zyrnicki W (1994) J Mol Spectrosc 166: 251
BrCl	$^{79}Br^{35}Cl$	2.13605328(67)	B_e	Uehara H, Konno T, Ozaki Y, Horiai K, Nakagawa K, Jones JWC (1994) Can J Phys 72: 1145–1154
BrCs	$^{133}Cs^{79}Br$	3.0679745(12)	B_e	Honerjäger R, Tischer R (1974) Z naturforsch A 29: 819–821
BrCu	CuBr	2.17345355(31)	B_e	Hirao T, Bernath PF (2001) Can J Phys 79: 299–343
BrF		1.75897319(29)	U_{01}	Birk H (1993) Z Naturforsch A 48: 581
BrGa	$^{69}Ga^{79}Br$	2.3524907(82)	Y_{01}	Nair KPR, Schütze-Pahlmann HW, Hoeft J (1981) Chem Phys Lett 80: 349
BrH	HBr	1.4144292(1)	DPF	Coxon JA, Hajigeorgiou PG (2015) JQSRT 151: 133–154
BrI	IBr	2.468985986(44)	DPF	Nelander B, Sablinskas V, Dulick M, Braun V, Bernath PF (1998) Mol Phys 93: 137–144
BrIn	$^{113}In^{79}Br$	2.5431492(88)	Y_{01}	Hoeft J, Nair KPR (1989) Chem Phys Lett 164: 33
BrK	$^{39}K^{79}Br$	2.8200766(12)	B_e	Rusk JR, Gordy W (1962) Phys Rev 127: 817
BrLa	LaBr	2.6520742(3)	B_e	Rubinoff DS, Evans CJ, Gerry MCL (2003) J Mol Spectrosc 218: 169–179
BrLi	LiBr	2.17043	U_{01}	Brazier R, Oliphant NH, Bernath PF (1989) J Mol Spectrosc 134: 421–432
BrNa	$^{23}Na^{79}Br$	2.50204034(56)	B_e	Uehara H, Horiai K, Kerim A, Ozaki Y, Konno T (1993) Chem Phys Lett 213: 101
BrRb	$^{85}Rb^{79}Br$	2.94474381(66)	Y_{01}	Tiemann E, Hölzer B, Hoeft J (1971) Z Naturforsch A 32: 123
BrSc	$Sc^{79}Br$	2.3808453(10)	B_e	Lin W, Evans CJ, Gerry MCL (2000) Phys Chem Chem Phys 2: 43–46
BrSc	$Sc^{81}Br$	2.3808439(10)	B_e	Lin W, Evans CJ, Gerry MCL (2000) Phys Chem Chem Phys 2: 43–46
$BrSi^+$	$^{28}Si^{79}Br^+$	2.095456(5)	Y_{01}	Ishiguro M, Okabayashi T, Tanimoto M (1995) J Mol Struct 352/353: 317

(continued)

Table 3.6 (continued)

Molecule		r_e(Å)	Method[a]	Reference
BrTl	TlBr	2.6181114(15)	$U_{01} + V$	Brault JW, Davis SP (1982) Phys Scr 25: 268
BrY	YBr	2.53462(1)	B_e	Walker KA, Gerry MCL (1998) J Chem Phys 109: 5439–5445
Br_2		2.2810272(7)	DPF	Yukiya T, Nishimiya N, Samejima Y, Yamaguchi K, Suzuki M, Boone CD, Ozier I, Le Roy RJ (2013) J Mol Spectrosc 283: 32–43
CCl^+	$^{12}C^{35}Cl^+$	1.537754(13)	B_e	Gruebele M, Polak M, Blake GA, Saykally RJ (1986) J Chem Phys 85: 6276
CF^+		1.154089	U_{01}	Cazzoli G, Cludi L, Puzzarini C, Gauss J (2010) A&A 509: A1/1–5
CH^+		1.1284625(58)	DPF	Cho Y-S, Le Roy RJ (2016) J Chem Phys 144: 024311
CNi	NiC	1.6273(2)	B_e	Brugh DJ, Morse MD (2002) J Chem Phys 117: 10703
CO		1.128230(1)	U_{01}	Authier N, Bagland N, Le Floch A (1993) J Mol Spectrosc 160: 590
CPt	PtC	1.679	r_0	Steimle TC, Jung KY, Li BZ (1995) J Chem Phys 103: 1767
CRu	RuC	1.605485(2)	B_e	Wang F, Steimle TC, Adam AG, Cheng L, Stanton JF (2013) J Chem Phys 139: 174318
CS		1.53482211(41)	U_{01}	Uehara H, Horiai K, Sakamoto T (2015) J Mol Spectrosc 313: 19–39
CSe	$^{12}C^{80}Se$	1.676086(64)	B_e	Lovas FJ, Tiemann E (1974) J Phys Chem Ref Data 3: 609
C_2		1.242440	B_e	Douay M, Nietmann R, Bernath PF (1988) J Mol Spectrosc 131: 250–260
CaO	$^{40}Ca^{16}O$	1.82223067(32)	Y_{01}	Blom CE, Hedderich HG, Lovas FJ, Suenram RD, Maki AG (1992) J Mol Spectrosc 152: 109
CaS		2.31770(14)	B_e	Melville TC, Coxon JA (2002) J Phys Chem A 106: 8271–8275
Ca_2		4.27781(3)	DPF	Le Roy RJ, Henderson RDE (2007) Mol Phys 105: 663–677
CdNe	CdNe	4.26(5)	B_e	Kvaran A, Funk DJ, Kowalski A, Breckenridge WH (1988) J Chem Phys 89: 6069–6080
Cd_2		3.76(4)	B_e	Strojecki M, Ruszczak M, Lukomski M, Koperski J (2007) Chem Phys 340: 171–180
ClCs	$^{133}Cs^{35}Cl$	2.90627295(81)	B_e	Honerjäger R, Tischer R (1974) Z Naturforsch A 29: 819–821
ClCu	CuCl	1.22576663(53)	B_e	Parekunnel T, O'Brien LC, Kellerman TL, Hirao T, Elhanine M, Bernath PF (2001) J Mol Spectrosc 206: 27–32

(continued)

Table 3.6 (continued)

Molecule		r_e(Å)	Method[a]	Reference
ClF	$^{35}Cl^{19}F$	1.6283410(40)	B_e	Willis RE Jr (1979) Ph.D Duke University
ClGa	$^{69}Ga^{35}Cl$	2.2016892(76)	Y_{01}	Hoeft J, Nair KPR (1986) Z Phys D 4: 189
ClH	HCl	1.27454677(6)	DPF	Coxon JA, Hajigeorgiou PG (2015) JQSRT 151: 133–154
ClI	ICl	2.3208757(19)	U_{01}	Hedderich HG, Bernath PF (1992) J Mol Spectrosc 155: 384–392
ClIn	$^{113}In^{35}Cl$	2.4011658(83)	Y_{01}	Hoeft J, Nair KPR (1989) Chem Phys Lett 155: 273
ClK	KCl	2.66663129(32)	U_{01}	Caris M, Lewen F, Müller HSP, Winnewisser G (2004) J Mol Struct 695–696: 243–251
ClLa	LaCl	2.4980231(3)	B_e	Rubinoff DS, Evans CJ, Gerry MCL (2003) J Mol Spectrosc 218: 169–179
ClLi	LiCl	2.02067150(10)	U_{01}	Bittner DM, Bernath PF (2018) Astrophys J J Supp Series 235:8
ClLu	LuCl	2.3733088(47)	U_{01}	Cooke SA, Krumrey C, Gerry MCL (2005) Phys Chem Chem Phys 7: 2570–2578
ClNa	NaCl	2.360789166(10)	U_{01}	Cabezas C, Cernicharo J, Quintana-Lacaci G, Peña I, Agundez M, Velilla Prieto L, Castro-Carrizo A, Zuñiga J, Bastida A, Alonso JL, Requena A (2016) Astrophys J 825: 150
ClRb	$^{35}Cl^{85}Rb$	2.7867248(17)	U_{01}	Ogilvie JF, Uehara H, Horiai K (2000) Bull Chem Soc Japan 73: 321–327
ClSb	SbCl	2.335472(13)	U_{01}	Cooke S, Gerry MCL (2005) J Mol Spectrosc 234: 195–203
ClSc	ScCl	2.23029317(95)	B_e	Lin W, Beaton SA, Evans CJ, Gerry MCL (2000) J Mol Spectrosc 199: 275–283
$ClSi^+$	$SiCl^+$	1.9439729(10)	U_{01}	Takeda K, Harada K, Cabezas C, Endo Y (2018) J Mol Spectrosc 345: 39–45
ClTl	TlCl	2.484739(10)	$U_{01} + V$	Maki AG (1989) J Mol Spectrosc 137: 147
ClY	$^{89}Y^{35}Cl$	2.382533(34)	B_e	Xin J, Edvinsson G, Klynning L (1991) J Mol Spectrosc 148: 59
Cl_2		1.9885(1)	B_e	Edwards HGM, Long DA, Mansour HR (1978) J Chem Soc Farad Trans II 74: 1200
CrMo	$^{52}Cr^{98}Mo$	1.8182(15)	B_e	Spain EM, Behm JM, Morse MD (1991) Chem Phys Lett 179: 411
Cr_2	$^{52}Cr_2$	1.671(10)	r_0	Michalopoulos DL, Geusic ME, Hansen SG, Powers DE, Smalley RE (1982) J Phys Chem 86: 3914
CsF		2.34535322(40)	Y_{01}	Honerjäger R, Tischer R (1974) Z Naturforsch A 29: 819–821

(continued)

Table 3.6 (continued)

Molecule		r_e(Å)	Method[a]	Reference
CsH	^{133}CsH	2.4942(1)	Y_{01}	Magg U, Jones H (1988) Chem Phys Lett 148: 6
CsI		3.315095(18)	DPF	Braun V, Guo B, Zhang K-Q, Dulick M, Bernath PF, McRae GA (1994) Chem Phys Lett 228: 633–640
CsK	KCs	4.284(7)	B_e	Ferber R, Klincare I, Nikolayeva O, Tamanis M, Knöckel H, Tiemann E, Pashov A (2008) J Chem Phys 128: 244316
CsRb	^{85}Rb^{133}Cs	4.427136(39)	Y_{01}	Gustavsson T, Amiot C, Vergès J (1988) Mol Phys 64: 279
Cs_2	$^{133}Cs_2$	4.647676(64)	Y_{01}	Amiot C, Demtröder W, Vidal CR (1988) J Chem Phys 88: 5255
CuF	^{63}Cu^{19}F	1.7449302(13)	Y_{01}	Jakob P, Sugawara K, Wanner J (1993) J Mol Spectrosc 160: 596
CuH		1.46254378(24)	DPF	Seto JY, Morbi Z, Charron F, Lee SK, Bernath PF (1999) J Chem Phys 110: 11756–11767
CuI	^{63}Cu^{127}I	2.3383516(10)	Y_{01}	Manson EL, DeLucia FC, Gordy W (1975) J Chem Phys 62: 4796
CuLi	^{7}Li^{63}Cu	2.700(3)	B_e	Russon LM, Rothschopf GK, Morse MD (1997) J Chem Phys 107: 1079
CuY	^{89}Y^{63}Cu	2.6197(6)	B_e	Arrington CA, Brugh DJ, Morse MD, Doverstal M (1995) J Chem Phys 102: 8704
Cu_2	$^{63}Cu_2$	2.21927(3)	B_e	Ram RS, Jarman CN, Bernath PF (1992) J Mol Spectrosc 155: 468–486
FGa	GaF	1.77434086(129)	U_{01}	Uehara H, Horiai K, Katsuie S (2016) J Mol Spectrosc 325: 20–28
FGe^+	GeF^+	1.6648210(38)	U_{01}	Tanaka K, Akiyama Y, Tanaka T, Yamada C, Hirota E (1990) Chem Phys Lett 171: 175
FH	HF	0.91683897(4)	DPF	Coxon JA, Hajigeorgiou PG (2015) JQSRT 151: 133–154
FI	^{127}I^{19}F	1.909756896(68)	B_e	Magg U, Birk H, Nair KPR, Jones H (1989) Z Naturforsch A 44: 313
FIn	InF	1.985397301(21)	DPF	Karkanis T, Dulick M, Morbi Z, White JB, Bernath PF (1994) Can J Phys 72: 1213–1217
FK	KF	2.17145967(23)	DPF	Frohman DJ, Bernath PF, Brooke JSA (2016) J Mol Spectrosc 169: 104–110
FLa	LaF	2.02335060(8)	B_e	Rubinoff DS, Evans CJ, Gerry MCL (2003) J Mol Spectrosc 218: 169–179
FLi	LiF	1.563864240(62)	U_{01}	Bittner DM, Bernath PF (2018) Astrophys J Supp Series 235:8
FLu	LuF	1.91711815(11)	B_e	Cooke SA, Krumrey C, Gerry MCL (2005) Phys Chem Chem Phys 7: 2570–2578
FNa	NaF	1.92594654(12)	DPF	Frohman DJ, Bernath PF, Brooke JSA (2016) J Mol Spectrosc 169: 104–110

(continued)

Table 3.6 (continued)

Molecule		r_e(Å)	Method[a]	Reference
FRb	$^{85}Rb^{19}F$	2.2703191(11)	B_e	Veazey SE, Gordy W (1965) Phys Rev A 138: 1303
FSb	SbF	1.917584(28)	U_{01}	Cooke S, Gerry MCL (2005) J Mol Spectrosc 234: 195–203
FSc	$^{45}Sc^{19}F$	1.78733(45)	B_e	Shenyavskaya EA, Ross AJ, Topouzkhanian A, Wannous G (1993) J Mol Spectrosc 162: 327
FSi^+	SiF^+	1.526494836(56)	U_{01}	Petrmichl RH, Peterson KA, Woods RC (1988) J Chem Phys 89: 5454
FTl	TlF	2.0843517(39)	U_{01}	Ogilvie JF (1994) J Phys B 27: 47
FY	$^{89}Y^{19}F$	1.925462(33)	B_e	Kaledin LA, Shenyavskaya EA (1991) Mol Phys 72: 1203
F_2	$^{19}F_2$	1.412642(10)	B_e	Martínez RZ, Bermejo D, Santos J, Cancio P (1994) J Mol Spectrosc 168: 343
GaH		1.660150(1)	DPF	Rey M, Tyuterev VG, Coxon JA, Le Roy RJ (2006) J Mol Spectrosc 238: 260–263
GaI		2.5746263(4)	U_{01}	Tiemann E, Arnst H, Stieda WU, Törring T, Hoeft J (1982) J Chem Phys 67: 133
GeH^+		1.58(2)	B_e	Tsuji M, Shimada S, Nishimura Y (1982) Chem Phys Lett 89: 75
GeO		1.62457750(84)	U_{01}	Bizzocchi L, Degli Esposti C, Dore L, Gauss J, Puzzarini C (2014) Mol Phys 113: 801–807
GeS		2.01204279(34)	U_{01}	Uehara H, Horiai K, Ozaki Y, Konno T (1995) J Mol Struct 352/353: 395–405
GeSe		2.1346291210(24)	DPF	Giuliano BM, Bizzocchi L, Sanchez R, Villanueva P, Cortijo V, Sanz ME, Grabow J-U (2011) J Chem Phys 135: 084303
GeTe		2.340167287(70)	DPF	Giuliano BM, Bizzocchi L, Sanchez R, Villanueva P, Cortijo V, Sanz ME, Grabow J-U (2011) J Chem Phys 135: 084303
HHe^+		0.774334277(3)	U_{01}	Coxon JA, Hajigeorgiou PG (1999) J Mol Spectrosc 193: 306–318
HI		1.6290588(4)	DPF	Coxon JA, Hajigeorgiou PG (2015) J Quant Spectr Rad Trans 151: 133–154
HIn	InH	1.8377415(5)	DPF	Shayesteh A, Ghazizadeh E (2016) J Mol Spectrosc 330:72–79
HK	KH	2.240164(10)	DPF	Uehara H, Horiai K, Konno T (1997) J Mol Struct 414–414: 457–462
HKr^+	KHr^+	1.4211904(5)	U_{01}	Linnartz H, Zink LR, Evenson KM (1997) J Mol Spectrosc 184: 56
HLa	^{139}LaH	2.031969(20)	B_e	Ram RS, Bernath PF (1996) J Chem Phys 104: 6444–6451

(continued)

Table 3.6 (continued)

Molecule		r_e(Å)	Method[a]	Reference
HLi	LiH	1.59491150(15)	DPF	Dulick M, Zhang K-Q, Guo B, Bernath PF (1998) J Mol Spectrosc 188: 14–26
HNa	NaH	1.887023(15)	DPF	Walji S-D, Sentjens KM, Le Roy RJ (2015) J Chem Phys 142: 044305
HNe^+	NeH^+	0.991109(2)	DPF	Coxon JA, Hajigeorgiou PG (2016) J Mol Spectrosc 330: 63–71
HO^-	OH^-	0.964094(37)	B_e	Matsushima F, Yonezu T, Okabe T, Tomaru K, Moriwaki Y (2006) J Mol Spectrosc 235: 261–264
HRb	RbH	2.365318(59)	U_{01}	Essig K, Urban R-D, Birk H, Jones H (1993) Z Naturforsch A 48: 1111
HS^-	SH^-	1.3432329(37)	B_e	Elhanine M, Farrenq R, Guelachvili G, Morillon-Chapey (1938) J Mol Spectrosc 129: 240
HS^-	SD^-	1.343173(15)	B_e	Zelinger Z, Bersch A, Petri M, Urban W, Civis S (1995) J Mol Spectrosc 171: 579
HSc	ScH	1.775427(8)	B_e	Ram RS, Bernath PF (1996) J Chem Phys 105: 2668
HSi^+	SiH^+	1.50436155(30)	B_e	Doménech JL, Schlemmer S, Asvany O (2017) Astrophys J 849: 60/1–4
HTl	TlH	1.8726434(71)	U_{01}	Urban RD, Jones H (1992) Chem Phys Lett 190: 609
HXe^+	XeH^+	1.6028114(13)	U_{01}	Peterson KA, Petrmichl RH, McClain RL, Woods RC (1991) J Chem Phys 95: 2352
HY	^{89}YH	1.922765(8)	B_e	Ram RS, Bernath PF (1996) J Chem Phys 105: 2668
HY	YD	1.920404(4)	B_e	Ram RS, Bernath PF (1994) J Chem Phys 101: 9283
H_2		0.74153(7)		Edwards HGM, Long DA, Mansour HR, Najm KAB (1979) J Raman Spectr 8: 251–254
H_2	$^1H^3H$	0.74149(1)	B_e	Edwards HGM, Long DA, Mansour HR, Najm KAB (1979) J Raman Spectr 8: 251–254
H_2	$^2H^3H$	0.74151(2)	B_e	Edwards HGM, Long DA, Mansour HR, Najm KAB (1979) J Raman Spectr 8: 251–254
H_2	$^3H_2$	0.74151(2)	B_e	Edwards HGM, Long DA, Mansour HR (1978) J Chem Soc Faraday Trans II 74:1203–1207
HfO	$^{180}Hf^{16}O$	1.7231481	Y_{01}	Lesarri A, Suenram RD, Brugh D (2002) J Chem Phys 117: 9651–9662
HfS		2.156141(4)	U_{01}	Cooke SA, Gerry MCL (2002) J Mol Spectrosc 216: 122–130
Hg_2		3.605(9)	B_e	Koperski J, Qu X, Meng H, Kenefick R, Fry ES (2008) Chem Phys 348: 103–112
IIn	InI	2.7536497(2)	DPF	Yoo JH, Köckert H, Mullaney JC, Stephens SL, Evans CJ, Walker NR, Le Roy RJ (2016) J Mol Spectrosc 330: 80–88
IK	$^{39}K^{127}I$	3.0478427(58)	Y_{01}	Bluhm H, Tiemann E (1994) J Mol Spectrosc 163: 238

(continued)

Table 3.6 (continued)

Molecule		r_e(Å)	Method[a]	Reference
ILa	LaI	2.8788547(2)	B_e	Rubinoff DS, Evans CJ, Gerry MCL (2003) J Mol Spectrosc 218: 169–179
ILi	LiI	2.3919810(34)	U_{01}	Guo B, Dulick M, Yost S, Bernath PF (1997) Mol Phys 91: 459–469
INa	$^{23}Na^{127}I$	2.71145380(81)	Y_{01}	Lindner J, Bluhm H, Fleisch A, Tiemann E (1994) Can J Phys 72: 1137
IRb	$^{85}Rb^{127}I$	3.1768663(39)	B_e	Rusk JR, Gordy W (1962) Phys Rev 127: 817
ISc	ScI	2.6078	B_e	Liao Z, Xia Y, Chan M-C, Cheung ASC (2013) Chem Phys Lett 570: 33–36
ITl	TlI	2.813614(1)	$U_{01} + V$	Tiemann E, Arnst H, Stieda WU, Törring T, Hoeft J (1982) J Chem Phys 67: 133
IY	YI	2.7637565(24)	B_e	Norman L, Evans CJ, Gerry MCL (2000) J Mol Spectrosc 199: 311–313
I_2		2.66637	Y_{01}	Luc P (1980) J Mol Spectrosc 80: 41–55
IrN	IrN	1.6068276(32)	B_e	Ram RS, Bernath PF (1999) J Mol Spectrosc 193: 363–375
KLi	KLi	3.31914(5)	DPF	Tiemann E, Knöckel H, Kowalczyk P, Jastrzebski W, Pashov A, Salami H, Ross AJ (2009) Phys Rev A 79: 042716
KNa	$^{23}Na^{39}K$	3.4990348(15)	Y_{01}	Yamada C, Hirota E (1992) J Mol Spectrosc 153: 91
K_2	$^{39}K_2$	3.92433(35)	Y_{01}	Amiot C (1991) J Mol Spectrosc 146: 370
KrNe	NeKr	3.64801653	DPF	Xu Y, Jäger W, Djauhari J, Gerry MCL (1995) J Chem Phys 103: 2827
KRb		4.06851(17)	Y_{01}	Ross AJ, Effantin C, Crozet P, Boursey E (1990) J Phys B: At Mol Opt Phys 23: L247-L251
KrXe		4.200903	DPF	Xu Y, Jäger W, Djauhari J, Gerry MCL (1995) J Chem Phys 103: 2827
Kr_2		4.017(12)	B_e	LaRocque PE, Lipson RH, Herman PR, Stoicheff BP (1986) J Chem Phys 84: 6627
LiNa	NaLi	2.888563(50)	DPF	Steinke M, Knöckel H, Tiemann E (2012) Phys Rev A 85: 042720
Li_2		2.672993(2)	DPF	Le Roy RJ, Dattani NS, Coxon JA, Ross AJ, Crozet P, Linton C (2009) J Chem Phys 131: 204309
MgO	$^{24}Mg^{16}O$	1.7481722(9)	B_e	Mürtz P, Thümmel H, Pfelzer C, Urban W (1995) Mol Phys 86: 513
MgNe	$^{24}Mg^{20}Ne$	4.40(15)	B_e	Wallace I, Breckenridge WH (1993) J Chem Phys 98: 2768

(continued)

Table 3.6 (continued)

Molecule		r_e(Å)	Method[a]	Reference
MgS	$^{24}Mg^{32}S$	2.14257249(85)	B_e	Walker KA, Gerry MCL (1997) J Mol Spectrosc 182: 178
MgXe	$^{24}Mg^{132}Xe$	4.56(12)	B_e	McCaffrey JG, Funk DJ, Breckenridge WH (1993) J Chem Phys 99: 9472
Mg_2		3.89039	DPF	Knöckel H, Rühmann S, Tiemann E (2013) J Chem Phys 138: 094303
Mo_2	$^{98}Mo_2$	1.940(9)	B_e	Hopkins JB, Langridge-Smith PRR, Morse MD, Smalley RE (1983) J Chem Phys 78: 1627
NP	$^{31}P^{14}N$	1.490866(2)	B_e	Cazzoli G, Lcudi L, Puzzarini C (2006) J Mol Struct 780–781: 260–267
NSb	SbN	1.8355630(20)	U_{01}	Cooke SA, Gerry MCL (2004) Phys Chem Chem Phys 6: 4579–4585
NSc	$^{45}Sc^{14}N$	1.687232(26)	B_e	Ram RS, Bernath PF (1992) J Chem Phys 96: 6344–6347
NSi^-	SiN^-	1.604(5)	PES	Meloni G, Sheehan SM, Ferguson MJ, Neumark DM (2004) J Phys Chem A 108: 9750–9754
NTa	TaN	1.6830999(88)	B_e	Ram RS, Liévin J, Bernath PF (2002) J Mol Spectrosc 215: 275–284
NY	$^{89}Y^{14}N$	1.80405(50)	B_e	Ram RS, Bernath PF (1994) J Mol Spectrosc 165: 97–106
N_2		1.097679(1)	DPF	Le Roy RJ, Huang Y, Jary C (2006) J Chem Phys 125:164310
NaRb	$^{23}Na^{85}Rb$	3.6435	B_e	Kasahara S, Ebi T, Tanimura M, Ikoma H, Matsubara K, Baba M, Kato H (1996) J Chem Phys 105: 1341–1347
Na_2	$^{23}Na_2$	3.0791(56)	B_e	Babaky O, Hussein K (1989) Can J Phys 67: 912
NeXe	NeXe	3.88691028	DPF	Xu Y, Jäger W, Djauhari J, Gerry MCL (1995) J Chem Phys 103: 2827
NeZn	ZnNe	4.16(10)	B_e	McCaffrey JG, Bellert D, Leung AWK, Breckenbridge WH (1999) Chem Phys Lett 302: 113–118
Ne_2		3.094(1)	DPF	Wüest A, Merkt F (2003) J Chem Phys 118: 8807–8812
OP^+	$^{31}P^{16}O^+$	1.4249927(4)	U_{01}	Petrmichl RH, Peterson KA, Woods RC (1991) J Chem Phys 94: 3504
OPb	$^{208}Pb^{16}O$	1.9217850(1)	B_e	Surkus A (1996) Spectrochim Acta A, 52: 1925–1927
OSi	SiO	1.50966468(15)	U_{01}	Müller HSP, Spezzano S, Bizzocchi L, Gottlieb CA, Degli Esposti C, McCarthy MC (2013) J Phys Chem A 117: 13843–13854
OSn	SnO	1.8324259(5)	U_{01}	Dewberry CT, Etchison KC, Grubbs GS, Powoski RA, Serafin MM, Peebles SA, Cooke SA (2008) J Mol Spectrosc 248: 20–25
OSr	$^{88}Sr^{16}O$	1.91985111(34)	B_e	Li H, Skelton R, Focsa C, Pinchemel B, Bernath PF (2000) J Mol Spectrosc 203: 188–195

(continued)

Table 3.6 (continued)

Molecule		r_e(Å)	Method[a]	Reference
OTh	ThO	1.8420327(23)	DPF	Schmitz JR, Kaledin LA, Heaven MC (2019) J Mol Spectrosc 360: 39–43
OZn	ZnO	1.7047(2)	B_e	Zack LN, Pulliam RL, Ziurys LM (2009) J Mol Spectrosc 256: 186–191
OZr	ZrO	1.71174527(74)	U_{01}	Beaton SA, Gerry MCL (1999) J Chem Phys 110: 10715–10724
O_2	$^{16}O_2$			See Table 3.7
PN	$^{31}P^{14}N$	1.49086694(2)	Y_{01}	Cazzoli G Cludi L Puzzarini C (2006) J Mol Struct 780-781: 260-267
PSb	SbP	2.205372(20)	U_{01}	Cooke SA, Gerry MCL (2004) Phys Chem Chem Phys 6: 4579–4585
PSc	ScP	2.1995	B_e	Liao Z, Xia Y, Chan M-C, Cheung AS-C (2012) Chem Phys Lett 551: 60–63
P_2		1.8934	B_e	Rao TVR, Laksham SVJ (1970) Ind J Pure Appl Phys 8: 617
PbS	PbS	2.2868195(16)	U_{01}	Knöckel H, Kröckertskothen T, Tiemann E (1985) Chem Phys 93: 349–358
PbSe		2.402232277(16)	$U_{01} + V$	Giuliano BM, Bizzocchi L, Cooke S, Banser D, Hess M, Fritzsche J, Grabow J-U (2008) Phys Chem Chem Phys 10: 2078–2088
PbTe		2.594975987(11)	$U_{01} + V$	Giuliano BM, Bizzocchi L, Cooke S, Banser D, Hess M, Fritzsche J, Grabow J-U (2008) Phys Chem Chem Phys 10: 2078–2088
Pb_2	$^{208}Pb_2$	2.9271(23)	B_e	Heaven MC, Miller TA, Bondybey VE (1983) J Phys Chem 87: 2072
PtSi		2.061206(42)	$U_{01} + V$	Cooke SA, Gerry MCL, Brugh DJ, Suenram RD (2004) J Mol Spectrosc 223: 185–194
Rb_2		4.2099515(70)	DPF	Seto JY, Le Roy RJ, Verges J, Amiot C (2000) J Chem Phys 113: 3067–3076
SSi	SiS	1.92926351(26)	U_{01}	Müller HSP, McCarthy MC, Bizzocchi L, Gupta H, Esser S, Lichau H, Caris M, Lewen F, Hahn J, Degli Esposti C, Schlemmer S, Thaddeus P (2007) Phys Chem Chem Phys 9: 1579–1586
SSn	SnS	2.2089829(22)	U_{01}	Tiemann E, Arnst H, Stieda WU, Törring T, Hoeft J (1982) J Chem Phys 67: 133
SSr	SrS	2.4397045(22)	U_{01}	Etchison KC, Dewberry CT, Cooke SA (2007) Chem Phys 342: 71–77
SZn	ZnS	2.0464	B_e	Zack LN, Ziurys LM (2009) J Mol Spectrosc 257: 213–216

(continued)

Table 3.6 (continued)

Molecule		r_e(Å)	Method[a]	Reference
SZr	ZrS	2.15652076(92)	U_{01}	Beaton SA, Gerry MCL (1999) J Chem Phys 110: 10715–10724
SeSn	SnSe	2.32557537(13)	U_{01}	Bizzocchi L, Giuliano BM, Hess M, Grabow J-U (2007) J Chem Phys 126: 114305
SeSi	SiSe	2.05828249(24)	DPF	Giuliano BM, Bizzocchi L, Grabow J-U (2008) J Mol Spectrosc 251: 261–267
SeTe	SeTe	2.35904044(14)	U_{01}	Banser D, Grabow J-U, Cocinero EJ, Lesarri A, Alonso JL (2006) J Mol Struct 795: 163–172
SiTe		2.27354785(24)	DPF	Giuliano BM, Bizzocchi L, Grabow J-U (2008) J Mol Spectrosc 251: 261–267
SnTe		2.52279640(7)	U_{01}	Bizzocchi L, Giuliano BM, Hess M, Grabow J-U (2007) J Chem Phys 126: 114305
Sn_2	$^{116}Sn_2$	2.748(3)	B_e	Bondybey VE, Heaven M, Miller TA (1983) J Chem Phys 78: 3593
Sr_2	$^{88}Sr_2$	4.446	Y_{01}	Gerber G, Möller R, Schneider H (1984) J Chem Phys 81: 1538
Xe_2		4.3773(49)	B_e	Wüest A, Hollenstein U, de Bruin KG, Merkt F (2004) Can J Chem 82: 750–761
Zn_2		4.19	B_e	Czajkowski M, Koperski J (1999) Spectrochin Acta A 55: 2221–2229

[a] B_e and Y_{01}, see Sect. 3.2. DPF, see Sect. 3.6. r_0, see Sect. 3.7 U_{01}, see Sect. 3.3. $U_{01} + V$, see Sect. 3.4.
see also Table 3.7

Table 3.7 Equilibrium structure of diatomic radicals in their ground electronic state

Molecule		State	r_e(Å)	Reference
AgS	^{107}AgS	$^2\Pi_i$	2.287792459(93)	Okabayashi T, Oya A, Yamamoto T, Mizuguchi D, Tanimoto M (2016) J Mol Spectrosc 329: 13–19
AgS	^{109}AgS	$^2\Pi_i$	2.287792511(64)	Okabayashi T, Oya A, Yamamoto T, Mizuguchi D, Tanimoto M (2016) J Mol Spectrosc 329: 13–19
AlC	AlC	$^4\Sigma^-$	1.95570(10)	Clouthier DJ, Kalume A (2016) J Chem Phys 144: 034305
AlO	AlO	$^2\Sigma^+$	1.61778520(40)	Breier AA, Waßmuth B, Büchling T, Fuchs GW, Gauss J, Giesen TF (2018) J Mol Spectrosc 350: 43–50
AlS	AlS	$^2\Sigma^+$	2.02823790(70)	Breier AA, Waßmuth B, Büchling T, Fuchs GW, Gauss J, Giesen TF (2018) J Mol Spectrosc 350: 43–50
Al_2	Al_2	$^3\Pi_u$	2.7010(20)	Fu Z, Lemire GW, Bishea GA, Morse MD (1990) J Chem Phys 93: 8420–8441
AsH	AsH	$^3\Sigma^-$	1.523200(60)	Beutel M, Setzer KD, Shestakov O, Fink EH (1996) J Mol Spectrosc 178: 165–171
ArXe		0^+	4.095773(29)	Piticco L, Merkt F, Cholewinski AA, McCourt FRW, Le Roy RJ (2010) J Mol Spectrosc 264: 83–93
AsH	AsD	$^3\Sigma^-$	1.522760(50)	Beutel M, Setzer KD, Shestakov O, Fink EH (1996) J Mol Spectrosc 178: 165–172
AsNa	NaAs	$^3\Sigma^-$ (0^+)	2.729750(10)	Setzer KD, Fink EH, Alekseyev AB, Liebermann H-P, Buenker RJ (2016) J Mol Spectrosc 320: 39–47
AsO	AsO	$^2\Pi_r$	1.6235271(28)	Ito F, Nakanaga T, Takeo H, Essig K, Jones H (1995) J Mol Spectrosc 174: 417–424
AuO	AuO	$^2\Pi_{3/2}$	1.84876171(23)	Okabayashi T, Koto F, Tsukamoto K, Yamazaki E, Tanimoto M (2005) Chem Phys Lett 403: 223–227
Au_2^-	Au_2^-	$^2\Sigma_u^+$	2.587	León I, Yang Z, Wang L-S (2013) J Chem Phys 138: 184304; erratum: J Chem Phys 139:089903
BC	BC	$^4\Sigma^-$	1.49116(34)	Fernando WTML, O'Brien LC, Bernath PF (1990) J Chem Phys 93: 8482–8487
BNi	BNi	$^2\Sigma^+$	1.686	Goudreau ES, Adam AG, Tokaryk DW, Linton C (2015) J Mol Spectrosc 314: 13–18
BO	$^{11}B^{16}O$	$^2\Sigma^+$	1.2045531(45)	Li G, Hargreaves J, Wang J-G, Bernath PF (2010) J Mol Spectrosc 263: 123–125
BPt	PtB	$^2\Sigma^+$	1.741(6)	Ng YW, Wong YS, Pang HF, Cheung AS-C (2012) J Chem Phys 137: 124302
B_2	B_2	$^3\Sigma^-$	1.58660(10)	Bredohl H, Dubois I, Mélen F (1987) J Mol Spectrosc 121: 128–134
BaBr	BaBr	$^2\Sigma^+$	2.8444949(99)	Ernst WE, Weiler G, Törring T (1985) Chem Phys Lett 121: 494–498

(continued)

Table 3.7 (continued)

Molecule		State	r_e(Å)	Reference
BaCl	BaCl	$^2\sum^+$	2.682764(10)	Ryzlewicz C, Schütze-Pahlmann, H-U, Heöft J, Törring T (1982) Chem Phys 71: 389
BaF	BaF	$^2\sum^+$	2.1592964(75)	Ryzlewicz C, Schütze-Pahlmann, H-U, Heöft J, Törring T (1982) Chem Phys 71: 389
BaH	BaH	$^2\sum^+$	2.23188651(19)	Ram RS, Bernath PF (2013) J Mol Spectrosc 283: 18–21
BaI	BaI	$^2\sum^+$	3.084761(11)	Törring T, Döbl K (1985) Chem Phys Lett 115: 328
BeH	BeH	$^2\sum^+$	1.342394(12)	Le Roy RJ, Appadoo DRT, Colin R, Bernath PF (2006) J Mol Spectrosc 236: 178–188
BeS^-	BeS-	$^2\sum^+$	1.8060(40)	Dermer AR, Green, ML, Mascaritolo KJ, Heaven MC (2017) J Phys Chem A 121: 5645–5650
Be_2^+	Be_2^+	$^2\Sigma_u^+$	2.2110(80)	Antonov IO, Barker BJ, Bondybey VE, Heaven MC (2010) J Chem Phys 133: 074309
BiBr	BiBr	0^+	2.609530(70)	Kuijpers P, Dymanus A (1976) Chem Phys Lett 39: 217
BiCl	BiCl	0^+	2.471550(70)	Kuijpers P, Törring T, Dymanus A (1976) Chem Phys 18: 401
BiF	BiF	$^3\sum^-$	2.0342760(10)	Fink EH, Setzer KD, Ramsay DA, Towle JP, Brown JM (1996) J Mol Spectrosc 178: 143–156
BiH	BiH	$^3\sum^-$	1.79530(20)	Setzer K-D, Fink EH, Hill C, Brown JM (2015) J Mol Spectrosc 312: 97–109
BiI	BiI	0^+	2.800530(80)	Kuijpers P, Törring T, Dymanus A (1976) Chem Phys 12: 309
BiLi	BiLi	$^3\sum^-$ (0^+)	2.66220(50)	Setzer KD, Fink EH, Liebermann H-P, Buenker RJ, Alekseyev AB (2015) J Mol Spectrosc 318: 38–45
BiNa	BiNa	$^3\sum^-$	2.9384870(80)	Setzer KD, Uibel C, Zyrnicki W, Pravilov AM, Fink EH, Lieberman H-P, Buenker RJ (2000) J Mol Spectrosc 204: 163–175
BrCa	Ca79Br	$^2\sum^+$	2.593585	Bernath PF, Field RW, Pinchemel B, Lefebvre Y, Schamps J (1981) J Mol Spectrosc 88: 175–193
BrCr	CrBr	$^6\sum^+$	2.337282(30)	Herman TJ, Ziurys LM (2019) J Chem Phys 151: 194301
BrH^+	HBr^+	$^2\prod$	1.447981(13)	Chanda A, Ho WC, Dalby FW, Ozier I (1995) J Mol Spectrosc 169: 108–147
BrMg	MgBr	$^2\sum^-$	2.3474320(10)	Walker KA, Gerry MCL (1997) J Chem Phys 107: 9835–9841
BrN	NBr	$^3\sum^-$	1.777990(40)	Sakamaki T, Okabayashi T, Tanimoto M (1998) J Chem Phys 109: 7169–7175
BrO	BrO	$^2\prod_{3/2}$	1.7172499	Drouin BJ, Miller CE, Müller HSP, Cohen EA (2001) J Mol Spectrosc 205: 128–138

(continued)

Table 3.7 (continued)

Molecule		State	r_e(Å)	Reference
BrP	PBr	$^3\sum^-$	2.172072(17)	Okabayashi T, Kawajiri H, Umeyama M, Ide C, Oe S, Tanimoto M (2008) J Chem Phys 129: 124301
BrSr	SrBr	$^2\sum^+$	2.7352168(95)	Törring T, Döbl K, Weiler G (1985) Chem Phys Lett 117: 539
BrTi	^{79}BrTi	$^4\Phi_{3/2}$	2.41448(17)	Adam AG, Hopkins WS, Sha W, Tokaryk DW (2006) J Mol Spectrosc 236: 42–51
BrYb	YbBr	$^2\sum^+$	2.6453860(40)	Dickinson CS, Coxon JA (2004) J Mol Spectrosc 224: 18–26
BrZn	ZnBr	$^2\sum^+$	2.26848(90)	Burton MA, Ziurys LM (2019) J Chem Phys 150: 034303
CCl	CCl	$^2\prod$	1.645218(16)	Endo Y, Saito S, Hirota E (1982) J Mol Spectrosc 94: 199
CF	CF	$^2\prod$	1.27197031(72)	Liu YY, Liu ZA, Davies PB (1995) J Mol Spectrosc 171: 402–419
CH	CH	$^2\prod$	1.1197888(58)	Bernath PF, Brazier CR, Olsen T, Hailey R, Fernando WTML, Woods C, Hardwick JL (1991) J Mol Spectrosc 147: 16–26
CH	^{13}CH	$^2\prod$	1.1197930(60)	Zachwieja M (1997) J Mol Spectrosc 182: 18–33
CH	CD	$^2\prod$	1.11886609(50)	Zachwieja M, SzajnaW, Hakalla R (2012) J Mol Spectrosc 275: 53–60
CN	CN	$^2\sum^+$	1.17180630(86)	Brooke JSA, Ram RS, Western CM, Li G, Schwenke DW, Bernath PF (2014) Ap J Suppl 210: 23
CNi^+	NiC^+	$^2\sum^+$	1.628	Chang YC, Shi X, Lau K-C, Yin Q-Z, Liou HT, Ng CY (2010) J Chem Phys 133: 054310
CO^+	CO^+	$^2\sum^+$	1.11517760(64)	Spezzano S, Brünken S, Müller HSP, Klapper G, Lewen F, Menten KM, Schlemmer S (2013) J Phys Chem A 117: 9814–9818
CP	CP	$^2\sum^+$	1.56197826(26)	Ram RS, Brooke SA, Western CM, Bernath PF (2014) JQSRT 138: 107–115
CP	CP	$^2\sum^+$	1.56197800(20)	Ram RS, Tam S, Bernath PF (1992) J Mol Spectrosc 152: 89–100
CSi	SiC	$^3\prod$	1.7190(10)	Bernath PF, Rogers SA, O'Brien LC, Brazier CR, McLean AD (1988) Phys Rev Lett 60: 197–199
CTi^+	TiC^+	$^2\sum^+$	1.6670(40)	Luo Z; Huang H, Chang Y-C, Zhang Z, Yin Q-Z, Ng CY (2014) J Chem Phys 141: 144307
CV^+	VC^+	$^3\Delta_1$	1.65490(30)	Chang YC, Luo Z, Pan Y, Zhang Z, Song Y-N, Kuang SY, Yin QZ, Lau K-C, Ng CY (2015) Phys Chem Chem Phys 17: 9780–9793
CaCl	CaCl	$^2\sum^+$	2.4367869(84)	Schütze-Pahlmann, H-U (1981) Dissertation Freie Universität Berlin

(continued)

Table 3.7 (continued)

Molecule		State	r_e(Å)	Reference
CaF	CaF	$^2\Sigma^+$	1.95163655(25)	Hou S, Bernath PF (2018) JQSRT 210: 44–51
CaH	CaH	$^2\Sigma^+$	1.957780(20)	Shayesteh A, Ram RS, Bernath PF (2013) J Mol Spectrosc 288: 46–51
CaI	CaI	$^2\Sigma^+$	2.828586(10)	Törring T, Döbl K (1985) Chem Phys Lett 115: 328
CaK	KCa	$^2\Sigma^+$	4.23968785	Gerschmann J, Schwanke E, Pashov A, Knöckel H, Ospelkaus S, Tiemann E (2017) Phys Rev A 96: 032505
CaLi	LiCa	$^2\Sigma^+$	3.35600(10)	Ivanova M, Stein A, Pashov A, Stolyarov AV, Knöckel H, Tiemann E (2011) J Chem Phys 135: 174303
CdH	^{114}CdH	$^2\Sigma^+$	1.7600980(30)	Shayesteh A, Le Roy RJ, Varberg TD, Bernath PF (2006) J Mol Spectrosc 237: 87–96
CeS	CeS	$\Omega = 3$	2.33522875(15)	Ram RS, Bernath PF (2014) J Mol Spectrosc 299: 6–10
ClCo	CoCl	$^3\Phi_4$	2.065599(13)	Ram RS, Gordon I, Hirao T, Yu S, Bernath PF (2007) J Mol Spectrosc 243: 69–78
ClCo	CoCl	$^3\Phi_3$	2.06707(25)	Ram RS, Gordon I, Hirao T, Yu S, Bernath PF (2007) J Mol Spectrosc 243: 69–78
ClCr	CrCl	$^6\Sigma^+$	2.193952(10)	Oike T, Okabayashi T, Tanimoto M (1998) J Chem Phys 109: 3501–3507
ClGe	GeCl	$^2\Pi_r$	2.163739(23)	Tanaka K, Honjou H, Tsuchiya M, Tanaka T (2008) J Mol Spectrosc 251: 369–373
ClHf	HfCl	$^2\Delta_{3/2}$	2.290532(57)	Ram RS, Adam AG, Tsouli A, Liévin J, Bernath PF (2000) J Mol Spectrosc 202:116–130
ClIr	IrCl	$^3\Phi_4$	2.156	Adam AG, Foran S, Linton C (2016) J Mol Spectrosc 319: 10–16
ClMg	MgCl	$^2\Sigma^+$	2.1963888(58)	Bogey M, Demuynck C, Destombes JL (1989) Chem Phys Lett 155: 265; errata 161: 92
ClMn	MnCl	$^7\Sigma^+$	2.2351512(75)	Halfen DT, Ziurys LM (2005) J Chem Phys 122: 054309
ClN	NCl	$^3\Sigma^-$	1.610705(19)	Yamada C, Endo Y, Hirota E (1986) J Mol Spectrosc 117: 134–137
ClNi	NiCl	$^2\Pi_{3/2}$	2.0614	Harms JC, Gipson CN, Grames EM, O'Brien JJ, O'Brien LC (2016) J Mol Spectrosc 321: 78–81
ClO	ClO	$^2\Pi$	1.5689335	Drouin BJ, Miller CE, Cohen EA, Wagner G, Birk M (2001) J Mol Spectrosc 207: 4–9
ClXe	XeCl	$^2\Sigma^+$	3.23	Sur A, Hui, AK, Tellinghuisen J (1979) J Mol Spectrosc 74: 465–479
ClP	PCl	$^3\Sigma^-$	2.014609(49)	Kanamori H, Yamada C, Butler JE, Kawaguchi K, Hirota E (1985) J Chem Phys 83: 4945

(continued)

Table 3.7 (continued)

Molecule		State	r_e(Å)	Reference
ClP^+	PCl^+	$^2\Pi$	1.9000(10)	Coxon JA, Naxakis S, Roychowdhury UK (1987) Can J Phys 65: 980–983
ClRh	RhCl	$^3\Pi_2$	2.070	Adam AG, Foran S, Linton C (2016) J Mol Spectrosc 319: 10–16
ClSb	SbCl	0^+	2.335472(13)	Cooke SA, Gerry MCL (2005) J Mol Spectrosc 234: 195–203
ClSr	SrCl	$^2\Sigma^+$	2.5758842(89)	Schütze-Pahlmann H-U, Ryzlewicz C, Hoeft J, Törring T (1982) Chem Phys Lett 93: 74–77
ClTa	TaCl	$\Omega = 0+$	2.236511(32)	Ram RS, Adam AG, Bernath PF (2005) J Mol Spectrosc 232: 358–368
ClTi	TiCl	$^4\Phi$	2.264642102(31)	Herbin H, Farrenq R, Guelachvili G, Pinchemel B, Picqué N (2004) J Mol Spectrosc 226: 103–111
$ClTi^+$	$TiCl^+$	$^3\Phi$	2.18320(10)	Focsa C, Pinchemel B, Féménias J-L, Huet TR (1998) J Chem Phys 107: 10365–10372
ClV	VCl	$^5\Delta$	2.2145	Ram RS, Liévin J, Bernath PF, Davis SP (2003) J Mol Spectrosc 217: 186–194
ClYb	YbCl	$^2\Sigma^+$	2.4882850(10)	Dickinson CS, Coxon JA, Walker NR, Gerry MCL (2001) J Chem Phys 115: 6979
ClZn	ZnCl	$^2\Sigma^+$	2.13003305(24)	Tenenbaum ED, Flory MA, Pulliam RL, Ziurys LM (2007) J Mol Spectrosc 244: 153–159
ClZr	ZrCl	$^4\Phi$	2.3661	Ram RS, Bernath PF (1997) J Mol Spectrosc 186: 335–348
Cl_2^+	Cl_2^+	$^2\Pi_{3/2(g)}$	1.891	Wu L, You S-P, Shao X-P, Chen G-J, Ding N, Wang Y-M, Yang X–H (2015) Chin Phys B 24: 083301
CoF	CoF	$^3\Phi_4$	1.7356980(80)	Ram RS, Bernath PF, Davis SP (1995) J Mol Spectrosc 173: 158–176
CoH	CoH	$^3\Phi_4$	1.532664(16)	Gordon IE, Le Roy RJ, Bernath PF (2006) J Mol Spectrosc 237: 11–18
CoH	CoD	$^3\Phi_4$	1.517531(29)	Gordon IE, Le Roy RJ, Bernath PF (2006) J Mol Spectrosc 237: 11–19
CoO	CoO	$^4\Delta$	1.6278620(10)	McLamarrah SK, Sheridan PM, Ziurys LM (2005) Chem Phys Lett 414: 301–306
CoS	CoS	$^4\Delta_{7/2}$	1.9786450(11)	Yu S, Gordon IE, Sheridan PM, Bernath PF (2006) J Mol Spectrosc 236: 255–259
CrF	CrF	$^6\Sigma^+$	1.784	Bencheikh M, Koivisto R, Launila O, Flament JP (1997) J Chem Phys 106: 6231–6239
CrH	CrH	$^6\Sigma^+$	1.6554111(57)	Bauschlicher CW, Ram RS, Bernath PF, Parsons CG, Galehouse D (2001) J Chem Phys 115: 1312–1318
CsO	CsO	$^2\Sigma^+$	2.300745(16)	Yamada C, Hirota E (1999) J Chem Phys 111: 9587–9592
CuS	CuS	$^2\Pi$	2.04988(11)	Thompsen JM, Ziurys LM (2001) Chem Phys Lett 344: 75–84

(continued)

Table 3.7 (continued)

Molecule		State	r_e(Å)	Reference
FFe	FeF	$^6\Delta$	1.78063	Kermode SM, Brown JM (2001) J Mol Spectrosc 207: 161–171
FGa^+	GaF^+	$^2\sum^+$	1.6828(26)	Grabandt O, Mooyman R, De Lange CA (1990) Chem Phys 143: 227–238
FH^+	HF^+	$^2\prod$	1.0016005(32)	Allen MD, Evenson KM, Brown JM (2004) J Mol Spectrosc 227: 13–22
FIr	IrF	$^3\Phi_4$	1.897	Adam AG, Granger AD, Downie LE, Tokaryk DW, Linton C (2009) Can J Phys 87: 557–565
FMg	MgF	$^2\sum^+$	1.74992986(12)	Hou S, Bernath PF (2017) JQSRT 203: 511–516
FMn	MnF	$^7\sum^+$	1.83584356	Sheridan PM, Ziurys LM (2003) Chem Phys Lett 380: 632–646
FN	NF	$^3\sum^-$	1.316979(29)	Kobayashi K, Saito S (1998) J Chem Phys 108: 6606–6610
FNi	NiF	$^2\prod_{3/2}$	1.73871	Benomier M, van Groenendael A, Pinchemel B, Hirao T, Bernath PF (2005) J Mol Spectrosc 233 (2005) 244–255
FO	FO	$^2\prod_{3/2}$	1.35411170(61)	Tamassia F, Brown JM, Saito S (2000) J Chem Phys 112: 5523–5526
FP	PF	$^3\sum^-$	1.589329(20)	Yamada C, Chang MC, Hirota E (1987) J Chem Phys 86: 3804–3806
FRh	RhF	$^3\prod_2$	1.964	Adam AG, Foran S, Linton C (2016) J Mol Spectrosc 319: 10–16
FS	SF	$^2\prod$	1.5962440(88)	Endo Y, Nagai K, Yamada C, Hirota E (1983) J Mol Spectrosc 97: 212–219
FSb	SbF	0^+	1.917584(28)	Cooke SA, Gerry MCL (2005) J Mol Spectrosc 234: 195–203
FSi	SiF	$^2\prod$	1.6009965(21)	Tanaka T, Tamura M, Tanaka K (1997) J Mol Struct 412–414: 153–166
FSr	SrF	$^2\sum^+$	2.0753910(70)	Schütze-Pahlmann H-U, Ryzlewicz C, Hoeft J, Törring T (1982) Chem Phys Lett 93: 74–77
FTa	TaF	$^3\sum^-$ (0+)	1.8184	Ng KF, Zou W, Liu W, Cheung AS-C (2017) J Chem Phys 146: 094308
FTi	TiF	$^4\Phi$	1.8311	Ram RS, Bernath PF (2005) J Mol Spectrosc 231: 165–170
FTi^+	TiF^+	$^3\Phi$	1.78000(10)	Focsa C, Pinchemel B, Collet D, Huet TR (1998) J Mol Spectrosc 189: 254–263
FV	VF	$^5\Delta$	1.7758	Ram RS, Bernath PF, Davis SP (2002) J Chem Phys 116: 7035–7039
FXe	XeF	$^2\sum^+$	2.3210(20)	Monts DL, Ziurys LM, Beck SM, Liverman MG, Smalley RE (1979) J Chem Phys 71: 4057–4065
FYb	YbF	$^2\sum^+$	2.0165140(10)	Dickinson CS, Coxon JA, Walker NR, Gerry MCL (2001) J Chem Phys 115: 6979

(continued)

Table 3.7 (continued)

Molecule		State	r_e(Å)	Reference
FZn	ZnF	$^2\Sigma^+$	1.76770(10)	Flory MA, McLamarrah SK, Ziurys LM (2006) J Chem Phys 125: 194304
FeS	FeS	$^5\Delta_4$	2.00900(10)	Zhang S, Zhen J, Zhang Q, Chen Y (2009) J Mol Spectrosc 255: 101–105
HHf	HfH	$^2\Delta_{3/2}$	1.830691(15)	Ram RS, Bernath PF (1994) J Chem Phys 101: 74–79
HIr	IrH	$^3\Phi_4$	1.602	Adam AG, Granger AD, Linton C, Tokaryk DW (2012) Chem Phys Lett 535: 21–25
HMg	MgH	$^2\Sigma^+$	1.72968540(70)	Henderson RDE, Shayesteh A, Tao J, Haugen CC, Bernath PF, Le Roy RJ (2013) J Phys Chem A 117: 13373–13387
HMn	MnH	$^7\Sigma^+$	1.7308601(47)	Gordon IE, Appadoo DRT, Shayesteh A, Walker KA, Bernath PF (2005) J Mol Spectrosc 229: 145–149
HN	NH	$^3\Sigma^-$	1.03606721(31)	Melosso M, Bizzocchi L, Tamassia F, Degli Esposti C, Canè E, Dore L (2019) Phys Chem Phys 21: 3564–3573
HNi	^{58}NiH	$^2\Delta$	1.45450(10)	Abbasi M, Shayesteh A, Crozet P, Ross AJ (2018) J Mol Spectrosc 349: 49–59
HNi	^{58}NiD	$^2\Delta$	1.45590(10)	Abbasi M, Shayesteh A, Crozet P, Ross AJ (2018) J Mol Spectrosc 349: 49–59
HO	OH	$^2\Pi_i$	0.9696160(90)	Martin-Drumel MA, Eliet S, Pirali O, Guinet M, Hindle F, Mouret G, Cuisset A (2012) Chem Phys Lett 550: 8–14
HO	OD	$^2\Pi_i$	0.969680(10)	Amano T (1984) J Mol Spectrosc 103: 436–454
HO$^+$	OH$^+$	$^3\Sigma^-$	1.02893	Rehfuss BD, Jagod M-F, Xu L-W, Oka T (1992) J Mol Spectrosc 151: 59–70
HP	PH	$^3\Sigma^-$	1.422179(16)	Ram RS, Bernath PF (1996) J Mol Spectrosc 176: 329–336
HP	PD	$^3\Sigma^-$	1.42182772(89)	Hughes RA, Brown JM (1997) J Mol Spectrosc 185: 197–201
HPb	PbH	$^2\Pi_{1/2}$	1.839290(20)	Setzer KD, Borkowska-Burnecka J, Zyrnicki W, Fink EH (2008) J Mol Spectrosc 252: 176–184
HPb	PbD	$^2\Pi_{1/2}$	1.838335(12)	Setzer KD, Borkowska-Burnecka J, Zyrnicki W, Fink EH (2008) J Mol Spectrosc 252: 176–184
HS	SH	$^2\Pi_i$	1.34061940(30)	Martin-Drumel MA, Eliet S, Pirali O, Guinet M, Hindle F, Mouret G, Cuisset A (2012) Chem Phys Lett 550: 8–14
HSb	SbH	$^3\Sigma^-$	1.70187	Beutel M, Setzer KD, Shestakov O, Fink EH (1996) J Mol Spectrosc 179: 79–84

(continued)

Table 3.7 (continued)

Molecule		State	r_e(Å)	Reference
HSb	SbD	$^3\Sigma^-$	1.70853	Beutel M, Setzer KD, Shestakov O, Fink EH (1996) J Mol Spectrosc 179: 79–84
HSe	SeH	$^2\Pi$	1.464319(64)	Ram RS, Bernath PF (2000) J Mol Spectrosc 203: 9–15
HSe	SeD	$^2\Pi$	1.4640715(25)	Brown JM, Fackerel AD (1982) Physica Scripta 25: 351
HSr	SrH	$^2\Sigma^+$	2.1460574(10)	Shayesteh A, Walker KA, Gordon I, Appadoo DRT, Bernath PF (2004) J Mol Struct 695–696: 23–37
HTa	TaH	$^3\Phi_2$	1.7569600(40)	Lee SY, Christopher CR, Manke KJ, Vervoort TR, Varberg TD (2014) Mol Phys 112: 2424–2432
HTe	TeH	$^2\Pi_{3/2}$	1.655890(30)	Fink EH, Setzer KD, Ramsay DA, Vervloet M (1989) J Mol Spectrosc 138: 19–28
HZn	^{65}ZnH	$^2\Sigma^+$	1.5934780(20)	Shayesteh A, Le Roy RJ, Varberg TD, Bernath PF (2006) J Mol Spectrosc 237: 87–96
H_2	HD^+	$^2\ \Sigma_g^+$	1.0580(30)	Stimson S, Evans M, Hsu C-W, Ng CY (2007) J Chem Phys 126: 164303
$HeKr^+$	$HeKr^+$	$\Omega = 1/2$	3.0599(76)	Carrington A, Pyne C, Shaw AM, Taylor SM, Hutson JM, Law MM (1996) J Chem Phys 105: 8602–8614
$He_2{}^+$	$He_2{}^+$	$^2\ \Sigma_u^+$	1.080610(60)	Raunhardt M, Schäfer M, Vanhaecke N, Merkt F (2008) J Chem Phys 128: 164310
HfN	HfN	$^2\Sigma^+$	1.724678(36)	Ram RS, Bernath PF (1997) J Mol Spectrosc 184: 401–412
IMg	MgI	$^2\Sigma^+$	2.57300(10)	Kilchenstein KM, Halfen DT, Ziurys LM (2017) J Mol Spectrosc 339: 1–5
INi	NiI	$^3\Sigma^-$	1.9653	Shestakov O, Gielen R, Setzer KD, Fink EH (1998) J Mol Spectrosc 192: 139–147
IO	IO	$^2\Pi$	1.87609130(30)	Miller CE, Cohen EA (2001) J Chem Phys 115: 6459–6470
IP	PI	$^3\Sigma^-\ (0^+)$	2.3812250(20)	Setzer KD, Beutel M, Fink EH (2003) J Mol Spectrosc 221: 19–22
IPb	PbI	$^2\Pi_{1/2}$	2.79760460(70)	Yoo JH, Köckert H, Mullaney JC, Stephens SL, Evans CJ, Walker NR, Le Roy RJ (2016) J Mol Spectrosc 330: 80–88
ISr	SrI	$^2\Sigma^+$	2.973641	Törring T, Döbl K, Weiler G (1985) Chem Phys Lett 117: 539
IYb	YbI	$^2\Sigma^+$	2.84830(50)	Noonan K, Melville TC, Dickinson CS, Coxon JA (2003) J Mol Spectrosc 222: 296–298
$I_2{}^+$	$I_2{}^+$	$^2\Pi_{3/2,\ g}$	2.584386(25)	Deng L–H, Zhu Y-Y, Li C-L, Chen Y-Q (2012) J Chem Phys 137: 054308
LaO	LaO	$^2\Sigma^+$	1.8259114(62)	Törring T, Zimmermann K, Hoeft J (1988) Chem Phys Lett 151: 520

(continued)

Table 3.7 (continued)

Molecule		State	r_e(Å)	Reference
LiO	LiO	$^2\prod$	1.68822160(20)	Yamada C, Fujitake M, Hirota E (1989) J Chem Phys 91: 137
LiPb	PbLi	$^4\sum^-$ (1/2)	2.70402	Setzer KD, Borkowska-Burnecka J, Zyrnicki W, Pravilov AM, Fink EH, Das KK, Liebermann H-P, Alekseyev AB, Buenker RJ (2003) J Mol Spectrosc 217: 127–141
LiS	LiS	$^2\prod$	2.1497	Brewster M, Ziurys LM (2001) Chem Phys Lett (2001) 349: 249–256
LiSb	LiSb	$^3\sum^-$ (0^+)	2.6046(10)	Setzer KD, Fink EH, Alekseyev AB, Liebermann H-P, Buenker RJ (2018) J Mol Spectrosc 347: 41–47
LiSr	$^7Li^{88}Sr$	$^2\sum^+$	3.5405	Schwanke E, Gerschmann J, Knöckel H, Ospelkaus S, Tiemann E (2020) J Phys B: At Mol Opt Phys 53: 065102
LiTe	LiTe	$^2\prod_{3/2}$	2.48964	Setzer KD, Fink EH, Alekseyev AB, Liebermann H-P, Buenker RJ (2001) J Mol Spectrosc 206: 181–197
LuO	LuO	$^2\sum^+$	1.79028770(40)	Cooke SA, Krumrey C, Gerry MCL (2011) J Mol Spectrosc 267: 108–111
MnS	MnS	$^6\sum^+$	2.06630(60)	Douay M, Pinchemel B, Dufour C (1985) Can J Phys 63: 1380–1388
NO	NO	$^2\prod$	1.1507886(10)	Saleck AH, Yamada KMT, Winnewisser G (1991) Mol Phys 72: 1135
NOs	OsN	$^2\Delta_{3/2}$	1.618023(91)	Ram RS, Liévin J, Bernath PF (1999) J Chem Phys 111: 3449–3456
NPt	PtN	$^2\prod$	1.682	Steimle TC, Jung KY, Li BZ (1995) J Chem Phys 103: 1767
NRe	^{185}ReN	0^+	1.637800(20)	Ram RS, Bernath PF, Balfour WJ, Cao J, Qian CXW, Rixon SJ (1995) J Mol Spectrosc 168: 350–362
NRu	RuN	$^2\sum^+$	1.5714269(60)	Ram RS, Bernath PF (2002) J Mol Spectrosc 213: 170–178
NS	NS	$^2\prod$	1.4939121(13)	Saleck AH, Ozeki H, Saito S (1995) Chem Phys Lett 244: 199–206
NSi	SiN	$^2\sum^+$	1.572066(41)	Saito S, Endo Y, Hirota E (1983) J Chem Phys 78: 6447
NaO	NaO	$^2\prod$	2.0515480(10)	Yamada C, Fujitake M, Hirota E (1989) J Chem Phys 90: 3033
NaSb	SbNa	$^3\sum^-$ (0^+)	2.901070(30)	Setzer KD, Fink EH, Liebermann H-P, Buenker RJ, Alekseyev AB (2015) J Mol Spectrosc 318: 1–11
NbO	NbO	$^4\sum^-$	1.68315	Kingston CT, Liao C-KD, Merer AJ, Tang SJ (2001) J Mol Spectrosc 207: 104–112
NbS	NbS	$^4\sum^-$	2.1645	Launila O (2005) J Mol Spectrosc 229: 31–38

(continued)

Table 3.7 (continued)

Molecule		State	r_e(Å)	Reference
NdO	NdO	$\Omega = 4$	1.7991	Shenyavskaya E, Bernard A, Vergès J (2003) J Mol Spectrosc 222: 240–247
NiO	NiO	$^3\sum^-$	1.6271291(10)	Namiki K, Saito S (1996) Chem Phys Lett 252: 343–347
NiS	NiS	$^3\sum^-$	1.962496(28)	Ram RS, Yu S, Gordon I, Bernath PF (2009) J Mol Spectrosc 258: 20–25
OIr	IrO	$^2\Delta_{5/2}$	1.7231	Pang HF, Ng YW, Cheung AS-C (2012) J Phys Chem A 116: 9739–9744
OP	PO	$^2\prod$	1.47637355(10)	Bailleux S, Bogey M, Demuynck C, Liu Y, Walters A (2002) J Mol Spectrosc 216: 465–471
OPt	PtO	$^3\sum^-$ (0^+)	1.724598(11)	Cooke SA, Gerry MCL (2005) Phys Chem Chem Phys 7: 2453–2459
ORb	RbO	$^2\sum^+$	2.254195	Yamada C, Fujitake M, Hirota E (1989) Annu Rev Inst for Molecular Science Japan 37
ORu	RuO	$^5\Delta_4$	1.714	Wang N, Ng YW, Cheung AS-C (2013) J Phys Chem A 117: 13279–13283
OS	SO	$^3\sum^-$	1.4810087(51)	Tiemann E (1982) J Mol Spectrosc 91: 60–71
OS^+	SO^+	$^2\prod$	1.4243980(20)	Milkman IW, Choi JC, Hardwick JL, Moseley JT (1988) J Mol Spectrosc 130: 20
OSc	ScO	$^2\sum^+$	1.66608(19)	Mukund S, Yarlagadda S, Bhattacharyya S, Nakhate SG (2012) J Qant Spectr Rad Tranf 113: 2004–2008
OSe	SeO	$^3\sum^-$	1.6394990(60)	Parent CR, Kuijpers PJM (1979) Chem Phys 40: 425
OTa	TaO	$^2\Delta_{3/2}$	1.6873430(29)	Ram RS, Bernath PF (1998) J Mol Spectrosc 191: 125–136
OTe	TeO	$^3\sum^-$ (0^+)	1.82500(10)	Setzer KD, Fink EH (2014) J Mol Spectrosc 295: 21–25
OTh^+	ThO^+	$^2\sum^+$	1.807	Goncharov V, Heaven MC (2006) J Chem Phys 124: 064312
OTi	TiO	$^3\Delta$	1.62009060(31)	Breier AA, Waßmuth B, Fuchs GW, Gauss J, Thomas F Giesen TF (2019) J Mol Spectrosc 355: 46–58
OTi^+	TiO^+	$^2\Delta_{3/2}$	1.583	Huang H, Luo Z, Chang YC, Lau K-C, Ng CY (2013) J Chem Phys 138: 174,309
OV^+	VO^+	$^3\sum^-$	1.5570(20)	Luo Z, Chang Y-C, Huang H, Ng CY (2015) J Phys Chem A 119: 11162–11169
OW	WO	$^3\sum^-$ (0^+)	1.6577613(64)	Krumrey C, Cooke SA, Russel DK, Gerry MCL (2009) Can J Phys 87: 567–573
OY	YO	$^2\sum^+$	1.7882710(60)	Hoeft J, Törring T (1993) Chem Phys Lett 215: 367–370

(continued)

Table 3.7 (continued)

Molecule		State	r_e(Å)	Reference
OZr^+	ZrO^+	$^2\Delta_{3/2}$	1.6910(20)	Luo Z, Chang Y-C, Zhang Z, Ng CY (2015) Mol Phys 113: 2228–2242
O_2	$^{16}O_2$	$^3\Sigma^-$	1.207524766(70)	Hajigeorgiou PG (2016) J Mol Spectrosc 330: 4–13
O_2^+	O_2^+	$^2\Pi_g$	1.116877(30)	Prasad CVV, Lacombe D, Walker K, Kong W, Bernath PF, Hepburn J (1997) Mol Phys 91: 1059–1074
OsSi	OsSi	$^3\Sigma^-$ (0^+)	2.1207(27)	Johnson EL, Morse MD (2015) J Chem Phys 143: 104303
PS	PS	$^2\Pi$	1.8977405(45)	Kawaguchi K, Hirota E, Ohishi M, Suzuki H, Takano S, Yamamoto S, Saito S (1988) J Mol Spectrosc 130: 81
PSi	SiP	$^2\Pi_i$	2.0775(17)	Jakubek ZJ, Nakhate SG, Simard B (2002) J Chem Phys 116: 6513–6520
PtS	PtS	$^3\Sigma^-$ (0^+)	2.038250(13)	Cooke SA, Gerry MCL (2004) J Chem Phys 121: 3486
SSc	ScS	$^2\Sigma^+$	2.135310(40)	Gengler J, Chen J, Steimle TC, Ram RS, Bernath PF (2006) J Mol Spectrosc 237: 36–45
STa	TaS	$^2\Delta$	2.1049	Wallin S, Edvinsson G, Taklif AG (1998) J Mol Spectrosc 192: 368–377
STi	TiS	$^3\Delta_r$	2.0827323(22)	Pulliam RL, Zack LN, Ziurys LM (2010) J Mol Spectrosc 264: 50–54
SV	VS	$^4\Sigma^-$	2.0501	Ran Q, Tam WS, Cheung AS-C, Merer AJ (2003) J Mol Spectrosc 220: 87–106
SY	YS	$^2\Sigma^+$	2.272	Zang J, Zhang Q, Zhang D, Qin C, Zhang Q, Chen Y (2015) J Mol Spectrosc 313: 49–53
SYb	YbS	$X0^+$	2.3591(2)	Melville TC, Coxon JA, Linton C (2000) J Chem Phys 113: 1771–1774
S_2	S_2	$^3\Sigma^-$	1.8892	Huber KP, Herzberg G (1979) Molecular Spectra and Molecular Structure IV, Van Nostrand, New York
Se_2	Se_2	$^3\Sigma_g^-$	2.163670(20)	Fink EH, Setzer KD, Ramsay DA, Zhu Q-S (1994) Can J Phys 72: 919–924
Si_2	Si_2	$^3\Sigma_g^-$	2.2462	Huber KP, Herzberg G (1979) Molecular Spectra and Molecular Structure IV, Van Nostrand, New York
Te_2	$^{128}Te_2$	$^3\Sigma_g^-$ (0_g^+)	2.55769(11)	Babaky O, Hussein K (1991) Can J Phys 69: 57

see also Table 3.6

References

1. Gordy W, Cook RL (1984) Microwave molecular spectra. Wiley, New York
2. Herzberg G (1950) Spectra of diatomic molecules. Van Nostrand, New York
3. Le Roy RJ (2011) Determining equilibrium structures and potential energy functions for diatomic molecules. In: Demaison J, Boggs JE, Császár AG (eds) Equilibrium molecular structures. From spectroscopy to quantum chemistry. CRC Press, Boca Raton, pp 159–203
4. Tiemann E (1992) Diatomic molecules. In: Hüttner W (ed) Molecular constants mostly from microwave, molecular beam and sub-doppler laser spectroscopy. Landolt-Börnstein, vol 24. Springer, Heidelberg
5. Demaison J, Vogt N (2020) Accurate Structure Determination of Free Molecules. Switzerland, Springer Nature, p 277
6. Dunham JL (1932) The energy levels of a rotating vibrator. Phys Rev 41:721–731
7. Watson JKG (1973) The isotope dependence of the equilibrium rotational constants in $^1\Sigma$ states of diatomic molecules. J Mol Spectrosc 45:99–113
8. Watson JKG (1980) The isotope dependence of diatomic Dunham coefficients. J Mol Spectrosc 80:411–421
9. Tiemann E, Knöckel H, Schlembach J (1982) Influence of the finite nuclear size on the electronic and rotational energy of diatomic molecules. Ber Bunsenges Phys Chem 86:821–824
10. Morse PM (1929) Diatomic molecules according to the wave mechanics. II. Vibrational levels. Phys Rev 34:57–64
11. Pekeris CL (1934) The rotation-vibration coupling in diatomic molecules. Phys Rev 45:98–103
12. Kuchitsu K, Morino Y (1965) Estimation of anharmonic potential constants. I. Linear XY_2 molecules. Bull Chem Soc Japan 38:805–813
13. Shayesteh A, Henderson RDE, Le Roy RJ, Bernath PF (2007) Ground state potential energy curve and dissociation energy of MgH. J Phys Chem A 111:12495–12505
14. Le Roy RJ, Dattani NS, Coxon JA, Ross AJ, Crozet P, Linton C (2009) Accurate analytic potentials for $Li_2(X\ ^1\Sigma_{g+})$ and $Li_2(A\ ^1\Sigma_{u+})$ from 2 to 90 Å, and the radiative lifetime of Li(2p). J Chem Phys 131:204309
15. Lovas FJ, Tiemann E (1974) Microwave spectral tables I. Diatomic molecules. J Phys Chem Ref Data 3:609–770
16. McGurk J, Tigelaar HL, Rock SL, Norris CL, Flygare WH (1973) Detection, assignment of the microwave spectrum and the molecular Stark and Zeeman effects in CSe, and the Zeeman effect and sign of the dipole moment in CS. J Chem Phys 58:1420–1424
17. Xu Y, Jäger W, Djauhari J, Gerry MCL (1995) Rotational spectra of the mixed rare gas dimers Ne–Kr and Ar–Kr. J Chem Phys 103:2827–2833
18. Hirota E (1985) High-resolution spectroscopy of transient molecules. Springer, Berlin, Heidelberg
19. Cooke SA, Gerry MCL (2005) The influence of nuclear volume and electronic structure on the rotational energy of platinum monoxide, PtO. Phys Chem Chem Phys 7:2453–2459
20. Nelis T, Brown JM, Evenson KM (1990) The rotational spectrum of the CH radical in its $a^4\Sigma^-$ state, studied by far-infrared laser magnetic resonance. J Chem Phys 90:4067–4076
21. Brown JM, Colbourn EA, Watson JKG, Wayne FD (1979) An effective Hamiltonian for diatomic molecules: ab initio calculations of parameters of HCl^+. J Mol Spectrosc 74:294–318
22. Brown JM, Merer AJ (1979) Lambda-type doubling parameters for molecules in Π electronic states of triplet and higher multiplicity. J Mol Spectrosc 74:488–494
23. Brown JM, Cheung AS-S, Merer AJ (1987) Λ-Type doubling parameters for molecules in Δ electronic states. J Mol Spectrosc 124:464–475
24. Demaison J, Hübner H, Włodarczak G (1998) Molecular constants mostly from microwave, molecular beam, and sub-doppler laser spectroscopy. In: Hüttner W (ed) Landolt-Börnstein—group II molecules and radicals, vol 24, Subvolume A. Springer, Berlin
25. Christen D (2021) Molecular constants mostly from microwave, molecular beam, and sub-doppler laser spectroscopy. Paramagnetic diatomic molecules (radicals), part 2. In: Hüttner W (ed) Landolt-Börnstein—group II molecules and radicals, vol 29. Springer, Berlin

26. Huber KP, Herzberg G (1979) Molecular spectra and molecular structure IV. Constants of diatomic molecules. Springer, New York
27. Lovas FJ, Tiemann E, Coursey JS, Kotochigova SA, Chang J, Olsen K, Dragoset RA (2005) Diatomic spectral database (version 2.1). http://physics.nist.gov/Diatomic

Chapter 4
Rotational Constants of a Polyatomic Molecule

Abstract This chapter is devoted to the determination of the rotational constants of a polyatomic molecule. Different experimental methods are briefly presented.

4.1 Introduction

The structure of many isolated molecules is derived from their rotational constants. They will be first defined. Then, the different ways to obtain them will be discussed. Particular emphasis will be put on their accuracy. Only the notions useful for a structure determination will be given. A comprehensive review of rotational spectroscopy is given by the reference books of Gordy and Cook [1], Townes and Schawlow [2], Demtröder [3], as well as the "Handbook of high-resolution spectroscopy" [4]. Chapter 4 of Demaison, Boggs, and Császár [5] and Chaps. 3 and 4 of Demaison and Vogt [6] review this aspect of molecular spectroscopy.

4.2 Rotational Energy

4.2.1 Definition of the Rotational Constants

We will consider a molecule of N atoms assumed to be point masses of mass m_α, and we will use a Cartesian coordinate system whose origin is at the center of mass of the molecule. The moment of inertia tensor $\mathbf{I}$ which is a 3×3 symmetric matrix is defined by its elements. The diagonal elements are (with g, g', $g'' = x, y, z$ by cyclic permutation)

$$I_{gg} = \sum_{\alpha=1}^{N} m_\alpha \left(g_\alpha'^2 + g_\alpha''^2 \right) \tag{4.1}$$

and the non-diagonal elements

N. Vogt and J. Demaison, *Equilibrium Structure of Free Molecules*, Lecture Notes in Chemistry 111, https://doi.org/10.1007/978-3-031-36045-9_4

$$I_{gg'} = -\sum_{\alpha=1}^{N} m_\alpha g_\alpha g'_\alpha \tag{4.2}$$

It is always possible to rotate the axis system so that the non-diagonal terms of **I** vanish. The new axes are called principal inertial axes, and the three remaining diagonal elements are called principal moments of inertia. By convention, they are noted $I_a \le I_b \le I_c$. The rotational constants are inversely proportional to the principal moments of inertia. They are given by

$$A = \frac{h}{8\pi^2 I_a} \ge B = \frac{h}{8\pi^2 I_b} \ge C = \frac{h}{8\pi^2 I_c} \tag{4.3}$$

where h is the Planck constant. The conversion factor is

$$\frac{h}{8\pi^2} = I_a(\text{u Å}^2) \cdot A(\text{MHz}) = 505379.005(50) \tag{4.4}$$

In classical physics, the energy (E_r) of a freely rotating rigid body is

$$E_\text{r} = (I_a\omega_a^2 + I_b\omega_b^2 + I_c\omega_c^2)/2 \tag{4.5}$$

$\omega_a, \omega_b, \omega_c$ are the angular velocities.

In Eq. (4.5), the angular momenta $J_g = I_g\omega_g (g = a, b, c)$ are introduced

$$E_\text{r} = \frac{1}{2}\left(\frac{J_a^2}{I_a} + \frac{J_b^2}{I_b} + \frac{J_c^2}{I_c}\right) \tag{4.6}$$

The quantum mechanical transcription

$$J_g \to -i\frac{h}{2\pi}\frac{\partial}{\partial g} \tag{4.7}$$

gives the Hamiltonian operator **H** whose eigenvalues are the rotational energies. There are different solutions for different cases:

- Linear molecule ($I_a = 0$, $I_b = I_c$)

$$E = BJ(J+1) \text{ where } J = 0, 1, 2, \ldots \tag{4.8}$$

- Prolate symmetric top ($I_a < I_b = I_c$)

$$E = BJ(J+1) + (A-B)K^2 \tag{4.9}$$

- Oblate symmetric top ($I_a = I_b < I_c$)

$$E = BJ(J+1) + (C-B)K^2 \tag{4.10}$$

In these equations, J is the rotational quantum number. It is an integer. In Eqs. (4.9–4.10) K is a new quantum number: $K = 0, \pm 1, \pm 2, \ldots, \pm J$.

- Asymmetric top ($I_a \neq I_b \neq I_c$)

This corresponds to the great majority of molecules, but there is no general formula for the energy levels. The energy levels are the eigenvalues of the Hamiltonian.

For asymmetric tops, there is a parameter called asymmetry parameter measuring the asymmetry

$$\kappa = \frac{2B - A - C}{A - C} \tag{4.11}$$

$\kappa = -1$ corresponds to the limiting case of the prolate symmetric top and $\kappa = +1$ to this of the oblate symmetric top.

Equations (4.8–4.10) are only valid for a completely rigid molecule. There are two phenomena to consider.

4.2.2 *Vibrational Effects*

The bonds are not rigid, and they are subject to vibrational motions. The vibrational energy is quantified, and each vibrational state has its rotational fine structure (usually, the rotational energy is several orders of magnitude smaller than the vibrational energy [7]). Thus, what can be determined is the rotational constants in a given vibrational state. For this reason, an index v is added to the rotational constants to indicate to which vibrational state it refers to. $v = 0$ means that the ground vibrational state is being studied. For a diatomic molecule, $v = 1$ means that the first excited state of the stretching vibration is being investigated.

A perturbational treatment gives [8]

$$B_v^\xi = B_e^\xi - \sum_k \alpha_k^\xi \left(v_k + \frac{d_k}{2}\right) + \sum_{i \geq j} \gamma_{ij}^\xi \left(v_i + \frac{d_i}{2}\right)\left(v_j + \frac{d_j}{2}\right) + \sum_{i \geq j} \gamma_{l_i l_j}^\xi \ell_i \ell_j + \cdots \tag{4.12}$$

where $\xi = a, b, c$, B_e^ξ is the equilibrium rotational constant and α_i^ξ and γ_{ij}^ξ are the vibration–rotation interaction constants of different orders. This expansion converges rapidly, and the α_i^ξ constants are about two orders of magnitude smaller than the B_e^ξ constants, see Sect. 5.2.

4.2.3 *Centrifugal Distortion Effects*

As a consequence of the centrifugal force, the chemical bonds are slightly stretched when the molecule rotates. To take into account this centrifugal distortion effect, Eqs. (4.8–4.10) are slightly modified.

- Linear molecule ($I_a = 0$, $I_b = I_c$)

$$E_{vJ} = B_v J(J+1) - D_v J^2(J+1)^2 + \ldots \quad (4.13)$$

- Prolate symmetric top ($I_a < I_b = I_c$)

$$E_{vJK} = B_v J(J+1) + (A_v - B_v)K^2 - D_{vJ} J^2(J+1)^2 - D_{vJK} J(J+1)K^2 - D_{vK} K^4 + \ldots \quad (4.14)$$

- Oblate symmetric top ($I_a = I_b < I_c$)

$$E_{vJK} = B_v J(J+1) + (C_v - B_v)K^2 - D_{vJ} J^2\left(J+1^2\right) - D_{vJK} J(J+1)K^2 - D_{vK} K^4 + \ldots \quad (4.15)$$

D_J, D_{JK}, and D_K are called quartic centrifugal distortion constants. Strictly speaking, they should also have an index v, but it is often dropped.

The ground state rotational parameters are given in Table 4.1 for some prolate and oblate symmetric top molecules. The vibrational dependence of the rotational constants is shown in Eq. (4.12), and their order of magnitude is discussed in Sect. 5.2, see Table 5.1.

There are higher-order centrifugal distortion constants (sextic, octic, ...), but they are much smaller.

For an asymmetric rotor, the situation is more complicated and will be discussed in Sect. 4.4.

4.3 Methods of Determination

Note that it is possible to determine the structure of a molecule without using the rotational constants. Some important methods are reviewed in Chaps. 7 and 8.

The rotational constants can be calculated from the geometrical structure or determined experimentally using different spectroscopic methods involving either the absorption or emission of a photon or the inelastic scattering of a photon (Raman scattering) or an electron (photoelectron spectroscopy). The calculation of the rotational constants from the structure is discussed in Sect. 4.10.

Table 4.1 Rotational constants (in MHz) and quartic centrifugal distortion constants (in kHz) for a few prolate/oblate symmetric top molecules

	A/C	B	D_J	D_{JK}	D_K	Ref.
CH_3F	155,352.56	25,536.148	60.217	439.751	2106.92	a
CH_3Cl	156,052.05	13,292.881	18.102	198.936	2187.44	b
$CH_3{}^{79}Br$	155,310.79	9568.193	9.873	128.655	2539.24	c
CH_3I	155,110.69	7501.2758	6.308	98.766	2627.14	d
SiH_3CN	84,740.50	4973.006	1.4399	63.702	711.5	e
CH_3SiF_3	4060.77	3717.904	0.851	2.526	-2.508	f
AsH_3	104,884.06	112,470.576	2924.7	-3715.5	3346.5	g
SbH_3	83,608.61	88,038.990	1884.66	-2377.2	2242.9	h
BiH_3	77,977.23	79,193.227	1612.98	-1919.8	2030	i
BF_3	5161.63	10,344.044	12.890	-22.760	10.647	j
$CHCl_3$	1713.55	3302.076	1.512	-2.518	1.144	k

[a] Papousek D, Hsu Y-C, Chen H–S, Pracna P, Klee S, Winnewisser M (1993) J Mol Spectrosc 159: 33–41
[b] Nikitin A, Champion JP (2005) J Mol Spectrosc 230: 168–173
[c] Kwabia Tchana F, Kleiner I, Orphal L, Lacôme N, Bouba O (2004) J Mol Spectrosc 228: 441–452
[d] Carocci S, Di Lieto A, De Fanis A, Minguzzi P, Alanko S, Pietilä J (1998) J Mol Spectrosc 191: 368–373
[e] Priem D, Cosléou J, Demaison J, Merke I, Stahl W, Jerzembeck W, Bürger H (1998) J Mol Spectrosc 191: 183–198
[f] Styger C, Ozier I, Wang S-X, Bauder A (2006) J Mol Spectrosc 239: 115–125
[g] Tarrago G, Dana V, Mandin J-Y, Klee S, Winnewisser BP (1996) J Mol Spectrosc 178: 10–21
[h] Fusina L, Di Lonardo G, De Natale P (1998) J Chem Phys 109: 997–1003
[i] Jerzembeck W, Bürger H, Constantin FL, Margulès L, Demaison J (2004) J Mol Spectrosc 226: 24–31
[j] Maki A, Watson JKG, Masiello T, Blake TA (2006) J Mol Spectrosc 238: 135–144
[k] Bialkowska-Jaworska E, Kisiel Z, Pszczólkowski L (2006) J Mol Spectrosc 238: 72–78

Most techniques are reviewed in the "Handbook of High-resolution Spectroscopy" edited by Quack and Merkt [4] as well as in Chap. 11 of Demtröder [3].

The most current ones are briefly described below.

4.3.1 Rotational Spectroscopy

If the molecule has a permanent electric dipole moment, it has a rotational spectrum (usually in the microwave and millimeter-wave ranges), the selection rules being: $\Delta J = 0, \pm 1$ and $\Delta K = 0$.

Thus, for a linear molecule the rotational frequency is

$$v = 2B_v(J+1) - 4D_v(J+1)^3 \tag{4.16}$$

whereas for a symmetric top, it is

$$v = 2B_v(J+1) - 4D_{vJ}(J+1)^3 - 2D_{vJK}(J+1)K^2 \tag{4.17}$$

The intensity of the transitions depends on the value of the dipole moment (or of the components of the dipole moment on the principal axes for an asymmetric molecule).

As the frequency range of the rotational transitions is quite large—from microwave to submillimeter-wave—different spectrometers are used.

The basic microwave spectrometer is composed of a source (klystron, carcinotron, or Gunn diode), a cell (for instance a waveguide), and a detector. To improve the sensitivity, it is possible to modulate the microwave frequency or the absorption frequency by applying an external ac electric field (Stark spectrometer [2]) or by using a double resonance technique [9].

Quite often, the rotational spectrum is measured with a molecular beam Fourier transform microwave spectrometer (MBFTMW) [10]. Due to the very low temperature of the beam, nearly all the molecules are in the vibrational ground state, and only the ground state rotational constants can be determined. The gas is excited by a series of regularly spaced microwave pulses. After each pulse, the molecules relax by spontaneous emission. As the detection of the signal takes place in the absence of microwave power, there is no source noise. Furthermore, there is no power nor modulation broadening. The signal after averaging is treated by fast Fourier transform. Although it is possible to perform spectroscopy in a waveguide cell [11], a significant improvement was to combine the technique of FTMW with a Fabry–Perot cavity and a pulsed molecular beam synchronized with the microwave pulse [12, 13]. The weak point of this method is the small bandwidth of the measurements. This difficulty was solved by Pate et al. [14, 15] by using a high-speed waveform generator to generate a "chirped" pulse that sweeps up to 12 GHz.

Table 4.2 compares the ground state rotational constants of the centrosymmetric molecule cyclopropane, c-C_3H_6, determined by different methods and confirms the high accuracy of the FTMW method (although cyclopropane has no dipole moment, it was possible to measure its distortion dipole microwave spectrum). Additional references may be found in Appendix 5.4.

At higher frequencies, different techniques are available: either frequency multiplication of lower frequencies or difference frequency modulation by mixing mid-IR or FIR laser radiations with microwaves. These methods are reviewed by De Lucia [16] and Blake et al. [17]. The millimeter-wave (MMW) and submillimeter-wave (SMM) spectroscopies are obviously useful to determine the rotational constants of light molecules, but they also permit to significantly improve the accuracy of the rotational constants of heavier molecules by more than one order of magnitude, see the example of dimethyl sulfoxide, $(CH_3)_2SO$, in Table 4.3.

Table 4.2 Ground state rotational constants (MHz) of cyclopropane, c-C_3H_6, determined by different methods

Method	B_0	C_0	Ref.
Raman	20,074.7(60)		a
Raman	20,093.8(45)	12,522.3(90)	b
Raman	20,094.5(22)		c
FTIR	20,093.317(30)		d
FTMW	20,093.3348(28)	12,555.7498(18)	e
fs-DFWM	20,093.322(12)		f

[a] Jones JWC, Stoicheff BP (1964) Can J Phys 42: 2259–2263
[b] Butcher RJ, Jones JWC (1973) J Mol Spectrosc 47: 64–83
[c] Rubin B, Steiner DA, McCubbin TK, Polo SR (1978) J Mol Spectrosc 72: 57–61
[d] Plíva J, Johnes JWC (1984) Can J Phys 62: 1369–1373
[e] Brupbacher T, Styler C, Vogelsanger B, Ozier I, Bauder A (1989) J Mol Spectrosc 138: 197–203
[f] Kummli DS, Frey HM, Keller M, Leutwyler S (2005) J Chem Phys 123: 05,430

Table 4.3 Rotational constants (MHz) in A-reduction of dimethyl sulfoxide, $(CH_3)_2SO$

Method	A_0	B_0	C_0	Ref
FTMW	7036.5798(78)	6910.8272(78)	4218.7770(83)	a
SMM	7036.576681(33)	6910.833844(26)	4218.778084(21)	b

[a] Fliege E, Dreizler H, Typke V. (1983) Z Naturforsch 38a: 668–675
[b] Margulès L, Motiyenko RA, Alekseev EA, Demaison J (2010) J Mol Spectrosc 260: 23–29

4.3.2 *Raman Spectroscopy [18, 19]*

When the molecule has no dipole moment, Raman spectroscopy may be used because the intensity of the transitions does not depend on the permanent dipole moment but on polarizability change. In this case, the selection rule is $\Delta J = 0, \pm 1, \pm 2$.

This method is based on the inelastic scattering of light by molecules. Typically, a laser beam illuminates the sample. This puts the molecules in a transient virtual energy state. Scattered photons are emitted. If the molecule relaxes back to its original state, it gives the Rayleigh scattering (i.e., elastic scattering) which is filtered out. If the molecule relaxes back to a higher state, it gives the Stokes scattering which is usually used in Raman spectroscopy. Finally, if the molecule relaxes back in a lower state, it gives the anti-Stokes scattering. For the total energy of the system to remain constant after the molecule moves to a new rovibronic state, the scattered photon shifts to a different energy. This energy difference is equal to that between the initial and final rovibronic states of the molecule.

Raman spectroscopy uses a single continuous wave. However, to improve the sensitivity, two lasers of frequencies ω_1 and ω_2 can be used such that the difference $\omega_1 - \omega_2$ is equal to a Raman active transition. This method is called coherent anti-Stokes Raman spectroscopy (CARS) [20], see also Sect. 4.3.4. A nice illustration of the possibilities of the CARS technique is the analysis of the rotational spectra of sulfur trioxide, SO_3, leading to the determination of the equilibrium structure [21].

4.3.3 *High-Resolution Infrared Spectroscopy*

When the infrared spectrum is measured in gas phase and when the resolution of the spectrometer is high enough, a rotational fine structure may appear because rotational transitions are occurring at the same time as vibrational transitions. The analysis of this fine structure permits to obtain rotational constants in vibrationally excited states as well as in the ground vibrational state. The intensity of an absorption band is proportional to the square of the derivative of the dipole moment with respect to the normal coordinate q.

The most used technique is Fourier transform spectroscopy using a Michelson interferometer [22, 23]. It has to be noted that a synchrotron source significantly increases the sensitivity [24].

Laser spectroscopy using a tunable infrared laser is another method. Many variants are available, see Chap. 11 of Ref. [3]. A particularly interesting technique is cavity-ringdown spectroscopy (CRDS). It is a very sensitive method. It is based on the measurements of the decay time of optical resonators filled with the absorbing gas. The resonator enhances weak signals by increasing the interaction time of photons with molecules [25]. A typical example of application is the accurate determination of the molecular parameters of acetylene [26].

4.3.4 *Rotational Coherence Spectroscopy (RCS) [27–30]*

A polarized ultrashort laser pulse tuned to a vibronic transition is sent on a supersonic molecular beam. It creates a transient alignment of molecules. There is a rephasing of the initial alignment due to the fact that the molecules rotate. The RCS rephasing of alignment is monitored.

Polarized fluorescence decay

A polarized picosecond pulse coherently prepares excited state rotational levels. This creates an initial alignment of excited molecules. By viewing the fluorescence, one can time resolve the phasing and dephasing of this alignment [31].

Time-resolved fluorescence depletion (TRFD)

A single polarized pulse tuned to a vibronic transition is split into two. One pulse is directly sent to the sample, and the other one is optically delayed before going to the sample. The total fluorescence is measured as a function of the delay between the pulses [32].

Time-resolved femtosecond degenerate four-wave mixing (fs-DFWM) [29, 33]

Three polarized pulses of same frequency are used. The first two pulses are directly sent to the sample. The third pulse is delayed. The three frequencies interact in a nonlinear medium and give rise to a fourth frequency. This process originates from a third-order polarization induced by Raman transitions. Its accuracy is comparable to FTMW, see Table 4.2 for the example of cyclopropane and Table 4.4. It is well adapted to large or short-lived molecules, and it has a high sensitivity.

Performances

The RCS techniques are particularly suitable for the determination of rotational constants of large molecules. A typical application is the measurement of both ground state and excited state rotational constants of para-cyclohexylaniline, $C_{12}H_{17}N$, with

Table 4.4 Comparison of the ground state rotational constants (in MHz) determined by fs-DFWM spectroscopy with other methods

Molecule	fs-DFWM	Ref.		Method	Ref.
CS_2	3271.549(18)	a	3271.5165(15)	FTIR	b
			3271.33(21)	Rotational Raman	c
$C^{34}S_2$	3175.060(21)	a	3175.0236(81)	FTIR	b
c-C_3H_6	20,093.322(12)	d	20,093.32(3)	FTIR	e
			20,093.3348(28)	FTMW	f
c-C_8H_8	2710.329(56)	g	2709.88(45)	Rotational Raman	h
c-C_4H_8	10,663.452(18)	i	10,666.92(90)	MBIR	j
c-$C_4H_8O_2$	$A = 5084.4(5)$	k	5083.2(3)	MBIR	l
	$B = 4684.(1)$		4684.4(3)		
	$C = 2744.7(8)$		2743.6(1)		
c-$C_6H_3F_3$	1759.410(4)	m	1761.04(26)	FTIR	n
			1758.312(15)	Rotational Raman	o

[a] Kummli DS, Frey HM, Leutwyler S (2006) J Chem Phys 124: 144307
[b] Ahonen T, Alanko S, Horneman V-M, Koivusaari M, Paso R, Tolonen A-M, Anttila R (1997) J Mol Spectrosc 181: 279–286
[c] Walker WJ, Weber A (1971) J Mol Spectrosc 39: 57–64
[d] Kummli DS, Frey HM, Keller M, Leutwyler S (2005) J Chem Phys 123: 054308
[e] Pliva J, Johns JWC (1985) J Mol Spectrosc 113: 175–185
[f] Brupbacher T, Styger C, Vogelsanger B, Ozier I, Bauder A (1989) J Mol Spectrosc 138: 197–203
[g] Kummli DS, Lobsiger S, Frey H-M, Leutwyler S, Stanton JF 2008) J Phys Chem A 112: 9134–9143
[h] Thomas PM, Weber A (1978) J Raman Spectrosc 7: 353–357
[i] Kummli DS, Frey HM, Leutwyler S (2007) J Phys Chem A 111: 11936–11942
[j] Li H, Cameron Miller C, Philips LA (1994) J Chem Phys 100: 8590–8601
[k] Den T, Menzi S, Frey H-M, Leutwyler S (2017) J Chem Phys 147: 074306
[l] Brown PR, Davies PB, Hansford GM, Martin NA (1993) J Mol Spectrosc 158: 468–478
[m] Kummli DS, Frey HM, Leutwyler S (2010 Chem Phys 367: 36 – 43
[n] Tavladorakis K, Parkin JE (1995) Spectrochim Acta A 51: 1469–1478
[o] Schlupf J, Weber A (1973) J Raman Spectrosc 1: 3–15

an accuracy better than one MHz [34]. For instance, the determined ground state rotational constants are (in MHz): $A_0 = 2406.4(6)$; $B_0 = 358.5(3)$; and $C_0 = 356.5(3)$. For this reason, it is also appropriate for the study of large molecular complexes such as the phenol dimer [35].

4.3.5 Photoelectron Spectroscopy [36]

A molecule is excited by a high-energy photon that ionizes the molecule. The energies of the ejected electrons depend of the original electronic state and also of the vibrational state and the rotational level.

It is mainly used to study the structure of molecular cations. It is now possible to resolve the rotational structure in the photoelectron spectra of polyatomic molecules and even to measure the fine and hyperfine structures of positively charged atoms and small molecules.

The equilibrium structure of several diatomic species has been determined by this method, see Tables 3.6 and 3.7. Polyatomic species have also been investigated. For instance, the ground state rotational constant of the perdeuterated ammonium ion, ND_4^+, has been determined to be $B_0 = 89{,}503(111)$ MHz [37].

4.4 Centrifugal Distortion of an Asymmetric Top [1, 3–7, 38]

The rotational Hamiltonian including centrifugal distortion may be written in the following way:

$$\mathbf{H} = A'\mathbf{J}_z^2 + B'\mathbf{J}_x^2 + C'\mathbf{J}_y^2 + \frac{\hbar^4}{4}\sum_{\alpha\beta\gamma\delta} \tau_{\alpha\beta\gamma\delta}\mathbf{J}_\alpha\mathbf{J}_\beta\mathbf{J}_\gamma\mathbf{J}_\delta \tag{4.18}$$

where α, β, γ, δ take the values of (x, y, z) of the molecule-fixed Cartesian system and $\mathbf{J}_\alpha$ is the α-component of total angular momentum. There are 81 quartic centrifugal distortion constants $\tau_{\alpha\beta\gamma\delta}$, but many constants are equivalent and many do not contribute to first order. Finally, using symmetry considerations and commutation rules, we arrive at only six groups of terms. Unfortunately, these six centrifugal distortion constants are not determinable from an experimental spectrum because the system of normal equations is ill-conditioned. By submitting the Hamiltonian to a unitary transformation, called reduction, Watson [38] has shown that of the six quartic constants, only five combinations are determinable.

Watson [38] proposed two different reductions, one called A-reduction (for asymmetric tops) and another one called S-reduction (better for molecules close to the symmetric top, although it also gives satisfactory results for very asymmetric tops).

In addition to these two reductions, there are six different ways to identify the (*x*, *y*, *z*) reference system with the (*a*, *b*, *c*) principal axes system. In practice, only two representations are used: I^r where $x = b$, $y = c$ and $z = a$ which is adapted to prolate tops and III^r where $x = a$, $y = b$ and $z = c$ which is believed to be better for oblate tops. In conclusion, there are four different sets of quartic centrifugal distortion constants. As an example, the rotational and quartic centrifugal distortion constants of dimethyl sulfoxide, $(CH_3)_2SO$, in the symmetric and asymmetric reductions and the representation I^r and III^r are given in Table 4.5 [39]. Although the values of the centrifugal distortion constants are small compared to the rotational constants, the successive transformations of the Hamiltonian slightly modify the rotational constants. The following linear combinations can be determined from the analysis of the spectra:

$$B_z = B_z^{(A)} + 2\Delta_J = B_z^{(S)} + 2D_J + 6d_2 \tag{4.19a}$$

$$B_x = \begin{array}{l} B_x^{(A)} + 2\Delta_J + \Delta_{JK} - 2\delta_J - 2\delta_K \\ B_x^{(S)} + 2D_J + D_{JK} + 2d_1 + 4d_2 \end{array} \tag{4.19b}$$

$$B_y = \begin{array}{l} B_y^{(A)} + 2\Delta_J + \Delta_{JK} + 2\delta_J + 2\delta_K \\ B_y^{(S)} + 2D_J + D_{JK} - 2d_1 + 4d_2 \end{array} \tag{4.19c}$$

$B_g^{(A)}$ ($g = x, y, z$) are the experimental constants in the *A*-reduction, $B_g^{(S)}$ the experimental constants in the *S*-reduction, B_g the determinable constants, the Δ the quartic centrifugal distortion constants in the *A*-reduction, and the *D* the quartic centrifugal distortion constants in the *S*-reduction.

However, the determinable constants are still affected by the centrifugal distortion. The "unperturbed rigid rotor" constants, B_x', are given by

Table 4.5 Rotational constants (MHz) and quartic centrifugal distortion constants for the ground state of dimethyl sulfoxide, $(CH_3)_2SO$[a]

Repres	Symmetric reduction			Asymmetric reduction	
	I^r	III^r		I^r	III^r
A	7036.579634(32)	7036.582334(26)	A	7036.576681(33)	7036.49312(10)
B	6910.830104(26)	6910.830327(25)	B	6910.833844(26)	6910.92455(10)
C	4218.779498(21)	4218.776781(40)	C	4218.778084(21)	4218.77257(13)
D_J	3.463079(14)	6.089201(16)	Δ_J	4.050907(14)	6.62679(10)
D_{JK}	−0.925893(71)	−8.937244(53)	Δ_{JK}	−4.452825(45)	−12.15739(44)
D_K	3.767553(89)	3.989320(37)	Δ_K	6.706545(70)	6.67255(37)
d_1	−1.4548584(69)	0.164061(11)	δ_J	1.4548677(66)	−0.164790(83)
d_2	−0.2939137(42)	−0.2717127(48)	δ_K	1.285585(18)	−46.8755(53)

[a] Margulès L, Motiyenko RA, Alekseev EA, Demaison J (2010) J Mol Spectrosc 260: 23–29, reprinted by permission of the publisher Elsevier

$$B'_x = B_x + \frac{1}{2}\left(\tau_{yyzz} + \tau_{xyxy} + \tau_{xzxz}\right) + \frac{1}{4}\tau_{yzyz} \tag{4.20}$$

where B'_y and B'_z are obtained by cyclic permutation of x, y, and z and $\tau_{\alpha\beta\gamma\delta}$ are the quartic centrifugal distortion constants as defined by Kivelson and Wilson [40]. The problem is that these $\tau_{\alpha\beta\gamma\delta}$ constants are approximately determinable only for planar molecules thanks to so-called planarity relations. Fortunately, they can be easily calculated from the harmonic force field.

This centrifugal distortion correction is small, except for light molecules, see Table 4.6. However, it is often significantly larger than the standard deviation of the rotational constants. For this reason, it is better to take it into account, especially since the derived structure is sensitive to the true accuracy of the rotational constants [41]. The structure of sulfine, CH_2SO, is given as example in Table 4.7 [42]. Although the effect of the centrifugal correction is small, it often non-negligible. Similar results were found for SO_2 [43]. It is worth noting that both Eqs. (4.19a, 4.19b, 4.19c) and (4.20) contribute significantly to the correction.

Table 4.6 Centrifugal distortion contribution to the ground state rotational constants for some molecules (in MHz)

Molecule		Exp	Det.–Exp.[a]	Unpert.–Det[b]	Ref.
NH_2	*A*	712,634.609(5)	63.329	−71.312	1
	B	388,213.098(9)	−145.871	−96.806	
	C	244,843.726(9à	22.438	201.733	
HCOOH	*A*	77,512.2354(11)	0.0200	−0.0643	2
	B	12,055.1065(2)	−0.1556	0.0154	
	C	10,416.1151(2)	0.0231	0.1375	
$CH_2S=O$	*A*	40,404.470(26)	0.0130	−0.0298	3
	B	9394.8290(62)	−0.0916	0.0003	
	C	7607.1156(51)	−0.0101	0.1082	

[a] *Det.* determinable constants, Eqs. (4.19a, 4.19b, 4.19c)
[b] *Unpert.* unperturbed constants, Eq. (4.20)
[1] Müller HSP, Klein H, Belov SP, Winnewisser G, Morino I, Yamada KMT, Saito S (1999) J Mol Spectrosc 195: 177–184
[2] Baskakov OI, Alekseev EA, Motiyenko RA, Lohilahti J, Horneman V-M, Alanko S, Winnewisser BP, Medvedev IR, De Lucia FC (2006) J Mol Spectrosc 240: 188–201
[3] Demaison J, Vogt N, Ksenafontov DN (2020) J Mol Struct 1206: 127676

Table 4.7 Semiexperimental equilibrium structure of sulfine, $CH_2 = S = O$[a]

	With distortion	Without distortion
Δ_e [b]/uÅ^2	0.0001	0.0013
C=S/Å	1.60838(6)	1.60864(8)
S=O/Å	1.46504(5)	1.46549(7)
CH_{cis}/Å	1.0797(2)	1.0798(3)
CH_{trans}/Å	1.0817(4)	1.0791(6)
∠(SCO)/degree	114.801(1)	114.784(1)
∠(SCH_{cis})/degree	123.047(8)	122.96(1)
∠(SCH_{trans})/degree	115.19(4)	115.40 (5)

[a] Demaison J, Vogt N, Ksenafontov DN (2020) J Mol Struct 1206: 127676, reprinted by permission of the publisher Elsevier
[b] Equilibrium inertial defect

4.5 Anharmonic Resonances

When vibrational states have the same symmetry and are not too far away, they may interact through an anharmonic potential term. This resonance affects the rotational constants, and even if the perturbation is small, the accuracy of the derived structure may be significantly worsened.

In the simple case of two interacting states *i* and *j*, the perturbed energies are obtained by diagonalizing the 2 × 2 matrix

$$\begin{pmatrix} E_{RV}^{(0)}(i) & V \\ V & E_{RV}^{(0)}(j) \end{pmatrix} \tag{4.21}$$

$E_{RV}^{(0)}$ are the unperturbed energies, and V is the potential term linking the two states. The eigenfunctions may be written as

$$\left[\begin{array}{l} \psi_i = c\psi_i^{(0)} + s\psi_j^{(0)} \\ \psi_j = -s\psi_i^{(0)} + c\psi_j^{(0)} \end{array}\right. \tag{4.22}$$

The rotational energy may be written as

$$\begin{aligned} E_R\left(A_i', B_i', C_i'\right) &= \langle \psi_i | \mathbf{H}_R | \psi_i \rangle \\ &= c^2 \left\langle \psi_i^{(0)} | \mathbf{H}_R | \psi_i^{(0)} \right\rangle + s^2 \left\langle \psi_j^{(0)} | \mathbf{H}_R | \psi_j^{(0)} \right\rangle \\ &= E_R(A_i, B_i, C_i) + s^2 \left[E_R\left(A_j, B_j, C_j\right) - E_R(A_i, B_i, C_i) \right] \end{aligned} \tag{4.23}$$

where the prime, ‘, indicates the rotational constants affected by the resonance. If the resonance is not too strong, the rotational constants may be approximately written

$$B_i' = B_i + s^2(B_j - B_i) \tag{4.24}$$

A_i' and C_i' are obtained by cyclic permutation.

There are different types of resonances. The most common are when the harmonic frequencies of two states are close: (i) the Fermi resonance either $2\omega_r \approx \omega_s$ or $\omega_r + \omega_s \approx \omega_t$ and (ii) the Darling-Dennison resonance: $2\omega_r \approx 2\omega_s$.

A typical example of Fermi resonance is found in the rotational spectrum of F_2O. The distance between the states $v_1 = 1$ and $v_2 = 2$ is only 6.8 cm^{-1}. For the state $v_1 = 1$, the values of the rotational constants (in MHz) neglecting the resonance are: $A_1 = 59{,}213.58(100)$; $B_1 = 10{,}824.35(35)$; and $C_1 = 9128.09(50)$, whereas the values of the unperturbed constants are: $A_1^{(0)} = 58{,}744.08(100)$; $B_1^{(0)} = 10{,}830.66(50)$; and $C_1^{(0)} = 9160.23(50)$ [44]. For instance, $\alpha_1^A = -430.95$ MHz when the Fermi resonance is neglected but 38.55 MHz when it is taken into account. The differences are several orders of magnitude larger than the accuracy and affect significantly the structure. Another well-known example of Fermi resonance is OCS for which two Fermi resonances are not negligible: between the states ($1\ 0^0\ 0$) and ($0\ 2^0\ 0$) and between the states ($0\ 0^0\ 1$) and ($0\ 2^0\ 0$) [45, 46] (the frequencies of the normal vibrations are in cm^{-1}: $\nu_1 = 2062$, $\nu_2 = 520$, $\nu_3 = 859$). When the resonances are neglected, r_e(C = O) = 1.1545(2) Å and, when it is taken into account, r_e(C = O) = 1.15617(14) Å. The presence of resonances is a general phenomenon even for small molecules, and their analysis is not straightforward. It is one of the many difficulties of an experimental structure determination. This difficulty disappears when the semiexperimental method is used, see Sect. 5.5.

4.6 Coriolis Interaction [1, 5–8, 47]

The Coriolis interaction is another interaction between vibrational states which may complicate the analysis of the spectra. It is caused by the coupling of the total angular momentum J_g and the vibrational angular momentum. According to Jahn's rule [48], Coriolis interaction takes place if the product of the symmetry species of the two vibrational modes contains the species of rotation.

Although the vibration–rotation-interaction constant, or α-constant, introduced in Eq. (4.12) is only a small percentage of the rotational constant, it plays a critical role in an accurate structure determination. The α-constant can be determined experimentally by determining the rotational constants of the vibrationally excited states, but it sometimes fails. This is due to the Coriolis interaction. These α-constants can be computed from the cubic force field and are the sum of three terms, see Sect. 5.2.

$$\alpha_k^{\xi} = \alpha_k^{\xi}(\text{harm}) + \alpha_k^{\xi}(\text{Coriolis}) + \alpha_k^{\xi}(\text{anharm}) \tag{4.25}$$

The problem comes from the Coriolis term which may be calculated by a second-order perturbation calculation

$$\alpha_k^\xi(\text{Coriolis}) = = -2\frac{\left(B_e^\xi\right)^2}{\omega_k}\sum_l \frac{\left(3\omega_k^2+\omega_l^2\right)\left(\zeta_{kl}^\xi\right)^2}{\omega_k^2-\omega_l^2} \tag{4.26}$$

$\zeta_{kl}^\xi = -\zeta_{lk}^\xi$ is a Coriolis coupling constant. Note that the denominator can be very small. If the numerator is different from zero, it gives rise to an interaction between the vibrational modes k and l called Coriolis interaction. When this interaction is large, it complicates the analysis of the spectra, and the states k and l cannot be treated separately. The Hamiltonian for the coupled v''/v' vibrational states is

$$\mathbf{H} = \begin{pmatrix} \mathbf{H}_{v''v''} & \mathbf{H}_{v''v'} \\ \mathbf{H}_{v''v'} & \mathbf{H}_{v'v'} + \Delta E_{v''v'} \end{pmatrix} \tag{4.27}$$

$\mathbf{H}_{v''v''}$ and $\mathbf{H}_{v'v'}$ are the Watson's semirigid rotor Hamiltonians for the v'' and v' states, respectively, $\Delta E_{v''v'}$ is the energy difference between the v'' and v' states, and the coupling term is

$$\begin{aligned} H_{v''v'}^{\text{Cor}} &= i\left(G_\alpha + G_\alpha^J J^2 + G_\alpha^K J_z^2 + \cdots\right) J_\alpha \\ &+ \left(F_{\beta\gamma} + F_{\beta\gamma}^J J^2 + F_{\beta\gamma}^K J_z^2 + \cdots\right)\left(J_\beta J_\gamma + J_\gamma J_\beta\right) + \cdots \end{aligned} \tag{4.28}$$

where $(\alpha, \beta, \gamma) = (a, b, c)$.

Alternative notations exist.

A typical example of the difficulty to analyze a spectrum in presence of Coriolis interaction is the fundamental band ν_1 of NO_2 [49]. The (1 0 0) state at 1319.77 cm^{-1} is in Coriolis interaction with the states (0 2 0) at 1498.35 cm^{-1} and (0 0 1) at 1616.85 cm^{-1}. Table 4.8 shows that the differences between the different determinations are much larger than the standard deviations. Another example is the Coriolis interaction between the states $\nu_7 = 1$ and $\nu_9 = 1$ of *trans* formic acid, HCOOH [50]. The distance between the two states is small: 14.559 cm^{-1} making the analysis of the spectra difficult. The consequence is that the value of the constant α_1^A varies between −31.7(14) MHz and −416.943(4) MHz depending on the determinations.

4.7 Electronic Correction

It is not enough to take into account the anharmonic resonances and to make a centrifugal distortion correction (for asymmetric tops). There is another (often) small effect to consider. To calculate the moments of inertia, the usual method is to use atomic masses, i.e., to assume that all the mass is concentrated on the nucleus which is assumed to be a point mass. However, in most cases, the distribution of the electronic clouds around the atoms is non-spherical. A small correction has to be taken into account [1, 5, 6, 51, 52]. The rotational constants calculated from the atomic masses, B^ξ, are obtained by

Table 4.8 Rotational constants (MHz) of the $v_1 = 1$ state of NO_2. $\Delta B_g = B_g(v_1 = 1) - B_g(v = 0)$ [a]

ΔA	ΔB	ΔC	Ref.	Year
2663.60(1829)	−57.63(162)	−78.14	b	1982
2731.55(23)	−70.03(4)	−78.37(4)	c	1983
2645.04(627)	−72.87(48)	−75.66(52)	d	1983
2720.05(78)	−69.97(11)	−65.22(60)	e	1984
2734.4(2)	−70.04(3)	−78.42(6)	f	1986
2726.19(26)	−71.064(21)	−34.782(12)	g	1992

[a] Ground-state constants: [a]
$A_0 = 239{,}904.5163(57)$; $B_0 = 13{,}002.20066(39)$; $C_0 = 12{,}304.75510(39)$
[a] Liu Y, Liu X, Liu H, Guo Y (2000) J Mol Spectrosc 202: 306–308
[b] Camy-Peyret C, Flaud, J-M, Perrin A, Rao KN (1982) J Mol Spectrosc 95: 72–79
[c] Morino Y, Tanimoto M, Saito S, Hirota E, Awata R, Tanaka T (1983) J Mol Spectrosc 98: 331–348 (1983)
[d] Hardwick JL (1983) J Mol Spectrosc.99: 239–242 (1983)
[e] Perrin A, Mandin J-Y, Camy-Peyret C, Flaud J-M, Chevillard J-P, Guelachvili G (1984) J Mol Spectrosc 103: 417–435
[f] Morino Y, Tanimoto M (1986) J Mol Spectrosc 115: 442–445
[g] Perrin A, Flaud J-M, Camy-Peyret C, Vasserot A-M, Guelachvili G, Goldman A, Murcray FJ, Blatherwick RD (1992) J Mol Spectrosc 154: 391–406

$$B^{\xi} = B^{\xi}_{\text{eff}}\left(1 - \frac{m_{\text{e}}}{M_{\text{p}}} g^{\xi}\right) \tag{4.29}$$

where $\xi = a, b, c$, B^{ξ}_{eff} are the experimental constants, m_{e} the electron mass, M_{P} the proton mass, and g^{ξ} the molecular rotational g-factors in the principal axis system. The g-factors can be obtained experimentally from the analysis of the Zeeman hyperfine structure of the rotational transitions [52], and they are tabulated in the Landolt-Börnstein tables (see for instance Ref. [53] and previous volumes), but it is now much easier to calculate them ab initio [54].

The correction is large for light molecules, but it rapidly decreases when the rotational constant decreases, but it may become large when there is an excited electronic state close to the ground state. A typical example is ozone, O_3, for which $g_{aa} = -2.9777$; $g_{bb} = -0.2295$; and $g_{cc} = -0.076$ [55]. Using the rotational constants [56], the corrections to the rotational constants are (in MHz): $\Delta A = 173.28$; $\Delta B = 1.68$, and $\Delta C = 0.46$. Although the correction is large for A, its effect on the structure is small, albeit non-negligible in a very accurate determination. Indeed, neglecting this electronic correction for SO_2 decreases r(SO) by 0.0003 Å and increases $\angle$(OSO) by 0.038°. However, this good behavior is due to the fact that the structure is obtained from only one isotopologue (more precisely, its A_{e} and B_{e} rotational constants). Therefore, the system is well conditioned and not sensitive to errors, see Sect. 4.11.

4.8 Variational Calculations

It would be tempting to have a method similar to the direct potential fit which is used for diatomic molecules (see Sect. 3.6). One of the main difficulties is the number of degrees of freedom which is much larger for a polyatomic molecule than for a diatomic molecule. However, the variational method has been used by several authors, but most of them are mainly interested in the calculation of the rovibrational energy levels. This aspect was for instance reviewed by Tennyson [57]. However, Handy et al. [58] determined the structure of several small molecules. Their work leads to the development of a computer program called MULTIMODE [59–61]. The kinetic energy operator is a Watson Hamiltonian [62], and for a triatomic molecule, the potential energy may be written

$$V = \sum\nolimits_{ijk} K_{ijk} (\Delta r_{12})^i (\Delta r_{13})^j \Delta\theta^k \tag{4.30}$$

The basis functions are harmonic oscillator functions.

This method has been used to determine the equilibrium structure of several triatomic molecules (such as HCN) [63] and tetratomic molecules: H_2CO [64], H_2CS [65], HC≡CH [66], and H_2O_2 [67].

See also MORBID in Sect. 6.5.

4.9 Large-Amplitude Motions

See Chap. 6.

4.10 Calculation of the Rotational Constants from the Structure

Usually, the structure of a molecule is given in internal coordinates: bond lengths, bond angles, and dihedral angles. When the structure is simple, the moments of inertia can be calculated directly from these internal coordinates. For instance, for a linear triatomic molecule XYZ, the moment of inertia is

$$I = \frac{1}{M}\left[m_{\mathrm{X}} m_{\mathrm{Y}} d_{\mathrm{XY}}^2 + m_{\mathrm{Y}} m_{\mathrm{Z}} d_{\mathrm{YZ}}^2 + m_{\mathrm{X}} m_{\mathrm{Z}} d_{\mathrm{XZ}}^2\right] \tag{4.31}$$

For a bent triatomic molecule, XY_2, the values of the moments of inertia are

$$I_x = 2 m_{\mathrm{Y}} d_{\mathrm{XY}}^2 \sin^2\left(\frac{\theta}{2}\right) \tag{4.32a}$$

$$I_y = 2\frac{m_X m_Y}{M} d_{XY}^2 \cos^2\left(\frac{\theta}{2}\right) \tag{4.32b}$$

In these equations, M is the total mass of the molecule and d_{ij} is the distance between the atoms i and j. In Eqs. (4.32a), (4.32b), x corresponds to the C_2 axis, with the z axis perpendicular to the molecular xy plane, and θ is the $\angle$(YXY) bond angle.

A more complete list of expressions may be found in Table 13.1 of Gordy and Cook [1] or in Appendix IV.2 of Demaison, Boggs, and Császár [5] or in Appendix 2 of chap. 4 of Demaison and Vogt [6].

More generally, the moments of inertia of a molecule are calculated in two steps. First, the internal coordinates are converted into Cartesian coordinates. There are many computer programs available for this calculation. In particular, Thompson has proposed a general method for this conversion [68]. Once the Cartesian coordinates are available, Eqs. (4.1–4.2) are used to calculate the inertia tensor which is subsequently diagonalized in order to have the principal moments of inertia.

4.11 Determination of the Rotational Constants and of the Structure from the Rotational Constants

Only a summary is given in this section. A more detailed presentation may be found in [69] and Chap. 9 of Ref. [6].

The rotational constants (as well as the centrifugal distortion constants) are obtained by a (generally weighted) least-squares fit of the experimental rotational or rovibrational spectrum. The least-squares fit method is also used to determine the geometrical structure from the rotational constants although for small molecules there is an analytical expression of the moments of inertia as a function of the structural parameters. This permits to determine the structure. However, this is an exception. In most cases, the structure is obtained by a nonlinear least-squares fit. As input data, the rotational constants may be used [70], but it is much more usual to use the moments of inertia, see Eqs. (4.3–4.4).

An iterative procedure is used. The correction vector $\Delta\mathbf{r}_k$ for the p geometrical parameters at each iteration k is found by solving the linear system

$$\mathbf{J}_k' \Delta\mathbf{r}_k = \Delta\mathbf{I}_k' \tag{4.33}$$

$\Delta\mathbf{I}_k'$ is the vector of the residuals of the n moments of inertia $\mathbf{I}$ at the previous iteration, and $\mathbf{J}_k'$ is the Jacobian matrix.

Quite often, the moments of inertia have different uncertainties, and their variance is

$$\mathbf{V}(\mathbf{I}) = \sigma^2 \mathbf{W}^{-1} \tag{4.34}$$

where $\mathbf{W}$ is the diagonal weight matrix with $W_{ii} = 1/\sigma^2(I_i)$. Left-multiplying Eq. (4.33) by $\mathbf{W}^{1/2}$ gives

$$\mathbf{W}^{1/2}\mathbf{J}'_k\Delta\mathbf{r}_k = \mathbf{W}^{1/2}\Delta\mathbf{I}'_k \tag{4.35}$$

The substitution $\mathbf{J}_k = \mathbf{W}^{1/2}\mathbf{J}'_k$ and $\Delta\mathbf{I}_k = \mathbf{W}^{1/2}\Delta\mathbf{I}'_k$ allows us to solve the system by the ordinary least-squares method because the variance–covariance of the new variable $\Delta\mathbf{I}_k$ is $\sigma^2\mathbf{I}_n$.

One of the best methods to solve the system of normal equations is the singular value decomposition (SVD) [71]. First, the columns of the Jacobian matrix $\mathbf{J}_k$ are scaled to have unit length, then the singular values $\mathbf{D}$ of the scaled $\mathbf{J}_k$ are calculated

$$\mathbf{J}_k = \mathbf{U}_k\mathbf{D}_k\mathbf{V}_k^{\mathrm{T}} \tag{4.36}$$

where $\mathbf{U}_k$ is a $n \times p$ matrix, $\mathbf{D}_k$ is a diagonal $p \times p$ matrix containing the singular values μ_i, and $\mathbf{V}_k$ is a $p \times p$ orthonormal matrix containing the eigenvectors of $\mathbf{J}^{\mathrm{T}}\mathbf{J}$. The solution is

$$\Delta\mathbf{r}_k = \mathbf{V}_k\mathbf{D}_k^{-1}\mathbf{U}_k^{\mathrm{T}}\Delta\mathbf{I}_k \tag{4.37}$$

And the updated solution is

$$\mathbf{r}_{k+1} = \mathbf{r}_k + \Delta\mathbf{r}_k \tag{4.38}$$

There are many diagnostics available to check the quality of the least-squares fit, such as the standard deviation of the fit. There are two diagnostics which are particularly important: the condition number and the leverage. The condition number κ is defined from the largest and smallest singular values of the scaled Jacobian matrix

$$\kappa = \frac{\mu_{\max}}{\mu_{\min}} \tag{4.39}$$

It is an error amplifier and is used to determine whether the system of normal equations is ill-conditioned or not. If we consider a perturbation $\delta\mathbf{I}$ in $\mathbf{I}$ and $\delta\mathbf{J}$ in $\mathbf{J}$, it induces a perturbation $\delta\Delta\mathbf{r}_k$

$$\left\| \frac{\delta\Delta r_k}{\Delta r_k} \right\| \leq \kappa R^{-1}\left[2 + \kappa\sqrt{1 - R^2}\right]\max\left(\frac{[\![\delta\mathbf{J}]\!]}{\|\mathbf{J}\|}, \frac{[\![\delta\mathbf{I}]\!]}{\|\mathbf{I}\|}\right) \tag{4.40}$$

where R is the multiple correlation coefficient. This equation shows that not only the errors δI_i on the experimental data affect the accuracy of the solution but also the errors $\delta\mathbf{J}_k$ on the Jacobian $\mathbf{J}_k$. It is known that numerical differentiation is an unstable procedure prone to truncation and rounding errors. In other words, it is desirable to calculate the Jacobian with the highest possible accuracy. However, in most cases, the dominant error is due to δI_i.

Another important diagnostic to perform is an analysis of leverages by calculating the diagonal elements of the square hat matrix **H** of dimensions $n \times n$

$$\mathbf{H} = \mathbf{J}\left(\mathbf{J}^{\mathrm{T}}\mathbf{J}\right)^{-1}\mathbf{J}^{\mathrm{T}} = \mathbf{U}\mathbf{U}^{\mathrm{T}} \tag{4.41}$$

This matrix connects computed and measured input data. It may be shown that $0 \leq h_{ii} \leq 1$. A value of h_{ii} close to one indicates that a small change in the input value ΔI_i causes a large change in the solution, which is an indication of problem with the ith measurement. It is obviously desirable that none of the parameters is determined by a single data because its disproportionate influence would diminish the balancing effect of the least-squares fit.

4.12 Experimental Data

The known experimental rotational parameters (rotational and centrifugal distortion constants, dipole moments, etc.) are tabulated in Landolt-Börnstein tables: molecular constants mostly from microwave, molecular beam, and sub-Doppler laser spectroscopy [72–77].

The constants for many molecules are also compiled by the National Institute of Standards and Technology (NIST) and published in the J Phys Chem Ref Data, see Refs. [78–82].

Experimental rotational transitions may be found in several databases:

- The Jet Propulsion Laboratory *Submillimeter, Millimeter, and Microwave Spectral Line Catalog* is accessible via anonymous ftp at spec.jpl.nasa.gov or via their home page at http://spec.jpl.nasa.gov.
- The **C**ologne **D**atabase for **M**olecular **S**pectroscopy is available online at http://www.astro.uni-koeln.de/cdms/.
- Toyama microwave Atlas for spectroscopists and astronomers. See: http://www.sci.u-toyama.ac.jp/phys/4ken/atlas/.

References

1. Gordy W, Cook RL (1984) Microwave molecular spectra. Wiley, New York
2. Townes CH, Schawlow AL (1975) Microwave spectroscopy. Dover, New York
3. Demtröder W (2018) Atoms, molecules and photons. An introduction to atomic- molecular- and quantum physics. Springer, Berlin
4. Quack M, Merkt F (2011) Handbook of high-resolution spectroscopy. Wiley, Chichester, UK
5. Perrin A, Demaison J, Flaud J-M, Lafferty W, Sarka K (2011) Spectroscopy of polyatomic molecules: determination of the rotational constants. In: Demaison J, Boggs JE, Császár AG (eds) Equilibrium molecular structures. From spectroscopy to quantum chemistry. CRC Press, Boca Raton, pp 89–124

6. Demaison J, Vogt N (2020) Accurate structure determination of free molecules. Springer Nature, Switzerland, p 277
7. Wilson EB Jr, Decius JC, Cross PC (1955) Molecular vibrations: the theory of infrared and Raman vibrational spectra. McGraw Hill, New York
8. Mills IM (1972) Vibration-rotation structure in asymmetric and symmetric molecules. In: Rao KN, Mathews CW (eds) Molecular spectroscopy: modern research, vol 1. Academic Press, New York
9. Baker JG (1979) Microwave-microwave double resonance. In: Chantry GW (ed) Modern aspects of microwave spectroscopy. Academic Press, London, pp 65–122
10. Caminati W, Grabow J-U (2018) Advancements in microwave spectroscopy. In: Laane J (ed) Frontiers and advances in molecular spectroscopy. Elsevier, Amsterdam, pp 569–598
11. Dreizler H (1986) Experiences with microwave Fourier transform spectroscopy of molecular gases. Mol Phys 59:1–28
12. Balle TJ, Flygare WH (1981) Fabry-Perot cavity pulsed Fourier transform microwave spectrometer with a pulsed nozzle particle source. Rev Sci Instrum 52:33–45
13. Grabow J-U (2011) Fourier transform microwave spectroscopy measurement and instrumentation. In: Quack M, Merkt F (eds) Handbook of high-resolution spectroscopy. Wiley, Chichester, Chapter 19
14. Brown GG, Dian BC, Douglass KO, Geyer SM, Shipman ST, Pate BH (2008) A broadband Fourier transform microwave spectrometer based on chirped pulse excitation. Rev Sci Inst 79:053103
15. Shipman ST, Pate BH (2011) New techniques in microwave spectroscopy. In: Quack M, Merkt F (eds) Handbook of high-resolution spectroscopy. Wiley, Chichester, Chapter 20
16. De Lucia FC (2010) The submillimeter: a spectroscopist view. J Mol Spectrosc 261:1–17
17. Blake GA, Laughin KB, Cohen RC, Busarow KL, Gwo DH, Schmuttenmaer CA, Steyert DW, Saykally RJ (1991) Tunable far infrared laser spectrometers. Rev Sci Instr 62:1693–1700
18. Gardiner DJ (1989) Practical Raman spectroscopy. Springer, Berlin
19. Weber A (2011) High-resolution Raman spectroscopy of gases. In: Quack M, Merkt F (eds) Handbook of high-resolution spectroscopy. Wiley, Chichester, Chapter 30
20. Lang T, Motzkus M, Frey HM, Beaud P (2001) High resolution femtosecond coherent anti-Stokes Raman scattering: determination of rotational constants, molecular anharmonicity, collisional line shifts, and temperature. J Chem Phys 115:5418–5426
21. Masiello T, Vulpanovici N, Barber J, Chrysostome ETH, Nibler JW, Maki A, Blake TA, Sams RL, Weber A (2004) Analysis of high-resolution infrared and CARS spectra of $^{32}S^{18}O_3$. J Mol Spectrosc 227:50–59
22. Davis SP, Abrams MC, Brault JW (2001) Fourier transform spectrometry. Academic press, San Diego
23. Albert S, Keppler K, Quack M (2011) High resolution Fourier transform infrared spectroscopy. In: Quack M, Merkt F (eds) Handbook of high-resolution spectroscopy. Wiley, Chichester, Chapter 26
24. Brubach J-B, Manceron L, Rouzières M, Pirali O, Balcon D, Kwabia Tchana F, Boudon V, Tudorie M, Huet T, Cuisset A, Roy P (2010) Performance of the AILES THz-infrared beamline at SOLEIL for high resolution spectroscopy. AIP Conf Proc 1214:81–84. https://doi.org/10.1063/1.3326356
25. Romanini D, Ventrillard I, Méjean G, Morville J, Kerstel E (2014) Introduction to cavity enhanced absorption spectroscopy. In: Gagliardi G, Loock HP (eds) Cavity enhanced spectroscopy and sensing. Springer, Berlin, pp 1–60
26. Amyay B, Herman M, Fayt A, Campargue A, Kassi S (2011) Acetylene, $^{12}C_2H_2$: refined analysis of CRDS spectra around 1.52 μm. J Mol Spectrosc 267:80–91
27. Felker PM (1992) Rotational coherence spectroscopy: studies of the geometries of large gas-phase species by picosecond time-domain methods. J Phys Chem 96:7844–7857
28. Felker PM, Zewail AH (1994) Molecular structures from ultrafast coherence spectroscopy. In: Manz J, Wöste L (eds) Femtochemistry VCH, Weinheim, chap 5 VCH. Weinheim

29. Frey H-M, Kummli D, Lobsiger S, Leutwyler S (2011) High-resolution rotational Raman coherence spectroscopy with femtosecond pulses. In: Quack M, Merkt F (eds) Handbook of high-resolution spectroscopy. Wiley, Chichester, Chapter 31
30. Riehn C (2002) High-resolution pump–probe rotational coherence spectroscopy—rotational constants and structure of ground and electronically excited states of large molecular systems. Chem Phys 283:297–329
31. Baskin JS, Felker PM, Zewail AH (1986) Doppler free time resolved polarization spectroscopy of large molecules: measurement of excited state rotational constants. J Chem Phys 84:4708–4710
32. Côté MJ, Kauffman JF, Smith PG, McDonald JD (1989) Picosecond fluorescence depletion spectroscopy. I. Theory and apparatus. J Chem Phys 90:2865–2873
33. Brown EJ, Zhang Q, Dantus M (1999) Femtosecond transient-grating techniques: population and coherence dynamics involving ground and excited states. J Chem Phys 110:5772–5788
34. Riehn C, Weichert A, Lommatzsch U, Zimmermann M, Brutschy B (2000) High-resolution rotational coherence spectroscopy of para-cyclohexylaniline. J Chem Phys 112:3650–3661
35. Connell LL, Ohline SM, Joireman PW, Corcoran TC, Felker PM (1992) Rotational coherence spectroscopy and structure of phenol dimer. J Chem Phys 96:2585–2593
36. Merkt F, Willitsch S, Hollenstein U (2011) High-resolution photoelectron spectroscopy. In: Quack M, Merkt F (eds) Handbook of high-resolution spectroscopy. Wiley, Chichester, Chapter 43. https://doi.org/10.1002/9780470749593.hrs071
37. Signorell R, Palm H, Merkt F (1997) Structure of the ammonium radical from a rotationally resolved photoelectron spectrum. J Chem Phys 106:6523–6533
38. Watson JKG (1977) Aspect of quartic and sextic centrifugal effects on rotational energy levels. In: Durig JR (ed) Vibrational spectra and structure, vol 6. Elsevier, Amsterdam, pp 2–89
39. Margulès L, Motiyenko RA, Alekseev EA, Demaison J (2010) Choice of the reduction and of the representation in centrifugal distortion analysis: a case study of dimethyl sulfoxide. J Mol Spectrosc 260:23–29
40. Kivelson D, Wilson EB Jr (1952) Approximate treatment of the effect of centrifugal distortion on the rotational energy levels of asymmetric-rotor molecules. J Chem Phys 20:1575–1579
41. Rudolph HD, Demaison J, Császár AG (2013) Accurate determination of the deformation of the benzene ring upon substitution: equilibrium structures of benzonitrile and phenylacetylene. J Phys Chem A 117:12969–12982
42. Demaison J, Vogt N, Ksenafontov DN (2020) Accuracy of semiexperimental equilibrium structures: sulfine as an example. J Mol Struct 1206:127676
43. Demaison J, Liévin J (2022) Accuracy of the equilibrium structure of sulfur dioxide. Mol Phys 120:e1950857
44. Morino Y, Saito S (1966) Microwace spectrum of oxygen difluoride in vibrationally excited states; $\nu_1 - 2\nu_2$ Fermi resonance. J Mol Spectrosc 19:435–453
45. Morino Y, Matsumura C (1967) Microwave spectra and equilibrium structure of carbonyl sulfide. Bull Chem Soc Japan 40:1095–1100
46. Lahaye J-G, Vendenhaute R, Fayt A (1987) CO_2 laser saturation stark spectra and global rovibrational analysis of the main isotopic species of carbonyl sulfide ($OC^{34}S$, $O^{13}CS$, and ^{18}OCS). J Mol Spectrosc 123:48–83
47. Papousek D, Aliev MR (1982) Molecular vibrational-rotational spectra. Elsevier, Amsterdam
48. Jahn HA (1939) Note on the coriolis coupling in polyatomic molecules. Phys Rev 56:680–683
49. Perrin A, Flaud J-M, Camy-Peyret C, Vasserot A-M, Guelachvili G, Goldman A, Murcray FJ, Blatherwick RD (1992) The ν_1, $2\nu_2$, and ν_3 interacting bands of $^{14}N^{16}O_2$: lines positions and intensities. J Mol Spectrosc 154:391–406
50. Demaison J, Herman M, Liévin J (2007) Anharmonic force field of *cis*- and *trans*-formic acid from high-level ab initio calculations, and analysis of resonance polyads. J Chem Phys 126:164305
51. Flygare WH (1974) Magnetic interactions in molecules and an analysis of molecular electronic charge distribution from magnetic parameters. Chem Rev 74:653–687

52. Sutter DH, Flygare WH (1976) The molecular Zeeman effect. In: Boschke FL (ed) Topics in current chemistry, vol 63. Springer, Berlin, pp 91–196
53. Hüttner W (2002) Molecular constants mostly from microwave, molecular beam, and sub-doppler laser spectroscopy. Dipole moments, quadrupole coupling constants, hindered rotation and magnetic interaction constants of diamagnetic molecules, vol 24C. Springer, Berlin
54. Gauss J, Ruud K, Helgaker T (1996) Perturbation-dependent atomic orbitals for the calculation of spin-rotation constants and rotational *g*-tensors. J Chem Phys 105:2804–2812
55. Meerts WL, Stolte S, Dymanus A (1977) Electric and magnetic properties of ozone by molecular beam electric resonance spectroscopy. Chem Phys 19:467–472
56. Colmont J-M, Bakri B, Demaison J, Mäder H, Willaert F, VlG T, Barbe A (2005) Microwave Fourier transform, millimeterwave, and submillimeterwave spectra of ozone in its vibrational ground state. J Mol Spectrosc 233:293–296
57. Tennyson J (2000) Variational calculations of rotation-vibration spectra. In: Jensen P, Bunker PR (eds) Computational molecular spectroscopy. Wiley, Chichester, pp 305–323
58. Whitehead RJ, Handy NC (1975) Variational calculation of vibration-rotation energy levels for triatomic molecules. J Mol Spectrosc 55:356–376
59. Carter S, Bowman JM, Handy NC (1998) Extensions and tests of multimode: a code to obtain accurate vibration/rotation energies of many-mode molecules. Theor Chem Acc 100:191–198
60. Handy NC, Carter S (2004) Large vibrational variational calculations using multimode and an iterative diagonalization technique. Mol Phys 102:2201–2205
61. Carter S, Sharma AR, Bowman JM, Rosmus P, Tarroni R (2009) Calculations of rovibrational energies and dipole transition intensities for polyatomic molecules using multimode. J Chem Phys 131:224106
62. Watson JKG (1968) Simplification of the molecular vibration-rotation hamiltonian. Mol Phys 15:479–490
63. Carter S, Mills IM, Handy NC (1992) The equilibrium structure of HCN. J Chem Phys 97:1606–1607
64. Carter S, Handy NC, Demaison J (1997) The rotational levels of the ground vibrational state of formaldehyde. Mol Phys 90:729–738
65. Carter S, Handy NC (1998) The geometry and force field of thioformaldehyde. J Mol Spectrosc 192:263–267
66. Carter S, Handy NC (2002) The geometry and force field of acetylene. Mol Phys 100:681–698
67. Koput J, Carter S, Handy NC (1998) Potential energy surface and vibrational-rotational energy levels of hydrogen peroxide. J Phys Chem A 102:6325–6330
68. Thompson HB (1967) Calculation of cartesian coordinates and their derivatives from internal molecular coordinates. J Chem Phys 47:3407–3410
69. Demaison J (2011) The method of least-squares. In: Demaison J, Boggs JE, Császár AG (eds) Equilibrium molecular structures. From spectroscopy to quantum chemistry. CRC Press, Boca Raton, pp 89–124
70. Rudolph HD (1991) Contribution to the systematics of r_0-derived molecular structure determinations from rotational parameters. Struct Chem 2:581–588
71. Golub GH, Reinsch C (1970) Singular value decomposition and least squares solutions. Numer Math 14:403–420
72. Starck B (1967) Molecular constants from microwave spectroscopy. In: Hellwege K-H, Hellwege AM (eds) Landolt-Börnstein—group II molecules and radicals, vol 4. Springer, Berlin
73. Demaison J, Hüttner W, Starck B, Buck I, Tischer R, Winnewisser M (1974) Molecular constants mostly from microwave, molecular beam, and sub-doppler laser spectroscopy. In: Hellwege K-H, Hellwege AM (eds) Landolt-Börnstein—group II molecules and radicals, vol 6. Springer, Berlin
74. Demaison J, Dubrulle A, Hüttner W, Tiemann E (1982) Molecular constants mostly from microwave, molecular beam, and sub-doppler laser spectroscopy. In: Hellwege K-H, Hellwege AM (eds) Landolt-Börnstein—group II molecules and radicals, vol 14. Springer, Berlin

75. Demaison J, Hüttner W, Vogt J, Włodarczak G (1992) Molecular constants mostly from microwave, molecular beam, and sub-doppler laser spectroscopy. In: Hüttner W (ed) Landolt-Börnstein—group II molecules and radicals, vol 19. Springer, Berlin
76. Demaison J, Vogt J, Włodarczak G (2002) Molecular constants mostly from microwave, molecular beam, and sub-doppler laser spectroscopy. In: Hüttner W (ed) Landolt-Börnstein—group II molecules and radicals, vol 24. Springer, Berlin
77. Demaison J, Vogt J (2010) Molecular constants mostly from microwave, molecular beam, and sub-doppler laser spectroscopy. In: Hüttner W (ed) Landolt-Börnstein—group II molecules and radicals, vol 29. Springer, Berlin
78. Lovas FJ, Coursey JS, Kotochigova SA, Chang J, Olsen K, Dragoset RA (2003) Triatomic spectral database (version 2.0). http://physics.nist.gov/Triatomic
79. Lovas FJ, Suenram RD, Coursey JS, Kotochigova SA, Chang J, Olsen K, Dragoset RA (2004) Hydrocarbon spectral database (version 2.1). http://physics.nist.gov/Hydrocarbon
80. Lovas FJ, Dragoset RA, Olsen KJ (2009) NIST recommended rest frequencies for observed interstellar molecular microwave transitions. https://doi.org/10.18434/T4JP4Q
81. Jacox ME (1998) Vibrational and electronic energy levels of polyatomic transient molecules. Supplement A. J Phys Chem Ref Data 27:115–393
82. Jacox ME (2003) Vibrational and electronic energy levels of polyatomic transient molecules. Supplement B. J Phys Chem Ref Data 32:1–441

Chapter 5
Equilibrium Structures of Semirigid Molecules from the Rotational Constants

Abstract The determination of the equilibrium geometry of a semirigid molecule from its rotational constants is reviewed in detail. First, the experimental method is critically examined. Then, the semiexperimental method is described in detail. Several empirical methods are also presented. The accuracy of all these methods is discussed in detail.

5.1 Introduction

There are different ways to determine the geometrical structure of a molecule by spectroscopy, but the most usual one is to use the rotational constants. Structure determinations by gas-phase electron diffraction will be reviewed in Chap. 7 and by some other methods will be discussed in Chap. 8. The present chapter deals with semirigid molecules, i.e., molecules for which the amplitude of vibrations is small. The case of molecules with a large-amplitude motion will be treated in the next chapter.

The equilibrium rotational constants will first be defined and the different procedures do obtain them will be discussed: either the purely experimental methods or the semiexperimental methods. Then, it will be shown how these equilibrium rotational constants can be used to derive an equilibrium structure.

It is also possible to obtain an approximate equilibrium structure only using the ground state rotational constants. The r_z and mass-dependent methods will be presented, and the empirical methods will be briefly reviewed.

The book of Gordy and Cook [1] is still an essential reference but there are many other references on the subject [2–7].

N. Vogt and J. Demaison, *Equilibrium Structure of Free Molecules*, Lecture Notes in Chemistry 111, https://doi.org/10.1007/978-3-031-36045-9_5

5.2 Equilibrium Rotational Constants

What is determinable experimentally is the rotational constants in a given vibrational state. The difference between the equilibrium rotational constants and the experimental constants is called the rovibrational contribution and may be written as a series expansion. The effective rotational constant about the ξ-axis ($\xi = a, b, c$) in a vibrational state characterized by the vibrational quantum numbers $v = (v_1, v_2, \ldots)$ with degeneracies $d_1, d_2, \ldots$ (for linear and symmetric molecules) is given by [1–8]

$$B_v^\xi = B_e^\xi - \sum_k \alpha_k^\xi \left(v_k + \frac{d_k}{2} \right) + \sum_{i \geq j} \gamma_{ij}^\xi \left(v_i + \frac{d_i}{2} \right) \left(v_j + \frac{d_j}{2} \right) + \sum_{i \geq j} \gamma_{l_i l_j}^\xi \ell_i \ell_j + \cdots \quad (5.1)$$

B_e^ξ is the equilibrium rotational constant, and α_i^ξ and γ_{ij}^ξ are the vibration–rotation interaction constants of different order. The summation is over all the normal modes. The last term $\gamma_{l_i l_j}^\xi$ is different from zero only for degenerate modes (i.e., for linear and symmetric molecules). The convergence of the series expansion is usually fast, α_i^ξ being about two orders of magnitude smaller than B_e^ξ and γ_{ij}^ξ two orders of magnitude smaller than α_i^ξ; see Table 5.1 where the values are given for a few diatomic molecules (the advantage of diatomic molecules is that there is only one vibrational mode which avoids the difficulty of interactions or resonances between the vibrations). One sees that α decreases more rapidly than B_e (in agreement with the fact that B_e is homogeneous of degree minus one in the masses whereas α is homogeneous of degree minus 3/2 [9]) and that γ is always less than 1% of α. For this reason, the γ-terms are generally neglected, except for light molecules. This will be discussed in Sect. 5.3. Of course, the previous conclusions are only valid for a semirigid molecule, i.e., for a molecule without large-amplitude motion. For a weakly-bound complex, the convergence of Eq. (5.1) is much slower as shown for instance in Table 5.1 for the van der Waals complex $Ne \cdots H^+$. The first-order vibration–rotation interaction constants (also called α-constants) can be derived by standard perturbation theory [8]. They are the sum of three terms

$$\alpha_k^\xi = \alpha_k^\xi \text{ (harm)} + \alpha_k^\xi \text{ (Coriolis)} + \alpha_k^\xi \text{ (anharm)} \quad (5.2)$$

The first term is a function of the harmonic force field

$$\alpha_k^\xi \text{ (harm)} = -2 \frac{\left(B_e^\xi\right)^2}{\omega_k} \sum_{\gamma = a,b,c} \frac{3\left(a_k^{\xi\gamma}\right)^2}{4 I_e^\gamma} \quad (5.3)$$

In this equation, ω_k is the harmonic frequency of the normal mode k, I_e^ξ is the equilibrium moment of inertia, $a_k^{\xi\gamma} = \left(\partial I^{\xi\gamma} / \partial Q_k\right)_e$ is the derivative of the (ξ, γ) element of the inertia tensor with respect to the normal coordinate Q_k at equilibrium.

Table 5.1 Rovibrational contributions to the rotational constants (in MHz)

	$Y_{01} \approx B_e$	$-Y_{11} = \alpha_e$	$Y_{21} = \gamma_e$	α_e/Be (%)	γ_e/α_e(%)	Ref.
HCl	317,580.97(23)	9208.96(57)	53.04(28)	2.90	0.58	a
HBr	253,850.6(12)	6997.58(30)	23.6(16)	2.76	0.34	b
HI	195,213.69(39)	5107.6(11)	−8.53(69)	2.62	0.17	c
BF	48,335.128(26)	625.916(48)	2.019(28)	1.29	0.32	d
CS	24,584.2874(13)	127.4275(15)	−0.02947(39)	0.52	0.02	e
^{74}GeS	5593.1008(19)	22.4580(5)	−0.00123(1)	0.40	0.01	f
IBr	1703.79816(63)	5.90465(66)	−0.01340(30)	0.35	0.23	g
NeH^+	536,170.42	32,880.6	280.9	6.13	0.85	h

[a] Guelachvili G, Niay P, Bernage P (1981) J Mol Spectrosc 85: 271–281
[b] Nishimiya N, Yukiya T, Ohtsuka T, Suzuki M (1997) J Mol Spectrosc 182: 309–314
[c] Guelachvili G, Niay P, Bernage P (1981) J Mol Spectrosc 85: 253–270
[d] Zhang KQ, Guo B, Braun V, Dulick M, Bernath PF (1995) J Mol Spectrosc 170: 82–93
[e] Ram RS, Bernath PF, Davis SP (1995) J Mol Spectrosc 173: 146–157
[f] Uehara H, Horiai K, Ozaki Y, Konno T (1995) J Mol Struct 352/353: 395–405
[g] Appadoo DRT, Bernath PF, Le Roy RJ (1994) Can J Phys 72: 1265–1272
[h] Ram RS, Bernath PF, Brault JW (1985) J Mol Spectrosc 113: 451–457

The second term is the Coriolis term

$$\alpha_k^\xi(\text{Coriolis}) = -2\frac{\left(B_e^\xi\right)^2}{\omega_k}\sum_l \frac{\left(3\omega_k^2+\omega_l^2\right)\left(\zeta_{kl}^\xi\right)^2}{\omega_k^2-\omega_l^2} \tag{5.4}$$

$\zeta_{kl}^\xi = -\zeta_{lk}^\xi$ is a Coriolis coupling constant. Note that the denominator can be very small. If the numerator is different from zero, it gives rise to an interaction between the vibrational modes k and l called Coriolis interaction. When this interaction is large, it complicates the analysis of the spectra and the states k and l cannot be treated separately, and Eq. (5.4) must be modified [10]. This is sometimes a problem for an experimental structure determination. Fortunately, it can be avoided as will be seen below.

The last term is the anharmonic contribution which is far from negligible.

$$\alpha_k^\xi\ (\text{anharm}) = -2\frac{\left(B_e^\xi\right)^2}{\omega_k}\pi\sqrt{\frac{c}{h}}\sum_l \phi_{kll}a_k^{\xi\xi}\frac{\omega_k}{\omega_k^{3/2}} \tag{5.5}$$

ϕ_{kll} is a cubic force constant in the dimensionless normal coordinate representation, c is the speed of light, and h is Planck's constant.

Equations (5.3–5.5) are appropriate when the rotational constants are in units of cm^{-1} because the force constants ϕ_{kll} are usually expressed in cm^{-1}; it has to be multiplied by the speed of light in cm s^{-1} to obtain it in units of Hz.

A simple way to avoid difficulties due to the Coriolis interaction is to perform a summation of the α-constants which eliminates the (possibly small) denominator in the Coriolis term, Eq. (5.4). The rovibrational correction may be written

$$B_0^\xi - B_e^\xi = \sum_k \alpha_k^\xi \frac{d_k}{2}$$
$$= \left(B_e^\xi\right)^2 \left\{ \left[\sum_{k\gamma} \frac{3\left(a_k^{\xi\gamma}\right)^2}{4\omega_k I_e^\gamma} - \sum_{k<l} \frac{(\omega_l - \omega_k)^2 \left(\zeta_{kl}^\xi\right)^2}{\omega_k \omega_l (\omega_k + \omega_l)} \right] + \pi \sqrt{\frac{c}{h}} \sum_{kl} \frac{\phi_{kkl} a_l^{\xi\xi}}{\omega_l^{3/2}} \right\} \tag{5.6}$$

In this equation, the denominator $\omega_k^2 - \omega_l^2$ disappears. The calculation of this rovibrational correction is a more robust numerical procedure than the calculation of the individual α-constants.

5.3 Determination of the Rovibrational Constants α and γ

The α-constants can be easily calculated using a cubic force field either experimental or ab initio. Several quantum chemistry programs calculate the force field, the α-constants, as well as $B_0^\xi - B_e^\xi$ (to use this later correction, it is assumed that the γ-constants are negligible).

If the γ-constants are negligible, the α-constants may be easily obtained from the ground state constants and the constants of a vibrationally excited state, v_k, assuming that it is not perturbed

$$\alpha_k^\xi = \frac{B_0^\xi - B_{v_k}^\xi}{v_k} \tag{5.7}$$

When the γ-constants are not negligible, the α-constants and the γ-constants may be obtained experimentally by a least-squares fit to Eq. (5.1) using the rotational constants of the ground state and of the excited vibrational states (fundamental, combinations, and overtones) as input data. As the excited states are often perturbed by Coriolis interactions and/or anharmonic resonances, it is required to properly treat these interactions. This is not a trivial task because it requires a careful study of many spectra. This is the reason why a full set of vibration–rotation interaction constants has been experimentally determined for a few small molecules, mainly triatomic molecules of C_{2v} symmetry. Table 5.2 gives an example of such a determination for a linear molecule, HCN, and for an asymmetric molecule, SO_2. In these two examples, the contribution of the γ-constants is at least one order of magnitude larger than the standard deviation of the corresponding rotational constant and have, therefore, a small effect on the accuracy of the structure. For instance, for $^{32}S^{16}O_2$,

Table 5.2 Rotational constants and vibration–rotation interaction constants for HCN and SO_2 (in MHz)

	HCN[a]	SO_2 [b]		
	B	A	B	C
B_0	44 315.97570(40)	60,778.5526(39)	10,318.07332(60)	8799.70375(54)
α_1	312.68(10)	−33.33(45)	50.023(90)	42.328(91)
α_2	−107.042(64)	−1130.36(55)	−2.624(33)	15.685(47)
α_3	299.955(77)	616.57(23)	34.954(39)	32.405(46)
γ_{11}	−4.405(56)	−0.60(22)	0.049(43)	0.104(44)
γ_{22}	1.393(27)	6.86(28)	0.087(55)	0.404(56)
γ_{33}	−0.8758(98)	−6.25(40)	0.137(73)	−0.851(80)
γ_{12}	5.821(33)	24.77(38)	−0.1728(59)	−0.0791(59)
γ_{13}	5.81(10)	−17.85(40)		−0.422(80)
γ_{23}	−3.444(88)	3.88(16)	−0.0351(31)	0.0767(31)
$\gamma_{\ell\ell}$	−5.849(19)			
$\sum\alpha$	199.275(93)	−273.57(45)	41.177(52)	45.209(56)
$\sum\gamma$	2.714(62)	2.70(19)	0.0161(25)	−0.192(33)
$\sum\alpha + \sum\gamma$	196.56(11)	−276.20(49)	41.160(58)	45.401(65)
B_e	44 512.447(59)	60,502.21(26)	10,359.234(38)	8845.105(40)

[a] Maki AG, Mellau GC, Klee S, Winnewisser M, Quapp W (2000). J Mol Spectrosc 202: 67–82
[b] Flaud JM, Lafferty WJ (1993) J Mol Spectrosc 161: 396–402

neglecting the γ-constants gives $r_e(SO) = 1.4307800$ Å and $\angle(OSO) = 119.33006°$ instead of 1.4307885(33) Å and 119.32902(25)° [11]. It must be emphasized that an accurate determination of the γ-constants is still much more difficult than that of the α-constants. A typical example is SO_2 for which two independent determinations of $\sum\gamma/4$ were made at the same time. Flaud and Lafferty [12] give (in MHz) 2.70(19); 0.0161(25); −0.192(33) for A, B, C, respectively, whereas the results of Morino and Tanimoto [13] are 1.91(27); 0.105(30); −0.044(27). For this reason, the γ-constants are only taken into account for the very accurate structure determination of light molecules (i.e., with large rotational constants).

5.4 Experimental Equilibrium Structure

There are several ways to obtain an experimental equilibrium structure. The most usual one is to use the experimental vibration–rotation interaction constants (α-constants and, if possible, γ-constants).

5.4.1 *With the Experimental Vibration–Rotation Interaction Constants*

There are as many non-redundant internal coordinates as normal modes. For a molecule without symmetry, it is $3n - 6$, n being the number of atoms. As an isotopologue has at most three rotational constants (see Chap. 4), a complete structure determination requires the rotational constants of several isotopologues (in the frame of the Born–Oppenheimer approximation, all the isotopologues have the same equilibrium structure, see Chap. 2). Table 5.3 gives the minimum number of isotopologues necessary to determine a structure. Actually, it is recommended to substitute all the atoms (one at a time) in order to improve the conditioning, see Sect. 4.11. One of the difficulties is that an isotopic substitution induces a small shift in the rotational constants. Therefore, they have to be quite accurate because the system of normal equations of the least-squares fit is not well conditioned. Another difficulty is the Coriolis interactions and/or anharmonic resonances.

It is why different determinations of the same structure are sometimes not compatible. An illuminating example of ill-conditioning is the structure of FCP because neither F nor P have isotopes, and the C atom is close to the center of mass (the substitution $^{12}C \rightarrow {}^{13}C$ induces a very small shift). The C–F bond length was found to be 1.276(1) Å or 1.284 Å [14], see Sect. 5.5.

Typical examples of the importance to treat correctly the interactions between the vibrational states are the C–H bond length in HCO^+ whose published values are (in Å): 1.0919(9) or 1.09725(4) [15] and the C–H bond length in HCP whose published values are 1.0702(10) or 1.0660(1) [16].

In conclusion, the determination of an equilibrium structure from the experimental vibration–rotation interaction constants is time-consuming and extremely difficult task. For this reason, it has been performed for only a few small molecules listed in Appendix 5.4.

Table 5.3 Minimum number of rotational spectra to analyze in order to determine an experimental equilibrium structure

Molecule	Parameters	n_{vib} [a]	n_{isotop} [b]	$n_{spectra}$ [c]
Diatomic	1	1	1	1
Bent XY_2	2	3	1	4
Pyramidal XY_3	2	4	1	5
Linear XYZ	2	3	2	8
Bent XYZ	3	3	2	8
Planar X_2XY	3	6	2	14
C_{3v} XY_3Z	3	6	2	14

[a] Number of fundamental vibrations
[b] Number of necessary isotopologues
[c] Minimum number of spectra to analyze: $n_{spectra} = (n_{vib} + 1) \times n_{isotop}$

Table 5.4 Equilibrium structure of sulfur dioxide, SO_2 (bond length in Å, angle in degree)

r(SO)	∠(OSO)	Method	Year	Ref.
1.4308(2)	119.32(3)	MW	1964	1
1.43076(13)	119.33(1)	MW	1969	2
1.43080(1)	119.329(2)	MW + IR	1988	3
1.430782(15)	119.32598(24)	MW + IR	1993	4
1.4308302(17)	119.33042(29)	MW + IR	1995	5
1.4307932(40)	119.32898(24)	MW + IR	2009	6
1.4307897(99)	119.32893(27)	MW + IR[a]	2021	7
1.43079396(93)	119.328783(55)	MW + IR[b]	2021	7
1.4311	119.389	Ab initio[c]	2021	7

[a] Mean value for $^{32}S^{16}O_2$ and $^{34}S^{16}O_2$
[b] Using an experimental cubic force field and the rotational constants of 17 isotopologues
[c] CCSD(T)_ae/aug-cc-pwCVQZ
[1] Morino Y, Kikuchi Y, Saito S, Hirota E (1964) J Mol Spectrosc 13: 95–118
[2] Saito S (1969) J Mol Spectrosc 30: 1–16
[3] Morino Y, Tanimoto M, Saito S (1988) Acta Chem Scand A 42: 346–351
[4] Flaud JM, Lafferty WJ (1993) J Mol Spectrosc 161: 396–402
[5] Morino Y, Tanimoto M (1994) J Mol Spectrosc 166: 310–320
[6] Lafferty WJ, Flaud JM, Ngom EHA, Sams RL (2009) J Mol Spectrosc 253: 51–54
[7] Demaison J, Liévin J (2022) Mol Phys 120: e1950857

Another problem with the experimental method is that the standard deviation of the structural parameters is calculated using the law of propagation of errors, i.e., it neglects the systematic errors originating for instance from an approximate treatment of anharmonic resonances, etc. For this reason, the standard deviations are often too small. A typical example is the structure of SO_2 which was determined many times [11]. At first glance, it is a very favorable case because there are only two independent structural parameters and they can be obtained from the rotational constants of only one isotopologue. However, as shown in Table 5.4, the standard deviations are often too small [11].

5.4.2 With the Experimental Anharmonic Force Field

An easier way to determine the equilibrium structure of a small molecule is to calculate the rovibrational correction from the force field up to cubic terms using Eq. (5.6) [17]. This avoids the problems of Coriolis interactions and anharmonic resonances. Another big advantage is that it allows us to obtain at the same time the rovibrational corrections of all isotopologues without the need of further measurements (only the ground state rotational constants of the isotopologues are necessary). The force field is determined mainly using the known α-constants, the vibrational band centers, the quartic and sextic centrifugal distortion constants, and the anharmonicity constants.

However, the number of cubic constants increases rapidly with the number of atoms and the loss of symmetry of the molecule. For instance, the number of cubic constants is six for a linear triatomic molecule and a bent XY_2 but it is already ten for a bent XYZ molecule and it increases to 22 for a planar X_2XY molecule (such as H_2CO), and 38 for a pyramidal XY_3Z molecule. For this reason, the method is limited to very small molecules. Nevertheless, in this case, it often gives more reliable results than the direct experimental method. A typical example of application is the structure of SO_2 [11]. Some further examples are listed in Table 4 of Ref. [18] and in appendix of Ref. [19].

The determination of an experimental anharmonic force field is limited to small molecules because the number of cubic constants increases rapidly with the number of atoms and the loss of symmetry of the molecule (62,835 constants up to quartic terms for proline, $C_5H_9NO_2$ [20]). The joint use of electron diffraction and high-resolution spectroscopic data permits to slightly extend the range of the method [21]. For instance, an approximate anharmonic force field could be determined for 1,1-dichloroethene [22].

5.5 Semiexperimental (SE) Equilibrium Structure

The determination of an experimental cubic force field is a difficult task, furthermore limited to small molecules. However, the calculation of a cubic force field by quantum-chemical methods is now a relatively easy task. This is the reason why most equilibrium structures are now determined by the semiexperimental method whereby the equilibrium rotational constants are derived from experimental ground state rotational constants and rovibrational corrections derived from a quantum-chemical force field.

This method avoids the very difficult labor of determining reliable excited state rotational constants. Another advantage is that it is often enough to compute the force field with medium level methods as MP2 or DFT. The SE method was first used by Pulay et al. [23] for methane, CH_4. Although the force field was obtained at a very modest Hartree–Fock level of theory, the C–H bond length was determined with a remarkable accuracy: $r_e(\text{C–H}) = 1.0862(5)$ Å to be compared with the most accurate determination, 1.08595(30) Å [24]. The robustness of the SE method was confirmed by the determination of the structure of 1,1-difluoroethene, $CH_2{=}CF_2$. The force field was calculated at five different levels of theory, from HF/6-31G* to MP2/cc-pVQZ [25]. Although the individual rotation-vibration interaction constants are sensitive to the level of theory, the derived structures are almost identical.

It is thus possible to determine the equilibrium structure of rather large molecules, whereas the purely experimental method is limited to very small molecules. Chapter 9 gives a list of molecules whose semiexperimental structure was determined, and Table 5.5 compares the SE structure of several molecules to the BO one: pseudopelletierine, $C_9H_{15}NO$ (40 independent structural parameters), proline, $C_5H_9NO_2$ (45 independent structural parameters) [26], and 2-dexoxyribose, $C_5H_9O_4$ (51 independent structural parameters). The semiexperimental structure of fructose, $C_6H_{12}O_6$ (66 independent structural parameters), was also determined [27].

The accuracy of the SE method has been studied by many authors (for a review, see Ref. [7]). In particular, Bak et al. [28] and Pawłowski et al. [29] made a systematic analysis of the SE structure of a nearly 20 small molecules and concluded that the equilibrium bond distances determined by the SE method surpasses the accuracy obtainable either by purely experimental techniques (except for the smallest systems such as diatomic molecules) or by ab initio methods). Actually, it is difficult to estimate the accuracy of a SE structure. One way is to use as many isotopologues as possible. A nice example is given by the structure of methane, CH_4. Using the semiexperimental equilibrium rotational constants of Ref. [30] (rovibrational corrections calculated at the CCSD(T)_ae/cc-pCVQZ level and electronic corrections calculated at the CCSD(T)_fc/aug-cc-pVQZ level), it is possible to determine the C–H bond length for each isotopologue, see Table 5.6. The mean and the median are identical, indicating the absence of outlier and the value of the mean is r_e(C–H) = 1.08600(2) Å in perfect agreement with one of the most accurate ab initio structures: r_e(C–H) = 1.08605 Å, CCSD(T)_fc/cc-pV6Z + core(cc-pCVQZ) value [31]. A more direct way to estimate the accuracy of the SE structure is to compare the SE bond lengths to the experimental ones for diatomic molecules. It is shown in Table 5.7. It confirms that the accuracy of the SE structure is better than 0.001 Å for these molecules but inferior to that of the experimental structure. Inspection of Appendix 5.4 confirms that the experimental method is more accurate for very small molecules (mainly with 2 independent structural parameters) but, for larger molecules the SE method seems to be more accurate. In particular, for HCN, HNC, HCO^+, and PH_3, the experimental method seems to be more accurate. However, for HCCH, SiH_3X (X = F, Cl), and $SiHF_3$, the SE is more accurate.

One may fear that the accuracy deteriorates for molecules with very heavy atoms. There are not many data available but the results for SbH_3 and BiH_3 are encouraging, see Table 5.8. In particular, the SE structure of bromofluoromethane was also determined and found in good agreement with the ab initio structure [32], see also Table 5.8.

The semiexperimental method has another advantage: because the rovibrational correction is a small correction to the rotational constants, it is much less sensitive to non-dynamical correlation than an ab initio structure optimization, see Sect. 2.7 and Ref. [33].

In some cases, the quality of the ab initio force field is not good enough but it is easy to refine an ab initio force field using experimental data [19] as was done for instance for SO_2 [11], $CHClF_2$ [34], SF_6 [35], CdH_2 [36], HgH_2 [37] ZnH_2 [38], and many other molecules (see Appendix 5.4) including van der Waals complexes, see next chapter, Sect. 6.6.

There are, however, cases where the SE method does not give satisfactory results: when there are not enough isotopologues available, when some atoms have small Cartesian coordinates [7, 39, 40] and more generally when the system of normal equations is ill-conditioned [7, 40]. In such a case, the mixed estimation method

Table 5.5 Structure of the heavy-atom skeleton of some large molecules (distances in Å, angles in degrees)

	r_e^{se}	r_e^{BO}		r_e^{se}	r_e^{BO}
Pseudopelletierine[a]			Proline (cont)		
C1N9	1.461(1)	1.4622	C2C3C4	102.304(80)	102.54
C2C3	1.512(1)	1.5127	C3C4C5	102.121(81)	102.06
C6C7	1.526(1)	1.5279	N1C2C6O8	2.50(31)	2.37
C1C2	1.528(1)	1.5284	N1C2C6O7	−177.36(35)	−177.59
C1C8	1.544(1)	1.5446	H9N1C2C3	121.09(17)	122.02
C1N9C5	110.70(8)	110.74	N1C2C3C4	26.47(18)	25.96
C6C7C8	110.70(9)	110.70	C2C3C4C5	−40.23(17)	−39.93
C1C8C7	111.84(7)	111.88	2-desoxyribose[c]		
N9C1C2	112.35(8)	112.42	C1C2	1.5182(14)	1.5174
C1C2C3	113.08(8)	113.04	C2C3	1.5258(14)	1.5246
C1N9C10	113.34(6)	113.28	C3C4	1.5226(14)	1.5240
C2C3C4	115.070(8)	115.17	C4C5	1.5133(16)	1.5148
C2C3O11	122.45(4)	122.4	C1O6	1.4183(14)	1.4170
N9C1C8	108.58(8)	108.53	C1O1	1.4072(15)	1.4049
N9C1C2C3	49.6(1)	49.54	C3O3	1.4145(15)	1.4128
N9C5C6C7	56.79(9)	56.74	C4O4	1.4262(15)	1.4249
C1C2C3O11	141.9(2)	141.97	C1C2C3	111.747(81)	112.2
C10N9C5C6	166.3(1)	166.29	C2C3C4	109.559(85)	110.12
Proline[b]			C3C4C5	110.107(93)	109.7
N1-C2	1.4785(18)	1.4784	O6C1C2	111.755(83)	111.56
C2-C6	1.5321(18)	1.5338	O6C1O1	111.47(11)	111.56
C6-O8	1.3343(19)	1.334	C2C3O3	111.33(13)	111.16
C6-07	1.2027(18)	1.2023	C3C4O4	109.43(10)	109.73
C2-C3	1.5307(18)	1.532	C1C2C3C4	49.67(12)	49.05
C3-C4	1.5262(17)	1.5269	C2C3C4C5	−53.26(14)	−52.77
C4-C5	1.5215(15)	1.5237	C5O6C1C2	58.93(14)	58.45
N1C2C6	110.09(12)	110.12	O6C1C2C3	−51.98(14)	−51.26
C2C6O8	114.17(21)	113.86	C1C2C3O3	172.65(15)	171.97
C2C6O7	122.41(21)	122.51	C2C3C4O4	69.21(15)	69.31
N1C2C3	105.79(13)	105.73			

[a] Vallejo-López M, Écija P, Vogt N, Demaison J, Lesarri A, Basterretxea FJ, Cocinero EJ (2017) Chem Eur J 23: 16491–16496

[b] Vogt N, Demaison J, Krasnoshchekov SV, Stepanov NF, Rudolph HD (2017) Mol Phys 115: 942–951

[c] Vogt N, Demaison J, Cocinero EJ, Écija P, Lesarri A, Rudolph HD, Vogt J (2016). Phys Chem Chem Phys 18: 15555–15563

Table 5.6 Semiexperimental bond length (in Å) of methane[a]

Isotopologue	r_e(C–H)	Ref. (B_0)
CH_4	1.08601	b
CH_3D, *A*	1.08600	c
CH_3D, *B*	1.08603	c
CHD_3, *A*	1.08599	d
CHD_3, *C*	1.08600	d
CD_4	1.08599	e
$^{13}CH_4$	1.08601	f
$^{13}CD_4$	1.08599	g
CH_2D_2, *A*	1.08599	h
CH_2D_2, *B*	1.08601	h
CH_2D_2, *C*	1.08604	h
Mean	1.08600	
Median	1.08600	
Range	0.00005	
Standard deviation	0.00002	

[a] Rovibratiional corrections from: Puzzarini C, Heckert M, Gauss J (2008) J Chem Phys 128: 194,108
[b] Tarrago G, Dang-Nhu M, Poussigue G, Guelachvili G, Amiot C (1975) J Mol Spectrosc 57: 246–263
[c] Tarrago G, Delaveau M, Fusina L, Guelachvili G (1987) J Mol Spectrosc 126 : 149–158
[d] Roche C, Champion JP, Valentin A (1993) J Mol Spectrosc 160 : 517–523
[e] Oldani M, Bauder A, Loete M, Champion JP, Pierre G, Hilico JC, Robiette AG (1985) J Mol Spectrosc 113 : 229–242
[f] Bobin B, Hilico JC, Guelachvili G (1982) J Mol Spectrosc 93: 253–270
[g] McDowell RS, Buchwald MI, Sorem MS, Robiette AG, Deeley CM, Kreiner WA (1985) J Mol Spectrosc 112: 363–376
[h] Deroche JC, Guelachvili G (1975) J Mol Spectrosc 56: 76–87

may be used with success [7, 40, 41]. In this method, structural parameters are fitted concurrently to predicate parameters (with appropriate uncertainties) and moments of inertia (with uncertainties). This method can be illustrated by a simple example: the structure of FCP [14]. The structure has to be determined using the semiexperimental equilibrium rotational constants of $F^{12}C{\equiv}P$ and $F^{13}C{\equiv}P$ because that neither F nor P have isotopologues. Furthermore, as the carbon atom is close to the center of mass, the rotational constant of $F^{13}C{\equiv}P$ does not bring much new information. Indeed, an attempt to determine the structure using the two rotational constants gives r_e(C–F) = 1.272 Å and r_e(C≡P) = 1.548 Å. As anticipated, statistical diagnostics (see Sect. 4.11) indicate that these two parameters are fully correlated. It is confirmed by an experimental determination which gives incompatible parameters: r_e(C–F) = 1.284 Å and r_e(C≡P) = 1.538 Å. The solution is to use predicate values for the two bond lengths. They were calculated at the CCSD(T)_fc/6-311 + G(2d)

Table 5.7 Comparison of the accuracy of the experimental and semiexperimental equilibrium structures of diatomic molecules

Molecule	Exp.[a]	SE[b]	AI[c]
F_2	1.412642(10)	1.41268(1)	1.410784[c]
HF	0.91683897(4)	0.916879(12)	0.917084[c]
N_2	1.097679(1)	1.09773(5)	1.097512[c]
CO	1.128230(1)	1.12836(4)	1.128427[c]
PH	1.42182772(89)		1.4214[d]
PF	1.589329(20)		1.5903[d]
PCl	2.014609(49)		2.0157[d]
BH	1.232180(10)		1.2293[e]

[a] For references, see Table 3.6
[b] Rovibrational corrections from a CCSD(T)_ae/cc-pCVQZ cubic force field. Pawlowski F, Jørgensen P, Olsen J, Hegelund F, Helgaker T, Gauss J, Bak, KL, Stanton JF (2002). J Chem Phys 116: 6482–6496
[c] CCSD(T)_fc/cc-pV∞Z + core(cc-pCVQZ) + corrections using CCSDT/cc-pVTZ +and CCSDTQ/cc-pVDZ. Puzzarini C, Heckert M, Gauss J (2008) J Chem Phys 128: 194108/1-9
[d] CCSD(T)-F12b_ae/cc-pCVQZ-F12 + corrections using CCSDT/cc-pVTZ +and CCSDTQ/cc-pVDZ + relativistic correction. Breidung J, Thiel W (2019). J Phys Chem A, 123: 5600–5612
[e] CCSD(T)_fc/cc-pV6Z + core correction. Pawlowski F, Halkier A, Jørgensen P, Bak KL, Helgaker T, Klopper W (2003) J Chem Phys 118: 2539–2549

level of theory and submitted to an offset correction using the known structures of HC≡P and FC≡N: r_e(C–F) = 1.2770(20) Å and r_e(C≡P) = 1.5455(20) Å. The fit gives: r_e(C–F) = 1.2761(7) Å and r_e(C≡P) = 1.5445(6) Å in good agreement with the ab initio structure calculated at the (CCSD(T)_fc/cc-pV∞Z + core-valence correction) level of theory: 1.2757 Å and r_e(C≡P) = 1.5431 Å [42]. As a rule, when there are several small Cartesian coordinates, it is not possible to determine them accurately without the mixed estimation method as illustrated by the structure of hexatriene [43]. Another illuminating example of the advantage of the mixed estimation method is the semiexperimental equilibrium structure of BF_2OH [44].

The mixed estimation method is also extremely useful when the available experimental rotational constants do not permit a determination of a complete structure. In such a frequent case, the semiexperimental rotational constants are supplemented by structural parameters from ab initio calculations [45]. It is much more accurate than using the Kraitchman substitution method (see Sect. 5.9) to determine a partial structure [46].

The mixed estimation method is also valuable to determine the structure of an oblate molecule. In such a case, the semiexperimental method gives unsatisfactory results, a typical example being 1,3,4-oxadiazole, c-$C_2H_2N_2O$ [47]. An attempt to determine its semiexperimental structure gave r_e^{se} (CO) = 1.256 Å using the MP2/cc-pVTZ force field but 1.447 Å with a B3LYP/6–311 + G(3df,2pd) force field. The failure of the semiexperimental method is due to two problems: (i) the rotational constants of the ^{18}O isotopologue were not available and (ii) the isotopic substitutions

Table 5.8 Accuracy of the semiexperimental equilibrium structure of molecules with a heavy atom

Molecule	Parameter	SE	Exp	AI	Ref.
CH_2BrF[a]	C–F	1.35753(22)		1.3578	1
	C–Br	1.92842(22)		1.9256	
	C–H	1.08302(8)		1.0833	
	∠FCBr	110.148(55)		110.14	
	∠HCF	109.546(17)		109.51	
	∠HCBr	107.251(15)		107.27	
	∠HCH	113.036(11)		113.09	
Stibine[b, c]					
$^{121}SbH_3$	Sb–H	1.70042	1.69994	1.7015	2
	∠HCH	91.559	91,558	91.76	
$^{121}SbD_3$	Sb–D	1.70112	1.70087		
	∠DCD	91.676	91.671		
$^{123}SbH_3$	Sb–H	1.70031	1.69996		3
	∠HCH	91.567	91.560		
$^{123}SbD_3$	Sb–D	1.70112	1.70086		
	∠DCD	91.677	91.672		
$^{209}BiH_3$ [b]	Bi–H	1.77904	1.77834(50)	1.7814	4, 5
	∠HCH	90.278	90.322(10)	90.09	

[a] Rovibrational corrections from a CCSD(T)_fc/cc-pVTZ force field
Ab initio structure: CCSD(T)_fc/cc-pV∞Z + core(cc-pCVQZ) + corrections using CCSDT/cc-pVTZ +and CCSDTQ/cc-pVDZ + relativistic correction
[b] Structure and force field at the CCSD(T)/SDB-aug-cc-pVQZ level of theory. The electronic correction was taken into account to calculate the experimental and semiexperimental structures
[c] The electronic correction was taken into account to calculate the experimental and semiexperimental structures
[1] Puzzarini C, Cazzoli G, Baldacci A, Baldan A, Michauk C, Gauss J (2007) J Chem Phys 127: 164302/1-10
[2] Canè E, Di Lonardo G, Fusina L, Jerzembeck W, Bürger H, Breidung J, Thiel W (2006) J Mol Struct 780–781: 98–110
[3] Canè E, Di Lonardo G, Fusina L, Jerzembeck W, Bürger H, Breidung J, Thiel W (2005) Mol Phys 103: 557–577
[4] Jerzembeck W, Bürger H, Constantin FL, Margulès L, Demaison J (2004) J Mol Spectrosc 226: 24–31
[5] Jerzembeck W, Bürger H, Breidung J, Thiel W (2004) J Mol Spectrosc 226: 32–44

$^{12}C \rightarrow {}^{13}C$, $^{14}N \rightarrow {}^{15}N$, and $H \rightarrow D$ induced a very large rotation of the principal axis system (PAS), a behavior typical of oblate molecules (asymmetry parameter, $\kappa = +0.952$) [48]. In those cases, the rovibrational corrections change significantly upon isotopic substitution, causing large errors in the structure because of ill-conditioned normal equations (for instance, at the B3LYP/6-311 + G(3df,2pd) level of theory, $\Delta A = A_e - A_0 = 93.0$ MHz for the parent species and 74.6 MHz for the d_1 species).

Table 5.9 Equilibrium structure of 1.3.4-oxadiazole (distances in Å, angles in degrees)

Ref.	1	1	2	2	2
Structure	Semiexperimental		Mixed[a]	Ab initio	Extrapol[b]
Method	MP2	B3LYP		CCSD(T)_ae	
Basis set	cc-pVTZ	6–311 + G(3df,2pd)		cc-pwCVQZ	cc-pwCV(T,Q)Z
CO	1.2560	1.4830	1.3548(16)	1.3540	1.3537
CN	1.3504	1.2103	1.2840(11)	1.2854	1.2851
CH	1.0691	1.0655	1.0734(13)	1.0737	1.0736
NN	1.3799	1.4812	1.4058(13)	1.4064	1.4057
COC	110.13	93.38	101.29(12)	101.36	101.39
OCN	110.37	117.07	113.782(2)	113.78	113.75
CNN	104.57	106.24	105.572(60)	105.54	105.55
OCH	125.27	117.86	118.081(2)	118.08	118.10
NCH	124.36	125.07	128.137(3)	128.14	128.15

[a] Mixed regression with two predicates: ∠(OCH) = 118.08(30)° and ∠(OCN) = 113.78(30)°, see Sect. 5.5

[b] Extrapolation with the rovibrational corrections calculated at the B3LYP:6-311 + G(3df,2pd), see Sect. 5.6

[1] Demaison J, Császár AG, Margulès L, Rudolph HD (2011) J Phys Chem A 115: 14,078–14,091

[2] Demaison J, Jahn MK, Cocinero EJ, Lesarri A, Grabow J-U, Guillemin J-C, Rudolph HD (2013) J Phys Chem A 117: 2278—2284

Finally, a more accurate value was obtained from a mixed fit: 1.3548(16) Å in good agreement with the extrapolated value: 1.3537 Å [47]. See Table 5.9.

In conclusion, it is often said that the semiexperimental method is the best way to determine an accurate structure. Indeed, as the rovibrational corrections are a small percentage of the rotational constants, it is not necessary to use very large basis sets and high-level methods. Furthermore, the semiexperimental method is not very sensitive to multireference effects. However, it requires a full set of isotopologues, and it is sensitive to bad conditioning. Thus, when the goal is to compare similar structures, it may be better to make an ab initio optimization of the structures. A typical example is the CH bond lengths in pyrrole, C_4H_5N [49]. The semiexperimental equilibrium structure gives r_e(C2–H) = 1.0753(1) Å and r_e(C3–H) = 1.0753(3) Å, whereas a high-level ab initio optimization gives r_e(C2–H) = 1.0746 Å and r_e(C3–H) = 1.0756 Å. In conclusion, contrarily to the CCSD(T)_ae/cc-pwCVQZ optimization, the SE method does not allow us to determine which bond length is longer.

5.6 "Extrapolation" Method

The "extrapolation" method was developed by Leutwyler et al. [50–53]. It is a variant of the semiexperimental method. However, there are two differences: indeed, for the extrapolation method, the ground state rotational constants of only one isotopologue are necessary which is a big advantage but, in order to have reliable results, it requires to optimize the structure at two highlevels of theory.

The basic idea is to extrapolate (or interpolate) the ab initio structure in order to reproduce the experimental rotational constants. The ab initio equilibrium values of the rotational constants of the parent species X_e ($X = A, B, C$) are first calculated. For this goal, the ab initio structure is optimized for instance with the CCSD(T) method with a correlation-consistent polarized weighted core-covalence n-uple zeta (cc-pwCVnZ) basis set (n = D, T, Q) (see Sect. 2.4), all electrons being correlated (ae). To these X_e(T) and X_e(Q) values, the computed rovibrational correction $\Delta X_0 = X_o - X_e$ is added giving X_0(T) and X_0(Q) under the assumption that this rovibrational correction is not much sensitive to the level of theory, which is generally a good assumption. Then, the rotational constants X_0(T) and X_0(Q) are extrapolated as a function of the computed structural parameters r_e(T) and r_e(Q). The intersection of the line with the experimental X_0 of the parent species gives the extrapolated r_e. Generally, a linear extrapolation is accurate enough. However, when a rotational constant is large, a parabolic extrapolation gives better results. Furthermore, in some particular cases, it is advantageous to use the moments of inertia instead of the rotational constants [11].

The method was first applied to cyclopropane, C_3H_6 [50]. The equilibrium rotational constants were calculated at the MP2/cc-pVnZ, MP2/aug-cc-pVnZ, and CCSD(T)_ae/cc-pVnZ levels of theory where n = D, T. They were subsequently converted into ground state rotational constants using the rovibrational corrections calculated at the same level of theory. These computed rotational constants were plotted as the function of the bond lengths r(CC) and r(CH) calculated at the same level of theory. All the lines intersect at the same point, the horizontal lines representing the experimental ground state rotational constants giving for instance r_e(CC) = 1.5036(3) Å in perfect agreement with the semiexperimental value, see Appendix 5.4.

This method was already applied with success to many molecules [7], see also Appendix 5.1 which gives a list of molecules whose structure was determined by this method. It was found that the accuracy is about 0.001 Å for the bond lengths and 0.1° for the angles.

In Table 5.10, the results of the extrapolation method are compared to the semiexperimental and CCSD(T)_ae/cc-pwCVQZ structures confirming the accuracy of the extrapolation method.

Another advantage of the extrapolation method is that it permits to obtain an accurate equilibrium structure even when the non-dynamical correlation is non-negligible [33]. A typical example is the structure of ClNO. The CCSD(T)_ae/aug-cc-pwCVQZ value of the N–Cl bond length is 1.962 Å, whereas the experimental

Table 5.10 Equilibrium parameters (in Å or degree) between heavy atoms for some molecules. Comparison of the extrapolation method with other methods

	Extrapolation	r_e^{se}	Ab initio[a]
4-cyanopyridine[b]			
N1C2	1.3354	1.3351(4)	1.3361
C2C3	1.3893	1.3890(5)	1.3896
C3C4	1.3921	1.3921(5)	1.3925
C4C7	1.4366	1.4368(7)	1.4371
C7N8	1.1575	1.1573(6)	1.1583
N1C2C3	123.76	123.81(5)	123.81
C2C3C4	118.02	118.01(5)	118.02
C3C4C5	119.08	119.06(6)	119.07
1,3,4-oxadiazole[c]			
CO	1.3537	1.3548(16)	1.3540
CN	1.2851	1.2840(11)	1.2854
NN	1.4057	1.4058(13)	1.4064
COC	101.39	101.29(12)	101.36
OCN	113.75	113.782(2)	113.78
CNN	105.55	105.57(6)	105.54
Thiophene[d]			
CS	1.7102	1.7102(1)	1.7107
C2C3	1.3663	1.3663(2)	1.3665
CSC	92.027	92.079(8)	92.059
Trans-hexatriene[e]			
C1C2	1.3389	1.3390(8)	1.3394
C2C3	1.4495	1.4494(7)	1.4501
C3C4	1.3455	1.3461(14)	1.3460
C1C2C3	123.70	123.6(4)	123.72
C2C3C4	123.70	123.67(8)	123.71
Carbon disulfide CS	1.55448[f]	1.55252(6)[g]	1.553302[h]
Cyclohexane			
CC	1.526[i]	1.5258(6)[j]	1.5267[j]

(continued)

Table 5.10 (continued)

	Extrapolation	r_e^{se}	Ab initio[a]
CCC	111.04	111.114(9)	111.09
CCCC	55.98	55.73(3)	55.80

[a] CCSD(T)_ae/cc-pwCVQ Z
[b] Vogt N, Nair KPR, Grabow J-U, Demaison J (2018) Mol Phys 116: 3530–3537
[c] Demaison J, Jahn MK, Cocinero EJ, Lesarri A, Grabow J-U, Guillemin J-C, Rudolph HD (2013) J Phys Chem A 117: 2278–2284
[d] Dindić C, Ludovicy J, Terzi V, Lüchow A, Vogt N, Demaison J, HVL Nguyen (2022) Phys Chem Chem Phys 24: 3804–3815
[e] Craig NC, Demaison J, Groner P, Rudolph HD, Vogt N (2015) J Phys Chem A 119: 195 – 204
[f] Kummli DS, Frey HM, Leutwyler S (2006). J Chem Phys 124: 144,307
[g] Walrand J, Humblet V, Blanquet G (1988) J Mol Spectrosc 127: 304–323
[h] Coriani S, Marchesan D, Gauss J, Hättig C, Helgaker T, Jørgensen P (2005) J Chem Phys 123: 184107
[i] Kummli DS, Frey HM, Keller M, Leutwyler S (2005) J Chem Phys 123: 054308
[j] Demaison J, Craig NC, Groner P, Ecija P, Cocinero EJ, Lesarri A, Rudolph HD (2015) J Phys Chem A 119: 1486–1493

equilibrium value is 1.9723(7) Å. Furthermore, a semiexperimental determination gives 1.9731(3) Å. Although these two values seem to be extremely precise, a detailed inspection of the fits shows that the system of normal equations is ill-conditioned and that the three structural parameters are fully correlated. In such a case, a meaningful confidence interval requires at least ten standard deviations. The interpolated value gives 1.9706(1) Å, which is inside the confidence interval of the experimental and semiexperimental values.

5.7 Zero-Point Average Structure, r_z

The r_z distance is the distance between average nuclear positions in the vibrational ground state at 0 K [7, 22, 54–57]. It has a clear physical meaning permitting comparisons between molecules. It is now rarely used by spectroscopists, in particular because it cannot compete with the semiexperimental method. However, it has an historical interest because the r_z structure of many molecules was determined and was used to estimate an approximate equilibrium structure.

The r_z structural parameters are obtained from the B_z^ξ rotational constants

$$B_z^\xi = B_0^\xi + \sum_k \alpha_k^\xi(\text{harm}) + \alpha_k^\xi(\text{Coriolis}) = B_e^\xi - \sum_k \alpha_k^\xi(\text{anharm}) \qquad (5.8)$$

Thus, it is enough to have a harmonic force field to calculate the correction, see Eqs. (5.3–5.4). One difficulty is that the r_z structure is not isotopically invariant. In order to obtain a reliable structure, isotopic differences in bond lengths, δr, have to be taken into account while isotopic differences in bond angles may be neglected, see for instance Table 7 of Ref. [18]. The δr was usually estimated by a formula derived by Kuchitsu [58, 59]

$$\delta r = \frac{3}{2} a \delta \left(u^2 \right) - \delta K \tag{5.9}$$

where u^2 is the mean-square amplitude for the bond concerned, and K the mean-square perpendicular amplitude correction, both obtained from the harmonic force field, while a is the Morse anharmonicity parameter for the stretching of the bond. The a parameters are often assumed to be equal to those of the corresponding diatomic molecule and have been tabulated by Kuchitsu and Morino [60]. This "diatomic approximation" was carefully checked in many different molecules and found to be a reasonable one [21, 22, 56]. Note that it is now possible to get more accurate results using an ab initio anharmonic force field. For instance, the r_e and r_z bond lengths are listed in several ab initio packages such as Gaussian [61].

The r_z bond length may be used to estimate the equilibrium value by Kuchitsu's equation

$$r_e = r_z - \frac{3}{2} a u^2 + K \tag{5.10}$$

Note that it would be now more accurate to use an ab initio anharmonic force field. There is a particular case which is interesting: if the r_z value has been obtained without isotopic substitution (i.e., only using one isotopologue): the mass-dependence method of Nakata et al. [62]. The r_e bond length is obtained from the following equation

$$r_e = r_z + \sum_i \left\{ \frac{m_i^* + m_i}{m_i^* - m_i} + \frac{1}{2} \right\} \delta r_z(i) \tag{5.11}$$

where $\delta r_z(i)$ is the isotopic difference when the ith atom of mass m_i is substituted by one of its isotopic species of mass m_i^*. The sum is taken over all constituent atoms.

For a molecule with an X–H bond, the r_e structure can also be estimated from the r_z structure using the diatomic approximation as done by Oka and Morino to determine the structure of H_2Se [63]. In this method, the r_z parameter is a linear function of $(\mu_i)^{-1/2}$ where μ_i is the reduced mass of the X–H or X–D bond. The equation is

$$r_e = \frac{(r_z)_H \sqrt{\mu_H} - (r_z)_D \sqrt{\mu_D}}{\sqrt{\mu_H} - \sqrt{\mu_D}} \tag{5.12}$$

This equation was used to determine the equilibrium structure of PH_3 [64] and AsH_3 [65]. The results are satisfactory. For instance, for PH_3, using Eq. (5.12), one finds r_e(P–H) = 1.4114 Å [66] to be compared with the best experimental structure, 1.4116 Å (see Appendix 5.4).

Table 5.11 Structure of methyl cyanide, CH_3CN (distances in Å, angles in degree)

	r_z [a]	$r_z - r_e$ [b]	r_e [c]	r_e^{se} [d]	r_e^{BO} [e]	$r_m^{(1)}$ [f]
C–C	1.4617(2)	0.0043	1.4574	1.4586(1)	1.4590	1.457(2)
C≡N	1.1567(6)	−0.0009	1.1576	1.1554(1)	1.1553	1.156(2)
C–H	1.0947(24)	0.0051	1.0896	1.0865(1)	1.0865	1.087(3)
∠(CCH)	109.851(96)			109.85(1)	109.84	110.1(3)

[a] Demaison J, Dubrulle A, Boucher D, Burie J, Typke V (1979) J Mol Spectrosc 76: 1–16
[b] Using the harmonic force field of: Halonen L, Mills IM (1978) J Mol Spectrosc 73: 494–502, and for the Morse anharmonicity parameter 2.131; 2.268; and 1.981 Å for CC, CN, and CH, respectively
[c] Derived from the r_z structure
[d] Puzzarini C, Cazzoli G (2006) J Mol Spectrosc 240: 260–264. Force field calculated at the CCSD(T)_ae/cc-pwCVTZ level
[e] CCSD(T)_fc/cc-pV6Z + CV correction at the CCSD(T)/cc-pwCVQZ level, see footnote d for reference
[f] Le Guennec M, Włodarczak G, Burie J, Demaison J (1992) 154: 305–323

For the angles, it was found that θ_z is close to the θ_e value, the difference being generally smaller than 0.2° [18]. However, it has be noted that the value of the θ_z angle is not invariant upon isotopic substitution, and the change might be non-negligible when a hydrogen atom is involved. However, these changes seem to be quite small, even when hydrogen atoms are involved. For instance, Duncan [67] found for formaldehyde, H_2CO, $\angle(HCH)_z = 116.23(10)°$ and $\angle(DCD)_z = 116.32(10)°$. He concluded that these variations are in agreement with those observed for H_2O, H_2Se, and NH_3.

The accuracy of the r_e bond lengths derived from the r_z structure is more limited, being perhaps not better than a few thousandths of an Å. It cannot be better than the accuracy of the r_z structure, which can be rather poor, in particular for large molecules. However, a comparison of the structures of 39 small molecules indicates that the standard deviation (calculated from the median absolute deviation) of the r_e structure derived from the r_z one, Eq. (5.10) is as low as 0.0013 Å with very few outliers, see Appendix 5.2. This pleasing result is probably due to the fact that the r_z structure of all molecules of this table are accurate because the number of structural parameters is small. For larger molecules, a significant decrease of the accuracy should be expected, mainly because the accuracy of the r_z structure is not high enough. However, when the r_z structure is accurate, it is possible to derive a r_e structure whose accuracy is satisfactory, see for instance the example of CH_3CN in Table 5.11.

5.8 Effective Structures (r_0)

This method is the simplest one but it gives an empirical structure: the structural parameters are directly fitted to the ground state moments of inertia. The rovibrational correction is indeed smaller than 1% for most molecules. Thus, if we neglect this correction for a diatomic molecule, the systematic error on the bond length is also smaller than 1%. It remains true for polyatomic molecules whose structure may be determined from only one isotopologue, i.e., the molecules of formula XY_n, see Table 5.12.

It may be noted that the r_0 structure appears a bit expanded compared to the equilibrium structure. This is due to the fact that the equilibrium rotational constants are predominantly larger than the ground state ones. What it interesting it that the offset $p_0 - p_e$ ($p = r$ or $\angle$) is remarkably constant with 0.0025(12) Å for a bond between two heavy atoms, 0.0084(9) Å for a bond between a heavy atom and hydrogen, and 0.0058(7) Å for a bond between a heavy atom and deuterium. For the bond angles, the offset is $\angle_0 - \angle_e = 0.22(4)$ degree [68]. In other words, using these offsets, the equilibrium structure of a XY_n molecule can be determined with an acceptable accuracy from the ground state rotational constants, see Table 5.13.

Table 5.12 Comparison of the r_0 and r_e structures for some XY_2 molecules (distances in Å, angles in degree)[a]

	θ_0	θ_e	r_0	r_e	$\theta_0 - \theta_e$ (%)	$r_0 - r_e$ (%)
ClO_2	117.66	117.40	1.4725	1.4698	0.2	0.2
F_2O	103.31	103.07	1.4087	1.4053	0.2	0.2
SF_2	98.27	98.05	1.5889	1.5875	0.2	0.1
Cl_2O	111.10	110.88	1.6986	1.6959	0.2	0.2
CF_2	104.95	104.86	1.3002	1.2972	0.1	0.2
SiF_2	100.98	100.76	1.5913	1.5901	0.2	0.1
GeF_2	97.37	97.15	1.7331	1.7321	0.2	0.1
NF_2	103.33	103.13	1.3494	1.3469	0.2	0.2
NO_2	134.25	133.86	1.1964	1.1939	0.3	0.2
SO_2	119.54	119.33	1.4322	1.4308	0.2	0.1
SeO_2	114.02	113.83	1.6088	1.6076	0.2	0.1
O_3	116.97	116.78	1.2761	1.2717	0.2	0.3
Cl_2O	111.10	110.88	1.6986	1.6959	0.2	0.2
CCl_2	109.34	109.22	1.7140	1.7113	0.1	0.2
$SiCl_2$	101.47	101.32	2.0667	2.0653	0.1	0.1
$GeCl_2$	100.03	99.88	2.1704	2.1695	0.1	0.0

[a] Demaison J, Vogt N (2022) J Mol Spectrosc 387: 111,661, reprinted with permission of the publisher Elsevier

Table 5.13 Equilibrium structures of SCl_2, PO_2, and PF_2 (distances in Å, angles in degrees)[a]

Molecule	Parameter	From r_0 [b]	Semiexperimental[c]	Basis set[d]
SCl_2	r_e(S–Cl)	2.0100(12)	2.0111(20)	6-311 + G(3df)
	$\angle_e$(ClSCl)	102.589(42)	102.676(89)	
PO_2	r_e(P=O))	1.4620(12)	1.46424(7)	aug-pcS-3
	$\angle_e$(OPO)	135.296(42)	135.119(14)	
PF_2	r_e(PF)	1.5750(12)	1.57605(8)	aug-pcS-3
	$\angle_e$(FPF)	98.329(42)	98.332(9)	
ZrO_2	r_e(Zr=O)	1.7685(12)	1.76955(3)	def2-qzvppd
	$\angle_e$(OZrO)	107.891(42)	107.969(4)	
Br_2O	r_e(Br–O)	1.8379(12)	1.8389(4)	aug-pc-3
	$\angle_e$(BrOBr)	112.212(42)	112.206(4)	

[a] Demaison J, Vogt N (2022) J Mol Spectrosc 387: 111661, reprinted with permission of the publisher Elsevier
[b] r_0 structure – offset
[c] An uncertainty of 10% is assumed for the rovibrational corrections
[d] Used to calculate the force field, B3LYP method, except for SCl_2 where it is B2PLYPD3

Table 5.14 Different r_0 structures of HCN (in Å)

Data[a]	r_e(C–H)	r_e(C≡N)
r_e[a]	1.0650	1.1533
Set H[b]	1.0692	1.1555
Set D[c]	1.0673	1.1551
Set T[d]	1.0657	1.1554
Sets H + D[b,c]	1.0624	1.1568

[a] Ground state rotational constants and equilibrium structure from: Maki AG, Mellau GC, Klee S, Winnewisser M, Quapp W (2000). J Mol Spectrosc 202: 67–82
[b] Set H: $B_0(H^{12}C^{14}N)$; $B_0(H^{13}C^{14}N)$; $B_0(H^{12}C^{15}N)$; $B_0(H^{13}C^{15}N)$
[c] Set D: $B_0(D^{12}C^{14}N)$; $B_0(D^{13}C^{14}N)$; $B_0(D^{12}C^{15}N)$; $B_0(D^{13}C^{15}N)$
[d] Set T: $B_0(T^{12}C^{14}N)$; $B_0(T^{13}C^{14}N)$; $B_0(T^{12}C^{15}N$

Unfortunately, when several isotopologues are necessary to determine the structure, the system of normal equations of the least-squares fit is often ill-conditioned, and the small errors are amplified. A typical example is the linear molecule FCP discussed in Sect. 5.5. Another difficulty is that the results depend of the data used. See the example of HCN in Table 5.14. The r_0 structure can be determined using either all the rotational constants of the hydrogenated species, or all the rotational constants of the deuterated species, or all the rotational constants of the triatiated species, or all the rotational constants of the hydrogenated and deuterated species. Each method gives different results, and the worst result is obtained when the rotational constants of the hydrogenated and deuterated species are mixed. This may be explained by the fact that the r_0 C–H and C–D bonds have different lengths.

As a rule, when several isotopologues are used, the r_0 structure is not accurate. It is especially true for the X–H bond length [5, 7, 18]. In favorable cases, the bond lengths in polyatomic molecules are determined with a MAD of about 0.003 Å (corresponding to a standard deviation of 0.004 Å), the r_0 being generally (but not always, typical counterexamples are FCN and ClCN) larger than the r_e one. However, in some cases, large deviations are observed, up to 0.014 Å, see Table V.10 of Ref. [5]. For the C–H bond, the difference $|r_e - r_0|$ is larger, about 0.005 Å, and its sign is positive for ≡C(sp)–H bonds and negative when an sp^2 or sp^3 carbon is involved [7]. For the angles, the standard deviation (from the MAD) is 0.3° but with a maximum deviation of 1.7°, see Table V.2 of Ref. [5].

5.9 Substitution Structures

5.9.1 Kraitchman's Equations

As the accuracy of the effective structure is often unsatisfactory, more sophisticated methods have been developed. The most used one is the substitution method. At first sight, assuming that the rovibrational correction remains constant upon isotopic substitution should be better than neglecting it as in the r_0-method. Indeed, in many cases, the range of $\varepsilon^{\xi} = I_0^{\xi} - I_e^{\xi}$ ($\xi = a, b, c$) is small compared to its mean value. For instance, for OCSe (27 isotopologues), the mean value is 0.349 uÅ^2 and the range only 0.026 uÅ^2 ($\varepsilon/I_0 = 0.3\%$). Likewise, for HCN (11 isotopologues), the mean value is 0.049 uÅ^2 and the range 0.007 uÅ^2 (ε/I0 = 0.4%) [7]. For this reason, Costain [69] proposed to use the Kraitchman's equations [70] to determine the Cartesian coordinates of the substituted atoms. In these equations, the differences of the moments of inertia between the parent and the substituted isotopologues are used.

For a linear molecule, the equation is quite simple. The moment of inertia I_e may be written

$$I_e = \sum_i m_i z_i^2 \tag{5.13}$$

where z_i is the Cartesian coordinate of atom I relatively to the center of mass and m_i its mass. If an atom α is substituted, its new mass is $m_\alpha + \Delta m$, and the moment of inertia of the substituted isotopologue is

$$I_e^{\alpha} = \sum_i m_i z_i^2 + \mu z_\alpha^2 \tag{5.14}$$

with

$$\mu = \frac{M \Delta m}{M + \Delta m} \tag{5.15}$$

where M is the mass of the parent isotopologue. The Cartesian coordinate of the substituted atom α is

$$z_\alpha = \left[\frac{I_e^\alpha - I_e}{\mu} \right]^{1/2} \tag{5.16}$$

For a general asymmetric top molecule, the equations are more complicated, and they are given in the Appendix 5.3.

When this method was applied to a few small molecules, the results were encouraging, the r_s structure being close to the r_e one [1, 69, 71, 72]. However, problems became visible quite soon, in particular for large molecules. The origin is easy to see. The ground state moment of inertia I_0 may be written

$$I_0 = I_e + \varepsilon \tag{5.17}$$

ΔI_0 is used in Kraitchman's equations

$$\Delta I_0 = \Delta I_e + \Delta\varepsilon \tag{5.18}$$

The method will give accurate results if $\Delta\varepsilon \ll \Delta I_e$. Unfortunately, when an atom has a small Cartesian coordinate, $\Delta\varepsilon$ is not negligible. A typical example is the molecule N_2O for which $I_0(^{14}N^{15}N^{16}O) - I_0(^{14}N^{14}N^{16}O) = 0.0028$ uÅ^2, and the corresponding value for $\Delta\varepsilon$ is -0.0022 uÅ^2 [73]. This difficulty is quite common for large molecules for which some atoms have small Cartesian coordinates, all the more so as the value $\Delta\varepsilon$ increases with the mass of the molecule [74]. An example of the poor performance of Kraitchman's equations is the structure of cyclohexane where r_s(C2–C3) = 1.514 Å, whereas the corresponding r_e value is 1.5312(7) Å [75]. Similar results were found for ethynylcyclohexane [76].

It may be concluded that the Kraitchman's equations are ill-conditioned, and even when semiexperimental equilibrium moments of inertia are used, the results may be disappointing [46, 77].

5.9.2 Variants

Kraitchman's equations can be used when only one atom is substituted. Chutjian [78] has developed equations when symmetry-equivalent atoms are substituted. This is useful because the preparation of these multisubstituted species may be easier. Furthermore, it avoids the effect of large axis rotation which can further deteriorates the accuracy [48]. For more details, see [1, 2, 5, 7].

Pierce [79] proposed to locate atoms near a principal axis by using a double substitution method. However, this method requires very accurate moments of inertia for many isotopologues, which is not easy. Furthermore, the results are sometimes disappointing [80].

The Kraitchman's and Chutjian's equations are useful for a partial structure determination. However, their accuracy can be disappointing. When the number of isotopologues is large enough, it may be better to directly fit the ε^{ξ} together with the structural parameters using the ground state moments of inertia as input [81]. This r_s-fit method assumes that the ε^{ξ} are isotopically invariant and is expected to be more balanced. However, Watson et al. [82] proposed a still better method that will be presented in the next section.

5.10 Mass-Dependent Structures

5.10.1 *Original* r$_m$ *Method [9]*

In order to simplify the notations, the simple case of a linear molecule will be considered, but the validity of the results is general.

The Cartesian coordinate of a substituted atom i of mass m_i may be calculated using Kraitchman's equations, see Eq. (5.16)

$$[z_s(i)]^2 = \frac{1}{\mu}[I_0(m_i + \Delta m_i) - I_0(m_i)] = \frac{1}{\mu}(\Delta I_e + \Delta\varepsilon) \tag{5.19}$$

In this equation, μ is the reduced mass, and $\varepsilon = I_0 - I_e$ is the vibrational contribution to I_0. Watson [9] assumed that it is possible to expand $\Delta\varepsilon$ in Taylor series in Δm_i

$$[z_s(i)]^2 = [z_e(i)]^2 + \frac{\partial\varepsilon}{\partial m_i} + \left(\frac{1}{2}\frac{\partial^2\varepsilon}{\partial m_i^2} + \frac{1}{M}\frac{\partial\varepsilon}{\partial m_i}\right)\Delta m_i + \cdots \tag{5.20}$$

In this equation, M is the total mass of the molecule.

When the Cartesian coordinates of all atoms have been determined using Kraitchman's equations, it is possible to define the substitution moment of inertia I_s, neglecting the last term in the expansion, Eq. (5.20). It gives

$$I_s = \sum_{i=1}^{n} m_i[z_s(i)]^2 = I_e + \sum_{i=1}^{n} m_i\frac{\partial\varepsilon}{\partial m_i} \tag{5.21}$$

As ε is a homogeneous function of the masses m_i of degree 1/2.

Table 5.15 Different structures of OCS (in Å)

Method	r(C=O)	r(C=S)	Ref.
r_e	1.15617(14)	1.56140(14)	1
r_0	1.1568(11)	1.5645(9)	2
r_s	1.1605	1.5596	2
r_m	1.1587(13)	1.5593(10)	2
r_c	1.1547	1.5623	3
r_m^ρ	1.1551(11)	1.5621(9)	2
$r_m^{(1)}$	1.1576(7)	1.5605(11)	4
$r_m^{(2)}$	1.15648(2)	1.56107(3)	4

[1] Lahaye J-G, Vendenhaute R, Fayt A (1987) J Mol Spectrosc 123: 48–83
[2] Harmony MD, Taylor WH (1986) J Mol Spectrosc 118: 163–173
[3] Nakata M, Kuchitsu K (1994) J Mol Struct 320: 179–192
[4] Watson JKG, Roytburg A, Ulrich W (1999) J Mol Spectrosc 196: 102–119

$$\sum_{i=1}^{n} m_i \frac{\partial \varepsilon}{\partial m_i} = \frac{1}{2}\varepsilon \tag{5.22}$$

It follows that

$$2I_s = 2I_e + \varepsilon = I_e + I_0 \tag{5.23}$$

In conclusion, if the last term (dependent on Δm_i) in the expansion is neglected, it is possible to have a good approximation of I_e

$$I_e \approx I_m = 2I_s - I_0 \tag{5.24}$$

Although this approximate relation is interesting (see also Sect. 3.7), it cannot be used for an accurate structure determination because the last term in the expansion, Eq. (5.20), is not negligible, see for instance Table 5.15 which compares different structures of OCS. As shown by Smith and Watson [83], a more accurate relation is

$$I_m = 2I_s - I_0 = I_e + \frac{1}{M}\sum_i m_i \Delta m_i \frac{\partial^2 (M\varepsilon)}{\partial m_i^2} \tag{5.25}$$

To obtain an accurate approximation of I_e, it would be necessary to extrapolate the substitution coordinates to $\Delta m_i = 0$, which is unpractical.

5.10.2 The r_c Method

To improve the r_m method, Nakata et al. [84, 85] suggested to use complementary sets of isotopologues in such a way that the last term proportional to Δm_i is compensated.

For instance, for OCS, the two complementary sets correspond to the following isotopologues

- $^{16}O^{12}C^{32}S$, $^{18}O^{12}C^{32}S$, $^{16}O^{13}C^{32}S$, $^{16}O^{12}C^{34}S$
- $^{18}O^{13}C^{34}S$, $^{16}O^{13}C^{34}S$, $^{18}O^{12}C^{34}S$, $^{18}O^{13}C^{32}S$.

A least-squares fit to the eight moments I_m gives the r_c structure which is a better approximation of the r_e structure than the r_m structure, see Table 5.15. However, the large number of isotopologues required precludes a wide application of the r_c method, except for small molecules.

This method is reviewed by Demaison et al. [18].

5.10.3 The r_m^ρ Method

Harmony et al. [86–89] proposed an empirical method to improve the accuracy of the r_m method. Their method is based on the fact that the ratio $\rho(i) = I_s(i)/I_0(i)$ hardly depends on the isotopologues and can hence by replaced by that of the parent species $\rho(1)$. This ratio can be used to correct the I_0 of all isotopologues by scaling them by the fixed factor $\rho(1)$. For a nonlinear molecule,

$$I_\rho^{(\xi)} = (2\rho_\xi - 1) I_0^{(\xi)} \tag{5.26}$$

The structure is obtained by a least-squares fitting of the corrected moments of inertia $I_\rho^{(\xi)}$.

For a H → D substitution, the scaling is imperfect. To obviate this imperfection, Harmony et al. [87] introduced an approximate correction for the non-negligible bond-length change resulting from the deuteriation: The $I_\rho^{(\xi)}$ value for the deuterated species is corrected by an amount equivalent to an elongation of the C–D bond by an amount $\delta r = 0.0028$ Å [88, 90], see also Sects. 5.8 and 5.10.4. The r_m^ρ method, also reviewed by Demaison et al. [18], has permitted to obtain many reliable structures, see for instance the example of OCS in Table 5.15. However, as noted by Cooksy et al. [91], the correction $(2\rho_\xi - 1)$ has not the correct mass dependence because ε would have a dependence of the order of 1 instead of the order ½ as it should be. The method presented in the next section should be more satisfactory.

5.10.4 Mass-Dependent $r_m^{(1)}$ or $r_m^{(2)}$ Methods

Actually, there is a better approximation than assuming that the rovibrational corrections, ε^{ξ} ($\xi = a, b, c$), are isotopically invariant. It is known that, for a diatomic molecule, ε^{ξ} is proportional to $I_e^{1/2}$, and, for a polyatomic molecule, this proportionality remains approximately valid [92] because ε^{ξ} is a homogeneous function of atomic masses of degree 1/2 [9].

Therefore, the ground-state moment of inertia may be approximated by

$$I_0^{\xi} = I_m^{\xi} + c_{\xi}\sqrt{I_m^{\xi}} \tag{5.27}$$

In this equation, I_m^{ξ} is an approximation of the equilibrium value, I_e^{ξ}. It is thus possible to calculate an approximate equilibrium structure by fitting together the molecular parameters and the parameters c_{ξ}. It gives the $r_{\mathrm{m}}^{(1)}$ structure [82]. It gives satisfactory results in many cases, but it fails when a substituted atom is close to a principal axis or when a hydrogen atom is substituted by deuterium.

The first problem is illustrated by the central C atom in the linear molecule OCSe [93] whose Cartesian coordinate with respect to the center of mass is only 0.24 Å. Although ε increases with I when ^{16}O and ^{78}Se are substituted by heavier isotopes, it decreases when ^{12}C is substituted by ^{13}C. Similar results are obtained for many molecules such as OCS, N_2O, and CO_2, due to the fact that the derivative $\partial\varepsilon/\partial m_Y < 0$, m_Y being the mass of the atom Y with a small coordinate. An analysis of several triatomic linear molecules XYZ where the central atom Y has a small coordinate showed that ε varies as $m_X m_Z/M$ where M is the total mass of the molecule. Watson et al. [82] generalized the method by including an additional empirical term to Eq. (5.27), giving the $r_{\mathrm{m}}^{(2)}$ structure

$$I_0^{\xi} = I_m^{\xi} + c_{\xi}\sqrt{I_m^{\xi}} + d_{\xi}\left(\frac{m_1 m_2 \ldots m_n}{M}\right)^{1/(2n-2)} \tag{5.28}$$

In this equation, n is the number of atoms, m_i their respective masses and c_{ξ} and d_{ξ} fitting parameters. This equation does not take into account the possible rotation of axes upon isotopic substitution, which may be important in some cases, in particular for oblate molecules. In this case, mixed coefficients c_{ab}, c_{bc}, and c_{ca} multiplying the non-diagonal elements of the square root of the inertial tensor have to be added. Structures obtained in this way are called $r_{\mathrm{m}}^{(1r)}$ or $r_{\mathrm{m}}^{(2r)}$.

The problem of the substitution of a hydrogen atom by deuterium is more complicated. Laurie [94] noticed a contraction of the X–H bond upon deuteration because of the large change of mass. For a number of molecules, he found that the difference $r_0(X\text{–H}) - r_0(X\text{–D})$ was in the range of 0.002–0.004 Å, see also Ref. [68]. This shrinkage is explained by the fact that the heavier deuterium atom lies lower in the asymmetric potential well of the X–H stretching vibration. Watson et al. [82] proposed the introduction of this mass dependence in the $r_{\mathrm{m}}^{(1)}$ and $r_{\mathrm{m}}^{(2)}$ methods by using an effective X–H bond parameter

$$r_{\mathrm{m}}^{\mathrm{eff}}(X\!-\!\mathrm{H/D}) = r_{\mathrm{m}}(X\!-\!\mathrm{H}) + \delta_{\mathrm{H}}\left[\frac{M_{\mathrm{H/D}}}{m_{\mathrm{H/D}}(M_{\mathrm{H/D}} - m_{\mathrm{H/D}})}\right]^{1/2} \tag{5.29}$$

The mass coefficient of δ_{H} is the square root of the reduced mass for the vibration of either H or D against the rest of the molecule. Structures including this Laurie correction are called $r_{\mathrm{m}}^{(1\mathrm{L})}$ or $r_{\mathrm{m}}^{(2\mathrm{L})}$.

The r_{m} method gives satisfactory results for small molecules when the rotational constants of many isotopologues are available. It even applies to van der Waals molecules, see next chapter, Sect. 6.6.1.

Unfortunately, the quality of the fit deteriorates rapidly when the number of parameters to be fitted increases. In particular, when the Laurie correction is used, there is often a high correlation between $r_{\mathrm{m}}(X\text{–H})$ and δ_{H}. Actually, as for the semiexperimental method, it is better not to mix hydrogenated and deuterated molecules and to use instead the mixed estimation method (Sect. 5.5), the X–H bond lengths (and $\angle(WX\mathrm{H})$ bond angles being introduced as predicates. These predicates generally are the results of ab initio calculations, but the empirical relationship between the X–H bond length and the $\nu^{\mathrm{is}}(X\text{–H})$ isolated stretching frequency can also be used [95], see also Sect. 8.5 of Ref. [7], and Sect. 8.6.

The weak point of the mass-dependent method is that it requires the determination of three to six additional parameters, which rapidly leads to the ill-conditioning of the least-squares system of normal equations. However, it was shown in the cases of diphenyl disulfide [96], ethynylcyclohexane [76], diallyl disulfide [97], and other moderately large molecules such as fructose [27] that the predicate or mixed estimation method gives a structure whose quality is comparable to that of the traditional semiexperimental method.

For example, Table 5.16 compares the $r_{\mathrm{m}}^{(1)}$ or $r_{\mathrm{m}}^{(2)}$ structure of some large molecules to the equilibrium structure calculated ab initio with the CCSD(T) method with basis sets of quadruple zeta quality. The predicates were calculated at the MP2 level with basis sets of triple zeta quality and were given appropriate uncertainties, i.e., 0.0015 Å for the CH bond lengths, 0.0020 Å for the other bonds, 0.3° for the bond angles, and 0.7° for most dihedral angles. Another typical example confirming the results of Table 5.16 is proline, $C_5H_9NO_2$, for which the comparison with the equilibrium parameters gives a median absolute deviation (MAD) of 0.0004 Å for the bond lengths and 0.04° for the bond angles the mean absolute error being 0.10° [26]. This is a significant improvement compared to the predicate values.

5.11 Choice of the Method

The choice of the best method can be problematic even for a small molecule. A typical example is the nitric oxide, NO_2. An experimental equilibrium structure has been determined [98]. However, it is difficult to assess its accuracy because the excited vibrational states are perturbed by Coriolis interactions and anharmonic resonances. Therefore, the derived rotational constants may be inaccurate, see Sect. 4.6 and

Table 5.16 Comparison of the r_m and r_e structures for the heavy-atom skeleton of some molecules (bond lengths in Å and angles in degrees)

Parameter	r_e^{BO}	r_e^{se}	r_m [a]
Ethynylcyclohexane[b]			
C7C8	1.2074	1.2053(11)	1.2050(13)
C1C7	1.4637	1.4633(13)	1.4630(15)
C1C2	1.5349	1.5338(11)	1.5344(13)
C2C3	1.5256	1.5251(11)	1.5261(15)
C3Hq	1.0909	1.0907(17)	1.0909(18)
C3C4	1.5259	1.5248(10)	1.5268(14)
C1C7C8	178.55	178.39(21)	178.41(22)
C7C1Hx	107.95	107.95(17)	107.94(25)
C7C1C2	111.10	111.060(66)	111.126(85)
C1C2C3	110.98	110.929(98)	110.913(96)
C2C3C4	111.24	111.460(89)	111.35(10)
C3C4C5	110.99	110.940(84)	111.02(12)
C2C1C6	110.54	110.56(10)	110.65(11)
C8C7C1C2	−61.76	−61.735(64)	−61.84(10)
C7C1C2C3	179.90	179.88(17)	179.86(12)
C1C2C3C4	−56.19	−56.09(10)	−56.07(15)
Diphenyl disulfide[c]			
C2C3	1.389		1.3890(13)
C3C4	1.392		1.3914(14)
C4C5	1.391		1.3903(10)
C5C6	1.392		1.3915(11)
C1C6 cis S	1.392		1.3942(20)
C1S	1.7764		1.7771(15)
SS	2.020		2.0199(30)
Diallyl disulfide[d]			
C2C3	1.3341		1.3337(15)
C2C1	1.4891		1.4870(14)
C1S	1.8298		1.8325(10)
SS′	2.0328		2.0282(13)
S′C1′	1.8322		1.8332(15)
C1′C2′	1.4883		1.4851(16)
C3′C2′	1.3336		1.3314(12)
C3C2C1	123.40		123.50(16)
C2C1S	113.02		112.92(10)
C1SS′	102.56		102.284(92)

(continued)

Table 5.16 (continued)

Parameter	r_e^{BO}	r_e^{se}	r_m [a]
SS′C1′	103.44		103.910(35)
S′C1′C2′	112.26		112.24(15)
C1′C2′C3′	123.25		123.268(51)
SC1C2C3	−113.25		−114.43(14)
S′SC1C2	67.73		68.93(27)
C1SS′C1′	−92.95		−93.63(14)
SS′C1′C2′	−64.41		−63.44(16)
S′C1′C2′C3′	113.02		115.05(24)

[a] $r_m^{(1)}$ or $r_m^{(2)}$, see footnotes a, b, c

[b] $r_m^{(2)}$ from Vogt N, Demaison J, Rudolph HD, Juanes M, Fernández J, Lesarri A (2018) J Chem Phys 148: 064306

[c] $r_m^{(2)}$ from Demaison J,Vogt N, Saragi RT, Juanes M, Rudolph HD, Lesarri A (2019) ChemPhysChem 20: 366–373

[d] $r_m^{(1)}$ from: Demaison J, Vogt N, Saragi RT, Juanes M, Rudolph HD, Lesarri A (2019) Phys Chem Chem Phys 21: 19732–19736

Table 4.8. Moreover, the higher-order vibration–rotation interactions constants γ-terms in Eq. (5.1) are extremely large, several hundred of MHz for the *A* rotational constant [98] making the semiexperimental method inaccurate. Finally, the ab initio methods are not easy to use because NO_2 is a radical with excited electronic states close to the ground state [99]. In such a case, the empirical methods only using the ground state rotational constants permit to obtain reasonable results, as shown in Table 5.17. In this particular case, the r_m method gives a satisfactory result because it does not require the use of isotopologues.

Table 5.17 Equilibrium structure of NO_2 (distance in Å, angle in degrees)

Method	r_e (N=O)	∠(ONO)	Comment
Exp. (MW)	1.19455(3)	133.85(2)	a
Exp. (PES)	1.19464(15)	133.888(2)	b
Ab initio	1.1933	135.0	c
Exp. $r_z \rightarrow r_e$	1.193	133.80	See 5.7
Exp. $r_0 \rightarrow r_e$ ($^{14}N^{16}O_2$)	1.1939(12)	134.04(4)	See 5.8
Exp. $r_0 \rightarrow r_e$ ($^{15}N^{16}O_2$)	1.1938(12)	134.03(4)	See 5.8
Exp. r_m	1.1941(3)	133.90	See 5.10.3

[a] Spectroscopy: Morino Y, Tanimoto M (1984) Can J Phys 62: 1315–1322

[b] From an experimental potential energy surface: Hardwick JL, Brand JCD (1976). Can J Phys 54: 80–91

[c] From interpolated potential energy surfaces calculated at the icMRCI + Q/cc-pVQZ level: Schinke R, Grebenshchikov SY, Zhu H (2008) Chem Phys 346: 99–114

5.12 Accuracy

In most cases, the structure is determined using the least-squares method and the standard deviation, σ, of the fitted parameters is used to estimate their accuracy. However, when there are several determinations of the same structure, the confidence intervals built from the standard deviations are not compatible, see for instance Table 5.4. Another typical example is the structure of stibine, SbH_3. The experimental equilibrium structure, when determined from the rotational constants of $^{121}SbH_3$, is r_e(Sb–H) = 1.700013(1) Å and ∠HSbH) = 91.5586(1)° whereas, from $^{121}SbD_3$, one get r_e(Sb–D) = 1.700909(2) Å and ∠DSbD) = 91.6713(1)° [100]. In this particular case, the incompatibility seems to be due to the neglect of the higher-order γ-terms, see Eq. (5.1), as discussed in Appendix 5.4 of Ref. [7].

There are several origins to this incompatibility problem

- The number of degrees of freedom (df = number of input data – fitted parameters) is generally small. In such a case, following Student's distribution, the confidence interval is large. For instance, for df = 4, a 99% confidence interval is $\pm 4.6\sigma$
- The system of normal equations is often ill-conditioned. The Student's distribution cannot be used to define the confidence interval but a distribution which leads to much larger confidence intervals, see Sect. 7.8 of Ref. [41].
- The systematic errors are not taken into account in the calculation of the standard deviations, and they are often larger than the random errors, see Table 5.18 which compares the experimental and semiexperimental equilibrium rotational constants of ethyne and confirm that the semiexperimental rotational constants are affected by a systematic error of several MHz, two orders of magnitude larger than the accuracy of the experimental constants (a few kHz or smaller). The effect of systematic errors is often rather small, see for instance the example of 1,1-difluoroethene in Sect. 5.5 and Ref. [25]. It is easy to explain this behavior. For the ith internal coordinate, the difference between the exact equilibrium structure, r_e, and the optimized semiexperimental structure, r_e^{se}, is

Table 5.18 Experimental (exp) equilibrium rotational constants of ethyne and their comparison with corresponding semiexperimental (se) ones (in MHz)[a]

Isotopologue	B_e(exp)[b]	B_e(se)[c]	Exp—se
$H^{12}C{\equiv}^{12}CH$	35,450.82	35,446.97	3.85
$H^{12}C{\equiv}^{13}CH$	34,602.40	34,598.46	3.94
$H^{13}C{\equiv}^{13}CH$	33,732.87	33,729.33	3.54
$D^{12}C{\equiv}^{12}CH$	29,841.80	29,838.29	3.51
$D^{12}C{\equiv}^{12}CD$	25,495.674	25,493.02	2.65

[a] Sources: Liévin J, Demaison J, Herman M, Fayt A, Puzzarini C (2011) J Chem Phys 134: 064119; Tamassia F, Cané E, Fusina L, Di Lonardo G (2016) Phys Chem Chem Phys 18: 1937–1944
[b] The mean accuracy is about 0.02 MHz
[c] Rovibrational correction calculated at the CCSD(T)_ae/cc-pwCVQZ

$$r_e(i) - r_e^{se}(i) = \sum_j \frac{\partial r(i)}{\partial I_j} \delta I_j \tag{5.30}$$

where I_j is the jth moment of inertia and δI_j its error. Some $\partial r(i)/\partial I_j$ are positive, whereas the others are negative. If all atoms are substituted once, the sum nearly cancels. An illuminating example is the structure of HC_5N discussed in Sect. 9.10.3 of Ref. [7]. For instance, for the C≡N bond length, $\sum_j \partial r(\mathrm{CN})/\partial I_j = -0.0007$ which is only 2% of the mean value of the derivatives: Mean$[\partial r(i)/\partial I_j] = 0.0340$ (all values in u^{-1} Å^{-1}).

When the molecule has a symmetry element, it is possible to estimate the uncertainty of the semiexperimental rotational constants, using parameters which remain invariant upon isotopic substitution such as the inertial defect when the molecule is planar or a planar moment of inertia in a more general case. An interesting example is the molecule phenylacetylene, $C_6H_5C{\equiv}CH$ [101] because the rotational constants of 39 isotopologues are known, it permits a reliable statistical analysis. The rovibrational corrections were calculated at the B3LYP/6-311+G(3df,2pd) level of theory, and the mean value of the semiexperimental equilibrium value of the inertial defect Δ_{SE} was found to be −0.0134 uÅ^2 to be compared with the median value, −0.0133 uÅ^2 indicating the probable absence of outliers. The small variations of Δ_{SE} may be used to estimate its standard deviation which is approximately is 0.0012 uÅ^2. Using the law of propagation of errors, the standard deviation may be written

$$\sigma^2(\Delta_{SE}) = \sigma^2(I_a) + \sigma^2(I_b) + \sigma^2(I_c) \tag{5.31}$$

Assuming that the random error is proportional to the rovibrational correction gives $\sigma(I_c) \approx 1.6\,\sigma(I_b)$ and $\sigma(I_a) \approx \sigma(I_b)/1.9$. This allows us to determine $\sigma(I_g)$ with $g = a, b, c$, and the standard deviations of the semiexperimental rotational constants: $\sigma(A_{SE}) \approx 0.021$ MHz, $\sigma(B_{SE}) \approx \sigma(C_{SE}) \approx 0.003$ MHz. These very small values, due to the random errors, are confirmed by the results of the least-squares fit. There is another way to estimate the standard deviation of the A_{SE} constant because 15 isotopologues keep the C_{2v} symmetry of the parent species. Therefore, the equilibrium A rotational constants are not affected by a substitution on the $a$symmetry axis. The standard deviation of these 15 A_{SE} is 0.058 MHz, i.e., the same order of magnitude as the estimation from the inertial defect.

To estimate the systematic errors, it is possible to follow the same procedure, i.e., assume that the errors are roughly proportional to the rovibrational corrections but the systematic error of Δ_{SE} is

$$\delta\Delta_{SE} = \delta I_c - \delta I_b - \delta I_a \tag{5.32}$$

Giving finally: $\delta(A_{SE}) \approx -6.2$ MHz, $\delta(B_{SE}) \approx \delta(C_{SE}) \approx 0.85$ MHz. These systematic errors are several orders of magnitude larger than the random errors.

This confirms the conclusion that the accuracy of the semiexperimental equilibrium rotational constants is a few percent of the rovibrational correction [102].

Table 5.19 gives a few examples of apparently incompatible equilibrium structures confirming that the standard deviation of the parameters is not a good indicator of accuracy. These inconsistencies are mainly due to the amplification of the systematic errors by the ill-conditioning.

5.13 Software

The least-squares method used to determine the structure is also employed to derive the rotational constants from a fit of an experimental spectrum. It is described in Sect. 4.11. A more detailed presentation may be found in Chap. 9 of Ref. [7].

There are several computer programs available to fit of the internal coordinates to the moments of inertia. For instance, the program STRFIT has many options and is easy to use. It was developed by Z. Kisiel. It is available from http://www.ifpan.edu.pl/~kisiel/prospe.htm [103]. This program can fit different types of structure: r_0, r_z, r_e, and the different r_m structures.

One of the weak points of STRFIT is that it does not permit the weighting of data. A more general package, called MOMSTRUC, is from H. D. Rudolph. It is available from https://www.uni-ulm.de/~hrudolph/sw_archiv/MOMSTRUC/. This program is described in Appendix Va of Rudolph and Demaison [5].

A software package called molecular structure refinement (MSR) permitting the determination of equilibrium structures by means of the semiexperimental approach was described [104]. This last program permits the use of predicate parameters.

There are several quantum chemistry programs permitting the calculation of the force field and the rovibrational corrections. The most used seem to be Gaussian [61] and CFOUR [105]. Furthermore, there are several programs dedicated to the calculation of the rovibrational corrections from the force field in Cartesian coordinates up to cubic terms. For instance, SHRINK written by Sipachev [106]. A similar program was written by Groner and Warren [107].

5.14 Bibliography

The experimental equilibrium structures of polyatomic molecules are listed in Appendix 5.4. The equilibrium structures of polyatomic molecules are listed in Ref. [108] and in the Landolt-Börnstein Series Structure Data of Free Polyatomic Molecules, whose most recent volumes are [109–116], see also Chap. 9 for most recent values.

Table 5.19 Comparison of different determinations (distance in Å)

Molecule	Bond	Meth A	Value A	Ref. A	Meth B	Value B	Ref. B	Diff
HNC	r_e(NC)	Exp	0.9960643(29)	a	SE	0.99489(8)	b	0.00117(8)
HCO^+	r_e(CH)	Exp	1.09725(4)	c	Exp	1.092038(3)	d	0.00521(4)
CS_2	r_e(CS)	Exp	1.55252(6)	e	SE	1.55448(20)	f	0.0020(2)
1,3,4-oxadiazole	r_e(CO)	SE	1.483(8)	g	SE	1.3537(15)	g	0.129(8)
1,2-dichloro-benzene	r_e(C1C2)	SE	1.3921(6)	h	$r_m^{(2)}$	1.3676(23)	h	0.025(2)
1,2-dichloro-benzene	r_e(C2C3)	SE	1.3921(4)	h	$r_m^{(2)}$	1.3979(12)	h	0.006(2)
ClNO	r_e(ClN)	Exp	1.97453(25)	i	AI	1.965775	j	0.009(1)
HOF	r_e(OH)	SE	0.968619(80)	b	AI	0.966565	k	0.0021(1)
PH_3	r_e(PH)	Exp	1.411607(16)	l	Exp	1.4117846(16)	l	0.000178(16)
SbH_3	r_e(SbH)	Exp;	1.700013(1)	m	Exp	1.700909(2)	m	0.000896(2)

[a] Okabayashi T, Tanimoto M (1993) J Chem Phys 99: 3268–3271
[b] Pawlowski F, Jørgensen P, Olsen J, Hegelund F, Helgaker T, Gauss J, Bak, KL, Stanton JF (2002). J Chem Phys 116: 6482–6496
[c] Woods RC (1988) Phil Trans R Soc London A 324:141–146
[d] Dore L, Beninati S, Puzzarini C, Cazzoli G (2003) J. Chem. Phys. 118, 7857–7862
[e] Walrand J, Humblet V, Blanquet G (1988) J Mol Spectrosc 127: 304–323
[f] Kummli D, Frey HM, Leutwyler S (2006) J Chem Phys 124: 144307
[g] Demaison J, Jahn MK, Cocinero EJ, Lesarri A, Grabow J-U, Guillemin J-C, Rudolph HD (2013) J Phys Chem A 117: 2278 – 2284
[h] Kisiel Z, Białkowska-Jaworska E, Chen J, Pszczółkowski L, Gawryś P, Kosarzewski J (2020) J Mol Spectrosc 374:111380
[i] Cazzoli G, Degli Esposti C, Palmieri P, Simeone S (1983) Equilibrium structure and force field of nitrosyl chloride. J Mol Spectrosc 97: 165–185
[j] CCSD(T)_ae/cc-pCVQZ: Coriani S, Marchesan D, Gauss J, Hättig C, Helgaker T, Jørgensen P (2005). J Chem Phys 123: 184107
[k] CCSD(T)/cc-pV∞Z + core(cc-pCVQZ) + T/cc-pVTZ + Q/cc-pVDZ : Puzzarini C, Heckert M, Gauss J (2008). J Chem Phys 128: 194108
[l] Canè E, Fusina L, Bürger H, Jerzembeck W, Brünken S, Lewen F, Winnewisser G. (2002) J Mol Spectrosc 215: 1–9
[m] Canè E, Di Lonardo G, Fusina L, Jerzembeck W, Bürger H, Breidung J, Thiel W (2006) J Mol Struct 780–781: 98–110

5.15 Conclusion

Various methods permitting the determination of an equilibrium structure have been described in this chapter. The coexistence of several methods indicates that there is no ideal method. However, some conclusions can be drawn. The purely experimental methods (making use of experimental equilibrium rotational constants) are limited to very small molecules. On the other hand, the semiexperimental method can be used for much larger molecules provided that the mixed estimation (which also uses ab initio optimized structures as additional data) is employed, in particular to avoid the problem of ill-conditioning. For the same reason, the mixed estimation can also be used with success with the empirical mass-dependent methods, $r_{\mathrm{m}}^{(1)}$ or $r_{\mathrm{m}}^{(2)}$.

All the methods described in this chapter use the rotational constants as data. However, other spectroscopic parameters can furnish useful information on the structure.

- Nuclear quadrupole coupling constants, see Sect. 6.6.3.
- Spin–spin coupling constants: they permit to determine the distance between two atoms. However, the splitting induced by the spin–spin interaction is extremely small and often difficult to observe. Nevertheless, it could be measured for several complexes, in particular with HF. See Sects. 6.6.4 and 8.4 as well as Sect. 7.6.3 of Ref. [95].
- Internal rotation parameters, see Sect. 6.4.
- Electric dipole moment. The dipole moment vector is useful for obtaining information on the symmetry of the molecule. It also permits to identify the conformer when there are several possibilities. It is also useful in the determination of the structure of complexes, see Sect. 6.6.5. See also Sect. 7.2 of Ref. [95].

Appendix 5.1: Molecules Whose Equilibrium Structure Has Been Determined by the "Extrapolation" Method

CS_2 Kummli DS, Frey HM, Leutwyler S (2006) Femtosecond degenerate four-wave mixing of carbon disulfide: High-accuracy rotational constants. J Chem Phys 124: 144307

$C_2H_2N_2O$ Demaison J, Jahn MK, Cocinero EJ, Lesarri A, Grabow J-U, Guillemin J-C, Rudolph HD (2013) Accurate semiexperimental structure of 1,3,4-oxadiazole by the mixed estimation method. J Phys Chem A 117: 2278–2284

C_3H_6 Kummli DS, Frey HM, Keller M, Leutwyler S (2005) Femtosecond degenerate four-wave mixing of cyclopropane. J Chem Phys 123: 054308

C_4H_4S Dindić C, Ludovicy J, Terzi V, Lüchow A, Vogt N, Demaison J, Nguyen HVL (2022) Determination of the semiexperimental equilibrium structure of 2-acetylthiophene in the presence of methyl internal rotation and substituent effects compared to thiophene. Phys Chem Chem Phys 2022: 3804–3815

C_4H_8 Kummli DS, Frey HM, Leutwyler S (2007) High-accuracy structure of cyclobutane by femtosecond rotational Raman four-wave mixing. J Phys Chem A 111: 11936–11942

$C_4H_8O_2$ Den T, Menzi S, Frey H-M, Leutwyler S (2017) Accurate gas-phase structure of para-dioxane by fs Raman rotational coherence spectroscopy and ab initio calculations. J Chem Phys 147: 074306/1–074306/15

C_5H_{10} Kowalewski P, Frey H-M, Infanger D, Leutwyler S (2015) Probing the structure, pseudorotation, and radial vibrations of cyclopentane by femtosecond rotational Raman coherence spectroscopy. J Phys Chem A 119: 11215–11225

C_6F_6 Den TS, Frey, H-M, Leutwyler S (2014) Accurate rotational constant and bond lengths of hexafluorobenzene by femtosecond rotational Raman coherence spectroscopy and ab initio calculations. J Chem Phys 141: 194303/1-194303/9 (2014)

$C_6H_3F_3$ Kummli DS, Frey HM, Leutwyler S (2010). Accurate determination of the structure of 1,3,5-trifluorobenzene by femtosecond rotational Raman coherence spectroscopy and ab Initio calculations. Chem Phys 367: 36—43

$C_6H_4F_2$ Den T, Frey H-M, Felker PM, Leutwyler S (2015) Rotational constants and structure of para-difluorobenzene determined by femtosecond Raman coherence spectroscopy: A new transient type. J Chem Phys 143: 144306/1–144306/12

$C_6H_4N_2$ Vogt N, Nair KPR, Grabow J-U, Demaison J (2018) Microwave rotational spectrum and ab initio computations on 4-cyanopyridine: molecular structure and hyperfine interactions. Mol Phys 116: 3530–3537

C_6H_8 Craig NC, Demaison J, Groner P, Rudolph HD, Vogt N (2015) Electron delocalization in polyenes: A semiexperimental equilibrium structure for (3*E*)-1,3,5-hexatriene and theoretical structures for (3*Z*,5*Z*)-, (3*E*,5*E*)-, and (3*E*,5*Z*)-1,3,5,7-octatetraene. J Phys Chem A 119: 195–204

C_6H_{12} Brügger G, Frey,H-M, Steinegger P, Kowalewski P, Leutwyler, S (2011) Femtosecond rotational Raman coherence spectroscopy of cyclohexane in a pulsed supersonic jet. J Phys Chem A, 115: 12380–12389

C_8H_8 Kummli DS, Lobsiger S, Frey H-M, Leutwyler S, Stanton JF 2008) Accurate determination of the structure of cyclooctatetraene by femtosecond rotational coherence spectroscopy and ab initio calculations. J Phys Chem A 112: 9134–9143

C_8H_{12} Vogt N, Demaison J, Rudolph HD, Juanes M, Fernández J, Lesarri A (2018) Semiexperimental and mass-dependent structures by the mixed regression method: Accurate equilibrium structure and failure of the Kraitchman method for ethynylcyclohexane. J Chem Phys 148: 064306

SO_2 Demaison J, Liévin J (2022) Accuracy of the equilibrium structure of sulfur dioxide. Mol Phys. 120:e1959857 https://doi.org/10.1080/00268976.2021.1950857

Appendix 5.2: r_e Bond Lengths Derived from r_z Values (in Å)

Molecule	Bond	r_e(from r_z)	ref (r_z)	r_e	ref (r_e)	$r_e - r_e(z)$
$OCCl_2$	C–Cl	1.7381	a	1.7375	b	−0.0006
OCHCl	C–Cl	1.765	c	1.7628	d	−0.0022
CH_2Cl_2	C–Cl	1.766	e	1.7643	f	−0.0017
OCF_2	C–F	1.311	g	1.3100	b	−0.0010
$SeCF_2$	C–F	1.309	h	1.311	h	0.0020
CH_3F	C–F	1.383	i	1.3827	j	−0.0003
OCHCl	C–H	1.0897	k	1.0918	d	0.0021
CH_2Cl_2	C–H	1.080	l	1.0816	f	0.0016
HCOOH *trans*	C–H	1.091	m	1.0920	n	0.0010
H_2CO	C–H	1.1005	o	1.1005	p	0.0000
$H_2C{=}CH_2$	C–H	1.081	o	1.0809	q	−0.0001
CH_3F	C–H	1.086	i	1.0870	j	0.0010
H_2S	C–H	1.3362	r	1.3353	s	−0.0009
HCOOH *trans*	C–O	1.340	m	1.3406	n	0.0006
$H_2C{=}CH_2$	C=C	1.334	o	1.3307	q	−0.0033
OCF_2	C=O	1.170	g	1.1699	h	−0.0001
$OCCl_2$	C=O	1.1756	a	1.1759	b	0.0003
OCFCl	C=O	1.170	t	1.1724	u	0.0029
OCHCl	C=O	1.182	k	1.1820	d	0.0000
HCOOH *trans*	C=O	1.201	m	1.1980	n	−0.0030
H_2CO	C=O	1.2033	o	1.2046	p	0.0013
HNNH	N–H	1.028	v	1.0286	w	0.0006
HNNH	N=N	1.246	v	1.2457	w	−0.0003
HCOOH *trans*	O–H	0.969	m	0.9662	n	−0.0028
PCl_3	P–Cl	2.039	x	2.0394	x	0.0004
PH_2F	P–F	1.602	x	1.599	x	−0.0030
FPO	P–F	1.5725	y	1.5727	y	0.0002
FPS	P–F	1.5872	z	1.587	z	−0.0002
PHF_2	P–F	1.579	X	1.578	X	−0.0010
PH_2F	P–H	1.415	X	1.415	X	0.0000
PH_2Cl	P–H	1.411	x	1.412	x	0.0010
PHF_2	P–H	1.421	x	1.420	x	−0.0010
FPO	P=O	1.4531	y	1.4528	y	−0.0003
FPS	P=S	1.8884	z	1.8886	z	0.0002
SiF_4	Si–F	1.5523	aa	1.5517	bb	−0.0006

(continued)

(continued)

Molecule	Bond	r_e(from r_z)	ref (r_z)	r_e	ref (r_e)	$r_e - r_e$(z)
SiF_2	Si–F	1.5903	cc	1.5901	dd	−0.0002
Max						0.0029
Min						−0.0033
Median						−0.0001

[a] Yamamoto S, Nakata M, Kuchitsu K (1985) J Mol Spectrosc 112: 173–182
[b] Demaison J, Császár AG (2012) J Mol Struct 10123: 7–14
[c] Davis RW, Gerry MCL (1983) J Mol Spectrosc 97: 117–138
[d] Demaison J, Boggs, JE, Rudolph HD (2004) J Mol Struct 695–696: 145–163
[e] Duncan JL (1987) J Mol Struct 158: 169–177
[f] Vogt N, Demaison J, Rudolph HD (2014) Mol Phys 112: 2873–2883
[g] Nakata M, Kohata K, Fukuyama T, Kuchitsu K, Wilkins CJ (1980) J Mol Struct 68: 271–280
[h] Bürger H, Demaison J, Dréan, P, Jerzembeck W, Merke I, Stahl W (1998) Ber Bunsenges Phys Chem 102: 872–881
[i] Egawa T, Yamamoto S, Nakata M, Kuchitsu K (1987) J Mol Struct 156: 213–228
[j] Demaison J, Breidung J, Thiel W, Papousek D (1999) Struct Chem 10: 129–133
[k] Davis RW, Gerry MCL (1983) J Mol Spectrosc 97: 117–138
[l] Duncan JL (1987) J Mol Struct 158: 169–177
[m] Davis RW, Robiette AG, Gerry MCL, Bjarnov E, Winnewisser G (1980) J Mol Spectrosc 81: 93–109
[n] Demaison J, Herman M, Liévin J (2007) J Chem Phys 126: 164305/1–9
[o] Duncan JL (1974) Mol Phys 28: 1177–1191
[p] Morgan WJ, Matthews DA, Ringholm M, Agarwal J, Gong JZ, Ruud K, Allen WD, Stanton JF, Schaefer HF (2018) J Chem Theory Comput 14: 1333–1350
[q] Martin JML, Lee TJ, Taylor PR, François J-P (1995) J Chem Phys 103: 2589–2602
[r] Cook RL, De Lucia FC, Helminger P (1975) J Mol Struct 28: 237–246
[s] Coriani S, Marchesan D, Gauss J, Hättig C, Helgaker T, Jørgensen P (2005) J Chem Phys 123: 184107/1–12
[t] Heineking N, Jäger W, Gerry MCL (1993) J Mol Spectrosc 158: 69–81
[u] Demaison J, Perrin A, Bürger H (2003) J Mol Spectrosc 221: 47–56
[v] Demaison J, Hegelund F, Bürger H (1997) J Mol Struct 413–414: 447–456
[w] Bak KL, Gauss J, Jørgensen P, Olsen J, Helgaker T, Stanton JF (2001) J Chem Phys 114: 6548–6556
[x] Dréan P, Paplewski M, Demaison J, Breidung J, Thiel W, Beckers H, Bürger H (1996) Inorg Chem 35: 7671–7678
[y] Beckers H, Bürger H, Paplewski M, Bogey M, Demaison J, Dréan P, Walter A, Breidung J, Thiel W (2001) Phys Chem Chem Phys 3: 4247–4257
[z] Beckers H, Bogey M, Breidung M, Bürger H, Demaison J, Dréan P, Paplewski M, Thiel W, Walters A (2001) J Mol Spectrosc 210: 213–223
[aa] Breidung J, Demaison J, Margulès L, Thiel W (1999) Chem Phys Lett 313: 713–717
[bb] Boudon V, Richard C, Manceron L (2022) J Mol Spectrosc 383: 111549
[cc] Kirchhoff WH (1972) J Mol Spectrosc 41:333–380
[dd] Shoji H, Tanaka T, Hirota E (1973) J Mol Spectrosc 47: 268–274

Appendix 5.3: Kraitchman's Equations for an Asymmetric Top Molecule

For the treatment of the general asymmetric rotor, it is convenient to use the planar moment tensor **P** which is a 3 × 3 matrix. For the parent molecule in its principal axis system, **P** is diagonal, and its elements are

$$P_a = P_{aa} = \frac{1}{2}(-I_a + I_b + I_c) = \sum m_i a_i^2$$
$$P_{ab} = \sum m_i a_i b_i = 0$$

The other elements are obtained by cyclic permutation of *a*, *b*, *c*. Using the definition of the center of mass,

$$x_i' = x_i - \frac{\sum m_i x_i}{\sum m_i} \quad \text{with } x = a, b, c$$

one sees that the elements of the **P′** tensor of the substituted molecule with respect of the principal axis system of the parent molecule are (*a*, *b*, and *c* are the Cartesian coordinates of the substituted atom in the principal axis system)

$$P_{aa}' = P_{aa} + \mu a^2$$
$$P_{ab}' = \mu ab$$

The eigenvalues of **P′** give the planar principal moments P_x' ($x = a, b, c$) which are known experimental values

$$\begin{vmatrix} P_a + \mu a^2 - P' & \mu ab & \mu ac \\ \mu ab & P_b + \mu b^2 - P' & \mu bc \\ \mu ac & \mu bc & P_c + \mu c^2 - P' \end{vmatrix} = 0$$

Expanding the secular equation

$$\begin{aligned} &(P_a + \mu a^2 - P')(P_b + \mu b^2 - P')(P_c + \mu c^2 - P') \\ &+ 2\mu^3 a^2 b^2 c^2 - (P_a + \mu a^2 - P')\mu^3 b^2 c^2 \\ &- (P_b + \mu b^2 - P')\mu^3 a^2 c^2 - (P_c + \mu c^2 - P')\mu^3 a^2 c^2 \\ &= 0 \end{aligned}$$

Arranged according to the powers of *P'* leads to the following equations.

- Coefficient of $P'^2 = P_a' + P_b' + P_c'$
- Coefficient of $P' = P_a'P_b' + P_b'P_c' + P_c'P_a'$
- Constant term $= P_a'P_b'P_c'$.

Solving these three equations gives the Cartesian coordinates of the substituted atom

$$a^2 = \frac{1}{\mu}(P'_a - P_a)\frac{P'_b - P_a}{P_b - P_a} \times \frac{P'_c - P_a}{P_c - P_a} = \frac{\Delta P_a}{\mu}\left(1 + \frac{\Delta P_b}{I_a - I_b}\right)\left(1 + \frac{\Delta P_c}{I_a - I_c}\right)$$

The squared coordinates b^2 and c^2 are obtained by cyclic permutation.

In the bottom equation, the last two factors are near unity because the numerators ΔP_g are expected to be much smaller than the denominators $I_g - I_{g'}$. In conclusion, the coordinate g depends mainly on the isotopic difference ΔP_g.

Appendix 5.4: Experimental Equilibrium Structures of Polyatomic Molecules in Their Ground Electronic State

This section lists the experimental equilibrium structures of polyatomic molecules in their ground electronic state. When the semiexperimental structure is available, it is also given for the sake of comparison. When there is an existing CCSD(T)_ae/cc-p(w)CVQZ structure, it is also listed because it is often a good approximation of the equilibrium structure. When there is a higher-level calculation, it is also given.

Most data are taken from the handbooks "Structure Data of Free Polyatomic Molecules" (Landolt-Börnstein Series), see Chap. 9 for more details. For experimental equilibrium structures determined by gas-phase electron diffraction, see also review paper: Spiridonov VP, Vogt N, Vogt J (2001) Determination of molecular structure in terms of potential energy functions from gas-phase electron diffraction supplemented by other experimental and computational data. Struct Chem 12: 349–376.

Gross formula, chemical name and point-group symmetry are given as headlines.

ArBrH	Ar·HBr ($C_{\infty v}$)		Argon–hydrogen bromide (1/1)
Method	r_e(Ar···Br)	Ref.	
Exp.[a]	4.2113	1	
$r_m^{(1L)}$[b]	4.2440(9)	2	
Potential[c, d]	4.243	3	

[a] 6/12 Lennard-Jones potential by combining bending, stretching, and their coupling
[b] With r_e(HBr) = 1.41447 Å fixed at the experimental value
[c] CCSD(T)/aug-cc-pVTZ potential scaled, shifted, and dilated to generate three-dimensional morphed potential with r_e(H–Br) = 1.4085 Å
[d] The ground state of Ar·HBr is the Ar·HBr isomer, but Ar·BrH is the corresponding global minimum that is 12 ± 1 cm^{-1} more stable than the Ar·HBr local minimum, for the global minimum: r_e(ArBr) = 3.652 Å

[1] Keenan MR, Campbell EJ, Balle TJ, Buxton LW, Minton TK, Soper PD Flygare WH (1980). Rotational spectra and molecular structures of ArHBr and KrHBr. J Chem Phys 72: 3070–3080

[2] Kisiel, Z (2003). Least-squares mass-dependence molecular structures for selected weakly bound intermolecular clusters. J Mol Spectrosc 218: 58–67
[3] Willaert FF, McElmurry BA, Lucchese RR, Bevan, JW (2008). Probing the accuracy of the isomerization energy of the 3-D morphed potential of Ar–HBr. Chem Phys Lett 460: 525–530

ArClF	Ar·ClF ($C_{\infty v}$)		Argon–chorine fluoride (1/1)
Method	r_e(Ar···Cl)	Ref.	
Exp.[a]	3.286	1	
AI[b]	3.38	2	

[a] Obtained by fitting the experimental data to a one-dimensional Lennard-Jones potential
[b] MP2/6-31+G(2df) with counterpoise correction
[1] Harris SJ, Novick SE, Klemperer W, Falconer WE (1974). Intermolecular potential between an atom and a diatomic molecule: The structure of ArClF. J Chem Phys 61: 193–197
[2] Tao F, Klemperer W (1992). The van der Waals potential-energy surfaces and the structures of ArClF and $ArCl_2$. J Chem Phys 97 : 440–451

ArClH	Ar·HCl ($C_{\infty v}$)			Argon–hydrogen chloride (1/1)
Method	r_e(Ar···Cl)	R_e [a]	Ref.	
Microwave[b]	4.0061		1	
$r_m^{(1L)}$	4.02521(72)	4.0612	2	
H6(4,3,0)[c]		4.004	3	
AI[d]		4.02	4	
AI[e]		4.006	5	

[a] Distance of the Ar atom to the center of mass of HCl
[b] 6/12 Lennard–Jones potential by combining bending, stretching, and their coupling
[c] Potential for the state $v = 0$ obtained by fitting to results from high-resolution microwave, far-infrared, and infrared spectroscopy. There is a second local minimum corresponding to the isomer Ar·ClH with $R_e = 3.610$ Å
[d] From a force field calculated at the CCSD(T)/aug-cc-pVQZ-33211 level of theory
[e] CCSD(T)/CBS
[1] Keenan MR, Campbell EJ, Balle TJ, Buxton LW, Minton TK, Soper PD Flygare WH (1980). Rotational spectra and molecular structures of ArHBr and KrHBr. J Chem Phys 72: 3070–3080
[2] Marshall FE, Neill JL, Muckle MT, Pate BH, Kisiel Z, Grubbs II GS (2018) Observation of $^{36}ArH^{37}Cl$, $^{38}ArH^{35}Cl$, $^{38}Ar^{37}Cl$ in natural abundance using CP-FTMW spectroscopy. J Mol Spectrosc 344: 34–38
[3] Hutson JM (1992) Vibrational dependence of the anisotropic intermolecular potential of Ar-HCl. J Phys Chem 96: 4237–4247
[4] Jouypazadeh, H., Solimannejad, M., Farrokhpour, H. (2016). New potential energy surface and rovibrational spectra of Ar···HCl complex: An ab initio study. Comput Theo Chem 1083: 64–71
[5] Woon, D. E., Peterson, K. A., Dunning, T. H. (1998). Benchmark calculations with correlated molecular wave functions. IX. The weakly bound complexes Ar–H_2 and Ar–HCl. J Chem Phys 109: 2233–2241

ArFH	Ar·HF ($C_{\infty v}$)				Argon–hydrogen fluoride (1/1)
Method	r_e(Ar–F)	r_e(Ar–H)	R_e [a]	Ref.	
Exp.[b]			3.436(1)	1	
$r_m^{(1L)}$	3.4416(6)			2	
H6(4,3,2)[c]	3.485	2.56(1)	3.440	3, 4	
SAPT[d]			3.456	5	
AI[e]	3.482	2.564		6	

[a] Distance of the Ar atom to the center of mass of HF
[b] Using a Morse potential
[c] Potential obtained by fitting to results from high-resolution microwave, far-infrared, and infrared spectroscopy. There is a second local minimum corresponding to the isomer Ar·FH with R_e = 3.375 Å
[d] Many-body symmetry-adapted perturbation theory
[e] CCSD(T)/CBS with counterpoise correction

[1] Cousins BL, O'Brien SC, Lisy JM (1984) Argon isotope effect in the microwave spectra of argon hydrogen fluoride (ArHF). J Phys Chem 88: 5142–5143
[2] Kisiel Z (2003) Least-squares mass-dependence molecular structures for selected weakly bound intermolecular clusters. J Mol Spectrosc 218: 58–67
[3] Hutson JM (1992) Vibrational dependence of the anisotropic intermolecular potential of Ar-HF. J Chem Phys 96: 6752–6767
[4] Chang HC, Tao F, Klemperer W, Healey C, Jeremy M. Hutson JM (1993) The Ar–HF intermolecular potential: Overtone spectroscopy and ab initio calculations. J Chem Phys 99: 9337–9349
[5] Lotrich VF, Williams HL, Szalewicz K, Jeziorski B, Moszynski R, Wormer PES, van der Avoird A (1995) Intermolecular potential and rovibrational levels of Ar–HF from symmetry adapted perturbation theory. J Chem Phys 103: 6076–6092
[6] Van Mourik T, Dunning TH (1997). Ab initio characterization of the structure and energetics of the ArHF complex. J Chem Phys 107: 2451–2462

AsF_3	(C_{3v})			Trifluoroarsine
Method	r_e(As–F)	∠(FAsF)	Ref.	
Exp.[a]	1.7041(10)	95.77(12)	1	
Exp	1.7043(2)	95.88(4)	2	

[a] Using an experimental anharmonic force field

[1] Smith JG (1978) The anharmonic force field of arsenic trifluoride from the millimetre wave spectra of vibrationally excited states. Mol Phys 35: 461–475
[2] Bürger H, Ruland, H, Demaison J, Dréan P (2000) The $\nu_1 = 1$ and $\nu_3 = 1$ states of AsF_3 studied by high resolution IR and millimeterwave spectroscopy: the equilibrium structure of AsF_3. J Mol Struct 517–518: 105–117

AsH_2	(C_{2v}; ground electronic state $\tilde{X}^2B_1$)			Arsino
Method	r_e(As–H)	∠(HAsH)	Ref.	
Exp.[a]	1.5158(6)	90.79(8)	1	
Exp	1.5249	90.765	2	

(continued)

(continued)

AsH_2	(C_{2v}; ground electronic state $\tilde{X}^2B_1$)			Arsino
Method	r_e(As–H)	∠(HAsH)	Ref.	
Exp.[a]	1.5169(8)	90.7(2)	3	
AI[b]	1.506	90.7	4	
AI[c]	1.528	91.132	3	

[a] Derived from the r_z structure
[b] Multireference single- and double-excitation configuration interaction with the cc-pVQZ basis set for H and a relativistic core potential for As
[c] Minimum of the potential energy surface calculated at the CCSD(T)_fc/aug-cc-pVQZ level of theory
[1] Fujiwara H, Saito S (1998) Microwave spectrum of the AsD_2 (X^2B_1) radical: harmonic force field and molecular structure. J Mol Spectrosc 192: 399–405
[2] Zhao D, Qin C, Ji M, Zhang Q, Chen Y (2009). Absorption spectra of AsH_2 radical in 435–510 nm by cavity ringdown spectroscopy. J Mol Spectrosc 25: 192–197
[3] Grimminger RA, Clouthier DJ (2012). Toward an improved understanding of the AsH_2 free radical: Laser spectroscopy, ab initio calculations, and normal coordinate analysis. J Chem Phys 137: 224307
[4] Alekseyev AB, Buenker RJ, Liebermann H (2011). Ab initio calculations of the electronic states of AsH_2 including dissociation characteristics. J of Chem Phys 135: 24430

AsH_3	(C_{3v})			Arsine
Method	r_e(As–H)	∠(HAsH)	Ref.	
Exp.[a]	1.513(2)	92.08(7)	1	
Exp	1.51130(11)	92.0775(40)	2	
AI[b]	1.508	91.9	3	

[a] Derived from the r_z structure using the diatomic approximation
[b] Multireference single- and double-excitation configuration interaction with the cc-pVQZ basis set for H and a relativistic core potential for As
[1] McRae GA, Gerry MCL, Wong M, Ozier I, Cohen EA (1987) Microwave spectra of deuterated arsines: Distortion moment transitions of AsD_3, microwave spectra of AsH_2D and $AsHD_2$, and the structure of arsine. J Mol Spectrosc 123: 321–339
[2] Ulenikov ON, Bekhtereva ES, Yuhnik YB, Bürger H (2006) High-resolution infrared study of the $\nu_1 = 1$ and $\nu_3 = 1$ bands, and equilibrium structure of AsD_3. J Mol Struct 780–781: 115–123
[3] Alekseyev AB, Buenker RJ, Liebermann H (2012). AsH_3 ultraviolet photochemistry: An ab initio view. J Chem Phys 136: 224307

BClX	ClBX, X = O, S, Se ($C_{\infty v}$)				
Molecule	Name	Method	r_e(Cl–B)	r_e(B=X)	Ref.
ClBO	Chloroxoborane	r_c	1.683	1.205	1
		Exp	1.67668(26)	1.21308(26)	2
ClBS	Chlorothioxoborane	r_c	1.678	1.607	1
		Exp.	1.680567(89)	1.604923(90)	3

(continued)

(continued)

BClX	ClBX, X = O, S, Se ($C_{\infty v}$)				
Molecule	Name	Method	r_e(Cl–B)	r_e(B=X)	Ref.
		AI[a]	1.68205	1.60615	4
ClBSe	Chloroselenoborane	r_c	1.696	1.719	1

[a] CCSD(T)_ae/cc-pCVQZ

[1] Nakata M, Kuchitsu K (1994) Estimation of the equilibrium structures of polyatomic molecules using isotopic differences in vibrationally averaged structures. J Mol Struct 320: 179–192

[2] Hunt NT, Röpcke J, Davies PB (2000) Infrared diode laser spectrum of the ν_1 fundamental band of ClBO. J Mol Spectrosc 204: 120–124

[3] Bizzocchi L, Degli Esposti C, Puzzarini C (2002) Millimeter-wave spectroscopy of ClBS: an improved evaluation of the equilibrium structure of chlorothioborine. J Mol Spectrosc 216: 177–190

[4] Coriani S, Marchesan D, Gauss J, Hättig C, Helgaker T, Jørgensen P (2005) The accuracy of ab initio molecular geometries for systems containing second-row atoms. J Chem Phys 123: 184107/1-12

BFS	FBS ($C_{\infty v}$)			Fluorothioxoborane
Method	r_e(F–B)	r_e(B=S)	Ref.	
Exp	1.2762(2)	1.6091(2)	1	
CCSD(T)_ae/cc-pCVQZ	1.277466	1.609913	2	

[1] Bizzocchi L, Degli Esposti C (2001) Pyrolysis of sulfur tetrafluoride over boron: Excited-state rotational spectra and equilibrium structure of fluorothioborine (FBS). J Chem Phys 115: 7041–7050

[2] Coriani S, Marchesan D, Gauss J, Hättig C, Helgaker T, Jørgensen P (2005) The accuracy of ab initio molecular geometries for systems containing second-row atoms. J Chem Phys 123: 184107/1-12

BF_3	(D_{3h})		Trifluoroborane
Method	r_e(B–F)	Ref.	
Exp	1.3070(1)	1	
Exp	1.30731(10)	2	
Exp.[a]	1.30704(5)	3	

[a] Mean value for $^{10}BF_3$ and $^{11}BF_3$

[1] Zeisberger E, Ruoff A (1989) High resolution Fourier transform infrared spectrum of ^{10}B-labelled BF_3. J Mol Spectrosc 136: 295–316

[2] Maki A, Masiello T, Blake, TA (2005) The high-resolution infrared spectrum of $^{10}BF_3$ from 400 to 4600 cm^{-1}. J Mol Struct 742: 3–20

[3] Masiello T, Maki A, Blake TA (2007) The analysis of combination and overtone states of $^{11}BF_3$ from 1650 to 4600 cm^{-1}. J Mol Spectrosc 243: 16–31

BHO	HBO ($C_{\infty v}$)			Oxoborane
Method	r_e(B–H)	r_e(B=O)	Ref.	
Exp	1.16740(25)	1.20051(7)	1	
AI[a]	1.1663	1.2019	2	

[a] CCSD(T)_ae/cc-pVQZ

[1] Kawashima Y, Colarusso P, Zhang KQ, Bernath P, Hirota E (1998) Infrared and Microwave Spectra and Force Field of DBO: The Coriolis Interaction between the ν_1 and $\nu_2 + \nu_3$ States. J Mol Spectrosc 192: 152–161

[2] DeYonker NJ, Li S, Yamaguchi Y, Schaefer HF, Crawford TD, King RA, Peterson KA (2005). Application of equation-of-motion coupled-cluster methods to low-lying singlet and triplet electronic states of HBO and BOH. J Chem Phys 122: 234316

BHS	HBS ($C_{\infty v}$)			Thioxoborane
Method	r_e(B–H)	r_e(B=S)	Ref.	
Exp.[a]	1.1698(4)	1.5978(1)	1	
Exp	1.16962(2)	1.597788(5)	2	
CCSD(T)_ae/cc-pCVQZ	1.169857	1.599445	3	

[a] Rovibrational corrections from experimental anharmonic force field

[1] Turner P, Mills I (1982) The infrared spectrum, equilibrium structure and harmonic and anharmonic force field of thioborine, HBS. Mol Phys 46: 161–170

[2] Bizzocchi L, Degli Esposti C (2007) Millimeter- and submillimeter-wave spectroscopy of HBS and DBS in vibrationally excited states. J Mol Spectrosc 241: 67–74

[3] Coriani S, Marchesan D, Gauss J, Hättig C, Helgaker T, Jørgensen P (2005) The accuracy of ab initio molecular geometries for systems containing second-row atoms. J Chem Phys 123: 184107/1-12

B_2H_6	BH_3BH_3 (D_{2h})						Diborane
Method	r_e(B–H$_t$)[a]	r_e(B–H$_b$)[a]	$\angle(H_tBH_t)$	$\angle(H_bBH_b)$	$\angle(BH_bB)$	Ref.	
Exp.[b]	1.184(3)	1.314(3)	121.5(5)	96.9(5)		1	
AI[c]	1.188	1.316	122.2		84.42	2	

[a] b = bridge, t = terminal

[b] Derived from the r_z structure

[c] MP2/6-311G(d,p)

Dependent parameters:

$r_e(B\cdots B) = 1.743$

$r_e(B\cdots H_t) = 2.540$

[1] Duncan JL, Harper J (1984) The structure of the diborane molecule. Mol Phys 51: 371–380

[2] Tian SX (2005) Ab initio and electron propagator theory study of boron hydrides. J Phys Chem A 109: 5471–5480

BeH_2	($D_{\infty h}$)		Beryllium dihydride
Method	r_e(Be–H)	Ref.	
Exp	1.326407(3)	1	
CCSD(T)_ae/cc-pCVQZ	1.32648	2	

[1] Shayesteh A, Tereszchuk, K, Bernath PF (2003) Infrared emission spectra of BeH_2 and BeD_2. J Chem Phys 118: 3622–3627
[2] Koput J, Peterson KA (2006) Ab initio prediction of the potential energy surface and vibration–rotation energy levels of BeH_2. J Chem Phys 125: 044306/1-7

BeF_2	($D_{\infty h}$)		Beryllium difluoride
Method	r_e(Be–F)	Ref.	
Exp. (IR)	1.3729710(95)	1	
Exp	1.37298(12)	2	
Exp. (GED)	1.374(4)	3	
CCSD(T)_fc/cc-pV5Z[a]	1.374	2	

[a] Core correlation calculated at the CCSD(T)/cc-pCVQZ level
[1] Frum CI, Engleman R, Bernath PF (1991) Fourier transform emission spectroscopy of BeF_2 at 6.5 μm. J Chem Phys 95: 1435–1440
[2] Yu S, Shayesteh A, Bernath PF (2005) The vibration–rotation emission spectrum of hot BeF_2 J Chem Phys 123: 134303
[3] Vogt N, Girichev GV, Vogt J, Girichev AG (1995) Equilibrium structure of beryllium difluoride from least-squares analysis of gas-phase electron diffraction data. J Mol Struct 352/353:175–179

BiH_3	(C_{3v})			Bismuthine
Method	r_e(Bi–H)	∠(HBiH)	Ref.	
Exp	1.77834(50)	92.3215(100)	1	
AI[a]	1.77814	90.09	2	

[a] CCSD(T)_ae with ECP for Bi and aug-cc-pVQZ basis set for H
[1] Jerzembeck W, Constantin FL, Margulès L, Demaison J, Bürger H (2004) The ground state parameters of BiH3 from infrared and microwave spectra: ground state and equilibrium structures. J Mol Spectrosc 226: 24–31
[2] Breidung J, Thiel W, Figgen D, Stoll H (2004) A systematic ab initio study of the equilibrium geometry and vibrational wave numbers of bismuthine. J Chem Phys 120: 10404–10413

BrHKr	Kr·HBr ($C_{\infty v}$)			Krypton – hydrogen bromide (1/1)
Method	r_e(Kr···Br)[a]	R_e [b]	Ref.	
Exp.[c]	4.2969		1	
$r_m^{(1L)}$	4.3194(7)		2	
Potential[a, c]		4.27	3	

[a] Value for the ground state structure Kr·HBr, this isomer is 23(3) cm^{-1} less stable than the equilibrium structure Kr·BrH for which R_e = 3.88 Å [3]
[b] Distance of the Kr atom to the center of mass of HBr
[c] 6/12 Lennard-Jones potential by combining bending, stretching, and their coupling
[d] MP2/aug-cc-pVTZ potential scaled, shifted, and dilated to generate three-dimensional morphed potential
1 Keenan MR, Campbell EJ, Balle TJ, Buxton LW, Minton TK, Soper PD Flygare WH (1980). Rotational spectra and molecular structures of ArHBr and KrHBr. J Chem Phys 72: 3070–3080
2 Kisiel Z (2003) Least-squares mass-dependence molecular structures for selected weakly bound intermolecular clusters. J Mol Spectrosc 218: 58–67
3 Wang Z, Lucchese RR, Bevan JW (2004) A Kr–BrH Global minimum structure determined on the basis of potential morphing. J Phys Chem A 108: 2884–2892

BrHO	HOBr (C_s)				Hypobromous acid
Method	r_e(Br–H)	r_e(O–H)	∠(HOBr)	Ref.	
Exp	1.82800(1)	0.9640(3)	102.99(1)	1	
AI[a]	1.8287	0.9625	102.78	2	

[a] MRCI + Q/CBS
1 Cohen EA, Müller HSP, Tan TL, McRae GA (2010) High resolution spectroscopy of DOBr and molecular properties of hypobromous acid. J Mol Spectrosc 262: 30–36
2 Peterson KA (2000) An accurate global ab initio potential energy surface for the X $^1A'$ electronic state of HOBr. J Chem Phys 113: 4598–4612

BrH_3Si	SiH_3Br (C_{3v})					Bromosilane
	r_e(Si–H)	r_e(Si–Br)	∠(HSiH)	∠(HSiBr)	Ref.	
Exp	1.4696(10)	2.2070(10)	110.54(20)	108.38(20)	1	
r_m^{ρ}	1.4703(4)	2.2076(1)	110.65(3)	108.27(3)	1	
[a]	1.4699				This work	

[a] From isolated stretching frequency, see Sect. 8.6
1 Demaison J, Cosléou J, Bürger H, Mkadmi EB (1997) High-resolution infrared spectrum of $D_3Si^{79}Br$ in the ν_1/ν_4 region: the structure of silyl bromide. J Mol Spectrosc 185: 384–391

BrNO	NOBr (C_s)				Nitrosyl bromide
	r_e(N–Br)	r_e(N=O)	∠(BrNO)	Ref.	
Exp	2.144(19)	1.133(20)	114.1(11)	1	
CCSD(T)/TZ2P	2.173	1.142	114.4	2	

[1] Degli Esposti C, Tamassia F, Cazzoli G, Kisiel Z (1995) Millimeter-wave spectrum of nitrosyl bromide in the low-lying excited states: equilibrium structure and cubic force field. J Mol Spectrosc 170: 582–600
[2] Lee TJ (1995) Ab initio characterization of triatomic bromine molecules of potential interest in stratospheric chemistry. J Phys Chem 99: 15074–15080

BrO_2	(C_{2v}) Ground electronic state $\tilde{X}^2B_1$			Bromine dioxide
Method	r_e(Br–O)	∠(OBrO)	Ref.	
Exp.[a]	1.644	114.3	1	
CCSD(T)/TZ2P	1.660	114.8	2	
CCSD(T)/AREP/TZ(2df)	1.650	114.9	3	

[a] Calculated from the r_z structure
[1] Müller HSP, Miller CE, Cohen EA (1997) The rotational spectrum and molecular properties of bromine dioxide, OBrO. J Chem Phys 107: 8292–8302
[2] Miller CE, Nickolaisen SL, Francisco JS, Sander SP (1997) The OBrO $C(^2A_2) \leftarrow X(^2B_1)$ absorption spectrum. J Chem Phys 107: 2300–2307
[3] Pacios LF, Gómez PC (1997) Ab initio study of bromine dioxides OBrO and BrOO. J Phys Chem A 101: 1767–1773

Br_2O	(C_{2v})			Dibromine monoxide
Method	r_e(Br–O)	∠(BrOBr)	Ref.	
Exp.[a]	1.83786	112.249	1	
CCSD(T)/TZ2P	1.865	112.9	2	

[a] Calculated from the r_z structure
[1] Müller HSP, Cohen EA (1997) Dibromine monoxide, Br_2O: The rotational spectrum and molecular properties. J Chem Phys 106: 8344–8354
[2] Lee TJ (1995) Ab initio characterization of triatomic bromine molecules of potential interest in stratospheric chemistry. J Phys Chem 99: 15074–15080

CArO	Ar·CO (C_s)			Argon–carbon monoxide (1/1)
Method	R_e [a]	θ [b]	Ref.	
Exp.[c]	3.7052	95	1	
Exp.[c]	3.71	92	2	

(continued)

(continued)

CArO	Ar·CO (C_s)			Argon–carbon monoxide (1/1)
Method	R_e [a]	θ [b]	Ref.	
Exp.[c]	3.74	98	3	
Pot.[d]	3.63	77	4	

[a] Distance of the Ar atom to the center of mass of CO
[b] Angle between R_e and CO
[c] Ab initio potential energy surface refined with experimental data
[d] From a (6-exp) Buckingham form of the atom–atom potential
[1] Sumiyoshi Y, Endo Y (2015) Three-dimensional potential energy surface of Ar–CO. J Chem Phys 142: 024314.
[2] Havenith M, Schwaab GW (2005) Attacking a small beast: Ar–CO, a prototype for intermolecular forces. Z Phys Chem 219: 1053–1088
[3] Coudert LH, Pak I, Surin L (2004) The potential energy surface of the Ar·CO complex obtained by using high-resolution data. J Chem Phys 121: 4691–4698
[4] Mirsky K (1980) Carbon monoxide molecules in an argon matrix: empirical evaluation of the Ar·Ar, C·Ar and O·Ar potential parameters. Chem Phys 46: 445–455

CBrN	BrCN ($C_{\infty v}$)			Cyanogen bromide
Method	r_e(C–Br)	r_e(C≡N)	Ref.	
Exp.[a]	1.78611	1.15767	1	
Exp	1.7888(48)	1.1577(69)	2	
Exp	1.787499(8)	1.159510(11)	3	
Exp	1.7874902(42)	1.1595229(62)	4, 5	

[a] Direct diagonalization of the Hamiltonian and fit to observed frequencies and rotational constants
[1] Whiffen DH (1978) The force fields of cyanogen halides. III. Cyanogen bromide and cyanogen iodine. Spectrochim Acta A: 34: 1183–1192
[2] Le Guennec M, Wlodarczak G, Chen WD, Bocquet R, Demaison J (1992) Rotational spectrum and equilibrium structure of cyanogen bromide. J Mol Spectrosc 153: 117–132
[3] Tamassia F, Esposti CD, Dore L, Cazzoli G (1995) Millimeter-wave and infrared spectrum of $BrC^{15}N$: Equilibrium structure of cyanogen bromide. J Mol Spectrosc 174: 59–77
[4] Degli Esposti C, Tamassia F, Puzzarini C, Tarroni R, Zelinger Z (1996) Millimetre-wave and infrared spectroscopy of $Br^{13}CN$: anharmonic force field of cyanogen bromide from spectroscopic data and ab initio calculations. Mol Phys 88: 1603–1620
[5] Fayt A, Demaison J, Wlodarczak G, Bürger H (2003) Millimeter-wave and Fourier transform infrared spectra between 325 cm^{-1} and 4720 cm^{-1} of BrCN and global rovibrational analysis of the main isotopomers. Mol Phys 101: 675–695

CClFO	O=CClF (C_s)							Carbonyl chloride fluoride
	r_e(C=O)	r_e(C–F)	r_e(C–Cl)	∠(OCF)	∠(OCCl)	∠(ClCF)	Ref.	
Exp.[a]	1.1695	1.3190	1.7259	124.79	126.01	109.20	1	

(continued)

(continued)

CClFO	O=CClF (C_s)							Carbonyl chloride fluoride
	r_e(C=O)	r_e(C–F)	r_e(C–Cl)	∠(OCF)	∠(OCCl)	∠(ClCF)	Ref.	
SE[b]	1.1724(2)	1.3237(4)	1.7207(4)	123.98(4)	126.49(3)	109.54(3)	2	
AI[c]	1.1739	1.3235	1.7208	123.98	126.42	109.60	2	

[a] Derived from the r_Z structure
[b] Rovibrational corrections from a MP2/TZ2Pf cubic force field
[c] CCSD(T)_fc/cc-pV(Q + d)Z + MP2(ae)/cc-pwCVQZ – MP2(fc)/cc-pwCVQ Z
[1] Heineking N, Jäger W, Gerry MCL (1993) Isotopic and vibrational satellites in the rotational spectrum of carbonyl chloride fluoride. J Mol Spectrosc 158: 69–81
[2] Demaison J, Perrin A, Bürger H (2003) Ab initio anharmonic force field and equilibrium structure of carbonyl chlorofluoride. J Mol Spectrosc 221: 47–56

CClFS	S=CClF (C_s)						Thiocarbonyl chloride fluoride
Method	r_e(C=S)	r_e(C–Cl)	r_e(C–F)	∠(SCCl)	∠(SCF)	Ref.	
Exp	1.5927(13)	1.7133(13)	1.3216(11)	127.00(10)	123.99(10)	1	
Exp.[a]				127.28(9)	123.58(12)	2	

[a] r_Z value from gas-phase electron diffraction
[1] Hamm R, Kohrmann HJ, Günther H, Zeil W (1976) Microwave spectrum r_0-structure, r_Z- and r_e-structure of the isotopic species $^{12}C^{34}SF^{37}Cl$. Z Naturforsch A 31: 594–601
[2] Gleisberg F, Haberl A, Zeil W (1975) Strukturbestimmung von Chlorfluorthiocarbonyl durch Elektronenbeugung. Z Naturforsch A 30: 549–553

$CClF_3$	(C_{3v})				Clorotrifluoromethane
Method	r_e(C–F)	r_e(C–Cl)	∠(ClCF)	Ref.	
Exp. (GED)	1.3200(2)	1.7520(3)	110.1(1)	1	
CCSD(T)_ae/cc-pwCVQZ	1.3208	1.7495	110.27	2	

[1] Seo SS, Ewbank JD (2004) Equilibrium structure for $CClF_3$ using real-time and time-resolved gas electron diffraction. J Korean Chem Soc 48: 339–350
[2] Vogt N, Demaison J, Rudolph HD (2014) Accurate equilibrium structures of fluoro- and chloroderivatives of methane. Mol Phys 112: 2873–2883

CClN	ClCN ($C_{\infty v}$)			Cyanogen chloride
Method	r_e(C–Cl)	r_e(C≡N)	Ref.	
Exp.[a]	1.62995	1.15923	1	

(continued)

(continued)

CClN	ClCN ($C_{\infty v}$)			Cyanogen chloride
Method	r_e(C–Cl)	r_e(C≡N)	Ref.	
Exp	1.6290(24)	1.1606(28)	2	

[a] Direct diagonalization of the Hamiltonian and fit to observed frequencies and rotational constants

[1] Whiffen DH (1978) The force fields of cyanogen halides. II. Cyanogen chloride. Spectrochim Acta A 34: 1173–1181

[2] Cazzoli G, Favero PG, Degli Esposti C (1977) Microwave measurement of the α_3 vibration–rotation constant of ClCN and BrCN vibrational energy transfer from active nitrogen. Chem Phys Lett 50: 336–338

CCl_2O	Cl_2C=O (C_{2v})				Carbonyl dichloride
Method	r_e(C–Cl)	r_e(C=O)	∠(ClCCl)	Ref.	
Exp	1.7365(12)	1.1766(22)	111.91(12)	1	
SE[a]	1.7375(2)	1.1759(4)	111.85(2)	2	

[a] Rovibrational corrections from a MP2/cc-pV(T + d)Z force field

[1] Yamamoto S, Nakata M, Kuchitsu K (1985) Equilibrium structure and anharmonic potential constants of phosgene derived from rotational constants and electron diffraction intensity. J Mol Spectroc 112: 173–182

[2] Demaison J, Császár AG (2012) Equilibrium CO bond lengths. J Mol Struct 1023: 7–14

CCl_2S	Cl_2C=S (C_{2v})				Thiocarbonyl dichloride
Method	r_e(C=S)	r_e(C–Cl)	∠(ClCCl)	Ref.	
Exp	1.597(7)	1.722(4)	111.2(3)	1	
Exp.[a]			111.19(16)	2	

[a] r_z value

[1] Nakata M, Fukuyama T, Kuchitsu K (1982) Molecular structure of thiocarbonyl chloride as studied by gas electron diffraction. J Mol Struct 81: 121–129

[2] Carpenter JH, Rimmer DF, Smith JG, Whiffen DH (1975) Some molecular properties of thiocarbonyl chloride determined by microwave spectroscopy. J Chem Soc, Faraday Trans 2, 71: 1752–1762

CCl_4	(T_d)		Tetrachloromethane
Method	r_e(C–Cl)	Ref.	
Exp.[a]	1.761(2)	1	
CCSD(T)_ae/cc-pwCVQZ	1.7630	2	

[a] Derived from the r_g structure

[1] Vogt N, Rudert R, Rykov AN, Karasev NM, Shishkov IF, Vogt J (2011) Use of imaging plates (IPs) in the gas-phase electron diffraction (GED) experiments on the EG-100 M apparatus. The tetrachloromethane molecule as a test object. Struct Chem 22: 287–291

[2] Vogt N, Demaison J, Rudolph HD (2014) Accurate equilibrium structures of fluoro- and chloroderivatives. Mol Phys 112: 2873–2883

CFN	FCN ($C_{\infty v}$)			Cyanogen fluoride
Method	r_e(C–F)	r_e(C≡N)	Ref.	
Exp.[a]	1.26504	1.15594	1	
Exp	1.26405(74)	1.15680(81)	2	
Exp	1.264185(32)	1.156634(51)	3	
Exp.[b]	1.2640	1.1570	4	

[a] Direct diagonalization of the Hamiltonian and fit to observed frequencies and rotational constants
[b] Using an experimental potential surface
[1] Whiffen DH (1978) The force fields of cyanogen halides. I. Cyanogen fluoride. Spectrochim Acta A 34: 1165–1171
[2] Degli Esposti C, Favero PG, Serenellini S Cazzoli G (1982) Rotational spectra of excited vibrational states of $F^{12}C^{15}N$. Equilibrium structure of cyanogen fluoride J Mol Struct 82 : 221–236
[3] Farkhsi A, Bredohl H, Dubois I, Remy F, Fayt A (2000) FT Infrared spectra of FCN from 1200 to 1800 cm^{-1} and from 2800 to 7000 cm^{-1} and global rovibrational analysis of the main isotopomers. J Mol Spectrosc 201: 36–55
[4] Zúñiga J, Picón JAG, Bastida A, Requena A (2012) A spectroscopic potential energy surface for FCN. J Quant Spectrosc Radiat Transfer 113: 1155–1169

CFP	FCP ($C_{\infty v}$)			(Fluoromethylidyne)phosphine
Method	r_e(C–F)	r_e(C≡P)	Ref.	
Exp	1.2761(7)	1.5445(6)	1	
SE[a]	1.2759(4)	1.5445(2)	2	
CCSD(T)_fc/cc-pV∞Z + CV[b]	1.2757	1.5431	3	

[a] Rovibrational corrections obtained from a CCSD(T)_fc/cc-pV(n+d)Z (up to $n = 6$) cubic force field
[b] CV = core-valence correction
[1] Dréan P, Demaison J, Poteau L, Denis JM (1996) Rotational spectrum and structure of HCP. J Mol Spectrosc 176: 139–145
[2] Bizzocchi L, Degli Esposti C, Puzzarini C (2006) Millimetre-wave spectroscopy and ab initio calculations for fluorophosphaethyne (FCP). Mol Phys 104: 2627–2640
[3] Puzzarini C (2009) Extrapolation to the complete basis set limit of structural parameters: comparison of different approaches. J Phys Chem A 113: 14530–14535

CF_2	(C_{2v})			Difluoromethylene
Method	r_e(C–F)	∠(FCF)	Ref.	
Exp	1.2975	104.81	1	
SE[a]	1.2972	104.858	2	

[a] Rovibrational corrections obtained from a CCSD(T)_ae/cc-pwCVQZ force field
[1] Margulès L, Demaison J, Boggs JE (1999) Equilibrium C–F bond length and the structure of formyl fluoride, difluorocarbene, monofluoromethylene, and difluoromethane. J Phys Chem A 103: 7632–7638
[2] Demaison J, Margulès L, Martin JML, Boggs JE (2002) Anharmonic force field, structure, and thermochemistry of CF_2 and CCl_2. Phys Chem Chem Phys 4: 3282–3288

CF_2O	$F_2C{=}O$ (C_{2v})				Carbonyl difluoride
Method	r_e(C–F)	r_e(C=O)	∠(FCF)	Ref.	
Exp	1.311(2)	1.170(2)	107.6(8)	1	
SE[a]	1.3100(2)	1.1699(4)	107.80(3)	2	

[a] Rovibrational corrections from a MP2/cc-pVTZ force field

[1] Nakata M, Kohata K, Fukuyama T, Kuchitsu K, Wilkins CJ (1980) Molecular structure of carbonyl fluoride as studied by gas electron diffraction and microwave data. J Mol Struct 68: 271–280

[2] Demaison J, Császár AG (2012) Equilibrium CO bond lengths. J Mol Struct 1023: 7–14

CF_2S	$F_2C{=}S$ (C_{2v})				Thiocarbonyl difluoride
Method	r_e(C–F)	r_e(C–S)	∠(FCF)	Ref.	
Exp.[a]	1.3133	1.5836	106.89	1	
Exp.[a]	1.312(2)	1.585(4)	107.01(9)	2	
SE[b]	1.3093(2)	1.5900(3)	107.384(26)	3	
SE[b, c]	1.3084(1)	1.5913(4)	107.448(1)	3	
CCSD(T)_ae/ cc-pwCVQZ	1.3083	1.5910	107.449	3	

[a] r_e structure derived from r_z structure

[b] Rovibrational corrections from a B2PLYPD3/6-311 + G(3df) cubic force field

[c] Extrapolation method

[1] Xu Y, Gerry MCL, Joo DL, Clouthier DJ (1992) The microwave spectrum, spin-rotation coupling constants, and structure of thiocarbonyl fluoride, SCF_2, observed with a cavity microwave Fourier transform spectrometer. J Chem Phys 97: 3931–3939

[2] Bürger H, Demaison J, Dréan P, Jerzembeck W, Merke I, Stahl W (1998) High resolution infrared, microwave and millimeter wave spectra of $SeCF_2$. Structure determination assisted by ab initio calculations. Ber Bunsenges Phys Chem 102: 872–881

[3] Demaison J, Vogt N (2022) Effective and equilibrium structures of XY_n-type molecules: SCl_2, PF_2, PO_2, ZrO_2, and CF_2S as additional example and nonbonded distances F···F, Cl···Cl and O···O. J Mol Spectrosc 387: 111661

CF_2Se	$F_2C{=}Se$ (C_{2v})				Selenocarbonyl difluoride
Method	r_e(C–F)	r_e(C=Se)	∠(FCF)	Ref.	
Exp.[a]	1.309(2)	1.739(2)	107.74(4)	1	
Exp.[b]			107.5(4)	2	

[a] r_e structure derived from r_z structure

[b] r_z value grom gas-phase electron diffraction

[1] Bürger H, Demaison J, Dréan P, Jerzembeck W, Merke I, Stahl W (1998) High resolution infrared, microwave and millimeter wave spectra of $SeCF_2$. Structure determination assisted by ab initio calculations. Ber Bunsenges Phys Chem 102: 872–881

[2] Christen D, Oberhammer H, Zeil W, Haas A, Darmadi A (1980) The molecular structure of seleno- and thiocarbonyl difluoride. J Mol Struct 66: 203–210

CF_4	(T_d)		Tetrafluoromethane
Method	r_e(C–F)	Ref.	
Exp. (IR)	1.314860(21)	1	
Exp. (GED)	1.3151(4)	2	
AI[a]	1.31526	3	
CCSD(T)_ae/cc-pwCVQZ	1.3152	4	

[a] CCSD(T)_fc/cc-pVTZ adjusted

[1] Carlos M, Gruson O, Richard C, Boudon V, Rotger M, Thomas X, Maul C, Sydow C, Domanskaya A, Georges R, Soulard P, Pirali O, Goubet M, Asselin P, Huet TR (2017) High-resolution spectroscopy and global analysis of CF_4 rovibrational bands to model its atmospheric absorption. J Quant Spectrosc Radiat Transf 201: 75–93
See also Brodersen S (1991) The cubic potential function of CF_4 fitted directly to the experimental frequencies. J Mol Spectrosc 145: 331–351

[2] Seo SS, Ewbank JD, Ischenko AA (2005) Direct evaluation of equilibrium structure for tetrahedrally symmetric molecule using real-time gas electron diffraction. J Korean Chem Soc 49: 521–530

[3] Wang X-G, Sibert III EL, Martin JML (2000) Anharmonic force field and vibrational frequencies of tetrafluoromethane (CF_4) and tetrafluorosilane (SiF_4). J Chem Phys 112: 1353–1366

[4] Vogt N, Demaison J, Rudolph HD (2014) Accurate equilibrium structures of fluoro- and chloroderivatives of methane. Mol Phys 112: 2873–2883

CHClO	OCHCl (C_s)						Formyl chloride
Method	r_e(C–H)	r_e(C–Cl)	r_e(C=O)	∠(OCH)	∠(OCCl)	Ref.	
Exp.[a]	1.0897	1.765	1.182	126.49	123.07	1	
SE[b]	1.0915	1.7630(2)	1.1810(2)	126.56(4)	123.34(6)	2	

[a] Derived from the r_z structure

[b] Rovibrational corrections obtained from a MP2/TZ2Pf cubic force field

[1] Davis RW, Gerry MCL (1983) The microwave spectrum, harmonic force field, and structure of formyl chloride. J Mol Spectrosc 97: 117–138

[2] Demaison J, Boggs JE, Rudolph HD (2004) Ab initio anharmonic force field and ab initio and experimental equilibrium structures of formyl chloride. J Mol Struct 695–696: 145–153

$CHCl_3$	(C_{3v})					Trichloromethane
Method	r_e(C–H)	r_e(C–Cl)	∠(HCCl)	∠(ClCCl)	Ref.	
r_m^ρ	1.105(7)	1.755(2)	107.57(22)	111.30(20)	1	
r_e [a]	1.078(3)	1.761(2)	108.27(2)	110.65(2)	1	

(continued)

(continued)

$CHCl_3$	(C_{3v})					Trichloromethane
Method	r_e(C–H)	r_e(C–Cl)	∠(HCCl)	∠(ClCCl)	Ref.	
AI[b]	1.081(2)	1.757(3)	108.02	110.89	1	
SE[c]	1.0818(6)	1.7595(1)	108.080(16)	110.825(16)	2	

[a] Derived from r_z structure
[b] MP2 method with triple zeta basis sets and offset correction for the distances
[c] Rovibrational corrections from a MP2/TZ2Pf anharmonic force field
[1] Colmont J-M, Priem D, Dréan P, Demaison J, Boggs JE (1998) Rotational spectra of the isotopic species of chloroform: Experimental and ab initio structures. J Mol Spectrosc 191: 158–175
[2] Demaison J (2008) Equilibrium structure of chloroform. J Mol Spectrosc 251: 217–219

CHF_3	(C_{3v})				Trifluoromethane
Method	r_e(C–H)	r_e(C–F)	∠(FCF)	Ref.	
Exp	1.091(7)	1.3284(15)	108.58(17)	1	
SE[a]	1.0850	1.3304	108.41	2	
AI[b]	1.0860	1.3312	108.44	2	

[a] Rovibrational corrections obtained from a CCSD(T)_ae/TZ2Pf cubic force field
[b] CCSD(T)_fc/aug-cc-pVQZ + core correction from CCSD(T)/MT
[1] Kawashima Y, Cox AP (1978) Microwave *l*-type resonance transitions of the $v_6 = 1$ state in CHF_3 and CDF_3: Accidental degeneracy and molecular structure. J Mol Spectrosc 72: 423–429
[2] Breidung J, Cosléou J, Demaison J, Sarka K, Thiel W (2004) Ab initio anharmonic force field, molecular parameters, equilibrium structure and enthalpy of formation of fluoroform. Mol Phys 102: 1827–1841

CHN	HCN ($C_{\infty v}$)			Hydrogen cyanide
Method	r_e(C–H)	r_e(C≡N)	Ref.	
Exp. (MW + IR)	1.065112(48)	1.153283(11)	1	
SE[a]	1.06501(8)	1.15324(2)	2	
SE[b]	1.06528(12)	1.15336(14)	3	
CCSD(T)_ae/cc-pCVQZ	1.0655	1.1538	4	
AI[c]	1.06528	1.15301	5	
AI[d]	1.065188	1.153190	6	

[a] Rovibrational corrections obtained from a variational calculation of the vibration–rotation spectrum
[b] Rovibrational corrections obtained from a CCSD(T)_fc/cc-pVQZ force field
[c] CCSD(T)_fc/cc-pV6Z + core(cc-pCVQZ)
[d] CCSD(T)_fc/cc-pV∞Z + core(cc-pCVQZ) + corrections from: CCSDT/cc-pVTZ and CCSDTQ/cc-pVDZ
[1] Maki AG, Mellau GC, Klee S, Winnewisser M, Quapp W (2000) High-temperature infrared measurements in the region of the bending fundamental of $H^{12}C^{14}N$, $H^{12}C^{15}N$, and $H^{13}C^{14}N$. J Mol Spectrosc 202: 67–82
[2] Carter S, Mills IM, Handy NC (1997) The equilibrium structure of HCN. J Chem Phys 97: 1606–1607

[3] Pawlowski F, Jørgensen P, Olsen J, Hegelund F, Helgaker T, Gauss J, Bak KL, Stanton JF (2002). Molecular equilibrium structures from experimental rotational constants and calculated vibration–rotation interaction constants. J Chem Phys 116: 6482–6496
[4] Bak KL, Gauss J, Jørgensen P, Olsen J, Helgaker T, Stanton JF (2001) The accurate determination of molecular equilibrium structures. J Chem Phys 114: 6548–6556
[5] Heckert M, Kállay M, Gauss J (2005) Molecular equilibrium geometries based on coupled-cluster calculations including quadruple excitations. Mol Phys 103: 2109–2115
[6] Puzzarini C, Heckert M, Gauss J (2008) The accuracy of rotational constants predicted by high-level quantum chemical calculations. I. Molecules containing first-row atoms. J Chem Phys 128: 194108/1-9

CHN	HNC ($C_{\infty v}$)			Hydrogen isocyanide
Method	r_e(H–N)	r_e(N≡C)	Ref.	
Exp. (MW)	0.9960643(29)	1.1683506(16)	1	
SE[a]	0.99489(8)	1.16875(10)	2	
CCSD(T)_ae/cc-pCVQZ	0.9953	1.1693	3	
AI[b]	0.99532	1.16860	4	
AI[c]	0.995389	1.168693	5	

[a] Rovibrational corrections obtained from a CCSD(T)/cc-pVQZ force field
[b] CCSD(T)_fc/cc-pV6Z + core(cc-pCVQZ)
[c] CCSD(T)_fc/cc-pV∞Z + core(cc-pCVQZ) + corrections using CCSDT/cc-pVTZ and CCSDTQ/cc-pVDZ
[1] Okabayashi T, Tanimoto M (1993) The millimeter and submillimeter wave spectroscopy of HNC and DNC in the vibrationally excited states. J Chem Phys 99: 3268–3271
[2] Pawlowski F, Jørgensen P, Olsen J, Hegelund F, Helgaker T, Gauss J, Bak KL, Stanton JF (2002). Molecular equilibrium structures from experimental rotational constants and calculated vibration–rotation interaction constants. J Chem Phys 116: 6482–6496
[3] Bak KL, Gauss J, Jørgensen P, Olsen J, Helgaker T, Stanton JF (2001) The accurate determination of molecular equilibrium structures. J Chem Phys 114: 6548–6556
[4] Heckert M, Kállay M, Gauss J (2005) Molecular equilibrium geometries based on coupled-cluster calculations including quadruple excitations. Mol Phys 103: 2109–2115
[5] Puzzarini C, Heckert M, Gauss J (2008) The accuracy of rotational constants predicted by high-level quantum chemical calculations. I. Molecules containing first-row atoms. J Chem Phys 128: 194108/1-9

Rovibrational corrections obtained from

CHNO	HNCO (C_s)						Isocyanic acid
Method[a]	r_e(H–N)	r_e(N=C)	r_e(C=O)	∠(HNC)	∠(NCO)	Ref.	
SE	1.0059(4)	1.2147(4)	1.1644(4)	123.02(2)	171.51(10)	1	
CCSD(T)_ae/CVQZ	1.0032	1.2147	1.1642	123.27	172.39	2	

[a] Semirigid bender model
[1] Penocchio E, Mendolicchio M, Tasinato N, Barone V (2016). Structural features of the carbon-sulfur chemical bond: a semi-experimental perspective Can J Chem 94: 1065–1076
[2] Mladenović M, Elhiyani M, Lewerenz M (2009). Quasilinearity in tetratomic molecules: An ab initio study of the CHNO family. J Chem Phys: 130: 154109

CHNS	HNCS (C_s)						IsothIsothiocyanic acid
Method	r_e(H–N)	r_e(N=C)	r_e(C=S)	∠(NCS	∠(HNC)	Ref.	
Exp.[a]	1.00178(13)	1.20046(32)	1.57480(34)	174.399(12)	129.3525(52)	1	
SE[b]	1.0070(9)	1.2047(9)	1.5768(7)	172.23(18)	129.75(5)	2	
SE[c]	0.993(13)	1.205(17)	1.568(13)	187.3(26)	131.0(8)	3	

[a] Semirigid bender model
[b] Rovibrational corrections from a B2PLYP/cc-pVTZ anharmonic force field
[c] Rovibrational corrections from a CCSD(T)_fc/cc-pV(Q + d)Z anharmonic force field
[1] Ross SC, Niedenhoff M Yamada KMT (1994). Semirigid bender determination of the quasilinear bending potential energy function of HNCS from the ground state rotational spectra. J Mol Spectrosc 164: 432–444
[2] Penocchio E, Mendolicchio M, Tasinato N, Barone V (2016). Structural features of the carbon–sulfur chemical bond: a semi-experimental perspective. Can J Chem 94: 1065–1076
[3] McGuire BA, Martin-Drumel M-A, Thorwirth S, Brünken S, Lattanzi V, Neill JL, Spezzano S, Yu Z, P. Zaleski D, Remijan AJ, Pate BH, McCarthy MC (2016) Molecular polymorphism: microwave spectra, equilibrium structures, and an astronomical investigation of the HNCS isomeric family. Phys Chem Chem Phys 18: 22693–22705

CHO	HCO (C_s) Electronic ground state $^2A'$				Formyl
	r_e(C–H)	r_e(C=O)	∠(HCO)	Ref.	
Exp	1.1191(50)	1.1754(15)	124.43	1	
AI[a]	1.1182	1.1764	124.34	2	

[a] Ab initio potential energy surface calculated using the Multiconfiguration Time Dependent Hartree method
[1] Hirota E (1986) Third-order anharmonic potential constants and equilibrium structures of the formyl and hydroperoxyl radicals. J Mol Struct 146: 237–252
[2] Ndengué SA, Dawes R, Guo H.(2016). A new set of potential energy surfaces for HCO: Influence of Renner-Teller coupling on the bound and resonance vibrational states. J Chem Phys 144: 244301

CHO	HCO^+ ($C_{\infty v}$)			Oxomethylium
Method	r_e(C–H)	r_e(CO^+)	Ref.	
Exp	1.092038(4)	1.105575(1)	1	
MORBID	1.09780(35)	1.107035(61)	2	
SE[a]	1.0919(3)	1.1055(1)	3	
SE[b]	1.09198	1.10562	4	

[a] Mean value: Rovibrational corrections obtained from (i) a variational calculation and (ii) a scaled CASCF + MRCI force field
[b] Rovibrational corrections to the experimental rotational constants, $\Delta B = B_0 - B_e$, were computed variationally using the B_e corresponding to the minimum of the CCSD(T)/cc-pVQZ potential energy surface
[1] Dore L, Beninati S, Puzzarini C, Cazzoli G (2003) Study of vibrational interactions in DCO^+ by millimeter-wave spectroscopy and determination of the equilibrium structure of the formyl ion. J Chem Phys 118: 7857–7862
[2] Winterhoff G, Kempf SCG, Jensen P, Bunker PR (2018) Empirical potential energy surface and bending angle probability densities for the electronic ground state of HCO^+. J Mol Spectrosc 354: 71–82

[3] Puzzarini C, Tarroni R, Palmieri P, Carter S, Dore L (1996) Accurate ab initio prediction of the equilibrium geometry of HCO^+ and rovibration energy levels of DCO^+ Mol Phys 87 : 879–898
[4] Mladenović, M. (2017). Theoretical spectroscopic parameters for isotopic variants of HCO^+ and HOC^+. J of Chem Phys 147: 114111

CHP	HCP ($C_{\infty v}$)			Methylidynephosphine
Method	r_e(C–H)	r_e(C≡P)	Ref.	
Exp. (MW + IR)	1.0670	1.5402	1	
From isolated stretching frequency	1.070(1)		1	
SE[a]	1.0702(10)	1.5399(2)	2	
CCSD(T)_ae/cc-pCVQZ	1.070748	1.540530	3	

[a] Mean value: Rovibrational corrections obtained from (i) a variational calculation and (ii) a scaled CASCF + MRCI force field
[1] Dréan P, Demaison J, Poteau L, Denis J-M (1996) Rotational spectrum and structure of HCP. J Mol Spectrosc 176: 139–145
[2] Puzzarini C, Tarroni R, Palmieri P, Demaison J, Senent ML (1996) Rovibrational energy levels and equilibrium geometry of HCP. J Chem Phys 105: 3132–3141. See also Puzzarini C (2004) An improved determination of the equilibrium structure and molecular properties of XBS and XCP (X = H, F, Cl) molecules from ab initio calculations. Phys Chem Chem Phys 6: 344–351
[3] Coriani S, Marchesan D, Gauss J, Hättig C, Helgaker T, Jørgensen P (2005) The accuracy of ab initio molecular geometries for systems containing second-row atoms. J Chem Phys 123: 184107/1-12

CHS	HSC (C_s)				Mercaptomethylidyne
Method	r_e(H–S)	r_e(S–C)	∠(HSC	Ref.	
Exp.[a]	1.379(1)	1.6343(2)	104.2(1)	1	
AI[b]	1.367	1.656	102.7	2	

[a] Derived from the r_z structure
[b] CCSD(T)_ae/QZ2P
[1] Habara H, Yamamoto S (2000) Microwave spectrum and molecular structure of the HSC radical. J Chem Phys 112: 10905–10911
[2] Ochsenfeld C, Kaiser RI, Lee YT, Head-Gordon M (1999). Coupled-cluster ab initio investigation of singlet/triplet CH_2S isomers and the reaction of atomic carbon with hydrogen sulfide to HCS/HSC. J Chem Phys 110: 9982–9988

CH_2	(C_{2v}) Ground electronic state $\tilde{X}\,^3B_1$			Methylene
Method	r_e(C–H)	∠(HCH)	Ref.	
MORBID Hamiltonian	1.0753(1)	133.93(1)	1	
CCSD(T)_ae/cc-pCVQZ	1.0756	133.67	2	
AI [a]	1.07566	133.894	3	

[a] CCSD(T)_fc/cc-pV6Z + core(cc-pCVQZ) + corrections using CCSDT/cc-pVTZ and+ CCSDTQ/cc-pVDZ
[1] Jensen P, Bunker PR (1988) The potential surface and stretching frequencies of $\tilde{X}\ ^3B_1$ methylene (CH_2) determined from experiment using the Morse oscillator-rigid bender internal dynamics Hamiltonian. J Chem Phys 89: 1327–1332
[2] Peterson KA, Dunning TH Jr (1997) Benchmark calculations with correlated molecular wavefunctions. VIII. Bond energies and equilibrium geometries of the CH_n and C_2H_n ($n = 1$–4) series. J Chem Phys 106: 4119–4140
[3] Heckert M, Kállay M, Gauss J (2005) Molecular equilibrium geometries based on coupled-cluster calculations including quadruple excitations. Mol Phys 103: 2109–2115

CH_2Br_2	(C_{2v})					Dibromomethane
Method	r_e(C–H)	r_e(C–Br)	∠(HCH)	∠(BrCBr)	Ref.	
Exp.[a]	1.0845	1.9235	112.43	112.64	1	
Exp.[b]				112.527(9)	2	
AI[c]	1.0824	1.9305		112.90	2	

[a] Derived from the r_z structure
[b] r_z value
[c] MP2/6-31G**, the r_e(C–H) value is expected to be reliable
[1] Davis RW, Gerry MCL (1985) The microwave spectrum, centrifugal distortion constants, harmonic force field, and structure of dibromomethane. J Mol Spectrosc 109: 269–282
[2] Kisiel Z, Kraśnicki A, Pszczółkowski L, Shipman ST, Alvarez-Valtierra, L, Pate BH (2009) Assignment and analysis of the rotational spectrum of bromoform enabled by broadband FTMW spectroscopy. J Mol Spectrosc 257: 177–186

CH_2Cl_2	(C_{2v})					Dichloromethane
Method	r_e(C–H)	r_e(C–Cl)	∠(ClCCl)	∠(HCH)	Ref.	
Exp.[a]	1.080(3)	1.766(2)	111.96(10)	112.10(20)	1	
SE[b]	1.0816(2)	1.76425(3)	112.166(3)	111.714(4)	2	
CCSD(T)_ae/cc-pwCVQZ	1.0817	1.7643	112.391	111.590	2	

[a] Derived from the r_z structure
[b] Rovibrational corrections from a MP2/cc-pV(T + d)Z cubic force field
[1] Duncan JL (1987). The structure of the methylene chloride molecule. J Mol Struct 158: 169–177
[2] Vogt N, Demaison J, Rudolph HD (2014) Accurate equilibrium structures of fluoro- and chloroderivatives. Mol Phys 112: 2873–2883

CH_2F_2	(C_{2v})					Difluoromethane
Method	r_e(C–H)	r_e(C–F)	∠(FCF)	∠(HCH)	Ref.	
Exp	1.084(3)	1.3508(5)	108.49(6)	112.8(3)	1	
SE[a]	1.08703(3)	1.35323(1)	108.282(1)	113.442(9)	2	
CCSD(T)_ae/cc-pwCVQZ	1.0871	1.3518	108.480	113.127	2	

[a] Rovibrational corrections from a MP2/cc-pVTZ cubic force field
[1] Hirota E (1978) Anharmonic potential function and equilibrium structure of methylene fluoride. J Mol Spectrosc 71: 145–159
[2] Vogt N, Demaison J, Rudolph HD (2014) Accurate equilibrium structures of fluoro- and chloroderivatives. Mol Phys 112: 2873–2883

CH_2N^+	$HCNH^+$ ($C_{\infty v}$)				Methylidyneammonium
Method	r_e(C–H)	r_e(N–H)	r_e(C≡N)	Ref.	
Exp.[a]	1.0783	1.0099	1.1345	1	
SE[b]	1.0779(2)	1.0123(3)	1.1340(1)	2	
Potential[c]	1.0785	1.0127	1.1341	3	

[a] The α-constants of the isotopologues were obtained by scaling
[b] Rovibrational corrections from a CCSD(T)_fc/cc-pVQZ anharmonic force field
[c] Six-dimensional potential energy surface computed at the CCSD(T)-F12A_ae/cc-pCVTZ-F12
[1] Liu D-J, Lee S-T, Oka T (1988) The ν_3 fundamental band of $HCNH^+$ and the $2\nu_3 \leftarrow \nu_3$ and $\nu_2 + \nu_3 \leftarrow \nu_2$ hot bands of HCO^+. J Mol Spectrosc 128: 236–249
[2] Botschwina P, Heyl A, Horn M, Flügge J (1994). Calculated spectroscopic constants and the equilibrium geometry of $HCNH^+$. J Mol Spectrosc 163: 127–137
[3] Brites V, Jutier L (2012). New ab initio study of the spectroscopy of $HCNH^+$. J Mol Spectrosc 271: 25–32

CH_2O	(C_{2v})				Formaldehyde
Method	r_e(C–H)	r_e(C=O)	∠(HCH)	Ref.	
Exp	1.1005(20)	1.2033(10)	116.30(25)	1	
Exp	1.10046(16)	1.20461(19)	116.722(93)	2	
SE[a]	1.10072(60)	1.20465(70)	116.74(24)	3	
SE[b]	1.10064	1.20296	116.70	4	
CCSD(T)_ae/cc-pCVQZ	1.1008	1.2043	116.44	5	
AI[c]	1.100503	1.204541	116.6896	6	
CCSDTQ/CBS	1.10052	1.20457	116.694	7	

[a] Rovibrational corrections obtained from a CCSD(T)_fc/cc-pVQZ cubic force field
[b] Variational calculation
[c] CCSD(T)_fc/cc-pV∞Z + core(cc-pCVQZ) + corrections using CCSDT/cc-pVTZ and+ CCSDTQ/cc-pVDZ
[1] Duncan JL (1974) The ground-state average and equilibrium structures of formaldehyde and ethylene. Mol Phys 28: 1177–1191
[2] Lohilahti J (2006) Rotation-Vibration Spectroscopic Studies of Formaldehyde and Formic Acid Molecules. Ph.D. thesis, University of Oulu
[3] Pawlowski F, Jørgensen P, Olsen J, Hegelund F, Helgaker T, Gauss J, Bak, KL, Stanton JF (2002). Molecular equilibrium structures from experimental rotational constants and calculated vibration–rotation interaction constants. J Chem Phys 116: 6482–6496
[4] Carter S, Handy NC, Demaison J (1997) The rotational levels of the ground vibrational state of formaldehyde. Mol Phys 90: 729–738
[5] Bak KL, Gauss J, Jørgensen P, Olsen J, Helgaker T, Stanton JF (2001) The accurate determination of molecular equilibrium structures. J Chem Phys 114: 6548–6556

[6] Puzzarini C, Heckert M, Gauss J (2008) The accuracy of rotational constants predicted by high-level quantum chemical calculations. I. Molecules containing first-row atoms. J Chem Phys 128: 194108/1-9
[7] Morgan WJ, Matthews DA, Ringholm M, Agarwal J, Gong JZ, Ruud K, Allen WD, Stanton JF, Schaefer HF (2018) Geometric energy derivatives at the complete basis set limit: application to the equilibrium structure and molecular force field of formaldehyde. J Chem Theory Comput 14: 1333–1350

CH_2O_2	HCOOH *trans* conformer (C_s)			Formic acid
Method	Exp.[a]	SE[b]	AI[c]	
Ref.	1	2	2	
r_e(C–H)	1.091(5)	1.0920(1)	1.0921	
r_e(C=O)	1.201(5)	1.1980(1)	1.1974	
r_e(C–O)	1.340(5)	1.3406(1)	1.3405	
r_e(O–H)	0.969(5)	0.9662(1)	0.9660	
∠(HC=O)	123.3(15)	125.04(1)	125.07	
∠(OCO)	124.8(4)	124.83(1)	124.89	
∠(COH)	106.6(4)	106.97(1)	106.80	

[a] Derived from the r_z structure
[b] Rovibrational corrections obtained from a MP2/cc-pVTZ force field
[c] CCSD(T)_fc/cc-pVQZ + MP2/[(cc-pwCVQZ(ae) – cc-pwCVQZ(fc) + cc-pV5Z – cc-pVQZ]
[1] Davis RW, Robiette AG, Gerry MCL, Bjarnov E, Winnewisser G (1980) Microwave spectra and centrifugal distortion constants of formic acid containing ^{13}C and ^{18}O: Ref.inement of the harmonic force field and the molecular structure. J Mol Spectrosc 81: 93–109
[2] Demaison J, Herman M, Liévin J (2007) Anharmonic force field of *cis-* and *trans-*formic acid from high-level ab initio calculations, and analysis of resonance polyads. J Chem Phys 126: 164305/1-9

CH_2S	H_2C=S (C_{2v})				Thioformaldehyde
Method	r_e(C–H)	r_e(C=S)	∠(HCH)	Ref.	
Exp	1.0856(21)	1.6110(8)	116.30(25)	1	
SE[a]	1.0856	1.6110	116.24	2	
SE[b]	1.08531(2)	1.60909(1)	116.484(4)	3	
CCSD(T)_ae/cc-pCVQZ	1.085374	1.609677	116.294	4	

[a] Variational calculation
[b] Rovibrational corrections from a CCSD(T)_fc/cc-pV(Q + d)Z cubic force field
[1] Turner PH, Halonen L, Mills IM (1981) Fourier transform infrared spectra of H_2CS and D_2CS. J Mol Spectrosc 88: 402–419
[2] Carter S, Handy NC (1998) The geometry and force field of thioformaldehyde. J Mol Spectrosc 192: 263–267
[3] Müller HSP, Maeda A, Thorwirth S, Lewen F, Schlemmer S, Medvedev IR, Winnewisser M, De Lucia FC, Herbst E (2019) Laboratory spectroscopic study of isotopic thioformaldehyde, H_2CS, and determination of its equilibrium structure. Astron Astrophys 621: A143
[4] Coriani S, Marchesan D, Gauss J, Hättig C, Helgaker T, Jørgensen P (2005) The accuracy of ab initio molecular geometries for systems containing second-row atoms. J Chem Phys 123: 184107/1-12

CH_2Se	$H_2C{=}Se$ (C_{2v})				Selenoformaldeyde
Method	r_e(C–H)	r_e(C=Se)	∠(HCH)	Ref.	
Exp.[a]	1.085	1.751	117.04	1	
From isolated stretching frequency	1.0844(10)			2	

[a] r_e structure derived from r_z structure

[1] Bürger H, Demaison J, Dréan P, Jerzembeck W, Merke I, Stahl W (1998) High resolution infrared, microwave and millimeter wave spectra of $SeCF_2$. Structure determination assisted by ab initio calculations. Ber Bunsenges Phys Chem 102: 872–881

[2] Beckers H, Kim YS, Willner H (2008) Infrared spectrum of matrix isolated selenoformaldehyde, CH_2Se, and CD_2Se. Inorg Chem 47: 1693–1696

CH_3	(D_{3h}) Electronic ground state $X^2A''_2$		Methyl
Method	r_e(C–H)	Ref.	
Exp.[a]	1.0581(27)	1	
Exp.[b]	1.0767	2	
CCSD(T)_ae/ccpCVQZ	1.0762[c]	3	

[a] From an experimental potential energy curve

[b] Rovibrational correction from an approximate anharmonic force field

[c] The estimated accuracy is 0.0005 Å

[1] Špirko V, Bunker PR (1982). The potential function and rotation-vibration energy levels of the methyl radical CH_3. J Mol Spectrosc 95: 381–390

[2] Hirota E, Yamada C (1982) Intramolecular motions and molecular structure of the CH_3 radical. J Mol Spectrosc 96: 175–182

[3] Peterson KA, Dunning TH Jr (1997) Benchmark calculations with correlated molecular wave functions. VIII. Bond energies and equilibrium geometries of the CH_n and C_2H_n ($n = 1$–4) series. J Chem Phys 106: 4119–4140

CH_3Br	(C_{3v})				Bromomethane
Method	r_e(C–H)	r_e(C–Br)	∠(HCH)	Ref.	
Exp	1.086(3)	1.933(2)	111.17(42)	1	
Exp	1.0823	1.9340	111.157	2	

(continued)

(continued)

CH_3Br	(C_{3v})				Bromomethane
Method	r_e(C–H)	r_e(C–Br)	∠(HCH)	Ref.	
Exp	1.0825	1.9340	111.17	3	
AI[a]	1.08228	1.93218	111.155	4	

[a] CCSD(T)_fc/cc-pV∞Z + core(cc-pCVQZ) + corrections using CCSDT/cc-pVTZ and+ CCSDTQ/cc-pVDZ + relativistic correction

[1] Duncan JL (1970) The equilibrium geometries of the methyl halides. J Mol Struct 6: 447–456

[2] Graner G (1981) The methyl bromide molecule: A critical consideration of perturbations in spectra. J Mol Spectrosc 90: 394–438

[3] Demaison J, Margulès L, Boggs JE (2003) The equilibrium C–Cl, C–Br, and C–I bond lengths from ab initio calculations, microwave and infrared spectroscopies, and empirical correlations. Struct Chem 14: 159–174

[4] Puzzarini C, Cazzoli G, Baldacci A, Baldan A, Michauk C, Gauss J (2007) Rotational spectra of rare isotopic species of bromofluoromethane: Determination of the equilibrium structure from ab initio calculations and microwave spectroscopy. J Chem Phys 127: 164302/1-10

CH_3Cl	(C_{3v})				Chloromethane
Method	r_e(C–H)	r_e(C–Cl)	∠(HCH)	Ref.	
Exp	1.086(4)	1.778(2)	110.67(67)	1	
Exp	1.0872(50)	1.7756(20)	110.35(32)	2	
Exp	1.0842(2)	1.7768(2)	110.21(3)	3	
SE	1.083450(8)	1.777725(5)	110.5458(9)	4	
AI[a]	1.0834	1.7777	110.54	5	

[a] Potential energy surface computed by explicitly correlated coupled-cluster calculations with extrapolation to the CBS limit and taking also into account the contributions due to the CV correlation, higher-order (HO) electron correlation terms (up to the contributions due to the perturbative pentuples), the diagonal Born–Oppenheimer corrections (DBOC), and scalar relativistic (SR) effects

[1] Duncan JL (1970) The equilibrium geometries of the methyl halides. J Mol Struct 6: 447–456

[2] Imachi M, Tanaka T, Hirota E (1976). Microwave spectrum of methyl chloride in the excited vibrational states. J Mol Spectrosc 63: 265–280

[3] Demaison J, Margulès L, Boggs JE (2003) The equilibrium C–Cl, C–Br, and C–I bond lengths from ab initio calculations, microwave and infrared spectroscopies, and empirical correlations. Struct Chem 14: 159–174

[4] Pietropolli Charmet A, Stoppa P, De Lorenzi A, Melosso M, Achilli A, Dore L, Puzzarini C, Canè E, Filippo Tamassia F (2023) Computational, rotational and ro-vibrational experimental investigation of monodeuterated chloromethane. J Quant Spectr Radiat Tranfer, in press

[5] Owens A, Yurchenko SN, Yachmenev A, Tennyson J, Thiel W (2015) Accurate ab initio vibrational energies of methyl chloride. J Chem Phys 142: 244306

CH_3F	(C_{3v})					Fluoromethane
Method	r_e(C–H)	r_e(C–F)	∠(HCH)	∠(HCF)	Ref.	
Exp	1.095	1.382	110.45		1	
Exp.[a]	1.086(2)	1.383(1)	110.2(3)	108.8(3)	2	

(continued)

(continued)

CH_3F	(C_{3v})					Fluoromethane
Method	r_e(C–H)	r_e(C–F)	∠(HCH)	∠(HCF)	Ref.	
Exp	1.0870(5)	1.3827(4)	110.26(6)	108.67(6)	3	
AI[b]	1.0876	1.3827	110.19	108.74	3	

[a] Derived from the r_z structure
[b] CCSD(T)_fc/cc-pV(5,Q)Z corrected for core correlation and diffuse functions
[1] Duncan JL (1970) The equilibrium geometries of the methyl halides. J Mol Struct 6: 447–456
[2] Egawa T, Yamamoto S, Nakata M, Kuchitsu K (1987). Average and equilibrium structures of methyl flouride studied by electron diffraction. A joint analysis with rotational constants and cubic force constants. J Mol Struct 156: 213–228
[3] Demaison J, Breidung J, Thiel W, Papousek D (1999) The equilibrium structure of methyl fluoride. Struct Chem 10: 129–133

CH_3I	(C_{3v})				Iodomethane
Method	r_e(C–H)	r_e(C–I)	∠(HCH)	Ref.	
Exp	1.085(3)	2.133(2)	111.28(42)	1	
Exp	1.084(1)	2.132(1)	111.20(15)	2	
Exp	1.0819	2.1336	111.40	3	

[1] Duncan JL (1970) The equilibrium geometries of the methyl halides. J Mol Struct 6: 447–456
[2] Matsuura H, Overend J (1972). Equilibrium structure of methyl iodide. J Chem Phys 56: 5725–5727
[3] Demaison J, Margulès L, Boggs JE (2003) The equilibrium C–Cl, C–Br, and C–I bond lengths from ab initio calculations, microwave and infrared spectroscopies, and empirical correlations. Struct Chem 14: 159–174

CH_4	(T_d)		Methane
Method	r_e(C–H)	Ref.	
ED[a]	1.0856(17)	1	
IR (CH_4)	1.0859(5)	2	
IR (CD_4)	1.0864(5)	3	
IR	1.0858(10)	4	
SE[b]	1.0862(5)	5	
SE[c]	1.0859	6	
AI[d]	1.08580	6	
AI[e]	1.08605	7	
AI[f]	1.086153	8	

[a] Mean value for CH_4 and CD_4
[b] Corrected r_g value
[c] CCSD(T)_ae/cc-pCVTZ force field
[d] CCSD(T) complete basis set extrapolation
[e] CCSD(T)_fc/cc-pV6Z + core(cc-pCVQZ)

[f] CCSD(T)/cc-pV∞Z + corrections using CCSDT/cc-pVTZ and+ CCSDTQ/cc-pVDZ + core(cc-pCVQZ)

[1] Bartell LS, Kuchitsu K, DeNeui (1961) Mean and equilibrium molecular structures of methane and deuteromethane as determined by electron diffraction. J Chem Phys 35: 1211–1218
[2] Tarrago G, Dang-Nhu M, Poussigue G, Gulelachvili G, Amiot C (1975) The ground state of methane $^{12}CH_4$ through the forbidden lines of the ν_3 band. J Mol Spectrosc 57: 246–263
[3] Brodersen, S, Gray DL, Robiette AG (1977) Determination of *B*o from the v_3 Raman band of CD_4. Mol Phys 34: 617–628
[4] Gray DL, Robiette AG (1979) The anharmonic force field and equilibrium structure of methane. Mol Phys 37: 1901–1920
[5] Pulay P, Meyer W, Boggs JE (1978) Cubic force constants and equilibrium geometry of methane from Hartree–Fock and correlated wavefunctions. J Chem Phys 38: 5077–5085
[6] Stanton JF (1999) A refined estimate of the bond length of methane. Mol Phys 97: 841–845
[7] Heckert M, Kállay M, Gauss J (2005) Molecular equilibrium geometries based on coupled-cluster calculations including quadruple excitations. Mol Phys 103: 2109–2115
[8] Puzzarini C, Heckert M, Gauss J (2008) The accuracy of rotational constants predicted by high-level quantum chemical calculations. I. Molecules containing first-row atoms. J Chem Phys 128: 194108/1-9

CIN	ICN ($C_{\infty v}$)				Cyanogen iodide
	rMethod $_e$(C–I)	r_e(C≡N)	r_e(I···N)	Ref.	
Exp	1.99209(22)	1.16044(33)		1	
Exp.[a]	1.99241	1.15965		2	
Exp	1.992018(32)	1.160551(48)	3.152569	3	
AI[b]	1.9921	1.1602		4	

[a] Direct diagonalization of the Hamiltonian and fit to observed frequencies and rotational constants
[b] CCSD(T)_fc/SDB-cc-pVQZ + offset correction

[1] Cazzoli G, Esposti CD, Favero PG (1978) Equilibrium structure of cyanogen iodide. J Mol Struct 48: 1–8
[2] Whiffen DH (1978) The force fields of cyanogen halides. III. Cyanogen bromide and cyanogen iodine. Spectrochim Acta A: 34: 1183–1192
[3] Bizzocchi L, Degli Esposti C, Dore L (1998) An improved evaluation of the equilibrium structure of cyanogen iodide. J Mol Struct 443: 211–222
[4] Demaison J, Margulès L, Boggs JE (2003) The equilibrium C–Cl, C–Br, and C–I bond lengths from ab initio calculations, microwave and infrared spectroscopies, and empirical correlations. Struct Chem 14: 159–174

COPt	PtCO ($C_{\infty v}$)			Platinum monocarbonyl
Method	r_e(C–Pt)	r_e(C=O)	Ref.	
r_e [a]	1.76046	1.14354	1	
r_e [b]	1.7615	1.1438	This work	

[a] Derived from r_z structure
[b] Derived from r_0 structure, see Sect. 5.8

[1] Evans CJ, Gerry MCL (2001) Pure rotational spectrum and structure of platinum monocarbonyl, PtCO. J Phys Chem A 105: 9659–9663

COS	OCS ($C_{\infty v}$)				Carbonyl sulfide
Method	r_e(C=O)	r_e(C=S)	r_e(O···S)	Ref.	
Exp	1.15617(14)	1.56140(14)	2.717560(29)	1	
CCSD(T)_ae/ cc-pCVQZ	1.156183	1.562500		2	

[1] Lahaye J-G, Vendenhaute R, Fayt A (1987) CO_2 laser saturation Stark spectra and global rovibrational analysis of the main isotopic species of carbonyl sulfide ($OC^{34}S$, $O^{13}CS$, and ^{18}OCS). J Mol Spectrosc 123: 48–83

[2] Coriani S, Marchesan D, Gauss J, Hättig C, Helgaker T, Jørgensen P (2005) The accuracy of ab initio molecular geometries for systems containing second-row atoms. J Chem Phys 123: 184107/1-12

COSe	OCSe ($C_{\infty v}$)			Carbonyl selenide
Method	r_e(C=O)	r_e(C=Se)	Ref.	
Exp	1.15327(25)	1.70981(19)	1	
AI[a]	1.1536	1.7106	2	

[a] MRCI potential energy surface refined by a least-squares adjustment of spectroscopic data

[1] LeGuennec M, Wlodarczak G, Demaison J, Bürger H, Litz M, Willner H (1993) Millimeterwave and high-resolution infrared spectra of OCSe: determination of its structure. J Mol Spectrosc 157: 419–446

[2] Puzzarini C, de Lara-Castells MP, Tarroni R, Palmieri P, Demaison J (1999) Accurate ab initio prediction of the rovibrational energy levels and equilibrium geometry of carbonyl selenide (OCSe). Phys Chem Chem Phys 1: 3955–3960

CO_2	($D_{\infty h}$)		Carbon dioxide
Method	r_e(C=O)	Ref.	
Exp	1.15995	1	
SE[a]	1.16006(6)	2	
CCSD(T)_ae/cc-pCVQZ	1.1604	3	
AI[b]	1.160055	4	
AI[c]	1.16002	5	

[a] Rovibrational corrections obtained from a CCSD(T)_fc/cc-pVQZ force field

[b] CCSD(T)/cc-pV∞Z + core(cc-pCVQZ) + corrections using CCSDT/cc-pVTZ and+ CCSDTQ/cc-pVDZ

[c] Potential energy surface calculated up to CCSDTQ level

[1] Graner G, Rossetti C, Bailly D (1986) The carbon dioxide molecule. A test case for the r_0, r_e, and r_m structures. Mol Phys 58: 627–636

[2] Pawlowski F, Jørgensen P, Olsen J, Hegelund F, Helgaker T, Gauss J, Bak KL, Stanton JF (2002). Molecular equilibrium structures from experimental rotational constants and calculated vibration–rotation interaction constants. J Chem Phys 116: 6482–6496

[3] Bak KL, Gauss J, Jørgensen P, Olsen J, Helgaker T, Stanton JF (2001) The accurate determination of molecular equilibrium structures. J Chem Phys 114: 6548–6556

[4] Puzzarini C, Heckert M, Gauss J (2008) The accuracy of rotational constants predicted by high-level quantum chemical calculations. I. Molecules containing first-row atoms. J Chem Phys 128: 194108/1-9

[5] Sebald P, Stein C, Oswald R, Botschwina P (2013) Rovibrational States of N_3^- and CO_2 Up to High *J*: A Theoretical Study Beyond fc-CCSD(T). J Phys Chem A 117: 13806–13814
See also: Teffo J-L, Ogilvie JF (1993), Born–Oppenheimer effects and vibration-rotational spectra of linear molecules II. Application to CO_2. Mol Phys 80: 1507–1524

CS_2	($D_{\infty h}$)		Carbon disulfide
Method	r_e(C=S)	Ref.	
Exp.[a]	1.55252(6)	1, 2	
SE[b]	1.55448(20)	3	
CCSD(T)_ae/cc-pCVQZ	1.553302	4	

[a] From the experimental parameters of Ref.s. (1, 2)
[b] Extrapolation
[1] Walrand J, Humblet V, Blanquet G (1988). Diode-laser spectra of isotopic carbon disulfide $^{13}CS_2$ in the region of 12 μm. J Mol Spectrosc 127: 304–323
[2] Winther F, Heyne U, Guarneri A (1988) The infrared spectrum of CS_2 in the ν_3 band region. Z Naturforsch 43a: 215–218
[3] Kummli D, Frey HM, Leutwyler S (2006) Femtosecond degenerate four-wave mixing of carbon disulfide: High-accuracy rotational constants. J Chem Phys 124: 144307/1-7
[4] Coriani S, Marchesan D, Gauss J, Hättig C, Helgaker T, Jørgensen P (2005) The accuracy of ab initio molecular geometries for systems containing second-row atoms. J Chem Phys 123: 184107/1-12

CSe_2	($D_{\infty h}$)		Carbon diselenide
Method	r_e(C=Se)	Ref.	
Exp	1.692169(3)	1	
Exp.[a]	1.6911(13)	2	

[a] r_0 structure with offset correction
[1] Bürger H, Willner H (1988) The high-resolution Fourier transform spectrum of monoisotopic $C^{80}Se_2$ in the v_2, and $v_1 + v_3$ regions: equilibrium, ground, and excited state parameters determined by rovibrational analyses. J Mol Spectrosc 128: 221–235
[2] Demaison J, Vogt N (2022) Effective and equilibrium structures of XY_n-type molecules: SCl_2, PF_2, PO_2, ZrO_2, and CF_2S as additional example and nonbonded distances F···F, Cl···Cl and O···O. J Mol Spectrosc 387: 111661

C_2F_2	FC≡CF ($D_{\infty h}$)			Difluoroethyne
Method	r_e(C–F)	r_e(C≡C)	Ref.	
Exp	1.2767(15)	1.1973(30)	1	
CCSD(T)_fc/TZ2Pf	1.2894	1.1909	2	
AI/Exp.[a]	1.2832	1.1865	3	

[a] MP2/TZ2Pf structure corrected to reproduce the experimental equilibrium rotational constant
[1] Bürger H, Senzlober M, Sommer S (1994) Ground state and equilibrium structure of FCCF from high-resolution IR spectra of $F^{12}C^{13}CF$. J Mol Spectrosc 164: 570–573
[2] Breidung J, Schneider W, Thiel W, Lee TJ (1992). The vibrational frequencies of difluoroethyne. J Chem Phys 97: 3498–3499

[3] Bürger H, Schneider, W, Sommer S, Thiel W, Willner H. (1991) The vibrational spectrum and rotational constants of difluoroethyne FC≡CF. Matrix and high resolution infrared studies and ab initio calculations. J Chem Phys 95: 5660–5669

C_2HF	HC≡CF ($C_{\infty v}$)				Fluoroethyne
	r_e(C–H)	r_e(C≡C)	r_e(C–F)	Ref.	
Exp	1.0603(15)	1.1962(8)	1.2764(8)	1	
SE[a]	1.0591(5)	1.1961(2)	1.2765(2)	2	
AI[b]	1.0586	1.1969	1.2762	2	

[a] From the CCSD(T)_fc/cc-pVTZ cubic force field
[b] CCSD(T)_ae/cc-pVQZ
[1] Borro AF, Mills IM, Mose A (1995) Overtone spectra and anharmonic resonances in haloacetylenes. Chem Phys 190: 363–371
[2] Botschwina P, Oswald M, Flügge J, Heyl Ä, Oswald R (1993) Chem Phys Lett 209: 117–125

C_2H_2	HC≡CH ($D_{\infty h}$)			Ethyne
Method	r_e(C–H)	r_e(C≡C)	Ref.	
Exp	1.06167(14)	1.202866(72)	1	
SE[a]	1.06176(2)	1.20289(1)	2	
CCSD(T)_ae/cc-pwCVQZ	1.06204	1.20343	2	
Variational	1.0618	1.2028	3	
AI[b]	1.061663	1.203038	4	
AI[c]	1.06157	1.20268	2	

[a] Rovibrational corrections obtained from a CCSD(T)_ae/cc-pwCVQZ force field
[b] CCSD(T)_fc/cc-pV∞Z + core(cc-pCVQZ) + corrections using CCSDT/cc-pVTZ and+ CCSDTQ/cc-pVDZ
[c] CCSD(T)_fc/cc-pV∞Z + core(cc-pwCV5Z) + corrections using CCSDT/cc-pVTZ and+ CCSDTQ/cc-pVDZ + relativistic correction + diagonal Born–Oppenheimer correction
[1] Tamassia F, Cané E, Fusina L, Di Lonardo G (2016) The experimental equilibrium structure of acetylene. Phys Chem Chem Phys 18: 1937–1944
[2] Liévin J, Demaison J, Herman M, Fayt A, Puzzarini C (2011) Comparison of the experimental, semi-experimental and ab initio equilibrium structures of acetylene: Influence of relativisitic effects and of the diagonal Born–Oppenheimer corrections. J Chem Phys 124: 064119/1-8
[3] Carter S, Handy NC (2002) The geometry and force field of acetylene. Mol Phys 100: 681–698
[4] Puzzarini C, Heckert M, Gauss J (2008) The accuracy of rotational constants predicted by high-level quantum chemical calculations. I. Molecules containing first-row atoms. J Chem Phys 128: 194108/1-9

$C_2H_2Cl_2$	CH_2=CCl_2 (C_{2v})						1,1-Dichloroethene
Method	r_e(C–H)	r_e(C=C)	r_e(C–Cl)	∠(HCH)	∠(ClCCl)	Ref.	
Exp	1.079(12)	1.324(5)	1.721(3)	120.5(8)	114.0(3)	1	
Exp.[a]				121.4(7)	114.1(3)	2	
Exp.[b]	1.0793(11)					3	

[a] r_z values
[b] From isolated stretching frequency using ν(CH) = 2988.9 cm^{-1}
[1] Nakata M, Kuchitsu K (1982) Molecular structure of 1,1-dichloroethylene as studied by gas electron diffraction and microwave data. Estimation of equilibrium structure by an extended anharmonic model analysis. J Mol Struct 95: 205–214
[2] Davis RW, Gerry MCL (1983) The Microwave Spectrum, Harmonic Force Field, and Ground State Average Structure of 1,1 -Dichloroethylene. J Mol Spectrosc 102: 297–309
[3] Demaison J, Rudolph HD (2008) Ab initio anharmonic force field and equilibrium structure of propene. J Mol Spectrosc 248: 66–76

c-$C_2H_2N_2O$	(C_{2v})			1,2,5-oxadiazole
	Exp.[a]	Exp.[b]	AI[c]	
Ref.	1	2	1	
r_e(O–N)	1.3677(30)		1.3661	
r_e(N=C)	1.3044(20)		1.3053	
r_e(C–C)	1.4183(30)		1.4167	
r_e(C–H)	1.0729(10)	1.0743(11)	1.0749	
∠(NON')	111.30(7)		111.30	
∠(ONC)	105.45(6)		105.58	
∠(NCC)	108.90(5)		108.77	
∠(NCH)	120.88(8)		120.86	
∠(CCH)	130.22(3)		130.36	

[a] r_e extrapolated from r_z
[b] From isolated stretching frequency using ν(CH) = 3151 cm^{-1} from Ref. 3
[c] CCSD(T)_ae/cc-pVQZ + MP2(fc)/aug-cc-pVQZ – MP2(fc)/cc-pVQZ + correction for core correlation
[1] Vázquez J, Demaison J, López-González JJ, Boggs JE, Rudolph HD (2001) Experimental and ab initio equilibrium structure and harmonic force field of 1,2,5-oxadiazole. J Mol Spectrosc 207: 224–237
[2] Demaison J, Rudolph HD (2008) Ab initio anharmonic force field and equilibrium structure of propene. J Mol Spectrosc 248: 66–76
[3] Christensen DH, Jensen PW, Nielsen JT, Nielsen OF (1973) Vibrational spectra and fundamental frequencies of 1,2,5-oxadiazole and its deuterated species. Spectrochim Acta A 29: 1393–1405

C_2H_3N	CH_3NC (C_{3v})					Methyl isocyanide
Method	r_e(N≡C)	r_e(C–N)	r_e(C–H)	∠(HCN)	Ref.	
Exp.[a]	1.1666(6)	1.4259(3)	1.0902(20)	109.47(15)	1	
Exp.[b]				109.34	2	
Exp.[c]			1.0862(11)		3	
AI[d]	1.1693	1.4217	1.0862	109.47	1	

[a] r_e extrapolated from r_z
[b] r_z value for ∠(DCN).
[c] From isolated stretching frequency using $\nu(CH) = 2988.9\ cm^{-1}$ from Ref. 4
[d] CCSD(T)_fc/aug-cc-pV∞Z + MP2(ae)/cc-pwCVQZ – MP2(fc)/cc-pwCVQZ

[1] Margulès L, Demaison J, Rudolph HD (2001) Ab initio and experimental structures of CH_3NC. J Mol Struct 599: 23–30
[2] Halonen L, Mills IM (1978) Microwave spectrum of CHD_2CN and CHD_2NC, and the harmonic force field and r_z structure of methyl cyanide and isocyanide. J Mol Spectrosc 73: 494–502
[3] Demaison J, Rudolph HD (2008) Ab initio anharmonic force field and equilibrium structure of propene. J Mol Spectrosc 248: 66–76
[4] McKean DC, Duncan JL, Mackenzie MW (1977) The infrared spectrum of CHD2NC and the ground state structure of methyl isocyanide. J Mol Struct 42: 77–83

C_2H_4	$CH_2{=}CH_2$ (D_{2h})				Ethene
Method	r_e(C–H)	r_e(C=C)	∠(HCH)	Ref.	
Exp	1.081(2)	1.334(2)	117.37(33)	1	
SE[a]	1.0809(3)	1.3307(3)	117.12(6)	2, 3	
CCSD(T)_ae/cc_pCVQZ	1.0809	1.3312	117.10	4	

[a] Rovibrational corrections obtained from a CCSD(T)_fc/cc-pVTZ cubic force field[3]

[1] Duncan JL (1974) The ground-state average and equilibrium structures of formaldehyde and ethylene. Mol Phys 28: 1177–1191
[2] Martin JML, Taylor PR (1996) The geometry, vibrational frequencies, and total atomization energy of ethylene. A calibration study. Chem Phys Lett 248: 336–344
[3] Martin JML, Lee TJ, Taylor PR, François J-P (1995) The anharmonic force field of ethylene, C_2H_4, by means of accurate ab initio calculations. J Chem Phys 103: 2589–2602
[4] Bak KL, Gauss J, Jørgensen P, Olsen J, Helgaker T, Stanton JF (2001) The accurate determination of molecular equilibrium structures. J Chem Phys 114: 6548–6556

C_2H_6	CH_3CH_3 (D_{3h})				Ethane
	r_e(C–H)	r_e(C–C)	∠(CCH)	Ref.	
Exp.[a]		1.522(2)		1	
Exp.[b]	1.0866			2	
Exp.[c]			111.17(1)	3	

(continued)

(continued)

C_2H_6	CH_3CH_3 (D_{3h})				Ethane
	r_e(C–H)	r_e(C–C)	∠(CCH)	Ref.	
CCSD(T)_ae/wCVQZ	1.0892	1.5229	111.23	4	
AI[d]	1.0887(10)	1.5221(20)	111.22(10)	4	

[a] Estimation using the empirical r_m^ρ method
[b] From isolated stretching frequency
[c] r_z value
[d] CCSD(T)_fc/cc-pV∞Z + CCSD(T)_ae/cc-pwCVQZ – CCSD(T)_fc/cc-pwCVQZ

[1] Harmony MD (1990) The equilibrium carbon–carbon single-bond length in ethane. J Chem Phys 93: 7522–7523
[2] Demaison J, Rudolph HD (2008) Ab initio anharmonic force field and equilibrium structure of propene. J Mol Spectrosc 248: 66–76
[3] Hirota E, Endo Y, Saito S, Duncan JL (1981) Microwave spectra of deuterated ethanes: Internal rotation potential function and r_z structure. J Mol Spectrosc 89: 285–295
[4] Puzzarini C, Taylor PR (2004) An ab initio study of the structure, torsional potential energy function and electric properties of disilane, ethane, and their deuterated isotopomers. J Chem Phys 122: 054315/1-10

C_3	($D_{\infty h}$)		1,2-Propadiene-1,3-diylidene
Method	r_e(C=C)	Ref.	
Exp. r_m [a]	1.2940(9)	1	
MORBID	1.30906(54)	2	
SE[b]	1.29407(10)	3	
AI[c]	1.29397	3	

[a] Bent-structure model
[b] Variational calculation using a potential energy surface[c]
[c] CCSD(T)-F12b_fc/aug-cc-pV5Z + core correlation (CCSD(T)/aug-cc-pV5Z) + scalar relativity correction + CCSDTQP correction

[1] Breier AA, Giesen TF, Ross SC, Yamada KMT (2020) Improved bond length determination technique for C_3 and other linear molecules with a large amplitude bending vibration. J Mol Struct 1219: 128329
[2] Jensen P, Rohlfing CM, Almlöf J (1992) Calculation of the complete active space self-consistent field potential energy surface, the dipole moment surfaces, the rotation–vibration energies, and the vibrational transition moments for C_3 ($\tilde{X}^1\Sigma^+$ g). J Chem Phys 97: 3399–3411
[3] Schröder B, Sebald P (2016) High-level theoretical rovibrational spectroscopy beyond fc-CCSD(T): The C_3 molecule. J Chem Phys 144:044307

C_3H_4	CH_2=C=CH_2 (D_{2d})				Allene
Method	r_e(C–H)	r_e(C=C)	∠(HCH)	Ref.	
Exp.[a]	1.076(3)	1.3082(10)	118.2(5)	1	
SE[b]	1.0808	1.3069	118.35	2	

(continued)

(continued)

C_3H_4	$CH_2{=}C{=}CH_2$ (D_{2d})				Allene
Method	r_e(C–H)	r_e(C=C)	∠(HCH)	Ref.	
CCSD(T)_ae/cc-pCVQZ	1.0810	1.3074	118.21	2	

[a] Derived from r_g structure
[b] Rovibrational corrections derived from a SDQ-MP4/cc-pVTZ cubic force field
[1] Ohshima Y, Yamamoto S, Nakata M, Kuchitsu K (1987) Geometrical structure of allene studied by a joint analysis of electron-diffraction and spectroscopic data. J Phys Chem 91: 4696–4700
[2] Auer AA, Gauss J (2001) Equilibrium structure and fundamental frequencies of allene. Phys Chem Chem Phys 3: 3001–3005

C_3H_6	c-C_3H_6 (D_{3h})				Cyclopropane
Method	r_e(C–H)	r_e(C–C)	∠(HCH)	Ref.	
Exp.[a]	1.083(5)	1.501(4)	114.5(9)	1	
SE[b]	1.0786(10)	1.5030(10)	114.97(10)	2	
SE[c]	1.0775(5)	1.5034(3)	115.09(10)	3	
CCSD(T)_ae/cc-pVQZ	1.0781	1.5019	114.81	2	

[a] Derived from r_g structure
[b] Rovibrational corrections derived from a SDQ-MP4/cc-pVTZ cubic force field
[c] Extrapolation method
[1] Yamamoto S, Nakata M, Fukuyama T, Kuchitsu K (1985) Geometrical structure of cyclopropane as studied by gas electron diffraction and spectroscopic data. J Phys Chem 89: 3298–3302
[2] Gauss J, Cremer D, Stanton JF (2000) The r_e structure of cyclopropane. J Phys Chem A 104: 1319–1324
[3] Kummli DS, Frey HM, Keller M, Leutwyler S (2005) Femtosecond degenerate four-wave mixing of cyclopropane. J Chem Phys 123: 054308

C_6H_6	c-C_6H_6 (D_{6h})			Benzene
	r_e(C–C)	r_e(C–H)	Ref.	
Exp	1.3902(2)	1.0862(15)	1	
SE[a]	1.3914(10)	1.0802(20)	2	
CCSD(T)_ae/cc-pwCVQZ	1.3916	1.0811	3	

[a] Rovibrational corrections derived from a SDQ-MP4/cc-pVTZ cubic force field
[1] Pliva J, Johns JWC, Goodman L (1991) Infrared bands of isotopic benzenes: ν_{13} and ν_{14} of $^{13}C_6D_6$. J Mol Spectrosc 148: 427–435
[2] Gauss J, Stanton JF (2000) The equilibrium structure of benzene. J Phys Chem A 104: 2865–2868
[3] Demaison J, Rudolph HD, Császár AG (2013) Deformation of the benzene ring upon fluorination: equilibrium structures of all fluorobenzenes. Mol Phys 111: 1539–1562

ClFKr	Kr·ClF ($C_{\infty v}$)	Krypton – chlorine fluoride (1/1)

The equilibrium structure is assumed to be linear
r_e(Kr···Cl) = 3.356 obtained by fitting the data to a one-dimensional Lennard-Jones potential
r_e(Kr···Cl) = 3.3136 from an estimate of B_e

Novick SE, Harris SJ, Janda KC, Klemperer W (1975). Structure and bonding of KrClF: intermolecular force fields in van der Waals molecules. Can J Phys 53: 2007–2015
Klemperer W (1977) Rotational spectroscopy of van der Waals molecules. Farad Discuss Chem Soc 62: 179–184
Buxton LW, Campbell EJ, Flygare WH (1981)^{83}Kr nuclear quadrupole coupling in KrClF. Chem Phys 59: 55–59

$ClFO_2$	$FClO_2$ (C_s)						Chloryl fluoride
Method	r_e(Cl–O)	r_e(Cl–F)	∠(OClO)	∠(OClF)	r_e(O···O)	Ref.	
Exp.[a]	1.4199(20)	1.6907(20)	115.03(10)	101.69(10)	2.395	1, 2	
Exp.[b]					2.3948(6)	3	

[a] Derived from the r_z structure
[b] r_0 value + offset correction
[1] Robiette AG, Parent CR, Gerry MCL (1981) The harmonic force field and structure of chloryl fluoride. J Mol Spectrosc 86: 455–464
[2] Müller HSP (2000) Rotational and vibrational spectra of chloryl fluoride, $FClO_2$: nuclear quadrupole, spin-rotation, and shielding tensors and the harmonic force field. J Mol Struct 517–518: 335–350
[3] Demaison J, Vogt N (2022) Effective and equilibrium structures of XY_n-type molecules: SCl_2, PF_2, PO_2, ZrO_2, and CF_2S as additional example and nonbonded distances F···F, Cl···Cl and O···O. J Mol Spectrosc 387: 111661

$ClFO_2S$	SO_2ClF (C_s)			Sulfuryl chloride fluoride
Method	Exp.[a]	Exp.[b]	Exp.[c]	
Ref.	1	2	3	
r_e(S=O)	1.40543		1.406(3)	
r_e(S–Cl)	1.98571			
r_e(S–F)	1.54895			
∠(OSO)	124.039			
∠(ClSO)	108.979			
∠(FSO)	106.961			
r_e(O···O)	2.482	2.480(4)		

[a] Derived from r_z structure
[b] r_0 value with offset correction
[c] Using S=O isolated stretching frequency
[1] Müller HSP, Gerry MCL (1994) Microwave spectrum of sulfuryl chloride fluoride, SO_2ClF: structure, hyperfine constants and harmonic force field. J Chem Soc Faraday Trans 90: 2601–2610
[2] Demaison J, Vogt N (2022) Effective and equilibrium structures of XY_n-type molecules: SCl_2, PF_2, PO_2, ZrO_2, and CF_2S as additional example and nonbonded distances F···F, Cl···Cl and O···O. J Mol Spectrosc 387: 111661

[3] Brunvoll J, Hargittai I (1978) On the correlation of geometric and vibrational parameters of the SO_2 groups in sulfone molecules. Acta Chim Acad Sci Hung 96: 337–354

$ClFO_3$	$FClO_3$ (C_{3v})				Perchloryl fluoride
Method	r_e(Cl–O)	r_e(Cl–F)	∠(OClO)	Ref.	
Exp.[a]	1.3967	1.6105	115.80	1	
Exp	1.397(3)	1.610(5)	115.7(4)	2	

[a] Derived from the r_z structure
[1] Müller HSP, Gerry MCL (1996) Microwave Fourier transform spectroscopy of perchloryl fluoride: $^{19}F^{35/37}Cl^{16}O_3$ and $^{19}F^{35/37}Cl^{16}O_2{}^{18}O$. J Mol Spectrosc 175: 120–132
[2] Cané E, Fusina L, Burczyk K (2008) The ν_4 and $\nu_2 + \nu_5$ high resolution infrared bands of $F^{35}Cl^{18}O_3$ and $F^{37}Cl^{18}O_3$. J Mol Spectrosc 247: 57–63

ClFS	ClSF (C_s)				Sulfur chloride fluoride
Method	r_e(S–F)	r_e(S–Cl)	∠(ClSF)	Ref.	
Exp.[a]	1.60604	1.99401	100.661	1	
SE[b]	1.6040(1)	1.9933(1)	100.606(4)	This work	

[a] Derived from the r_z structure
[b] B3LYP/6-311+G(3df) cubic force field and rotational constants of Ref. (1)
[1] Preusser J, Gerry MCL (1997) Fourier-transform microwave spectrum, structure, harmonic force field, and hyperfine constants of sulfur chloride fluoride. J Chem Phys 106: 10037–10047

ClF_3	(C_{2v})				Chlorine trifluoride
Method	r_e(Cl–F_{eq})	r_e(Cl–F_{ax})	∠(F_{ax}ClF_{eq})	Ref.	
Exp.[a]	1.5967	1.69664	87.45	1	
AI[b]	1.605	1.703	87.1	2	

[a] Derived from the r_z structure
[b] SR-CCSD(T)/TZVP (SR = scalar relativistic)
[1] Müller HSP (2001) The rotational spectrum of chlorine trifluoride, ClF_3. Centrifugal distortion analysis, Cl nuclear magnetic shielding tensor, structure, and the harmonic force field. Phys Chem Chem Phys 3: 1570–1575
[2] Sergentu D-C, Amaouch M, Pilmé J, Galland N, Maurice R (2015) Electronic structures and geometries of the XF_3 (X = Cl, Br, I, At) fluorides. J Chem Phys 143: 114306

$ClGeH_3$	GeH_3Cl (C_{3v})					Germyl chloride
$FGeH_3$	GeH_3F (C_{3v})					Germyl fluoride
Molecule	Method	r_e(Ge–H)	r_e(Ge–X)[a]	∠(HGeX)[a]	Ref.	
GeH_3Cl	Exp	1.51559(63)	2.14470(17)	107.107(78)	1	

(continued)

(continued)

ClGeH$_3$	GeH$_3$Cl (C_{3v})					Germyl chloride
FGeH$_3$	GeH$_3$F (C_{3v})					Germyl fluoride
Molecule	Method	r_e(Ge–H)	r_e(Ge–X)[a]	∠(HGeX)[a]	Ref.	
GeH$_3$F	Exp	1.51453(13)	1.730945(40)	106.071(17)	2	

[a] $X = \text{Cl}$ or F

[1] Demaison J, Wlodarczak G, Burie J, Bürger H (1990) The rotational spectrum of germyl chloride: determination and comparison of the substitution, double substitution, and equilibrium structures. J Mol Spectrosc 140: 322–339

[2] LeGuennec M, Chen W, Wlodarczak G, Demaison J, Eujen R, Bürger H (1991) The millimeter-wave spectrum of germyl fluoride: determination and comparison of the effective, substitution, and equilibrium structures. J Mol Spectrosc 150: 493–510

ClHKr	Kr·HCl ($C_{\infty v}$)			Krypton – hydrogen chloride (1/1)
Method	r_e(Kr· · · Cl)	R_e [a]	Ref.	
[b]	4.1525		1	
[c]		4.0589(3)	2	

[a] Distance between the inert gas atom and the center of mass of the diatom

[b] 6/12 Lennard–Jones potential by combining bending, stretching and their coupling

[c] The potential is calculated at $\theta = 0°, 60°, 120°$ and $180°$ using a multipole expansion with Legendre polynomials, and interpolated between these angles using $P_0(\cos\theta)$ to $P_3(\cos\theta)$ at each R. θ is the angle at the diatom center of mass

[1] Keenan MR, Campbell EJ, Balle TJ, Buxton LW, Minton TK, Soper PD, Flygare WH (1980). Rotational spectra and molecular structures of ArHBr and KrHBr. J Chem Phys 72: 3070–3080

[2] Hutson JM, Barton AE, Langridge-Smith PRR, Howard BJ (1980) Intermolecular potential energy surfaces for Kr·Hal and Ar·HBr (1980) Chem Phys Lett 73: 218–223

ClHO	HOCl (C_s)				Hypochlorous acid
Method	r_e(O–H)	r_e(O–Cl)	∠(HOCl)	Ref.	
Exp	0.96437(13)	1.68897(2)	102.965(19)	1	
CCSD(T)_ae/ cc-pCVQZ	0.963125	1.690152	102.770	2	

[1] Escribano RM, Di Lonardo G, Fusina L (1996) Empirical anharmonic force field and equilibrium structure of hypochlorous acid, HOCl. Chem Phys Lett 259: 614–618

[2] Coriani S, Marchesan D, Gauss J, Hättig C, Helgaker T, Jørgensen P (2005) The accuracy of ab initio molecular geometries for systems containing second-row atoms. J Chem Phys 123: 184107/ 1-12

ClHSi	HSiCl (C_s)				Chlorosilylene
Method	r_e(Si–Cl	r_e(Si–H)	∠(HSiCl)	Ref.	
Exp	2.067(3)	1.525(5)	96.9(5)	1	
SE[a]	2.0724	1.5140	94.66	2	

(continued)

(continued)

ClHSi	HSiCl (C_s)				Chlorosilylene
Method	r_e(Si–Cl	r_e(Si–H)	∠(HSiCl)	Ref.	
CCSD(T)_ae/cc-pCVQZ	2.07122	1.51469	95.303	3	

[a] Rovibrational corrections obtained from a CCSD(T)cc-pCVTZ cubic force field.
[1] Harper WW, Clouthier DJ (1997) Reinvestigation of the HSiCl electronic spectrum: Experimental reevaluation of the geometry, rotational constants, and vibrational frequencies. J Chem Phys 106: 9461–9473
[2] Vázquez J, Stanton JF (2002) Theoretical investigation of the structure and vibrational spectrum of the electronic ground state $\tilde{X}(^1A')$ of HSiCl. J Phys Chem A 106: 4429–4434
[3] Coriani S, Marchesan D, Gauss J, Hättig C, Helgaker T, Jørgensen P (2005) The accuracy of ab initio molecular geometries for systems containing second-row atoms. J Chem Phys 123: 184107/1-12

ClHXe	Xe·HCl ($C_{\infty v}$)			Xenon–hydrogen chloride (1/1)
Method	r_e(Xe···Cl)	R [a]	Ref.	
[b]	4.2767		1	
[c]		4.274(22)	2	

[a] Distance between the inert gas atom and the center of mass of the diatom
[b] 6/12 Lennard-Jones potential by combining bending, stretching and their coupling
[c] From an experimental potential surface using the Maitland-Smith form, see Sect. 6.6.8
[1] Keenan MR, Buxton LW, Campbell EJ, Balle TJ, Flygare WH (1980). ^{131}Xe nuclear quadrupole coupling and the rotational spectra of XeHCl. J Chem Phys 73: 3523–3529
[2] Hutson JM, Howard BJ (1982) Anisotropic intermolecular forces. I. Rare-gas-hydrogen chloride systems. Mol Phys 45: 769–790

ClH_2^+	H_2Cl^+ (C_{2v})			Chloronium
Method	r_e(H–Cl)	∠(HClH)	Ref.	
Exp	1.30412(17)	94.243(27)	1	
AI[a]	1.3030	94.244	2	

[a] CEPA-1 potential energy surface with offset correction, accuracy: 0.001 Å
[1] Lee SK, Amano T, Kawaguchi K, Oldani M (1988). Difference-frequency laser spectroscopy of the ν_1 and ν_3 fundamental bands of H_2Cl^+: Determination of the equilibrium molecular structure. J Mol Spectrosc 130: 1–19
[2] Botschwina P (1988) Anharmonic potential-energy surfaces, vibrational frequencies and infrared intensities calculated from highly correlated wavefunctions J Chem Soc, Faraday Trans 2 84: 1263–1276

ClH_2P	PH_2Cl (C_s)				Chlorophosphane
Method	r_e(P–H)	r_e(P–Cl)	∠(HPH)	∠(HPCl)	
Exp.[a]	1.411(2)	2.063(2)	92.8(2)	96.4(2)	

(continued)

(continued)

ClH_2P	PH_2Cl (C_s)				Chlorophosphane
Method	r_e(P–H)	r_e(P–Cl)	∠(HPH)	∠(HPCl)	
AI[b]	1.412		92.6	96.7	

[a] r_e extrapolated from r_z
[b] CCSD(T)_fc/cc-pVTZ + empirical correction using the structure of PH_3
Dréan P, Paplewski M, Demaison J, Breidung J, Thiel W, Beckers H, Bürger H (1996) Millimeter-wave spectra, ab Initio calculations, and structures of fluorophosphane and chlorophosphane. Inorg Chem 35: 7671–7678

ClH_3Si	SiH_3Cl (C_{3v})				Silyl chloride
Method	r_e(Si–Cl)	r_e(Si–H)	∠(HSiCl)	Ref.	
Exp	2.0458	1.4684	108.425	1	
SE[a]	2.04584(2)	1.46878(5)	108.427(4)	2	
CCSD(T)_ae/cc-pCVQZ	2.0479	1.4691	108.521	1	
AI[b]	2.0463	1.4689	108.42	2	

[a] Rovibrational corrections from a CCSD(T)_fc/cc-pVTZ cubic force field.
[b] CCSD(T)/CBS + core + correction + corrections using CCSDT/cc-pVTZ and+ CCSDTQ/cc-pVDZ + relativistic correction
[1] Demaison J, Sormova H, Bürger H, Margulès L, Constantin FL, Ceausu-Velcescu A (2005) Pure rotational spectrum, ab initio anharmonic force field, and equilibrium structure of silyl chloride. J Mol Spectrosc 232: 323–330
[2] Cazzoli G, Puzzarini C, Gauss J (2012) Rotational spectrum of silyl chloride: hyperfine structure and equilibrium geometry. Mol Phys 110: 2359–2369

ClNO	(C_s)				Nitrosyl chloride
Method	r_e(N–Cl)	r_e(N=O)	∠(ClNO)	Ref.	
Exp	1.97263(7)	1.13571(7)	113.405(3)	1	
SE[a]	1.97308(32)	1.13566(32)	113.380(16)	2	
CCSD(T)_ae/ aug-cc-pwCVQZ[b]	1.96233	1.13620	113.3756	3	
Interpolation	1.9706(1)	1.13922(4)	113.4452(8)	3	

[a] Rovibrational corrections obtained from a CCSD(T)_ae/aug-cc-pVTZ cubic force field.
[b] The multireference character is not negligible
[1] Cazzoli G, Degli Esposti C, Palmieri P, Simeone S (1983) Equilibrium structure and force field of nitrosyl chloride. J Mol Spectrosc 97: 165–185
[2] Demaison J, Császár AG, Dehayem-Kamadjeu A (2006) The case of the weak N–X bond: Ab initio, semi-experimental, and experimental equilibrium structures of XNO (X = H, F, Cl, OH) and FNO2··· J Phys Chem A 110: 13609–13617
[3] Demaison J, Liévin J, Vogt N (2023) Accurate equilibrium structures of some challenging molecules: FNO, ClNO, HONO, FNO_2, and N_2O. J Mol Spectrosc 394: 111788

ClO_2	(C_s)			Chlorine dioxide
Method	r_e(O–Cl)	∠(OClO)	Ref.	
Exp. (MW)	1.469839(13)	117.403(27)	1	
Exp. (GED)	1.4707(16)	117.48(8)	2	

[1] Miyazaki, K, Tanoura M, Tanaka K, Tanaka T (1986) Microwave spectrum of chlorine dioxide in excited vibrational states. J Mol Spectrosc 116: 435–449
[2] Spiridonov VP, Gershikov AG (1986) Molecular potential functions from joint analysis of gas electron diffraction and spectroscopic data. J Mol Struct 140: 173–191

Cl_2Cu	$CuCl_2$ ($C_{\infty v}$) $^2\prod_{g(3/2)}$ ground electronic state		Copper dichloride
Method	r_e(Cu–Cl)	Ref.	
Exp	2.0353	1	
Exp	2.0341(3)	2	

[1] Beattie IR, Brown JM, Crozet P, Ross AJ, Yiannopoulou A (1997) On the geometry of the $CuCl_2$ molecule. Inorg Chem 36: 3207–3208
[2] Bosch E, Crozet P, Ross AJ, Brown JM (2000) Fourier transform spectra of the $E^2\prod_u$-$X^2\prod_{g(3/2)}$ system of $CuCl_2$: 2. Rovibronic levels of the ground state up to 4000 cm^{-1}. J Mol Spectrosc 202: 253–261

Cl_2Ge	$GeCl_2$ (C_{2v})			Germanium dichloride
Method	r_e(Ge–Cl)	∠(ClGeCl)	Ref.	
Exp	2.169452(15)	99.8825(15)	1	
Exp.[a]	2.1679(13)	99.81(4)	2	
AI[b]	2.191	100.5	3	

[a] r_0 structure with offset correction
[b] MRSDCI
[1] Tsuchiya MJ, Honju H, Tanaka K, Tanaka T (1995) Millimeter-wave spectrum of germanium dichloride $GeCl_2$. Equilibrium structure and anharmonic force field. J Mol Struct 352/353: 407–415
[2] Demaison J, Vogt N (2022) Effective and equilibrium structures of XY_n-type molecules: SCl_2, PF_2, PO_2, ZrO_2, and CF_2S as additional example and nonbonded distances F· · · F, Cl· · · Cl and O· · · O. J Mol Spectrosc 387: 111661
[3] Benavides-Garcia M, Balasubramanian K (1992) The singlet–triplet energy separations of $GeCl_2$, $GeBr_2$, GeI_2, GeHCl, GeHBr, and GeHI. J Chem Phys 97: 7537–7544

Cl_2H^-	$ClHCl^-$ ($D_{\infty h}$)	
Method	r_e(Cl–Cl)	Ref.
Exp	3.1122(9)	1
AI[a]	3.132	2

(continued)

(continued)

Cl_2H^-	$ClHCl^-$ ($D_{\infty h}$)	
Method	r_e(Cl–Cl)	Ref.
AI[b]	3.1153	3

[a] CEPA-1 method
[b] Adjusted CCSD(T)-F12/aug-cc-pV(5 + d)Z potential energy surface
[1] Kawaguchi K (1988) Gas-phase infrared spectroscopy of $ClHCl^-$. J Chem Phys 88: 4186–4189
[2] Botschwina P, Sebald P, Burmeister R (1988) Calculated spectroscopic properties for $ClHCl^-$. J Chem Phys 88: 5246–5248
[3] Sebald P, Oswald R, Botschwina P, Kawaguchi K (2013) Rovibrational states of $ClHCl^-$ isotopologues up to high J: a joint theoretical and spectroscopic investigation. Phys Chem Chem Phys 15: 6737–6748

Cl_2H_2Si	SiH_2Cl_2 (C_{2v})					Dichlorosilane
Method	r_e(Si–H)	r_e(Si–Cl)	∠(ClSiCl)	∠(HSiH)	Ref.	
Exp.[a]	1.4671	2.0316	109.67	112.45	1	
From isolated stretching frequency	1.4651(10)				2	

[a] Derived from the r_z structure
[1] Davis RW, Robiette AG, Gerry, MCL (1981) The harmonic force fields of methylene chloride and dichlorosilane from combined microwave and infrared data. J Mol Spectrosc 85: 399–415
[2] Christensen DH, Nielsen OF (1968) Infrared absorption spectra of SiH_2Cl_2, $SiDHCl_2$ and SiD_2Cl_2. J Mol Spectrosc 27: 489–498

Cl_2O	(C_{2v})			Dichorine monoxide
Method	r_e(Cl–O)	∠(ClOCl)	Ref.	
Exp	1.6959(1)	110.88(1)	1	
CCSD(T)_ae/cc-pCVQZ	1.696384	110.982	2	

[1] Sugie M, Ayabe M, Takeo H, Matsumura C (1995) Microwave spectra of dichlorine monoxide in its excited vibrational states. J Mol Struct 352/353: 259–265
[2] Coriani S, Marchesan D, Gauss J, Hättig C, Helgaker T, Jørgensen P (2005) The accuracy of ab initio molecular geometries for systems containing second-row atoms. J Chem Phys 123: 184107/1-12

Cl_2OS	$SOCl_2$ (C_s)				Thionyl chloride
	SE[a]	AI[b]	r_e [c]	r_e [d]	
Ref.	1	1	2	3	
r_e(S=O)	1.4330(2)	1.432	1.436(3)		
r_e(S–Cl)	2.0700(1)	2.071			

(continued)

(continued)

Cl_2OS	$SOCl_2$ (C_s)				Thionyl chloride
	SE[a]	AI[b]	r_e [c]	r_e [d]	
∠(ClSO)	107.703(2)	107.75			
∠(ClSCl)	96.929	96.903			
τ(ClSOCl)	103.58(1)	103.59			
r_e(Cl···Cl)	3.099			3.100(1)	

[a] Rovibrational corrections calculated from an anharmonic CCSD(T)/cc-pV(T + d)Z force field
[b] CCSD(T)_ae/cc-pwCVQZ values
[c] Using S=O isolated stretching frequency
[d] r_0 value + offset
[1] Martin-Drumel MA, Roucou A, Brown GG, Thorwirth S, Pirali O, Mouret, G, Hindle F, McCarthy MC, Cuisset A (2016) High resolution spectroscopy of six $SOCl_2$ isotopologues from the microwave to the far-infrared, J. Chem. Phys. 144: 084305
[2] Brunvoll J, Hargittai I (1978) On the correlation of geometric and vibrational parameters of the SO_2 groups in sulfone molecules. Acta Chim Acad Sci Hung 96: 337–354
[3] Demaison J, Vogt N (2022) Effective and equilibrium structures of XY_n-type molecules: SCl_2, PF_2, PO_2, ZrO_2, and CF_2S as additional example and nonbonded distances F···F, Cl···Cl and O···O. J Mol Spectrosc 387: 111661

Cl_2O_2	$ClClO_2$ (C_s)					Chloryl chloride
Method	r_e(Cl–Cl)	r_e(Cl–O)	∠(ClClO)	∠(OClO)	Ref.	
Exp.[a]	2.1921	1.4368	104.02	114.63	1	
CCSD(T)_fc/TZ2P[b]	2.279	1.468	104.1	114.9	2	

[a] Derived from the r_z structure
[b] Not corrected for core correlation
[1] Müller HSP, Cohen EA, Christen D (1999) The rotational spectrum of chloryl chloride $ClClO_2$, in its ground vibrational state. J Chem Phys 110: 11865–11875
[2] Lee TJ, Rohlfing CM, Rice JE (1992) An extensive ab initio study of the structures, vibrational spectra, quadratic force fields, and relative energetics of three isomers of Cl_2O_2. J Chem Phys 97: 6593–6605

Cl_2S	(C_{2v})			Sulfur dichloride
Method	r_e(Cl–S)	∠(ClSCl)	Ref.	
Exp.[a]		102.730	1	
Exp	2.0100(12)	102.589(7)	2	
SE[b]	2.0111(20)	102.676(89)	2	
CCSD(T)_ae/cc-pwCVQZ	2.0131	102.704	2	

[a] r_z value
[b] Rovibrational corrections obtained from a B2PLYPD3/6–311+G(3df) force field
[1] Davis RW, Gerry MCL (1977) The quadratic potential function and average structure of sulphur dichloride, SCl_2, from its microwave spectrum. J Mol Spectrosc 65: 455–473
[2] Demaison J, Vogt N (2022) Effective and equilibrium structures of XY_n-type molecules: SCl_2, PF_2, PO_2, ZrO_2, and CF_2S as additional example and nonbonded distances F···F, Cl···Cl and O···O. J Mol Spectrosc 387: 111661

Cl_2Si	(C_{2v})			Dichlorosilylene
Method	r_e(Cl–Si)	∠(ClSiCl)	Ref.	
Exp	2.065310(26)	101.3240(16)	1	
CCSD(T)_ae/cc-pCVQZ	2.067952	101.529	2	

[1] Fujitake M, Hirota E (1994) The microwave spectrum of dichlorosilylene ($SiCl_2$) in excited vibrational states. Spectrochim Acta A 50: 1345–1370
[2] Coriani S, Marchesan D, Gauss J, Hättig C, Helgaker T, Jørgensen P (2005) The accuracy of ab initio molecular geometries for systems containing second-row atoms. J Chem Phys 123: 184107/1-12

Cl_3OP	$POCl_3$ (C_{3v})						Phosphoryl chloride
Method	r_e(P–Cl)	r_e(O–P)	∠(OPCl)	∠(ClPCl)	r_e(Cl··· Cl)	Ref.	
Exp.[a]	1.9884	1.4447	114.9	103.5	3.124	1	
GED	1.9885	1.4458		103.3(2)	3.119	2	
Exp.[b]					3.123(1)	3	

[a] Derived from the r_z structure
[b] r_0 value + offset
[1] Firth S, Davis RW (1988) The microwave spectrum of oxygen-18-labeled phosphoryl chloride: Ref.inement of the harmonic force field and the molecular structure. J Mol Spectrosc 127: 209–221
[2] Kuchitsu K, Moritani T, Morino Y (1971) Molecular structures of phosphoryl fluoride, phosphoryl chloride, and thiophosphoryl chloride studied by gas electron diffraction. Inorg Chem 10: 344–350
[3] Demaison J, Vogt N (2022) Effective and equilibrium structures of XY_n-type molecules: SCl_2, PF_2, PO_2, ZrO_2, and CF_2S as additional example and nonbonded distances F··· F, Cl··· Cl and O··· O. J Mol Spectrosc 387: 111661

Cl_3P	PCl_3 (C_{3v})			Trichlorophosphine
Method	r_e(P–Cl)	∠(ClPCl)	Ref.	
Exp.[a]	2.039(3)	100.28(10)	1	
SE[b]	2.0394(10)	100.18(10)	2	
AI[c]	2.0395	100.19	2	

[a] r_e extrapolated from r_z
[b] Rovibrational corrections from CCSD(T)_fc/aug-cc-pV(Q + d)Z force field
[c] CCSD(T)-F12b_ae/cc-pCVQZ-F12 + corrections using CCSDT/cc-pVTZ and+ CCSDTQ/cc-pVDZ + relativistic correction
[1] Dréan P, Paplewski M, Demaison J, Breidung J, Thiel W, Beckers H, Bürger H (1996) Millimeter-wave spectra, ab Initio calculations, and structures of fluorophosphane and chlorophosphane. Inorg Chem 35: 7671–7678
[2] Breidung J, Thiel W (2019) Equilibrium structures of the phosphorous trihalides PF_3 and PCl_3 and the phosphoranes PH_3F_2, PF_5, PCl_3F_2, and PCl_5. J Phys Chem A 123: 5600–5612

Cl_3PS	$SPCl_3$ (C_{3v})						Thiophosphoryl chloride
Meth	r_e(P–Cl)	r_e(P=S)	∠(SPCl)	∠(ClPCl)	r_e(Cl···Cl)	Ref.	
Exp.[a]	2.0042	1.8805	116.348	101.8	3.111	1	
GED	2.0053	1.8815		101.8(2)		2	
Exp.[b]					3.111(1)	3	

[a] Derived from the r_z structure
[b] r_0 value + offset
[1] Firth S, Davis RW (1996) The microwave spectra of isotopic thiophosphoryl chlorides in the ground and $\nu_6 = 1$ vibrational states. J Mol Spectrosc 179: 32–40
[2] Kuchitsu K, Moritani T, Morino Y (1971) Molecular structures of phosphoryl fluoride, phosphoryl chloride, and thiophosphoryl chloride studied by gas electron diffraction. Inorg Chem 10: 344–350
[3] Demaison J, Vogt N (2022) Effective and equilibrium structures of XY_n-type molecules: SCl_2, PF_2, PO_2, ZrO_2, and CF_2S as additional example and nonbonded distances F···F, Cl···Cl and O···O. J Mol Spectrosc 387: 111661

CsHO	CsOH ($C_{\infty v}$)			Cesium hydroxide
Method	r_e(Cs–O)	r_e(O–H)	Ref.	
Exp	2.391(2)	0.960(10)	1	
Semirigid bender	2.393(2)	0.992	2	

[1] Lide DR, Matsumura (1969) Structure of the alkali hydroxides. IV. Interpretation of vibration–rotation interactions in CsOH and RbOH and refinement of structures. J Chem Phys 50: 3080–3086
[2] Brown RD, Godfrey PD, Kleiboemer B, McNaughton D (1994) Structures and flexibilities of the alkali hydroxides. J Mol Struct 327: 99–106

$FGeH_3$	GeH_3F (C_{3v}) See ClH_3Ge				Germyl fluoride
FHO	HOF (C_s)				Hypofluorous acid
	r_e(O–H)	r_e(O–F)	∠(HOF)	Ref.	
Exp	0.9657(16)	1.4350(31)	97.54(50)	1	
SE[a]	0.968619(80)	1.43447(11)	97.86(2)	2	
CCSD(T)_ae/cc-pCVQZ	0.9657	1.4326	97.85	3	
AI[b]	0.966565	1.432757	97.9596	4	

[a] Rovibrational corrections obtained from a CCSD(T)_fc/cc-pVQZ force field
[b] CCSD(T)_fc/cc-pV∞Z + core(cc-pCVQZ) + corrections using CCSDT/cc-pVTZ and+ CCSDTQ/cc-pVDZ
[1] Halonen L, Ha T-K (1988) Equilibrium structure and anharmonic force field of hypofluorous acid (HOF). J Chem Phys 89: 4885–4888
[2] Pawlowski F, Jørgensen P, Olsen J, Hegelund F, Helgaker T, Gauss J, Bak KL, Stanton JF (2002). Molecular equilibrium structures from experimental rotational constants and calculated vibration–rotation interaction constants. J Chem Phys 116: 6482–6496
[3] Bak KL, Gauss J, Jørgensen P, Olsen J, Helgaker T, Stanton JF (2001) The accurate determination of molecular equilibrium structures. J Chem Phys 114: 6548–6556

[4] Puzzarini C, Heckert M, Gauss J (2008) The accuracy of rotational constants predicted by high-level quantum chemical calculations. I. Molecules containing first-row atoms. J Chem Phys 128: 194108/1-9

FH_2P	PH_2F (C_s)				Fluorophosphane
Method	r_e(P–H)	r_e(P–F)	∠(HPH)	∠(HPF)	
Exp.[a]	1.415(2)	1.602(2)	92.0(2)	97.8(2)	
AI[b]	1.415	1.599	92.0	97.9	

[a] r_e extrapolated from r_z
[b] CCSD(T)_fc/cc-pVTZ + empirical correction using the structures of PH_3 and PF_3
Dréan P, Paplewski M, Demaison J, Breidung J, Thiel W, Beckers H, Bürger H (1996) Millimeter-wave spectra, ab initio calculations, and structures of fluorophosphane and chlorophosphane. Inorg Chem 35: 7671–7678

FH_3Si	SiH_3F (C_{3v})				Silyl fluoride
Method	r_e(Si–H)	r_e(Si–F)	∠(FSiH)	Ref.	
Exp	1.4675	1.5915	108.037	1	
SE[a]	1.46948(9)	1.59048(6)	108.304(9)	2	
AI[b]	1.4695	1.5911	108.30	2	

[a] Rovibrational corrections obtained from a CCSD(T)_ae/cc-pCVTZ cubic force field
[b] CCSD(T)/CBS + core + correction + corrections using CCSDT/cc-pVTZ and+ CCSDTQ/cc-pVDZ + relativistic correction
[1] Boulaftali N, Ben Sari-Zizi N, Graner G, Demaison J (2002) The equilibrium structure of silyl fluoride. J Mol Spectrosc 216: 284–291
[2] Puzzarini C, Cazzoli G, Gauss J (2010) Rotational spectra of isotopic species of silyl fluoride. Part II: Theoretical and semi-experimental equilibrium structure. J Mol Spectrosc 262: 37–41

FNO	(C_s)				Nitrosyl fluoride
Method	r_e(N–F)	r_e(N=O)	∠(FNO)	Ref.	
Exp	1.51658(25)	1.13155(23)	109.9220(72)	1	
SE[a]	1.51681(23)	1.13176(21)	109.9361(66)	2	
CCSD(T)_ae/ aug-cc-pCVQZ[b]	1.5120	1.1322	109.975	2	
Interpolation	1.5155(8)	1.1331(2)	109.969(2)	3	

[a] Rovibrational corrections obtained from a CCSD(T)_ae/aug-cc-pVTZ cubic force field
[b] The multireference character is not negligible
[1] Degli Esposti C, Cazzoli G, Favero PG (1985) Microwave spectrum of $^{18}O^{14}NF$ in excited vibrational states: Equilibrium structure of nitrosyl fluoride. J Mol Spectrosc 109: 229–238
[2] Demaison J, Császár AG, Dehayem-Kamadjeu A (2006) The case of the weak N–X bond: Ab initio, semi-experimental, and experimental equilibrium structures of XNO (X = H, F, Cl, OH) and FNO2... J Phys Chem A 110: 13609–13617
[3] Demaison J, Liévin J, Vogt N (2023) Accurate equilibrium structures of some challenging molecules: FNO, ClNO, HONO, FNO_2, and N_2O. J Mol Spectrosc 394: 111788

FOP	FPO (C_s)				Phosphenous fluoride
Method	r_e(PO)	r_e(PF)	∠(OPF)	Ref.	
Exp.[a]	1.4531	1.5725	110.16	1, 2	
SE[b]	1.4528(2)	1.5727(1)	110.16(2)	2	
AI[c]	1.4547	1.5747	110.13	2	
AI[d]	1.4532	1.7529	110.14	3	

[a] r_e extrapolated from r_z
[b] Rovibrational corrections obtained from a MP2)/cc-pVQZ + 1 cubic force field
[c] CCSD(T)_fc/cc-pV5Z + corrections for core-valence and diffuse functions
[d] CCSD(T)_fc/CBS + core correlation correction [CCSD(T)/wCVQZ] + scalar relativistic effects
[1] Gatehouse B, Brupbacher T, Gerry MCL (1999) Fourier transform microwave spectrum, geometry, and hyperfine coupling constants of phosphenous fluoride, OPF. J Phys Chem A 103: 560–567
[2] Beckers H, Bürger H, Paplewski P, Bogey M, Demaison J, Dréan P, Walter A, Breidung J, Thiel W (2001) Millimeter-wave spectroscopy, high resolution infrared spectrum, ab initio calculations, and molecular geometry of FPO. Phys Chem Chem Phys 3: 4247–4257
[3] Puzzarini C (2006) Theoretical study of XPO (X= H, F, Cl, Br) molecules. J Mol Struct 780–781: 238–246

FO_2	(C_s) electronic ground state 2A″				Fluoroperoxyl
Method	r_e(FO)	r_e(OO)	∠(FOO)	Ref.	
Exp	1.649(4)	1.200(4)	111.19(12)	1	
AI[a]	1.709	1.198	111.8	2	
AI[b]	1.632	1.192	110.91	3	
AI[c]	1.636	1.195	111.0	4	

[a] QCISD(T)/6-311G(d)
[b] CCSD(T)_fc/cc-pVQZ + core-valence correction
[c] iCAS-CI + Q (internally contracted, complete active space, multireference configuration interaction
[1] Yamada C, Hirota E (1984) The infrared diode laser spectrum of the ν_2 band of the FO_2 radical. J Chem Phys 80: 4694–4700
[2] Francisco JS, Zhao Y, Lester WA, Williams IH (1992) Theoretical studies of the structure and thermochemistry of FO_2 radical: Comparison of Möller–Plesset perturbation, complete-active-space self-consistent-field, and quadratic configuration interaction methods. J Chem Phys 96: 2861–2867
[3] Karton A, Parthiban S, Martin J ML (2009) Post-CCSD(T) ab initio thermochemistry of halogen oxides and related hydrides XOX, XOOX, HOX, XO_n, and HXO_n (X = F, Cl), and evaluation of DFT methods for these systems. J Phys Chem A 113: 4802–4816
[4] Feller D, Dixon, D A (2003) Coupled cluster theory and multireference configuration interaction study of FO, F_2O, FO_2, and FOOF. J Phys Chem A, 107: 9641–9651

FPS	(C_s)				Phosphenothious fluoride
Method	r_e(PS)	r_e(PF)	∠(FPS)	Ref.	
Exp.[a]	1.8884	1.5872	109.29	1	
SE[b]	1.8886(4)	1.5870(3)	109.28(2)	1	

(continued)

(continued)

FPS	(C_s)				Phosphenothious fluoride
Method	r_e(PS)	r_e(PF)	∠(FPS)	Ref.	
AI[c]	1.8907	1.5883	109.27	1	
AI[d]	1.8944	1.5882	109.40	2	

[a] r_e extrapolated from r_z
[b] Rovibrational corrections obtained from a MP2)/cc-pVQZ + 1 cubic force field
[c] CCSD(T)_fc/cc-pV5Z + corrections for core-valence and diffuse functions
[d] CCSD(T)_ae/cc-pVQZ + 1

[1] Beckers H, Bogey M, Breidung J, Bürger H, Demaison J, Dréan P, Paplewski P, Thiel W, Walters A (2001) Millimeter-wave spectroscopy, high-resolution infrared spectrum, ab initio calculations, and molecular geometry of FPS. J Mol Spectrosc 210: 213–223
[2] Beckers H, Bogey M, Breidung J, Bürger H, Dréan P, Paplewski P, Walters A (2000) FP=S in the gas-phase: Detection by rotationally resolved infrared and millimetre-wave spectra assisted by ab-initio calculations. Phys Chem Chem Phys 2: 2467–2469

F_2Ge	GeF_2 (C_{2v})				Germanium difluoride
Method	r_e(Ge–F)	∠(FGeF)	F···F	Ref.	
Exp	1.73209(1)	97.1478(4)	2.5975	1	
Exp.[a]			2.555(27)	2	
Exp.[b]	1.7306(12)	97.148(41)		3	

[a] From spin–spin interaction constant, see Sect. 8.4
[b] Extrapolated from the r_0 structure using the rotational constants of Ref. [1]

[1] Takeo H, Curl RF (1972) Microwave spectrum of GeF_2: Quadrupole coupling and centrifugal distortion. J Mol Spectrosc 43: 21–30
[2] Styger C, Gerry MCL (1993) ^{19}F spin rotation and spin–spin interactions of silicon difluoride and germanium difluoride prepared in an electric discharge, observed with a cavity microwave Fourier transform. J Mol Spectrosc 158: 328–338
[3] Demaison J, Vogt N (2022) Effective and equilibrium structures of XY_n-type molecules: SCl_2, PF_2, PO_2, ZrO_2, and CF_2S as additional example and nonbonded distances F···F, Cl···Cl and O···O. J Mol Spectrosc 387: 111661

F_2H^-	FHF^- ($D_{\infty h}$)	
Method	r_e(FF)	Ref.
Exp.[a]	2.27771(3)	1
Exp.[b]	2.27752(3)	2
AI[c]	2.27922	3

[a] For FHF^-
[b] For FDF^-
[c] Adjusted CCSD(T)-F12_fc/aug-cc-pV5Z potential energy surface

[1] Kawaguchi K, Hirota E (1987) Diode laser spectroscopy of the ν_3 and ν_2 bands of FHF^- in 1300 cm^{-1} region. J Chem Phys 87: 6838–6841
[2] Kawaguchi K, Hirota E (1995) Infrared diode laser spectroscopy of FDF^-. J Mol Struct 352–353: 389–394

[3] Sebald P, Bargholz A, Oswald R, Stein C, Botschwina P (2013). FHF– isotopologues: Highly anharmonic hydrogen-bonded systems with strong Coriolis interaction. J Phys Chem A 117: 9695–9703

F_2HP	PHF_2 (C_s)						Difluorophosphane
Method	r_e(P–H)	r_e(P–F)	∠(FPF)	∠(HPF)	r_e(F···F)	Ref.	
Exp.[a]	1.421(2)	1.579(2)	98.9(2)	95.7(2)	2.3996	1	
AI[b]	1.420	1.578	99.0	95.7		1	
Exp.[c]					2.3997(14)	2	

[a] r_e extrapolated from r_z
[b] CCSD(T)_fc/cc-pVTZ + empirical correction using the structures of PH_3 and PF_3
[c] r_0 + offset, using the rotational constants of Ref. (3)
[1] Dréan P, Paplewski M, Demaison J, Breidung J, Thiel W, Beckers H, Bürger H (1996) Millimeter-wave spectra, ab initio calculations, and structures of fluorophosphane and chlorophosphane. Inorg Chem 35: 7671–7678
[2] Demaison J, Vogt N (2022) Effective and equilibrium structures of XY_n-type molecules: SCl_2, PF_2, PO_2, ZrO_2, and CF_2S as additional example and nonbonded distances F···F, Cl···Cl and O···O. J Mol Spectrosc 387: 111661
[3] Kuczkowski RL (1968) The microwave spectrum, structure, and dipole moment of difluorophosphine. J Amer Chem Soc 90: 1705–1708

F_2H_2Si	SiH_2F_2 (C_{2v})					Difluorosilane
Method	r_e(Si–H)	r_e(Si–F)	∠(HSiH)	∠(FSiF)	Ref.	
Exp.[a]	1.4634	1.5760	114.35	107.77	1	
SE[b]	1.4615	1.5758	114.57	107.78	2	
AI[c]	1.4614	1.5760	114.21	107.88	2	

[a] r_e extrapolated from r_z
[b] Rovibrational corrections obtained from a MP2/cc-pVTZ + 1(Q) cubic force field
[c] CCSD(T)_fc/cc-pVQZ + 1 + offsets corrections obtained from SiH_4 and SiF_2
[1] Davis RW, Robiette AG, Gerry MCL (1980). The microwave spectrum, centrifugal distortion constants, harmonic force field, and molecular structure of difluorosilane. J Mol Spectrosc 83: 185–201
[2] D'Eu J-F, Demaison J, Bürger H (2003) Millimeter-wave and high-resolution FTIR spectroscopy of SiH_2F_2: the ground and $v_4 = 1$ states. J Mol Spectrosc 218: 12–21

F_2Kr	KrF_2 ($D_{\infty h}$)		Krypton difluoride
Method	r_e(^{84}Kr–F)	Ref.	
Exp	1.876930(23)	1	
Exp.[a]	1.880(1)	2	
AI[b]	1.9191	1	

[a] r_0 + offset
[b] CCSD(T)/6-31G* + ECP
[1] Bürger H, Kuna R, Ma S, Breidung J, Thiel W (1994) The vibrational spectra of krypton and xenon difluoride: high-resolution infrared studies and ab initio calculations. J Chem Phys 101: 1–14
[2] Demaison J, Vogt N (2022) Effective and equilibrium structures of XY_n-type molecules: SCl_2, PF_2, PO_2, ZrO_2, and CF_2S as additional example and nonbonded distances F···F, Cl···Cl and O···O. J Mol Spectrosc 387: 111661

F_2N	NF_2 (C_{2v}) Ground electronic state $^2B_1$			Nitrogen difluoride
Method	r_e(N–F)	∠(FNF)	Ref.	
Exp.[a]	1.34691(68)	103.132(20)	1	
SE[b]	1.3458	103.16	2	

[a] This is a mixed experimental/semiexperimental structure because the α_2 values were calculated ab initio
[b] Rovibrational corrections from a CASSCF force field
[1] Müller HSP, Löblein K, Hübner H, Hüttner W, Brown JM. (2008). The rotational spectrum of the NF_2 free radical: Determination of molecular structure. J Mol Spectrosc 251: 185–193
[2] Peterson KA; Mayrhofer RC, Woods RC (1990) The potential energy and dipole moment surfaces of NF_2 and O_3^- by complete active space self-consistent field. J Chem Phys 93: 5020–5028

F_2O	(C_{2v})			Oxygen difluoride
Method	r_e(F–O)	∠(FOF)	Ref.	
Exp	1.4053(4)	103.07(5)	1	
Exp.[a]	1.4052	103.07	2	
SE[b]	1.4054	103.09	3	

[a] Variational calculation of the potential energy surface
[b] Rovibrational correction from a CCSD(T)_ae/cc-pVTZ cubic force field
[1] Morino Y, Saito S (1966). Microwave spectrum of oxygen difluoride in vibrationally excited states; ν_1 - $2\nu_2$ Fermi resonance and equilibrium structure. J Mol Spectrosc 19: 435–453
[2] Saarinen T, Kauppi E, Halonen L (1990) Potential energy surface for F_2O. J Mol Spectrosc 142: 175–181
[3] Breidung J, Thiel W, Gauss J, Stanton JF (1999). Anharmonic force fields from analytic CCSD(T) second derivatives: HOF and F_2O. J Chem Phys 110: 3687–3696

F_2O_2S	SO_2F_2 (C_{2v})					Sulfuryl fluoride
Method	r_e(S=O)	r_e(S–F)	∠(OSO)	∠(FSF)	Ref.	
Exp.[a]	1.400	1.531	124.82	95.54	1	
Exp.[b]			125.1(2)		2	
AI[c]	1.401	1.532	124.91	95.53	1	

[a] r_e extrapolated from r_z
[b] Calculated value from gas-phase electron diffraction S=O and microwave O...O distances and has no strictly defined physical meaning
[c] CCSD(T)_fc/cc-pVQZ + 1 + MP2_ae/cc-pwCVQZ – MP2_fc/cc-pwCVQZ
[1] Merke I, Heineking N, Demaison J (2004) Rotational spectra of isotopic species of SO_2F_2: experimental and ab initio structures. J Mol Spectrosc 228: 308–313
[2] Brunvoll J, Hargittai I (1978) On the correlation of geometric and vibrational parameters of the SO_2 groups in sulfone molecules. Acta Chim Acad Sci Hung 96: 337–354

F_2P	PF_2 (C_{2v})			Phosphorous difluoride
Method	r_e(P–F)	∠(FPF)	Ref.	
Exp.[a]	1.5750(12)	98.329(42)	1	
SE[b]	1.57605(8)	98.332(9)	1	
SE[c]	1.57612(2)	98.338(2)	2	

[a] r_0 + offset
[b] Rovibrational correction from a B3LYP/aug-pcS-3 cubic force field
[c] Rovibrational correction from a CCSD(T)_fc/aug-cc-pVQZ cubic force field
[1] Demaison J, Vogt N (2022) Effective and equilibrium structures of XY_n-type molecules: SCl_2, PF_2, PO_2, ZrO_2, and CF_2S as additional example and nonbonded distances F···F, Cl···Cl and O···O. J Mol Spectrosc 387: 111661
[2] Breidung J, Thiel W (2019) Equilibrium Structures of the Phosphorus Trihalides PF_3 and PCl_3, and the Phosphoranes PH_3F_2, PF_5, PCl_3F_2, and PCl_5. J Phys Chem A 123: 5600–5612

F_2S	SF_2 (C_{2v})			Sulfur difluoride
Method	r_e(S–F)	∠(FSF)	Ref.	
Exp	1.58745(12)	98.048(13)	1	
Exp.[a]	1.5864(12)	98.05(4)	2	

[a] r_0 structure with offset correction
[1] Endo Y, Saito S, Hirota E, Chikaraishi T (1979) Microwave spectrum of sulfur difluoride in the first excited vibrational states. Vibrational potential function and equilibrium structure. J Mol Spectrosc 77: 222–234
[2] Demaison J, Vogt N (2022) Effective and equilibrium structures of XY_n-type molecules: SCl_2, PF_2, PO_2, ZrO_2, and CF_2S as additional example and nonbonded distances F···F, Cl···Cl and O···O. J Mol Spectrosc 387: 111661

F_2S_2	FSSF (C_2)					Difluorodisulfane
Method	r_e(S–F)	r_e(S–S)	∠(SSF)	τ(FSSF)	Ref.	
Exp.[a]	1.6295	1.8889	108.264	87.526	1	
AI[b]	1.630				2	
AI[c]		1.8895	108.73		3	
AI[d]				87.95	4	

[a] Derived from the r_z structure
[b] B3P86/cc-pVTZ value + offset
[c] The MP2/cc-pV(T + d)Z value is close to the r_e one, see Ref. (3)
[d] From an empirical correlation using MP2/cc-pV(T + d)Z and B3LYP/6–311 + G(3df) values, see Ref. (4)

[1] Davis RW, Firth S (1991) The microwave spectrum of the chain isomer of disulfur difluoride: FS–SF. J Mol Spectrosc 145: 225–235
[2] Altman JA, Handy NC, Ingamells VE (1997) A study of sulphur-containing molecules using Hartree–Fock, MP2 and DFT (hybrid) methodologies. Mol Phys 92: 339–352
[3] Demaison J, Vogt N, Saragi RT, Juanes M, Rudolph HD, Lesarri A (2019) The S–S Bridge: A mixed experimental-computational estimation of the equilibrium structure of diphenyl disulfide. Chem Phys Chem 20: 366–373
[4] Juanes M, Vogt N, Demaison J, León I, Lesarri A, Rudolph HD (2017) Axial–equatorial isomerism and semiexperimental equilibrium structures of fluorocyclohexane. Phys Chem Chem Phys 19: 29162–29169

F_2S_2	$S{=}SF_2$ (C_s)						Disulfur difluoride
Method	r_e(S–F)	r_e(S= S)	∠(FSF)	∠(SSF)	r_e(F···F)	Ref.	
Exp.[a]	1.6026	1.8543	91.72	108.02	2.300	1	
Exp.[b]					2.300(2)	2	

[a] Derived from the r_z structure
[b] r_0 value + offset correction

[1] Davis RW (1986) The microwave spectrum of the pyramidal isomer of disulfur difluoride: $S{=}SF_2$. J Mol Spectrosc, 116: 371–383
[2] Demaison J, Vogt N (2022) Effective and equilibrium structures of XY_n-type molecules: SCl_2, PF_2, PO_2, ZrO_2, and CF_2S as additional example and nonbonded distances F···F, Cl···Cl and O···O. J Mol Spectrosc 387: 111661

F_2Si	SiF_2 (C_{2v})			Silicon difluoride
Method	r_e(Si–F)	∠(FSiF)	Ref.	
Exp	1.59009(3)	100.758(3)	1	
CCSD(T)_ae/cc-pCVQZ	1.591931	100.911	2	

[1] Bizzocchi, L.; Degli Esposti, C. (2006) Millimeter and submillimeter-wave spectroscopy of silicon difluoride. J Mol Spectrosc 235: 117–124
[2] Coriani S, Marchesan D, Gauss J, Hättig C, Helgaker T, Jørgensen P (2005) The accuracy of ab initio molecular geometries for systems containing second-row atoms. J Chem Phys 123: 184107/1-12

F_2Xe	XeF_2 ($D_{\infty h}$)		Xenon difluoride
Method	r_e(^{136}Xe–F)	Exp	
Exp	1.974365(7)	1	
Exp.[a]	1.975(1)	2	
AI[b]	1.9988	3	

[a] r_0 + offset
[b] CCSD(T)/6-31G* + ECP
[1] Bürger H, Ma S (1993) The ground state and equilibrium structure of XeF_2 J Mol Spectrosc 157: 536–539
[2] Demaison J, Vogt N (2022) Effective and equilibrium structures of XY_n-type molecules: SCl_2, PF_2, PO_2, ZrO_2, and CF_2S as additional example and nonbonded distances F···F, Cl···Cl and O···O. J Mol Spectrosc 387: 111661
[3] Bürger H, Kuna R, Ma S, Breidung J, Thiel W (1994) The vibrational spectra of krypton and xenon difluoride: high-resolution infrared studies and ab initio calculations. J Chem Phys 101: 1–14

F_3HSi	$SiHF_3$ (C_{3v})				Trifluorosilane
Method	r_e(Si–H)	r_e(Si–F)	∠(FSiH)	Ref.	
Exp	1.4468(5)	1.5624(1)	110.64(3)	1	
SE[a]	1.44872(3)	1.56266(1)	110.6884(4)	2	

[a] Rovibrational corrections obtained from a MP2/cc-pVTZ + 1(Q). Q means that the basis set cc-pVQZ is used for Si
[1] Hoy AR, Bertram M, Mills IM (1973) Excited vibrational state microwave spectra, structure, and force-field of trifluorosilane. J Mol Spectrosc 46: 429–447
[2] Demaison J, Margulès L, Breidung J, Thiel W, Bürger H (1999) Ab initio anharmonic force field, spectroscopic parameters and equilibrium structure of trifluorosilane. Mol Phys 97: 1053–1067

F_3N	NF_3 (C_{3v})			Trifluoroamine
Method	r_e(N–F)	∠(FNF)	Ref.	
Exp	1.3673	101.88	1	
Exp	1.36757(58)	101.8513(10)	2	
SE[a]	1.3678	101.85	1	
CCSD(T)_ae/cc-pCVQZ	1.3662	101.95	1	

[a] Rovibrational corrections obtained from a CCSD(T)_ae/TZ2Pf cubic force field
[1] Breidung J, Constantin L, Demaison J, Margulès L, Thiel W (2003) Ground state rotational spectrum, $K = 3$ splittings, ab initio anharmonic force field and equilibrium structure of trifluoroamine. Mol Phys 101: 1113–1122
[2] Najib H (2015) Experimental values of the rotational and vibrational constants and equilibrium structure of nitrogen trifluoride. J Mol Spectrosc 312: 1–5

F_3P	PF_3 (C_{3v})			Trifluorophosphine
Method	r_e(P–F)	∠(FPF)	Ref.	
Exp	1.560986(43)	97.566657(64)	1, 2	
SE[a]	1.5610(10)	97.57(10)	3	
AI[b]	1.5614	97.56	3	

[a] Rovibrational corrections from CCSD(T)_fc/aug-cc-pV(Q + d)Z force field
[b] CCSD(T)-F12b_ae/cc-pCVQZ-F12 + full T/VTZ + (Q)/VDZ + relativistic correction
[1] Najib H (2014) Experimental rovibrational constants and equilibrium structure of phosphorus trifluoride. J Mol Spectrosc 305: 17–21
[2] Ben Sari-Zizi N, Bürger H, Litz M, Najib H, Radtke J (1996) High resolution FTIR study of PF_3 in the 800–950 cm^{-1} region. The ν_1 and ν_3 fundamental and $\nu_2 + \nu_4$ combination bands. J Mol Spectrosc 177: 46–57
[3] Breidung J, Thiel W (2019) Equilibrium structures of the phosphorous trihalides PF_3 and PCl_3 and the phosphoranes PH_3F_2, PF_5, PCl_3F_2, and PCl_5. J Phys Chem A 123: 5600–5612
See also Small CE, Smith JG (1978) The anharmonic force field of phosphorus trifluoride from the millimeter wave spectra of vibrationally excited states. J Mol Spectrosc 73: 215–233

F_4Si	SiF_4 (T_d)		Tetrafluorosilane
Method	r_e(Si–F)	Ref.	
Exp.[a]	1.5516985(30)	1	
Exp.[b]	1.5524(8)	2	
AI[c]	1.5523	2	

[a] Infrared spectroscopy
[b] Derived from the mean r_z value from gas-phase electron diffraction and spectroscopy
[c] CCSD(T)_ae/cc-pVQZ + 1 + offset correction from SiF_2
[1] Boudon V, Richard C, Manceron L (2022) High-Resolution spectroscopy and analysis of the fundamental modes of $^{28}SiF_4$. Accurate experimental determination of the Si–F bond length. J Mol Spectrosc 383: 111549
[2] Breidung J, Demaison J, Margulès L, Thiel W (1999) Equilibrium structure of SiF_4. Chem Phys Lett 313: 713–717

F_5P	PF_5 (D_{3h})			Phosphorous pentafluoride
Method	r_e(P–F_{eq})	r_e(P–F_{ax})	Ref.	
Exp.[a]	1.529(3)	1.576(3)	1	
SE[b]	1.5310(10)	1.5714(10)	2	
AI[c]	1.5313	1.5715	2	

[a] Derived from r_g structure
[b] Rovibrational corrections from CCSD(T)_fc/aug-cc-pV(Q + d)Z force field
[c] CCSD(T)-F12b_ae/cc-pCVQZ-F12 + higher order corrections using CCSDT/cc-pVTZ and CCSDTQ/cc-pVDZ + relativistic correction
[1] Kurimura H, Yamamoto S, Egawa T, Kuchitsu K (1986) Geometrical structure of PF_5 as studied by gas electron diffraction and spectroscopic data. J Mol Struct 140: 79–86
[2] Breidung J, Thiel W (2019) Equilibrium structures of the phosphorous trihalides PF_3 and PCl_3 and the phosphoranes PH_3F_2, PF_5, PCl_3F_2, and PCl_5. J Phys Chem A 123: 5600–5612

F_6S	SF_6 (O_h)		Sulfur hexafluoride
Method	r_e(S–F)	Ref.	
Exp.[a]	1.5561(8)	1	
Exp.[b]	1.5568(6)	2	
Exp	1.5560(1)	3	
AI + Exp.[b]	1.5568	4	

[a] Using an experimental anharmonic force field
[c] From gas-phase electron diffraction
[c] B3LYP/6-31G* force field refined using electron diffraction data

[1] Krohn BJ, Overend J (1984). Force-field model for the stretching anharmonicities of sulfur hexafluoride. J Phys Chem 88: 564–574
[2] Ischenko AA, Ewbank JD, Schaefer L (1994) Direct evaluation of equilibrium molecular geometries using real-time gas electron diffraction. J Phys Chem 98: 4287–4300
[3] Boudon V, Doménech JL, Bermejo D, Willner H (2004) High-resolution Raman spectroscopy of the ν_1 region and Raman–Raman double resonance spectroscopy of the $2\nu_1 - \nu_1$ band of $^{32}SF_6$ and $^{34}SF_6$. Determination of the equilibrium bond length of sulfur hexafluoride. J Mol Spectrosc 228: 392–400
[4] Kochikov IV, Tarasov YI, Spiridonov VP, Kuramshina GM, Saakjan AS, Yagola AG (2000) The use of ab initio anharmonic force fields in experimental studies of equilibrium molecular geometry. J Mol Struct 550–551: 429–438

F_6Se	SeF_6 (O_h)	Selenium hexafluoride
r_e(Se–F) = 1.6784(8)		

From gas-phase electron diffraction

Maggard P, Lobastov VA, Schaefer L, Ewbank JD, Ischenko AA (1995) Direct evaluation of equilibrium molecular geometries using real-time gas electron diffraction. 2. Selenium hexafluoride. J Phys Chem 99: 13115–13117

GeH_4	(T_d)		Germane
Method	r_e(Ge–H)	Ref.	
Exp	1.5158(16)	1	
Exp	1.51714(25)	2	
CCSD(T)_ae/cc-pCVQZ	1.5406	3	
CCSD(T)_ae/cc-pCVQZ-DK	1.51614	3	
SE[a]	1.51759	3	

[a] CCSD(T)_ae/cc-pCVQZ-DK value fine-tuned to better match rotation levels of the vibration ground state

[1] Ohno K, Matsuura H, Endo Y, Hirota E (1986) The microwave spectra of deuterated silanes, germanes, and stannanes. J Mol Spectrosc 118: 1–17
[2] Richard C, Boudon, V, Rizopoulos A, Vander Auwera J, Kwabia Tchana F (2021) Line positions and intensities for the ν_2/ν_4 bands of 5 isotopologues of germane near 11.5 μm. J Quant Spectrosc Radiat Transf 260: 107474/1-11
[3] Nikitin AV, Rey M, Rodina A, Krishna BM, Tyuterev VG (2016). Full-dimensional potential energy and dipole moment surfaces of GeH_4 molecule and accurate first-principle rotationally resolved intensity predictions in the infrared. J Phys Chem A 120: 8983–8997

HIO	HOI (C_s)				Hypoiodous acid
Method	r_e(O–H)	r_e(O–I)	∠(HOI)	Ref.	
Exp.[a]	0.959(8)	1.9874(3)	103.89(20)	1	
AI[b]	0.959	1.9688	104.71	2	

[a] Derived from the r_z structure
[b] From a CASPT2/ANO-RCC-VQZP potential energy surface
[1] Ozeki H, Saito S (2004) Submillimeter-wave spectra of hypoiodous acid. J Chem Phys 120: 5110–5116
[2] De Oliveira-Filho AGS, Aoto YA, Ornellas FR (2011) Full-dimensional analytical ab initio potential energy surface of the ground state of HOI. J Chem Phys 135: 044308

HKO	KOH ($C_{\infty v}$)			Potassium hydroxide
Method	r_e(O–H)	r_e(K–O)	Ref.	
Exp	0.960	2.196	1	
Semirigid bender	0.980(17)	2.1992(21)	2	

[1] Pearson EF, Winnewisser BP, Trueblood MB (1976) Millimeter wave spectrum of excited vibrational states of potassium hydroxide, KOH and KOD. Z Naturforsch 31a: 1259–1267
[2] Brown RD, Godfrey PD, Kleiboemer B, McNaughton D (1994) Structures and flexibilities of the alkali hydroxides. J Mol Struct 327: 99–106

HLiO	LiOH ($C_{\infty v}$)			Lithium hydoxide
Method	r_e(O–H)	r_e(Li–O)	Ref.	
Exp	0.953	1.584	1	
Semirigid bender	0.9691(21)	1.5816(10)	2	
SE[a]	0.949(2)	1.5776(4)	1	
CCSD(T)_ae/aug-cc-pCVQZ	0.9483	1.5805	1	

[a] Rovibrational corrections obtained from a CCSD(T)_ae/aug-cc-pCVQZ potential energy surface
[1] Higgins KJ, Freund SM, Klemperer W, Apponi AJ, Ziurys LM (2004) The rotational spectrum and dynamical structure of LiOH and LiOD: A combined laboratory and ab initio study. J Chem Phys 121: 11715–11730
[2] McNaughton D, Tack LM, Kleibömer B, Godfrey PD (1994) Rotational spectrum and bending potential of LiOH: A semirigid bender analysis. Struct Chem 5: 313–319

HMgO	MgOH ($C_{\infty v}$) Ground electronic state $^2\Sigma^+$			Magnesium hydoxide
Method	r_e(O–H)	r_e(Mg–O)	Ref.	
Exp	0.940	1.767	1	
MORBID	0.95	1.76087(81)	2	

[1] Apponi, AJ, Anderson MA, Ziurys L (1999) High resolution spectroscopy of MgOH ($X^2\Sigma^+$) in its ν_2 mode: Further evidence for quasilinearity. J Chem Phys 111: 10919–10925
[2] Bunker PR, Kolbuszewski M, Jensen P, Brumm M, Anderson MA, Barclay WL, Ziurys LM, Ni Y, Harris DO (1995). New rovibrational data for MgOH and MgOD and the internuclear potential function of the ground electronic state. Chem Phys Lett 239: 217–222

HNO	(C_s)				Nitrosyl hydride
Method	r_e(N–H)	r_e(N=O)	∠(HNO)	Ref.	
Exp	1.063(3)	1.206(3)	109.1(2)	1	
SE[a]	1.0536	1.2087	108.12	2	
CCSD(T)_ae/cc-pCVQZ	1.0524	1.2085	108.08	3	
AI[b]	1.051774	1.208070	108.2196	4	

[a] Rovibrational corrections obtained from a CCSD(T)/cc-pVTZ cubic force field
[b] CCSD(T)_ae/cc-pV∞Z + higher order corrections using CCSDT/cc-pVTZ and CCSDTQ/cc-pVDZ

[1] Hirota E (1986) Chem Soc Jpn 1986: 1438
[2] Demaison J, Császár AG, Dehayem-Kamadjeu A (2006) The case of the weak N–X bond: ab initio, semi-experimental, and experimental equilibrium structures of XNO ($X = $ H, F, Cl, OH) and FNO_2. J Phys Chem A 110: 13609–13617
[3] Bak KL, Gauss J, Jørgensen P, Olsen J, Helgaker T, Stanton JF (2001) The accurate determination of molecular equilibrium structures. J Chem Phys 114: 6548-655.
[4] Puzzarini C, Heckert M, Gauss J (2008) The accuracy of rotational constants predicted by high-level quantum chemical calculations. I. Molecules containing first-row atoms. J Chem Phys 128: 194108/1-9

HNOS	HNSO (C_s)						Thionyl imide
Method	r_e(N–H)	r_e(N=S)	r_e(S=O)	∠(HNS)	∠(NSO)	Ref.	
Exp.[a]	1.0247(12)	1.5117(15)	1.4498(13)	116.01	120.42	1	
AI[b]	1.0203	1.5079	1.4478	116.58	120.44	1	
AI[c]	1.0199					2	
AI[d]	1.0246					This work	

[a] r_e extrapolated from r_z
[b] CCSD(T)_ae/cc-pV∞Z + aug(N,O) – MP2_ae/cc-pVQZ + 1 + MP2_ae/cc-pwCVQZ
[c] CCSD(T)_ae/cc-pVQZ + MP2_ae/cc-pCVQZ – MP2_ae/cc-pVQZ
[d] MP2/cc-pVTZ + offset, see Sect. 2.8

[1] Demaison J, Margulès L, Boggs JE, Rudolph HD (2001) Experimental and ab initio equilibrium structures of *cis*-thionylimide, HNSO: estimation of the Laurie correction. Struct Chem 12: 1–13
[2] Demaison J, Margulès L, Boggs JE (2000) The equilibrium N–H bond length. Chem Phys 260: 65–81

HN_2^+	$(C_{\infty v})$			Diazenylium
Method	r_e(N–H)	r_e(N–N)	Ref.	
Exp	1.03359(43)	1.092766(92)	1	
Exp	1.03460(14)	1.092698(26)	2	
r_m^ρ	1.0347(1)	1.0929(0)	3	

[1] Owrutsky JC, Gudeman CS, Martner CC, Tack LM, Rosenbaum NH, Saykally RJ (1986) Determination of the equilibrium structure of protonated nitrogen by high resolution infrared laser spectroscopy. J Chem Phys 84: 605–617
[2] Amano T, Hirao T, Takano J (2005) Submillimeter-wave spectroscopy of HN_2^+ and DN_2^+ in the excited vibrational states. J Mol Spectrosc 234: 170–175

[3] Berry RJ, Harmony MD (1989) The use of scaled moments of inertia in experimental structure determination of polyatomic molecules. Struct Chem 1: 49–59

HOP	HPO (C_s) Ground electronic state $^1A'$				Oxophosphine
Method	r_e(P–H)	r_e(P=O)	∠(HPO)	Ref.	
Exp	1.455(7)	1.4800(9)	104.57	1	
AI[a]	1.452	1.490	104.1	2	

[a] CASPT2
[1] Ozeki H, Saito S (2003) Microwave spectra of HPO and DPO: molecular structure. J Mol Spectrosc 219: 305–312
[2] Luna A, Merchán M, Ross BO (1995) A theoretical analysis of the lowest excited states in HNO/NOH and HPO/POH. Chem Phys 196: 437–445

HORb	RbOH ($C_{\infty v}$)			Rubidium hydroxide
Method	r_e(O–H)	r_e(Rb–O)	Ref.	
Exp	0.957(10)	2.301(2)	1	
Semirigid bender	0.986(11)	2.3025(14)	2	

[1] Lide DR, Matsumura (1969) Structure of the alkali hydroxides. IV. Interpretation of vibration–rotation interactions in CsOH and RbOH and refinement of structures. J Chem Phys 50: 3080–3086
[2] Brown RD, Godfrey PD, Kleiboemer B, McNaughton D (1994) Structures and flexibilities of the alkali hydroxides. J Mol Struct 327: 99–106

HO_2	(C_s) Electronic ground state $^2A''$				Hydroperoxyl
Method	r_e(O–H)	r_e(O–O)	∠(HOO)	Ref.	
Exp	0.9707(20)	1.33054(85)	104.29(31)	1	
Exp	0.9708(7)	1.33051(31)	104.30(13)	2	
AI[a]	0.972	1.334	104.12	3	

[a] From a potential energy surface calculated at the MRCI + Q/aug-cc-pVQZ level
[1] Lubic KG, Amano T, Uehara H, Kawaguchi K, Hirota E (1984) The ν_1 band of the DO2 radical by difference frequency laser and diode laser spectroscopy: The equilibrium structure of the hydroperoxyl radical. J Chem Phys 81: 4826–4831
[2] Uehara H, Kawaguchi K, Hirota E (1985). Diode laser spectroscopy of the ν_3 and ν_2 fundamental bands of DO_2. J Chem Phys 83: 5479–5485
[3] Xu C, Xie D, Zhang DH, Lin SY, Guo H (2005) A new ab initio potential-energy surface of $HO_2(X^2A'')$ and quantum studies of HO_2 vibrational spectrum and rate constants for the H + O_2 ↔ O + OH reactions. J Chem Phys 122: 244305

HPS	(C_s)				Thioxophosphine
Method	r_e(P–H)	r_e(P=S)	∠(HPS)	Ref.	
Exp.	1.4321(2)	1.9287(1)	101.78(1)	1	

(continued)

(continued)

HPS	(C_s)				Thioxophosphine
Method	r_e(P–H)	r_e(P=S)	∠(HPS)	Ref.	
CCSD(T)_ae/ cc-pwCVQZ	1.4303	1.9293	101.84	1	
CCSD(T)-F12b/ cc-pVQZ-F12	1.4335	1.9331	101.8	2	
MRCI-F12/ cc-pCVQZ-F12	1.4210	1.9264	102.16	3	

[1] Halfen DT, Clouthier DJ, Ziurys LM, Lattanzi V, McCarthy MC, Thaddeus P, Thorwirth S (2011) The pure rotational spectrum of HPS ($\tilde{X}^1A'$): Chemical bonding in second-row elements. J Chem Phys 134: 134302/1-9
[2] Yaghlane SB, Cotton CE, Francisco JS, Linguerri R, Hochlaf M (2013) Ab initio structural and spectroscopic study of HPSx and HSPx (x = 0, +1, −1) in the gas phase. J Chem Phys 139: 174313
[3] Mok DKW, Lee EPF, Chau F, Dyke JM (2014) Simulation of the single-vibronic-level emission spectrum of HPS. J Chem Phys 140: 194311

HS_2	HSS (C_s) $^2A''$ ground electronic state				Thiosulfeno
Method	r_e(H–S)	r_e(S–S)	∠(HSS)	Ref.	
Exp.[a]	1.3523	1.9603	101.74	1	
AI[b]	1.350	1.968	101.4	2	

[a] Derived from the r_z structure
[b] CCSD(T)/aug-cc-pV5Z
[1] Tanimoto M, Klaus T, Müller HSP, Winnewisser G (2000) Rotational spectra of the thiosulfeno radical, HSS and DSS, between 0.3 and 0.9 THz. J Mol Spectrosc 199: 73–80
[2] Francisco JS (2007) Sulfur atom exchange in the reaction of SH radicals with S atoms. J Chem Phys 126: 214301

H_2Hg	HgH_2 ($C_{\infty v}$)		Mercury dihydride
Method	r_e(Hg–H)	Ref.	
Exp. for HgH_2	1.63324(1)	1	
Exp. for HgD_2	1.63315(1)	1	
AI[a]	1.633405	2	

[a] From a refined potential energy surface calculated at the CCSD(T)_ae/cc-pwCVQZ-PP with a spin–orbit correction
[1] Shayesteh A, Yu S, Bernath PF (2005) Infrared emission spectra and equilibrium bond length of gaseous HgH_2 and HgD_2. J Phys Chem A 109: 10280–10286
[2] Botschwina P, Sebald P, Figgen D, Stoll H (2007) Calculated spectroscopic properties of HgH_2. Mol Phys 105: 1193–1205

H_2N	NH_2 (C_{2v}; ground electronic state $\tilde{X}^2B_1$)			Amino
Method	r_e(N–H)	∠(HNH)	Ref.	
Exp	1.0254(12)	102.85(14)	1	
SE (NH_2)[a]	1.0239	103.105	2	
SE (ND_2)[a]	1.0238	103.114	2	
CCSD(T)_ae/cc-pA′CVQZ[b]	1.0242	103.098	2	

[a] Rovibrational corrections obtained from a CCSD(T)_ae/cc-pA′CVQZ[b] cubic force field.
[b] aug-cc-pCVQZ on N and cc-pVQZ on H
[1] Kobayashi K, Ozeki H, Saito S, Tonooka M, Yamamoto S (1997) The microwave spectrum of the NHD radical in the ground electronic state, $^2A''$. J Chem Phys 107: 9289–9296
[2] Demaison J, Margulès L, Boggs JE (2003) Equilibrium structure and force field of NH_2. Phys Chem Chem Phys 5: 3359–3363

H_2NS	(C_{2v}; ground electronic state $\tilde{X}^2B_1$)				Thionitroxyl
Method	r_e(N–H)	r_e(N–S)	∠(HNH)	Ref.	
Exp.[a]	1.005(2)	1.6398(4)	118.9(2)	1	
AI[b]	1.005	1.635	121.3	2	

[a] Derived from the r_z structure
[b] CCSD(T)_fc/ CBS + core correlation at the CCSD(T)/cc-pCVTZ
[1] Habara H, Yamamoto S, Saito S (1998) Microwave spectrum and molecular structure of the H_2NS radical. J Chem Phys 109: 2700–2707
[2] Jarraya M, Ben Yaghlane S, Feifel R, Linguerri R, Hochlaf M (2021) Thionitroxyl radical (H_2NS) isomers: Structures, vibrational spectroscopy, electronic states and photochemistry. Front Astron Space Sci 8: 641067

H_2N_2	HNNH (C_{2h})				*trans*-Diazene
Method	r_e(N–H)	r_e(N=N)	∠(HNN)	Ref.	
Exp.[a]	1.029(1)	1.247(1)	106.3(1)	1	
SE[b]	1.02883(60)	1.24575(70)	106.34(18)	2	
CCSD(T)_ae/cc-pCVQZ	1.0284	1.2467	106.17	3	
AI[c]	1.028296	1.245165	106.4122	4	

[a] r_e from r_z
[b] Rovibrational corrections from a CCSD(T)_ae/cc-pCVQZ cubic force field
[c] CCSD(T)_fc/cc-pV∞Z + core(cc-pCVQZ) + higher order corrections using CCSDT/cc-pVTZ and CCSDTQ/cc-pVDZ
[1] Demaison J, Hegelund F, Bürger H (1997) Experimental and ab initio equilibrium structure of *trans*-diazene HNNH. J Mol Struct 413–414: 447–456
[2] Pawlowski F, Jørgensen P, Olsen J, Hegelund F, Helgaker T, Gauss J, Bak KL, Stanton JF (2002). Molecular equilibrium structures from experimental rotational constants and calculated vibration–rotation interaction constants. J Chem Phys 116: 6482–6496
[3] Bak KL, Gauss J, Jørgensen P, Olsen J, Helgaker T, Stanton JF (2001) The accurate determination of molecular equilibrium structures. J Chem Phys 114: 6548–655
[4] Puzzarini C, Heckert M, Gauss J (2008) The accuracy of rotational constants predicted by high-level quantum chemical calculations. I. Molecules containing first-row atoms. J Chem Phys 128: 194108/1-9

H_2O	(C_{2v})			Water
Method	r_e(O–H)	∠(HOH)	Ref.	
Exp	0.9575	104.51	1	
SE[a]	0.957902(28)	104.40(9)	2	
CCSD(T)_ae/cc-pCVQZ	0.9571	104.22	3	
AI[b]	0.957746	104.5809	4	
Adiabatic (H_2O)	0.95785	104.50_0	5	
Adiabatic (D_2O)	0.95783	104.49_0	5	

[a] Rovibrational corrections obtained from a CCSD(T)/cc-pVQZ force field.
[b] CCSD(T)_fc/cc-pV∞Z + core(cc-pCVQZ) + higher order corrections using CCSDT/cc-pVTZ and CCSDTQ/cc-pVDZ
[1] Benedict WS, Gailar N, Plyler EK (1956) Rotation-vibration spectra od deuterated water vapor. J Chem Phys 24: 1139–1165
[2] Pawlowski F, Jørgensen P, Olsen J, Hegelund F, Helgaker T, Gauss J, Bak KL, Stanton JF (2002). Molecular equilibrium structures from experimental rotational constants and calculated vibration–rotation interaction constants. J Chem Phys 116: 6482–6496
[3] Bak KL, Gauss J, Jørgensen P, Olsen J, Helgaker T, Stanton JF (2001) The accurate determination of molecular equilibrium structures. J Chem Phys 114: 6548-655
[4] Puzzarini C, Heckert M, Gauss J (2008) The accuracy of rotational constants predicted by high-level quantum chemical calculations. I. Molecules containing first-row atoms. J Chem Phys 128: 194108/1-9
[5] Csaszár AG, Czakó G, Furtenbacher T, Tennyson J, Szalay V, Shirin SV, Zobov NF, Polyansky OL (2005) On equilibrium structures of the water molecule. J Chem Phys 122: 214305/1-10

H_2O^+	(C_{2v}) Electronic ground state $^2B_1$			Oxoniumyl
Method	r_e(O–H)	∠(HOH)	Ref.	
Exp	1.001	108.9	1	
Exp.[a]	1.0015(11)	108.86(10)	2	
Exp	0.9992(6)	109.30(10)	3	
AI[b]	1.0004	109.0745	4	

[a] Using an experimental anharmonic force field
[b] From a MRCI potential energy surface
[1] Dinelli BM, Crofton MW, Oka T (1988) Infrared spectroscopy of the ν_3 band of H_2O^+. J Mol Spectrosc 127: 1–11
[2] Kauppi E, Halonen L (1990). Equilibrium structure and potential-energy surface for the water cation. Chem Phys Lett 169: 393–398
[3] Huet TR, Pursell CJ, Ho WC, Dinelli BM, Oka T (1992) Infrared spectroscopy and equilibrium structure of $H_2O^+(\tilde{X}\,^2B_1)$. J Chem Phys 97: 5977–5987
[4] Weis B, Carter S, Rosmus P, Werner H-J, Knowles PJ (1989). A theoretical rotationally resolved infrared spectrum for $H_2O + (X\,^2B_1)$. J Chem Phys 91: 2818–2833

H_2OS	HSOH (C_s)			Oxadisulfane
Method	Exp.[a]	SE[b]	AI[c]	
Ref.	1	2	3	
r_e(S–O)	1.669514(30)	1.6616(1)	1.6619	
r_e(S–H)	1.3414	1.3420(2)	1.3414	
r_e(O–H)	0.9601	0.9606(3)	0.9601	
∠(OSH)	97.1784(17)	98.57(5)	98.55	
∠(SOH)	107.01	107.19(3)	107.01	
τ(HSOH)	84.659(21)	90.41(11)	91.29	

[a] Semirigid bender model
[b] CCSD(T)/cc-pVTZ cubic force field
[c] CCSD(T)_ae/cc-pCVQZ
[1] Ross SC, Yamada KMT, Ito F (2010). Torsion–rotation coupling and the determination of the torsional potential energy function of HSOH. Phys Chem Chem Phys 12: 11133–11150
[2] Winnewisser G, Lewen F, Thorwirth S, Behnke M, Hahn J, Gauss J, Herbst E (2003) Gas-phase detection of HSOH: Synthesis by flash vacuum pyrolysis of di-*tert*-butyl sulfoxide and rotational-torsional spectrum. Chem Eur J 9: 5501–5510
[3] Baum O, Esser S, Gierse N, Brünken S, Lewen F, Hahn J, Gauss J, Schlemmer S, Giesen TF (2006). Gas-phase detection of HSOD and empirical equilibrium structure of oxadisulfane. J Mol Struct 795 : 256–262

H_2O_2	(C_2)					Hydrogen peroxide
Method	r_e(O–H)	r_e(O–O)	∠(OOH)	τ(HOOH)	Ref.	
Exp.[a]	0.967	1.4556	102.32	113.70	1	
SE[b]	0.9617(2)	1.4524(3)	99.76(6)	113.6(3)	2	
CCSD(T)_ae/ cc-pCVQZ	0.9619	1.4497	99.99		3	
AI[c]	0.9623	1.4515	100.07	113.17	2	
AI[d]	0.9628	1.4509	100.09	112.81	4	

[a] Minimum of the torsional potential
[b] Rovibrational corrections obtained from a variational calculation with a potential energy surface from 4
[c] Extrapolation of optimized CCSD(T)_ae/cc-pCVTZ-5Z geometries supplemented by higher order perturbative quadruple correction calculated with the cc-pCVTZ basis set and the frozen core approximation
[d] CCSD(T)/aug-ccpV7Z with valence electrons correlated, corrections for core correlation, scalar relativistic correction, and higher-order valence-electron correlation effects
[1] Pelz G, Yamada KMT, Winnewisser G (1993) Torsional dependence of the effective rotational constants of H_2O_2 and H_2S_2. J Mol Spectrosc 159: 507–520
[2] Baraban J, Changala PB, Stanton JF (2018) The equilibrium structure of hydrogen peroxide. J Mol Spectrosc 343: 92–95
[3] Bak KL, Gauss J, Jørgensen P, Olsen J, Helgaker T, Stanton JF (2001) The accurate determination of molecular equilibrium structures. J Chem Phys 114: 6548–655
[4] Małyszek P, Koput J (2013) Accurate ab initio potential energy surface and vibration- rotation energy levels of hydrogen peroxide. J Comput Chem 34: 337–345

H_2S	(C_{2v})			Hydrogen sulfide
Method	r_e(S–H)	∠(HSH)	Ref.	
Exp	1.3362	92.06	1	
CCSD(T)_ae/cc-pCVQZ	1.335272	92.290	2	

[1] Cook RL, De Lucia FC, Helminger P (1975) Molecular force field and structure of hydrogen sulfide: recent microwave results. J Mol Struct 28: 237–246

[2] Coriani S, Marchesan D, Gauss J, Hättig C, Helgaker T, Jørgensen P (2005) The accuracy of ab initio molecular geometries for systems containing second-row atoms. J Chem Phys 123: 184107/1-12

H_2Se	(C_{2v})			Hydrogen selenide
	r_e(Se–H)	∠(HSeH)	Ref.	
Exp	1.459099(77)	90.958(11)	1	
Exp.[a]	1.459466(15)	90.821	2	

[a] From experimental potential energy surface

[1] Flaud J-M, Camy-Peyret C, Arcas P, Bürger H, Willner H (1994) $H_2{}^{80}Se$: high-resolution study of the $3\nu_2$, $\nu_1 + \nu_2$ and $\nu_2 + \nu_3$ bands and determination of equilibrium rotational constants and structure. J Mol Spectrosc 167: 383–399

[2] Ulenikov ON, Bekhtereva ES, Sanzharov NA, Jensen P (2004) A refined potential energy function for the electronic ground state of H_2Se. J Mol Spectrosc 227: 1–12

H_2Si	(C_{2v}) $^1A_1$ ground electronic state			Silylene
Method	r_e(Si–H)	∠(HSiH)	Ref.	
Exp	1.5137(3)	92.04(5)	1	
CCSD(T)_ae/cc-pCVQZ	1.512962	92.203	2	

[1] Kokkin DL, Ma T, Steimle T, Sears TJ (2016) Detection and characterization of singly deuterated silylene, SiHD, via optical spectroscopy. J Chem Phys 144: 244304

[2] Coriani S, Marchesan D, Gauss J, Hättig C, Helgaker T, Jørgensen P (2005) The accuracy of ab initio molecular geometries for systems containing second-row atoms. J Chem Phys 123: 184107/1-12

H_2Te	(C_{2v})			Hydrogen telluride
Method	r_e(Te–H)	∠(HTeH)	Ref.	
Exp	1.65145(10)	90.2635(90)	1	
Exp.[a]	1.652124(37)		2	

[a] From experimental potential energy surface, HTeH assumed to be 89.7365°

[1] Flaud J-M, Arcas P, Bürger H, Polanz O, Halonen L (1997) High-resolution study of the ν_2, $2\nu_1$, $\nu_1 + \nu_3$ and $2\nu_3$ bands of hydrogen telluride: determination of equilibrium rotational constants and structure. J Mol Spectrosc 183: 310–335.

[2] Gómez PC, Jensen P (1997) A potential energy surface for the electronic ground state of H_2Te derived from experiment. J Chem Phys 185: 282–289

H_2Zn	ZnH_2 ($C_{\infty v}$)		Zinc dihydride
Method	r_e(Zn–H)	Ref.	
Exp. for ^{64}Zn–H	1.52413(1)	1	
Exp. for ^{64}Zn–D	1.52394(1)	1	
AI[a]	1.52416	2	

[a] From a refined potential energy surface calculated at the CCSD(T)_ae/cc-pwCV5Z-PP

[1] Shayesteh A, Gordon IE, Appadoo DRT, Bernath PF (2005) Infrared emission spectra and equilibrium bond lengths of gaseous ZnH_2 and ZnD_2. Phys Chem Chem Phys 7: 3132–3142

[2] Sebald P, Vennekate H, Oswald R, Botschwina P, Stoll H (2010). A theoretical study of ZnH_2: a case of very strong Darling–Dennison resonance. Mol Phys 108: 487–499

H_3^+	(D_{3h})		Trihydrogen (1+) ion
	r_e(H–H)	Ref.	
H_3^+	0.873190(5)	1	
D_3^+	0.873183(5)	2	

[1] Majewski WA, McKellar ARW, Sadovskii D, Watson JKG (1994) New observations and analysis of the infrared vibration-rotation spectrum of H_3^+. Can J Phys 72: 1016-127.

[2] Amano T, Chan M-C, Civis S, McKellar ARW, Majewski WA, Sadovskii D, Watson JKG (1994) The infrared vibration–rotation spectrum of the molecular ion: extension to higher vibrational and rotational quantum numbers. Can J Phys, 72: 1007–1015.

H_3N	NH_3 (C_{3v})			Ammonia
Method	r_e(N–H)	∠(HNH)	Ref.	
Exp	1.0116(6)	106.7	1	
Exp.[a, b]	1.0141	107.49	2	
Exp.[a, c]	1.0149	107.40	2	
SE [d]	1.01139(60)	107.17(18)	3	
CCSD(T)_ae/cc-pCVQZ	1.0112	106.36	4	
AI[e]	1.010936	106.8147	5	
AI[f]	1.010990	106.7474	6	

[a] From an anharmonic potential function

[b] Better for the prediction of energies of excited states

[c] The best for the reproduction of inversion splittings

[d] Rovibrational corrections obtained from a CCSD(T)/cc-pVQZ force field

[e] CCSD(T)_fc/cc-pV∞Z + core(cc-pCVQZ) + higher order corrections using CCSDT/cc-pVTZ and CCSDTQ/cc-pVDZ

[f] CCSD(T)-F12a_ae/cc-pCVQZ-F12 + relativistic + DBOC + CCSDTQP

[1] Benedict WS, Plyler K (1957) Vibration–rotation bands of ammonia: II. The molecular dimensions and harmonic frequencies of ammonia and deuterated ammonia. Can J Phys 35: 1235–1241

[2] Spirko V, Kraemer WP (1989) Anharmonic potential function and effective geometries for the NH_3 molecule. J Mol Spectrosc 133: 331–344

[3] Pawlowski F, Jørgensen P, Olsen J, Hegelund F, Helgaker T, Gauss J, Bak KL, Stanton JF (2002). Molecular equilibrium structures from experimental rotational constants and calculated vibration–rotation interaction constants. J Chem Phys 116: 6482–6496

[4] Bak KL, Gauss J, Jørgensen P, Olsen J, Helgaker T, Stanton JF (2001) The accurate determination of molecular equilibrium structures. J Chem Phys 114: 6548–6556
[5] Puzzarini C, Heckert M, Gauss J (2008) The accuracy of rotational constants predicted by high-level quantum chemical calculations. I. Molecules containing first-row atoms. J Chem Phys 128: 194108/1-9
[6] Egorov O, Rey MM, Nikitin AV, Viglaska D (2021) New ab initio potential energy surfaces for NH_3 constructed from explicitly correlated coupled cluster methods. J Phys Chem A 125: 10568–10579

H_3N^+	$NH_3{}^+$ (D_{3h}) $^2A'_2$ ground electronic state		Ammoniumyl
Method	r_e(N–H)	Ref.	
Exp	1.014	1	
AI[a]	1.0243	2	
AI[b]	1.0205(2)	3	
AI[c]	1.0216	4	

[a] CEPA-1
[b] MRCI potential adjusted to experimental data
[c] CCSD(T)/aug-cc-pVQZ

[1] Bawendi MG, Rehfuss BD, Dinelli BM, Okumura M, Oka T (1989) Observation and analysis of the ν_3 band of $NH_3{}^+$. J Chem Phys 90: 5910–5917
[2] Botschwina P (1989) Ion and Cluster ion Spectrosopy and Structure, Maier JP (Ed), Elsevier, Amsterdam
[3] Kraemer WP, Špirko V (1992). Potential energy function and rotation-vibration energy levels of $NH_3{}^+$. J Mol Spectrosc 153: 276–284
[4] Léonard C, Carter S, Handy NC, Knowles PJ (2001) Theoretical determination of the vibrational levels of $NH_3{}^+$ and its isotopomers. Mol Phys 99: 1335–1346

H_3O^+	(C_{3v})			Oxonium
Method	r_e(O–H)	∠(HOH)	Ref.	
Exp	0.9758	111.3	1	
Exp	0.9702(89)	109.4(38)	2	
Exp	0.974(1)	113.6(1)	3	
CCSD(T)/CBS	0.9747	112.02	4	
AI[a]	0.9765	111.7	5	

[a] From a potential energy surface calculated at the CCSD(T)/aug-cc-pVQZ level

[1] Sears TJ, Bunker PR, Davies PB, Johnson SA, Špirko V (1985). Diode laser absorption spectroscopy of D_3O^+: Determination of the equilibrium structure and potential function of the oxonium ion. J Chem Phys 83: 2676–2685
[2] Araki M, Ozeki H, Saito S (1999) Microwave spectrum of the inversion–rotation transitions of the D_3O^+ ion: $\Delta k = \pm 3n$ interaction and equilibrium structure. Mol Phys 97: 177–183
[3] Tang J, Oka T (1999). Infrared Spectroscopy of H_3O^+: The ν_1 fundamental band. J Mol Spectrosc 196: 120–130
[4] Rajamäki T, Miani A, Halonen L (2003). Six-dimensional ab initio potential energy surfaces for H_3O^+ and NH_3: Approaching the sub-wave number accuracy for the inversion splittings. J Chem Phys 118: 10929–10938
[5] Huang X, Carter S, Bowman J (2003). Ab initio potential energy surface and rovibrational energies of H_3O^+ and its isotopomers. J Chem Phys 118: 5431–5441

H_3P	PH_3 (C_{3v})			Phosphine
Method	r_e(P–H)	∠(HPH)	Ref.	
Exp. (from PD_3)	1.4117846(16)	93.42523(19)	1	
Exp. (from PH_3)	1.411607(16)	93.4184(21)	1	
CCSD(T)_ae/cc-pCVQZ	1.411539	93.482	2	
SE (from PD_3)[a]	1.41185	93.383	3	
SE (from PH_3)[a]	1.41195	93.378	3	
AI[b]	1.4117	93.38	3	

[a] Rovibrational corrections from CCSD(T)_fc/aug-cc-pV(Q + d)Z force field
[b] CCSD(T)-F12b_ae/cc-pCVQZ-F12 + higher order corrections using CCSDT/cc-pVTZ and CCSDTQ/cc-pVDZ + relativistic correction
[1] Canè E, Fusina L, Bürger H, Jerzembeck W, Brünken S, Lewen F, Winnewisser G. (2002) The ground state spectroscopic parameters and equilibrium structure of PD_3. J Mol Spectrosc 215: 1–9
[2] Coriani S, Marchesan D, Gauss J, Hättig C, Helgaker T, Jørgensen P (2005) The accuracy of ab initio molecular geometries for systems containing second-row atoms. J Chem Phys 123: 184107/1-12
[3] Breidung J, Thiel W (2019) Equilibrium structures of the phosphorous trihalides PF_3 and PCl_3 and the phosphoranes PH_3F_2, PF_5, PCl_3F_2, and PCl_5. J Phys chem A 123: 5600–5612

H_3S^+	(C_{3v})			Sulfonium
Method	r_e(S–H)	∠(HSH)	Ref.	
Exp.[a]	1.350620(34)	94.2758(23)	1	
SE[b]	1.349872(13)	94.1405(26)	2	
CCSD(T)_ae/cc-pwCVQZ	1.34918	94.266	2	
AI[c]	1.3495(10)	94.26(5)	2	

[a] r_e from r_z
[b] Rovibrational corrections obtained from a CCSD(T)/cc-pwCVQZ cubic force field.
[c] CCSD(T)_fc/cc-pV6Z + + core(cc-pwCVQZ).
[1] Tinti F, Bizzocchi L, Degli Esposti C, Dore L (2006) Millimeter-wave spectroscopy of SHD_2^+ and its symmetric isotopologues: Determination of the molecular structure of the sulfonium ion. J Mol Spectrosc 240: 202–209.
[2] Puzzarini C (2007) Ab initio anharmonic force field and equilibrium structure of the sulfonium ion. J Mol Spectrosc 242: 70–75. See also Dore L, Bizzocchi L, Degli Esposti C (2009) Millimeter-wave spectroscopy of deuterated hydrogen sulfide, SH_2D^+. J Mol Spectrosc 254: 33–38

H_3Sb	SbH_3 (C_{3v})			Stibine
Method	r_e(Sb–H)	∠(HSbH)	Ref.	
Exp. (from $^{121}SbD_3$)	1.700909(2)	91.6713(2)	1	
Exp. (from $^{121}SbH_3$)	1.700013(1)	91.5586(1)	1	
Exp. (from $^{123}SbD_3$)	1.700897(3)	91.6728(3)	2	
Exp. (from $^{123}SbH_3$)	1.700013(1)	91.55910(12)	2	

(continued)

(continued)

H_3Sb	SbH_3 (C_{3v})			Stibine
Method	r_e(Sb–H)	∠(HSbH)	Ref.	
CCSD(T)_fc[a]	1.7015	91.76	2	

[a] aug-cc-pVQZ on H, SDB-aug-cc-pVQZ on Sb (with a large core pseudo-potential ECP46MWB and a core polarization potential)
[1] Canè E, Di Lonardo G, Fusina L, Jerzembeck W, Bürger H, Breidung J, Thiel W (2006) Rotation spectrum and high-resolution infrared spectra of the fundamental bands of $^{121}SbD_3$. Determination of the ground state and equilibrium structures. Ab initio calculations of the spectroscopic parameters. J Mol Struct 780–781: 98–110
[2] Canè E, Di Lonardo G, Fusina L, Jerzembeck W, Bürger H, Breidung J, Thiel W (2005) Rotation spectrum and infrared fundamental bands of $^{123}SbD_3$. Determination of molecular geometry and ab initio calculations of spectroscopic parameters. Mol Phys 103: 557–577

H_3Si^+	SiH_3^+ (D_{3h})		Silylium
Method	r_e(Si–H)	Ref.	
Exp	1.460	1	
AI (CEPA-2)	1.459(1)	2	

[1] Davies PB, Smith DM (1994) Diode laser spectroscopy and coupled analysis of the ν_2 and ν_4 fundamental bands of SiH_3^+. J Chem Phys 100: 6166–6174
[2] Botschwina P, Oswald M (1992) Ab initio calculation of the vibrational frequencies and the equilibrium geometry of SiH_3^+. J Chem Phys 96: 4044–4045

H_4N^+	NH_4^+ (T_d)		Ammonium
Method	r_e(N–H)	Ref.	
Exp	1.0208(20)	1	
CCSD(T)_ae/cc-pwCVQZ	1.0205	2	
AI[a]	1.0203	2	

[a] CCSD(T)_fc/cc-pV∞Z + CCSD(T)_ae/cc-pwCVQZ - CCSD(T)_fc/cc-pwCVQZ
[1] Crofton MW, Oka T (1987) Observation of forbidden transitions of ammonium ion (NH_4^+) ν_3 band and determination of ground state rotational constants. Observation of ν_3 band allowed transitions of ND_4^+. J Chem Phys 86: 5983–5988
[2] Demaison J, Margulès L, Boggs JE (2000) The equilibrium N–H bond length. Chem Phys 260: 65–81

H_4Si	SiH_4 (T_d)		Silane
Method	r_e(Si–H)	Ref.	
Exp	1.4741	1	
Exp	1.4734(3)	2	

(continued)

(continued)

H_4Si	SiH_4 (T_d)		Silane
Method	r_e(Si–H)	Ref.	
CCSD(T)_ae/cc-pCVQZ	1.474221	3	

[1] Ohno K, Matsuura H, Endo Y, Hirota E (1986) The microwave spectra of deuterated silanes: SiH_3D, SiH_2D_2, and $SiHD_3$. J Mol Spectrosc 111, 73–82
[2] Ohno K, Matsuura H, Endo Y, Hirota E (1986) The microwave spectra of deuterated silanes, germanes, and stannanes. J Mol Spectrosc 118: 1–17
[3] Coriani S, Marchesan D, Gauss J, Hättig C, Helgaker T, Jørgensen P (2005) The accuracy of ab initio molecular geometries for systems containing second-row atoms. J Chem Phys 123: 184107/1-12

H_4Sn	SnH_4 (T_d)		Stannane
Method	r_e(Sn–H)	Ref.	
Exp	1.6935(26)	1	
SE[a]	1.6954(6)	2	

[a] Rovibrational corrections calculated from a CCSD(T)/ANO-RCC force field
[1] Ohno K, Matsuura H, Endo Y, Hirota E (1986) The microwave spectra of deuterated silanes, germanes, and stannanes. J Mol Spectrosc 118: 1–17
[2] Fusina L, Nivellini GD, Salzillo T, Lamarra M, Tarroni R (2012) Ab initio quartic force field of stannane and rotational analysis of the ν_1 infrared band of $H^{120}SnD_3$. J Chem Phys 137: 204316

NO_2	(C_{2v}) Electronic ground state X^2A_1			Nitrogen dioxide
Method	r_e(NO)	∠(ONO)	Ref.	
Exp	1.19389(4)	133.86(3)	1	
Exp	1.19455(3)	133.85(2)	2	
Exp. (GED)	1.195(3)	133.7(2)	3	
Exp.[a]	1.19464(15)	133.888(2)	4	
MORBID	1.18724(470)	133.767(195)	5,6	
AI[b]	1.1933	135.0	7	

[a] From an experimental potential energy surface
[b] From interpolated potential energy surfaces calculated at the icMRCI + Q/cc-pVQZ level
[1] Morino Y, Tanimoto M, Saito S, Hirota E, Awata R, Tanaka T (1983) Microwave spectrum of nitrogen dioxide in excited vibrational states—Equilibrium structure. J Mol Spectrosc 98: 331–348
[2] Morino Y, Tanimoto M (1984) Vibration–rotation interaction constants and the cubic potential function of the NO_2 molecule in the small-amplitude approximation. Can J Phys 62: 1315–1322
[3] Borisenko KB, Kolonits M, Rozsondai B, Hargittai I (1997) Electron diffraction study of the nitrogen dioxide molecular structure at 294, 480, and 691 K. J Mol Struct 413–414: 121–131
[4] Hardwick JL, Brand JCD (1976) Anharmonic potential constants and the large amplitude bending vibration in nitrogen dioxide. Can J Phys 54: 80–91
[5] Tashkun SA, Jensen P (1994) The low-energy part of the potential function for the electronic ground state of NO_2 derived from experiment. J Mol Spectrosc 165: 173–184
[6] Schryber JH, Polyansky OL, Jensen P, Tennyson J (1997) On the spectroscopically determined potential energy surfaces for the electronic ground states of NO_2 and H_2O. J Mol Spectrosc 185: 234–243

[7] Schinke R, Grebenshchikov SY, Zhu H (2008) The photodissociation of NO_2 in the second absorption band: Ab initio and quantum dynamics calculations. Chem Phys 346: 99–114

N_2O	($C_{\infty v}$)					Dinitrogen monoxide
Method	r_e(NN)	r_e(NO)	$r(+)^a$	$r(-)^b$	Ref.	
Exp	1.127292	1.185089	2.312381(4)	0.05779(7)	1	
CCSD(T)_ae/cc-pwCVQZ	1.1267	1.1848			2	
SE[c]	1.12695(10)	1.18539(5)			3	

[a] $r(+) = r_e(NN) + r_e(NO)$
[a] $r(-) = r_e(NN) - r_e(NO)$
[c] Using an ab initio, potential surface calculated at the CCSD(T)-F12b/aug-cc-pV5Z + core-valence (CCSD(T)/aug-cc-pCV5Z), scalar relativity corrections, and corrections beyond CCSDTQP
[1] Teffo J-L, Chédin A (1989) Internuclear potential and equilibrium structure of the nitrous oxide molecule from rovibrational data. J Mol Spectrosc 135: 389–409
[2] Demaison J, Császár AG, Dehayem-Kamadjeu A (2006) The case of the weak N-X bond: Ab initio, semi-experimental, and experimental equilibrium structures of XNO (X = H, F, Cl, OH) and FNO_2. J Phys Chem 110: 13609–13617
[3] Schröder B, Sebald P, Stein C, Weser O, Botschwina P (2015) Challenging high-level ab initio rovibrational spectroscopy: The nitrous oxide molecule. Z Phys Chem 229: 1663–1690

OS_2	S_2O (C_s)				Disulfur monoxide
Method	r_e(SS)	r_e(SO)	∠(SSO)	Ref.	
Exp	1.88424(11)	1.45621(13)	117.876(4)	1	
CCSD(T)_ae/cc-pCVQZ	1.885254	1.455467	118.038	2	

[1] Lindenmeyer J, Rudolph HD, Jones H (1986) The equilibrium structure of disulfur monoxide: Diode laser spectroscopy of ν_1 and ν_3 of $S_2{}^{18}O$ and ν_3 of $S_2{}^{16}O$. J Mol Spectrosc 119: 56–67
[2] Coriani S, Marchesan D, Gauss J, Hättig C, Helgaker T, Jørgensen P (2005) The accuracy of *ab initio* molecular geometries for systems containing second-row atoms. J Chem Phys 123: 184107/1-12

O_2P	PO_2 (C_{2v}) Ground electronic state $^2A_1$			Phosphorous dioxide
Method	r_e(PO)	∠(OPO)	Ref.	
Exp.[a]	1.4645(12)	135.296(7)	1	
SE[b]	1.46424(7)	135.119(14)	1	
AI[c]	1.466	135.6	2	

[a] r_0 value + offset
[b] Rovibrational corrections calculated from the B3LYP/aug-pcS-3 force field
[c] SCF + correction using the known structure of PO
[1] Demaison J, Vogt N (2022) Effective and equilibrium structures of XY_n-type molecules: SCl_2, PF_2, PO_2, ZrO_2, and CF_2S as additional example and nonbonded distances F⋯F, Cl⋯Cl and O⋯O. J Mol Spectrosc 387: 111661

[2] Kabbadj Y, Lievin J (1989) Ab initio study of the electronic structure of the PO_2 radical. Physica Scripta, 40: 259–269

O_2S	SO_2 (C_{2v})			Sulfur dioxide
Method	r_e(SO)	∠(OSO)		
Exp	1.43076(13)	119.33(1)	1	
Exp	1.430782(15)	119.3297(30)	2	
Exp	1.4308302(17)	119.33042(29)	3	
Exp.[a]	1.4307897(99)	119.32893(27)	4	
Exp.[b]	1.43079396(93)	119.328783(55)	4	
CCSD(T)_ae/ aug-cc-pwCVQZ	1.4311	119.389	4	

[a] Mean value for $^{32}S^{16}O_2$ and $^{34}S^{16}O_2$
[b] Using an experimental cubic force field and the rotational constants of 17 isotopologues
[1] Saito S (1969) Microwave spectrum of sulfur dioxide in doubly excited vibrational states and determination of the γ constants. J Mol Spectrosc 30: 1–16
[2] Flaud J-M, Lafferty V (1993) $^{32}S^{16}O_2$: a refined analysis of the $3\nu_3$ band and determination of equilibrium rotational constants. J Mol Spectrosc 161: 396–402
[3] Morino Y, M. Tanimoto M (1994) Analysis of Darling-Dennison interactions and related phenomena in the SO_2 molecule for the force field study. J Mol Spectrosc 166: 310–320
[4] Demaison J, Liévin J (2022) Accuracy of the equilibrium structure of sulfur dioxide. Mol Phys 120: e1959857

O_2Se	SeO_2 (C_{2v})			Selenium dioxide
Method	r_e(SeO)	∠(OSeO)	Ref.	
Exp. (MW)	1.6076(6)	113.83(8)	1	
Exp. (GED)	1.6078	113.910(1)	2	

[1] Takeo H, Hirota E, Morino Y (1970) Equilibrium structure and potential function of selenium dioxide by microwave spectroscopy. J Mol Spectrosc 34: 370–382
[2] Spiridonov VP, Gershikov AG (1986) Molecular potential functions from joint analysis of gas electron diffraction and spectroscopic data. J Mol Struct 140: 173–191

O_3	(C_{2v})			Ozone
Method	r_e(OO)	∠(OOO)	Ref.	
Exp.[a]	1.27173(20)	116.783(33)	1	
Exp.[b]	1.27276(15)	116.7542(25)	2	
Exp. (extrapolation)	1.27163(80)	116.799(200)	3	
CCSD(T)_ae/aug-cc-pwCVQZ[c]	1.26599	117.2312	3	
i-MRCI/CBS	1.2751	116.78	4	

[a] Using the experimental rovibrational corrections
[b] Minimum of the experimental potential surface
[c] Coupled-cluster diagnostics: $T_1 = 0.024$; $D_1 = 0.078$

[1] Colmont J-M, Demaison J, Cosléou J (1995) Rotational spectra of all ^{17}O bisubstituted ozones: comparison between different structures. J Mol Spectrosc 171: 453–467
[2] Tyuterev VG, Tashkun S, Jensen P, Barbe A, Cours T (1999) Determination of the effective ground state potential energy function of ozone from high-resolution infrared spectra. J Mol Spectrosc 198: 57–76
[3] Demaison J, Liévin J, Vogt N (2023) Accurate equilibrium structures of some challenging molecules: FNO, ClNO, HONO, FNO_2, and N_2O. J Mol Spectres 394: 111788
[4] Tyuterev VG, Kochanov RV, Tashkun SA, Holka F, Szalay P (2013) New analytical model for the ozone electronic ground state potential surface and accurate ab initio vibrational predictions at high energy range. J Chem Phys 139: 134307/1-22

O_3S	SO_3 (D_{3h})		Sulfur trioxide
Method	r_e(SO)	Ref.	
Exp.($^{32}S^{16}O_3$)	1.4173403(13)	1	
Exp.($^{34}S^{16}O_3$)	1.4173470(68)	2	
Exp.($^{32}S^{18}O_3$)	1.417333(4)	3	
Exp.($^{34}S^{18}O_3$)	1.417339(3)	4	
Exp.[a]	1.4184(10)	5	
CCSD(T)_ae/cc-pCVQZ	1.417940	6	

[a] Using an experimental anharmonic force field

[1] Barber J, Chrysostome EtH, Masiello T, Nibler JW, Maki A, Weber A, Blake TA, Sams RL (2002) Analysis of ν_2, ν_4 Infrared Hot Bands of $^{32}S^{16}O_3$: Resolution of the puzzle of the ν_1 CARS spectrum. J Mol Spectrosc 216: 105–112
[2] Barber J, Chrysostome ETH, Masiello T, Nibler JW, Maki A, Weber A, Blake TA, Sams RL (2003) Analysis of the ν_2, ν_4 infrared hot bands and ν_1 CARS spectrum of $^{34}S^{16}O_3$. J Mol Spectrosc 218: 204–212
[3] Masiello T, Vulpanovici N, Barber J, Chrysostome EtH, Nibler JW, Maki A, Blake TA, Sams RL, Weber A (2004) Analysis of high-resolution infrared and CARS spectra of $^{32}S^{18}O_3$. J Mol Spectrosc 227: 50–59
[4] Masiello T, Barber J, Chrysostome EtH, Nibler JW, Maki A, Weber A, Blake TA, Sams RL (2004) Analysis of high-resolution infrared and CARS spectra of $^{34}S^{18}O_3$. J Mol Spectrosc 223: 84–95
[5] Dorney AJ, Hoy AR, Mills IM (1973) An anharmonic force field for SO_3. J Mol Spectrosc 45: 253–260
[6] Coriani S, Marchesan D, Gauss J, Hättig C, Helgaker T, Jørgensen P (2005) The accuracy of ab initio molecular geometries for systems containing second-row atoms. J Chem Phys 123: 184107/1-12

O_4Os	OsO_4 (T_d)		Osmium tetroxide
Method	r_e(OsO)	Ref.	
Exp.	1.70919(16)	1	
Exp.[a]	1.7092(13)	2	

[a] Derived from the r_0 structure

[1] Louviot M, Boudon V, Manceron L, Roy P, Bermejo D, Martinez RZ (2012) High-resolution spectroscopy and structure of osmium tetroxide. A benchmark study on $^{192}OsO_4$. Inorg Chem 51: 10356–10365

[2] Demaison J, Vogt N (2022) Effective and equilibrium structures of XY_n-type molecules: SCl_2, PF_2, PO_2, ZrO_2, and CF_2S as additional example and nonbonded distances F· · · F, Cl· · · Cl and O· · · O. J Mol Spectrosc 387: 111661

References

1. Gordy W, Cook RL (1984) Microwave molecular spectra. Wiley, New York
2. Groner P (1999) The quest for equilibrium structure of molecules. In: Durig JR (ed) Vibrational spectra and molecular structure, vol 24. Elsevier, Amsterdam, pp 165–252
3. Demaison J (2007) Experimental, semi-experimental and ab initio equilibrium structures. Mol Phys 105:3109–3138
4. Puzzarini C, Stanton JF, Gauss J (2010) Quantum-chemical calculation of spectroscopic parameters for rotational spectroscopy. Int Rev Phys Chem 29:273–367
5. Rudolph HD, Demaison J (2011) Determination of structural parameters from the inertial moments. In: Demaison J, Boggs JE, Császár AG (eds) Equilibrium molecular structures. From spectroscopy to quantum chemistry. CRC Press, Boca Raton, Boca Raton, pp 125–158
6. Vázquez J, Stanton JF (2011) Semiexperimental equilibrium structures: computational aspects. In: Demaison J, Boggs JE, Császár AG (eds) Equilibrium molecular structures. From spectroscopy to quantum chemistry. CRC Press, Boca Raton, pp 53–87
7. Demaison J, Vogt N (2020) Accurate structure determination of free molecules. Springer Nature, Switzerland, 277 p
8. Mills IM (1972) Vibration-rotation structure in asymmetric and symmetric molecules. In: Rao KN, Mathews CW (eds) Molecular spectroscopy: modern research, vol 1. Academic Press, New York
9. Watson JKG (1973) The estimation of equilibrium molecular structures from zero-point rotational constants. J Mol Spectrosc 48:479–502
10. Aliev MR, Watson JKG (1985) Higher order effects in the vibration rotation spectra od semirigid molecules. In: Rao KN (ed) Molecular spectroscopy; modern research, vol 3. Academic Press, Orlando, pp 1–67
11. Demaison J, Liévin J (2022) Accuracy of the equilibrium structure of sulfur dioxide. Mol Phys 120:e1950857
12. Flaud JM, Lafferty WJ (1993) $^{32}S^{16}O_2$: a refined analysis of the $3\nu_3$ band and determination of equilibrium rotational constants. J Mol Spectrosc 161:396–402
13. Morino Y, Tanimoto M (1994) Analysis of Darling-Dennison interactions and related phenomena in the SO_2 molecule for the force field study. J Mol Spectrosc 166:310–320
14. Dréan P, Demaison J, Poteau J, Denis J-M (1996) Rotational spectrum and structure of HCP. J Mol Spectrosc 176:139–145
15. Puzzarini C, Tarroni R, Palmieri P, Carter S, Dore L (1996) Accurate ab initio prediction of the equilibrium geometry of HCO+ and of rovibration energy levels of DCO+. Mol Phys 87:879–898
16. Puzzarini C, Tarroni R, Palmieri P, Demaison J, Senent ML (1996) Rovibrational energy levels and equilibrium geometry of HCP. J Chem Phys 105:3132–3141
17. Hoy AR, Mills IM, Strey G (1972) Anharmonic force constant calculations. Mol Phys 24:1265–1290

18. Demaison J, Wlodarczak G, Rudolph HD (1997) Determination of reliable structures from rotational constants. In: Hargittai M, Hargittai I (eds) Advances in molecular structure research, vol 3. JAI Press, Greenwich, pp 1–52
19. Sarka K, Demaison J (1988) Perturbation theory, effective Hamiltonians and force constants. In: Jensen P, Bunker PR (eds) Computational molecular spectroscopy. Wiley, New York, pp 255–303
20. Allen WD, Czinki E, Császár AG (2004) Molecular structure of proline. Chem Eur J 10:4512–4517
21. Kuchitsu K, Nakata M, Yamamoto S (1988) Joint use of electron diffraction and high-resolution spectroscopic data for accurate determination of molecular structure. In: Hargittai I, Hargittai M (eds) Stereochemical applications of gas-phase electron diffraction, vol A, chap 7. VCH Publishers, New York, p 227
22. Nakata M, Kuchitsu K (1982) Molecular structure of 1,1-dichloroethylene as studied by gas electron diffraction and microwave data. Estimation of equilibrium structure by an extended anharmonic model analysis. J Mol Struct 95:205–214
23. Pulay P, Meyer W, Boggs JE (1978) Cubic force constants and equilibrium geometry of methane from Hartree-Fock and correlated wavefunctions. J Chem Phys 38:5077–5085
24. Stanton JF (1999) A refined estimate of the bond length of methane. Mol Phys 97:841–845
25. Vogt N, Demaison J, Vogt J, Rudolph HD (2014) Why it is sometimes difficult to determine the accurate position of a hydrogen atom by the semiexperimental method: structure of molecules containing the OH or the CH_3 group. J Comput Chem 35:2333–2342
26. Vogt N, Demaison J, Krasnoshchekov SV, Stepanov NF, Rudolph HD (2017) Determination of accurate semiexperimental equilibrium structure of proline using efficient transformations of anharmonic force fields among the series of isotopologues. Mol Phys 115:942–951
27. Vogt N, Demaison J, Cocinero EJ, Écija P, Lesarri A, Rudolph HD, Vogt J (2016) The equilibrium molecular structures of 2-deoxyribose and fructose by the semiexperimental mixed estimation method and coupled-cluster computations. Phys Chem Chem Phys 18:15555–15563
28. Bak KL, Gauss J, Jørgensen P, Olsen J, Helgaker T, Stanton JF (2001) The accurate determination of molecular equilibrium structures. J Chem Phys 114:6548–6556
29. Pawłowski F, Jørgensen P, Olsen J, Hegelund F, Helgaker T, Gauss J, Bak KL, Stanton JF (2002) Molecular equilibrium structures from experimental rotational constants and calculated vibration–rotation interaction constants. J Chem Phys 116:6482–6496
30. Puzzarini C, Heckert M, Gauss J (2008) The accuracy of rotational constants predicted by high-level quantum-chemical calculations. I. Molecules containing first-row atoms. J Chem Phys 128:194108
31. Heckert M, Kállay M, Gauss J (2005) Molecular equilibrium geometries based on coupled-cluster calculations including quadruple excitations. Mol Phys 103:2109–2115
32. Puzzarini C, Cazzoli G, Baldacci A, Baldan A, Michauk C, Gauss J (2007) Rotational spectra of rare isotopic species of bromofluoromethane: determination of the equilibrium structure from *ab initio* calculations and microwave spectroscopy. J Chem Phys 127:164302/1-10
33. Demaison J, Liévin J, Vogt N (2023) Accurate equilibrium structures of some challenging molecules: FNO, ClNO, HONO, FNO_2, and N_2O. J Mol Spectrosc 394: 111788
34. Palmieri P, Tarroni R, Hühn MM, Handy NC, Willetts A (1995) An improved anharmonic force field of $CHClF_2$. Chem Phys 190:327–344
35. Kochikov IV, Tarasov YI, Spiridonov VP, Kuramshina GM, Saakjan AS, Yagola AG (2000) The use of ab initio anharmonic force fields in experimental studies of equilibrium molecular geometry. J Mol Struct 550–551:429–438
36. Sebald P, Oswald R, Botschwina P, Stoll H, Figgen D (2009) Accurate potential energy surface and calculated spectroscopic properties for CdH_2 isotopomers. J Phys Chem A 113:11772–11782
37. Botschwina P, Sebald P, Figgen D, Stoll H (2007) Calculated spectroscopic properties of HgH_2. Mol Phys 105:1193–1205

38. Sebald P, Vennekate H, Oswald R, Botschwina P, Stoll H (2010) A theoretical study of ZnH_2: a case of very strong Darling-Dennison resonance. Mol Phys 108:487–499
39. Demaison J, Margulès L, Rudolph HD (2010) Accurate determination of an equilibrium structure in the presence of a small coordinate: the case of dimethyl sulfide. J Mol Struct 978:229–233
40. Demaison J (2011) The method of least squares. In: Demaison J, Boggs JE, Császár AG (eds) Equilibrium molecular structures. From spectroscopy to quantum chemistry. CRC Press, Boca Raton, pp 29–52
41. Belsley DA (1991) Conditioning diagnostics. Collinearity and weak data in regression. Wiley, New York
42. Puzzarini C (2009) Extrapolation to the complete basis set limit of structural parameters: comparison of different approaches. J Phys Chem A 113:14530–14535
43. Craig NC, Demaison J, Groner P, Rudolph HD, Vogt N (2015) Electron delocalization in polyenes: a semiexperimental equilibrium structure for (3E)-1,3,5-hexatriene and theoretical structures for (3Z,5Z)-, (3E,5E)-, and (3E,5Z)-1,3,5,7-octatetraene. J Phys Chem A 119:195–204
44. Vogt N, Demaison J, Rudolph HD, Perrin A (2015) Interplay of experiment and theory: high-resolution infrared spectrum and accurate equilibrium structure of BF_2OH. Phys Chem Chem Phys 17:30440–30449
45. Demaison J, Craig NC (2011) Semiexperimental equilibrium structure for *cis, trans*-1,4-difluorobutadiene by the mixed estimation method. J Phys Chem A 115:8049–8054
46. Demaison J, Craig NC, Conrad AR, Tubergen MJ, Rudolph HD (2012) Semiexperimental equilibrium structure of the lower energy conformer of glycidol by the mixed estimation method. J Phys Chem A 116:9116–9122
47. Demaison J, Jahn MK, Cocinero EJ, Lesarri A, Grabow J-U, Guillemin J-C, Rudolph HD (2013) Accurate semiexperimental structure of 1,3,4-oxadiazole by the mixed estimation method. J Phys Chem A 117:2278–2284
48. Demaison J, Császár AG, Margulès LD, Rudolph HD (2011) Equilibrium structures of heterocyclic molecules with large principal axis rotations upon isotopic substitution. J Phys Chem A 115:14078–14091
49. Császár AG, Demaison J, Rudolph HD (2015) Equilibrium structures of three-, four-, five-, six-, and seven-membered unsaturated N-containing heterocycles. J Phys Chem A 119:1731–1746
50. Kummli DS, Frey HM, Keller M, Leutwyler S (2005) Femtosecond degenerate four-wave mixing of cyclopropane. J Chem Phys 123:054308
51. Kummli DS, Frey HM, Leutwyler S (2006) Femtosecond degenerate four-wave mixing of carbon disulfide: high-accuracy rotational constants. J Chem Phys 124:144307
52. Kummli DS, Frey HM, Leutwyler S (2007) High-accuracy structure of cyclobutane by femtosecond rotational raman four-wave mixing. J Phys Chem A 111:11936–11942
53. Frey H-M, Kummli D, Lobsiger S, Leutwyler S (2011) High-resolution rotational Raman coherence spectroscopy with femtosecond pulses. In: Quack M, Merkt F (eds) Handbook of high-resolution spectroscopy. Wiley, Chichester, Chap 31
54. Császár AG (2011) Structures averaged over nuclear motions. In: Demaison J, Boggs JE, Császár AG (eds) Equilibrium Molecular Structures. From Spectroscopy to Quantum Chemistry. CRC Press, Boca Raton, pp 233–262
55. Kuchitsu K (1992) The potential energy surface and the meaning of internuclear distances. In: Domenicano A, Hargittai I (eds) Accurate molecular structures. Oxford University Press, Oxford, p 14
56. Nakata M, Kohata K, Fukuyama T, Kuchitsu K (1980) Molecular structure of phosgene as studied by gas electron diffraction and microwave spectroscopy: the r_z structure and isotope effect. J Mol Spectrosc 83:105–117
57. Kuchitsu K, Oyanagi K (1977) Experimental determination of average and equilibrium structures of polyatomic gas molecules by diffraction and spectroscopic methods. Faraday Discuss 62:20–28

58. Kuchitsu K (1968) Comparison of molecular structures determined by electron diffraction and spectroscopy. Ethane and diborane. J Chem Phys 49:4456–4462
59. Kuchitsu K, Fukuyama T, Morino Y (1969) The role of isotopic differences in the determination of molecular structure. A supplementary note on the structure of acrolein. J Mol Struct 4:41–50
60. Kuchitsu K, Morino Y (1965) Estimation of anharmonic potential constants. I. Linear XY_2 molecules. Bull Chem Soc Japan 38:805–813
61. Frisch MJ, Trucks GW, Schlegel HB et al (2016) Gaussian 16. Gaussian, Inc., Wallingford CT, 2016. See https://gaussian.com/
62. Nakata M, Kuchitsu K, Mills IM (1984) Estimation of equilibrium structure by use of isotopic differences in the r_z structure: modification of the mass dependence method and application to H_2O, H_2CO, and H_2CS. J Phys Chem 88:344–348
63. Oka T, Morino Y (1962) Analysis of the microwave spectrum of hydrogen selenide. J Mol Spectrosc 8(1–6):300–314
64. McRae GA, Gerry MCL, Cohen EA (1986) The microwave spectra of PH_2D and PHD_2, and the harmonic force field and structure of phosphine. J Mol Spectrosc 116:58–70
65. McRae GA, Gerry MCL, Wong M, Ozier I, Cohen EA (1987) Microwave spectra of deuterated arsines: Distortion moment transitions of AsD3, microwave spectra of AsH_2D and $AsHD_2$, and the structure of arsine. J Mol Spectrosc 123:321–339
66. Dréan P, Paplewski M, Demaison J, Breidung J, Thiel W, Beckers H, Bürger H (1996) Millimeter-wave spectra, ab initio calculations, and structures of fluorophosphane and chlorophosphane. Inorg Chem 35:7671–7678
67. Duncan JL (1974) The ground-state average and equilibrium structures of formaldehyde and ethylene. Mol Phys 28:1177–1191
68. Demaison J, Vogt N (2022) Effective and equilibrium structures of XY_n-type molecules: SCl_2, PF_2, PO_2, ZrO_2, and CF_2S as additional example and nonbonded distances F···F, Cl···Cl and O···O. J Mol Spectrosc 387:111661
69. Costain CC (1958) Determination of molecular structures from ground state rotational constants. J Chem Phys 29:864–874
70. Kraitchman J (1953) Determination of molecular structure from microwave spectroscopic data. Am J Phys 21:17–24
71. Costain CC (1966) Further comments on the accuracy of r_s substitution structures. Trans Amer Cryst Ass 2:157–164
72. Van Eijck BP (1982) Influence of molecular vibrations on substitution coordinates. J Mol Spectrosc 91:348–362
73. Teffo J-L, Chédin A (1989) Internuclear potential and equilibrium structure of the nitrous oxide molecule from rovibrational data. J Mol Spectrosc 135:389–409
74. Demaison J, Rudolph HD (2002) When is the substitution structure not reliable? J Mol Spectrosc 215:78–84
75. Demaison J, Craig NC, Groner P, Écija P, Cocinero EJ, Lesarri A, Rudolph HD (2015) Accurate equilibrium structures for piperidine and cyclohexane. J Phys Chem A 119:1486–1493
76. Vogt N, Demaison J, Rudolph HD, Juanes M, Fernández J, Lesarri A (2018) Semiexperimental and mass-dependent structures by the mixed regression method: accurate equilibrium structure and failure of the Kraitchman method for ethynylcyclohexane. J Chem Phys 148:064306
77. Vogt N, Nair KPR, Grabow J-U, Demaison J (2018) Microwave rotational spectrum and ab initio computations on 4-cyanopyridine: molecular structure and hyperfine interactions. Mol Phys 116:3530–3537
78. Chutjian A (1964) Determination of structure by isotopic substitution in molecules with symmetrically equivalent atoms. J Mol Spectrosc 14:361–370
79. Pierce L (1959) Note on the use of ground state rotational constants in the determination of molecular structures. J Mol Spectrosc 3:575–580
80. Demaison J, Wlodarcak G, Burie J, Bürger H (1990) The rotational spectrum of germyl chloride: determination and comparison of the substitution, double substitution, and equilibrium structure. J Mol Spectrosc 140:322–339

81. Rudolph HD (1991) Contribution to the systematics of r_0-derived molecular structure determinations from rotational parameters. Struct Chem 2:581–588
82. Watson JKG, Roytburg A, Ulrich W (1999) Least-squares mass-dependence molecular structures. J Mol Spectrosc 196:102–119
83. Smith JG, Watson JKG (1978) A test if the r_m method of structure determination. J Mol Spectrosc 69:47–52
84. Nakata M, Fukuyama T, Kuchitsu K, Takeo H, Matsumura C (1980) Molecular structure of phosgene as studied by gas electron diffraction and microwave spectroscopy: The r_s, r_m, and r_e structures. J Mol Spectrosc 83:118–129
85. Nakata M, Sugie M, Takeo H, Matsumura C, Fukuyama T, Kuchitsu K (1981) Structure of dichlorine monoxide as studied by microwave spectroscopy. Determination of equilibrium structure by a modified mass dependence method. J Mol Spectrosc 86:241–249
86. Harmony MD, Taylor WH (1986) The use of scaled moments of inertia for structural calculations. J Mol Spectrosc 118:163–173
87. Berry RJ, Harmony MD (1990) The use of scaled moments of inertia in experimental structure determinations of polyatomic molecules. Struct Chem 1:49–59
88. Tam HS, Choe J-I, Harmony MD (1991) Molecular structure of polyatomic molecules using scaled moments of inertia. J Phys Chem 95:9267–9272
89. Harmony MD (1999) Molecular structure determination from spectroscopic data using scaled moments of inertia. In: Durig JR (ed) Vibrational Spectra and Molecular Structure, vol 24. Elsevier, Amsterdam, pp 1–83
90. Harmony MD (1990) The equilibrium carbon-carbon single-bond in ethane. J Chem Phys 93:7522–7523
91. Cooksy AL, Watson JKG, Gottlieb CA, Thaddeus P (1994) The structure of the HCCCO radical. Rotational spectra and hyperfine structure of monosubstituted isotopomers. J Chem Phys 101:178–186
92. Demaison J, Nemes L (1979) A correlation between the vibrational correction ε and the effective moment of inertia. J Mol Struct 55:295–299
93. Le Guennec M, Wlodarczak G, Demaison J, Bürger H, Litz M, Willner H (1993) Millimeterwave and high-resolution infrared spectrum of OCSe: determination of its structure. J Mol Spectrosc 157:419–446
94. Laurie VW (1958) Note on the determination of molecular structure from spectroscopic data. J Chem Phys 28:704–706
95. Legon AC, Demaison J (2011) Other spectroscopic sources of molecular properties: intermolecular complexes as examples. In: Demaison J, Boggs JE, Császár AG (eds) Equilibrium molecular structures. From spectroscopy to quantum chemistry. CRC Press, Boca Raton, Boca Raton, pp 205–231
96. Demaison J, Vogt N, Saragi RT, Juanes M, Rudolph HD, Lesarri A (2019) The S-S bridge: a mixed experimental-computational estimation of the equilibrium structure of diphenyl disulfide. ChemPhysChem 20:366–373
97. Demaison J, Vogt N, Saragi RT, Juanes M, Rudolph HD, Lesarri A (2019) How flexible is the disulfide linker? A combined rotational–computational investigation of diallyl disulfide. Phys Chem Chem Phys 21:19732–19736
98. Morino Y, Tanimoto M (1984) Vibration–rotation interaction constants and the cubic potential function of the NO_2 molecule in the small-amplitude approximation. Can J Phys 62:1315–1322
99. Persch G, Vedder H, Demtröder W (1987) The hyperfine structure in excited electronic states of NO_2: Analysis of $N' = 10 \leftarrow N'' = 00$ transitions in the system. J Mol Spectrosc 123:356–365
100. Canè E, Di Lonardo G, Fusina L, Jerzembeck W, Bürger H, Breidung J, Thiel W (2006) Rotation spectrum and high-resolution infrared spectra of the fundamental bands of $^{121}SbD_3$. Determination of the ground state and equilibrium structures. Ab initio calculations of spectroscopic parameters. J Mol Struct 780–781:98–110
101. Rudolph HD, Demaison J, Császár AG (2013) Accurate determination of the deformation of the benzene ring upon substitution: equilibrium structures of benzonitrile and phenylacetylene. J Phys Chem A 117:12969–12982

102. Vogt N, Demaison J, Vogt J (2011) Accuracy of the rotational constants. J Mol Struct 988:119–127
103. Kisiel Z (2003) Least-squares mass-dependence molecular structures for selected weakly bound intermolecular clusters. J Mol Spectrosc 218:58–67
104. Mendolicchio M, Penocchio E, Licari D, Tasinato N, Barone V (2017) Development and implementation of advanced fitting methods for the calculation of accurate molecular structures. J Chem Theory Comput 13:3060–3075
105. Stanton JF, Gauss J, Harding ME, Szalay PG (2011) CFour A quantum chemical program package with contributions from 39 authors. See http://www.cfour.de
106. Sipachev VA (2004) The use of quantum-mechanical third-order force constants in structural studies. J Mol Struct 693:235–240
107. Groner P, Warren RD (2001) Approximate r_e structures from experimental rotational constants and ab initio force fields. J Mol Struct 599:323–335
108. Vogt N, Vogt J (2019) Structure data of free polyatomic molecules. Springer Nature, Switzerland, 926 p
109. Graner G, Hirota E, Iijima T, Kuchitsu K, Ramsay DA, Vogt J, Vogt N (1998) Structure data of free polyatomic molecules (Kuchitsu K, ed), vol II-25A. Springer, Berlin, Germany
110. Graner G, Hirota E, Iijima T, Kuchitsu K, Ramsay DA, Vogt J, Vogt N (1999) Structure data of free polyatomic molecules (Kuchitsu K, ed), vol II-25B. Springer, Berlin, Germany
111. Graner G, Hirota E, Iijima T, Kuchitsu K, Ramsay DA, Vogt J, Vogt N (2000) Structure data of free polyatomic molecules (Kuchitsu K, ed), vol II-25C. Springer, Berlin, Germany
112. Graner G, Hirota E, Iijima T, Kuchitsu K, Ramsay DA, Vogt J, Vogt N (2003) Structure data of free polyatomic molecules (Kuchitsum K, ed), vol II-25D. Springer, Berlin, Germany
113. Hirota E, Iijima T, Kuchitsu K, Ramsay DA, Vogt J, Vogt N (2006) Structure data of free polyatomic molecules (Kuchitsu K, Vogt N, Tanimoto M, eds), vol II-28A. Springer, Berlin, Germany
114. Hirota E, Iijima T, Kuchitsu K, Ramsay DA, Vogt J, Vogt N (2006) Structure data of free polyatomic molecules (Kuchitsu K, Vogt N, Tanimoto M, eds), vol II-28B. Springer, Berlin, Germany
115. Hirota E, Iijima T, Kuchitsu K, Ramsay DA, Vogt J, Vogt N (2007) Structure data of free polyatomic molecules (Kuchitsu K, Vogt N, Tanimoto M, eds), vol II-28C. Berlin, Germany, Springer
116. Hirota E, Iijima T, Kuchitsu K, Ramsay DA, Vogt J, Vogt N (2007) Structure Data of Free Polyatomic Molecules. Kuchitsu K, Vogt N, Tanimoto M (Eds), vol II-28D. Springer, Berlin, Germany; Hirota E, Kuchitsu K, Steimle T, Vogt J, Vogt N (2014) Structure data of free polyatomic molecules (Kuchitsu K, Vogt N, Tanimoto M, eds), vol II-30A. Springer, Berlin, Germany; Hirota E, Kuchitsu K, Steimle T, Vogt J, Vogt N (2014) Structure Data of Free Polyatomic Molecules (Kuchitsu K, Vogt N, Tanimoto M, eds), vol II-30B. Springer, Berlin, Germany

Chapter 6
Structure of Non-rigid Molecules by Spectroscopic Methods

Abstract This chapter discusses the determination of the structure of a molecule when a large-amplitude motion is present. Different cases are discussed: inversion, ring-puckering, internal rotation, quasi-linear molecules, and weakly bound complexes.

6.1 Introduction

The basic assumption made in Chaps. 3–5 is that the amplitude of vibrations is small. When a large-amplitude motion is present, this assumption is no longer appropriate and the rovibrational Hamiltonian has to be modified, which may have consequences on the determination of structure. In this chapter, we will discuss how the Hamiltonian has to be modified and how to determine a structure.

There is a great diversity of large-amplitude motions (LAM) which will be discussed in this chapter: inversion, ring-puckering, internal rotation, quasi-linear molecules, weakly bound complexes where the presence of a very weak bond linking the monomers requires the use of specific methods to determine the structure of the complex, see Sect. 6.6.

In some cases, the equilibrium structure cannot be determined using the methods described in Chap. 5, but it remains generally possible to determine the minimum (or minima) of the potential energy surface (PES).

It is beyond the scope of this book to present a detailed theory of the large-amplitude motions. Only the notions useful to determine rotational constants and a structure will be discussed. Nevertheless, references to the theory will be given. See, for instance, Refs. [1, 2].

See also previous Sect. 4.8.

Some molecules have symmetrically equivalent equilibrium configurations separated by a potential barrier. Typical examples are: inversion, internal rotation, etc. If the barrier is high enough, the perturbation it produces is negligible and the molecules behave as semirigid molecules. When the barrier is lowered, some atoms are affected by a large-amplitude motion. As an example, consider a double-minimum potential which is typical for the inversion. If ψ_A is the wave function corresponding to one

N. Vogt and J. Demaison, *Equilibrium Structure of Free Molecules*, Lecture Notes in Chemistry 111, https://doi.org/10.1007/978-3-031-36045-9_6

equilibrium configuration and ψ_B the wave function for the other configuration, a better solution is obtained by using a linear combination of the original wave functions. In such case, two combinations are possible

$$\psi_+ = \frac{1}{\sqrt{2}}(\psi_A + \psi_B) \tag{6.1}$$

and

$$\psi_- = \frac{1}{\sqrt{2}}(\psi_A - \psi_B) \tag{6.2}$$

It engenders a splitting of the energy levels, the energy being

$$\begin{aligned} E_\pm &= \frac{1}{2}[\langle\psi_A|\mathbf{H}|\psi_A\rangle + \langle\psi_B|\mathbf{H}|\psi_B\rangle \pm \langle\psi_A|\mathbf{H}|\psi_B\rangle \pm \langle\psi_B|\mathbf{H}|\psi_A\rangle] \\ &= \langle\psi_A|\mathbf{H}|\psi_A\rangle \pm \langle\psi_A|\mathbf{H}|\psi_B\rangle \end{aligned} \tag{6.3}$$

If the splitting is detectable, the molecule should be considered as non-rigid.

The general method is to consider a zero-order Hamiltonian which takes into account the rotation and the large-amplitude motion and to use the perturbation theory to allow for the neglected terms arising from the effects of the small amplitude vibrations.

The mean-square amplitude of a molecular vibration is inversely proportional to the product of the reduced mass and the frequency of the vibration. If it has both a small reduced mass and a frequency that is low because of small force constants, the resultant amplitude of vibration will be large.

6.2 Inversion

In theory, the inversion motion is possible in any nonplanar molecule, but it was observed for only a very few. A typical example is ammonia, NH_3. At equilibrium the molecule is pyramidal but the N atom can pass through the plane of the hydrogen atoms giving rise to the phenomenon of inversion. The potential energy of the molecule as a function of the distance between the N atom and the H_3 plane has a double minimum corresponding to the two equivalent configurations. The energy levels are split into doublets and transitions between sublevels of the same vibrational state are called inversion transitions. The first transition to be measured in the microwave range was an inversion transition of NH_3 [3]. The frequency of the inversion transitions can be calculated using an empirical exponential function [4].

The $J = 1 \leftarrow 0$ rotational transition of $^{14}NH_3$ was measured at 572496.69(60) MHz by Helminger and Gordy [5], see Fig. 6.1. Because of spin statistics, it is a single line rather than an inversion doublet. To determine its "unsplit" position, it is

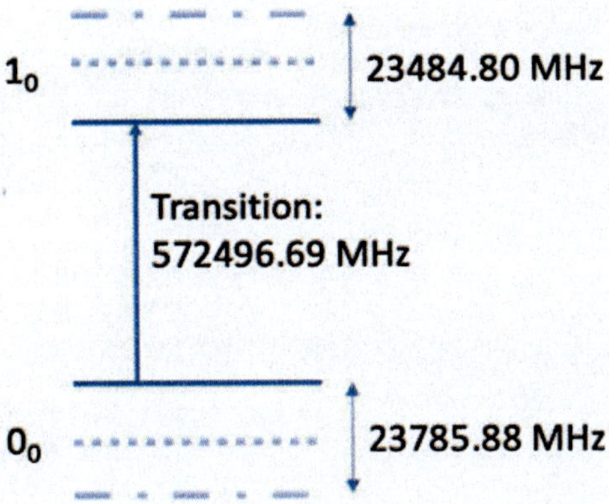

Fig. 6.1 Energy-level diagrams for the $J_K = 0_0$ and 1_0 states of $^{14}NH_3$ showing the observed transition and the hypothetical levels corresponding to no inversion

enough to add to its frequency half the sum of the inversion splittings of the $J = 0$, $K = 0$ (23785.88 MHz) and $J = 1$, $K = 0$ (23484.80 MHz). The inversion splittings are calculated using the empirical exponential function [4]. It gives 596132.03 MHz for the transition between the hypothetical levels assuming no inversion. This simple method permits to obtain all the unsplit frequencies and to derive the B rotational constant.

It is also possible to fit together the inversion and rotational transitions as well as the rovibrational transitions using a Hamiltonian in which the large-amplitude inversion is removed from the vibrational problem and treated together with the rotational problem. The theory was developed by Papoušek et al. [6] and Urban [7]. This work was reviewed and completed by Chen et al. [8].

There is also a recent review on the subject [9].

The equilibrium structure of ammonia, NH_3, although it is the subject of a large-amplitude inversion motion was determined from experimental as well as semiexperimental rotational constants, see Table 6.1. Benedict and Plyler [10] analyzed the infrared spectrum of ammonia and were able to determine a full set of α-constants, see Eq. (5.1), as well as the equilibrium rotational constants. Although their constants are not very accurate: $B_e = 9.965$ cm $^{-1}$ and $C_e = 6.341$ cm^{-1} compared to the best semiexperimental values: $B_e = 10.01589$ cm^{-1} and $C_e = 6.347373$ cm^{-1} [11], their final structure seems to be of good quality. A semiexperimental structure was also determined by Pawlowski et al. [12] assuming that all vibrations are of small amplitude and was found in good agreement with both the experimental structure and a high-level ab initio structure, see Table 6.1. This is a typical example confirming that it is not necessary to have accurate rovibrational corrections when the system of normal equations is well conditioned (this is generally the case when the structure is determined from the rotational constants of one single isotopologue). This is because these corrections to the rotational constants are small (0.75% for B and 1.94% for C).

The experimental equilibrium structure of a few other molecules with inversion splitting was determined, among them: H_3O^+ [13] and its isotopologue D_3O^+ [14, 15]. A six-dimensional potential energy surface was also calculated for H_3O^+ at the

Table 6.1 Equilibrium structure of ammonia, NH_3 (distances in Å and angles in degree)

Method	r_e(N–H)	∠(HNH)	Ref.
Exp	1.0116(6)	106.7	1
SE[a]	1.01139(60)	107.17(18)	2
AI[b]	1.010936	106.8147	3

[a] Rovibrational corrections obtained from a CCSD(T)_ae/cc-pVQZ force field
[b] CCSD(T)_fc/cc-pV∞Z + core(cc-pCVQZ) + corrections using CCSDT/cc-pVTZ and CCSDTQ/cc-pVDZ
[1] Benedict WS, Plyler K (1957) Can J Phys 35:
[2] Pawlowski F, Jørgensen P, Olsen J, Hegelund F, Helgaker T, Gauss J, Bak, KL, Stanton JF (2002) J Chem Phys 116: 6482–6496
[3] Puzzarini C, Heckert M, Gauss J (2008) J Chem Phys 128: 194,108/1–9

Table 6.2 Equilibrium structure of oxonium, H_3O^+ (distances in Å and angles in degree)

Method	r_e(O–H)	∠(HOH)	Ref.
Exp	0.9758	111.3	1
Exp	0.9702(89)	109.4(38)	2
Exp	0.974(1)	113.6(1)	3
CCSD(T)/CBS	0.9747	112.02	4
AI[a]	0.9765	111.7	5

[a] From a potential energy surface calculated at the CCSD(T)/aug-cc-pVQZ
[1] Sears TJ, Bunker PR, Davies PB, Johnson SA, Špirko V (1985) J Chem Phys 83: 2676–2685
[2] Araki M, Ozeki H, Saito S (1999) Mol Phys 97: 177–183
[3] Tang J, Oka T (1999) J Mol Spectrosc 196: 120–130
[4] Rajamäki T, Miani A, Halonen L (2003) J Chem Phys 118: 10929–10938
[5] Huang X, Carter S, Bowman J (2003) J Chem Phys 118: 5431–5441

CCSD(T)/CBS level of theory confirming the accuracy of the experimental equilibrium structure [16]. There is also another potential energy surface calculated at the CCSD(T)/aug-cc-pVQZ giving compatible results [17], see Table 6.2. Similar results were obtained for SiH_3^+ [18], and the inversion splitting was observed for PH_3^+ [19]. See also [9] for the amine derivatives.

6.3 Ring-Puckering

There are many reviews on this subject, see for instance Refs. [20, 21]. A short history of ring-puckering may be found in [22], and the theory is discussed by Harris et al. [23, 24].

A ring-puckering is an out-of-plane vibration of a cyclic molecule.

As the atoms of a three-membered ring define a plane, there is no out-of-plane vibration. Four-membered rings have one out-of-plane vibration, called ring-puckering. More generally, for an n-membered ring there are $n - 3$ out-of-plane skeletal modes. Experience shows that the fundamental of the puckering mode is usually much lower in frequency than the remaining vibrations and the puckering vibration is of large amplitude.

There are two opposite forces involved: the ring strain tends to make the molecule planar since puckering the ring further decreases the already highly strained ring angles and thus increases the potential energy of the molecule. Opposing the ring strain is the barrier to internal rotation which results in a torsional force about the ring bonds. This force tends to make the ring attain a puckered configuration. If the torsional force dominates, the potential function will have a double minimum with a barrier to the planar configuration.

6.3.1 Four-Membered Rings

For a four-membered ring, as usual it will be assumed that the low-frequency, large-amplitude vibration can be separated from the remaining high-frequency, small-amplitude vibrations and the Hamiltonian can be assumed one-dimensional. The coordinate usually chosen to describe the ring-puckering is x, the half-distance between the two diagonals of the ring and the Hamiltonian can be written

$$\mathbf{H} = \frac{P_x^2}{2\mu} + V(x) \tag{6.4}$$

where P_x is the vibrational momentum conjugate to x, μ the reduced mass of the vibration, and $V(x)$ the potential energy. If the ring in its planar configuration has a plane of symmetry, a quartic-quadratic potential has been shown to give good results,

$$V(x) = Ax^4 + Bx^2 \tag{6.5}$$

where A is a positive constant and B may be positive or negative. If B is positive, the potential function has only one minimum. If B is negative, a double minimum appears with a barrier height $B^2/4A$ separating the minima.

The Hamiltonian is first rewritten using a reduced coordinate z obtained by the following transformation [25]

$$x = kz \tag{6.6}$$

where k is a function of the reduced mass μ.

The Hamiltonian matrix can be set up in the harmonic oscillator basis and diagonalized to obtain the eigenvalues.

As the ring-puckering vibration is a low-frequency vibration, many excited states are sufficiently populated at room temperature and can be observed by rotational spectroscopy. The spectra can be analyzed the standard Watsonian, see Sect. 4.4 and Ref. [26]. For example, the B_v rotational constants can be written

$$\langle v|B|v\rangle = \beta_0 + \beta_2\langle v|z^2|v\rangle + \beta_4\langle v|z^4|v\rangle \tag{6.7}$$

β_0 is the rotational constant of the planar configuration. The parameters β_1 and β_2 permit to obtain the coefficients of the potential function. If a zigzag behavior of the rotational constants with the puckering quantum number v is observed, it means that the ring is not planar and that the potential function has a double minimum. This applies also for the centrifugal distortion constants.

It may happen that some ring-puckering states are not isolated and, therefore, affected by a Coriolis interaction (see Sect. 4.6). In such a case, a two-state Hamiltonian has to be used. A typical example is given by the rotational spectrum of 1,1-difluorocyclobutane [27].

If the planar ring does not have a plane of symmetry, odd-powers of x appear in the potential function. In this case, the minima are no more equivalent.

6.3.2 *Five-Membered-Rings*

There are two out-of-plane vibrations: puckering (bend) and twist. The theory assumes that these two-ring modes are separable from all other modes.

If both vibrations are symmetric, the potential energy function is

$$V(x_1, x_2) = a_1x_1^4 + b_1x_1^2 + a_2x_2^4 + b_2x_2^2 + cx_1^2x_2^2 \tag{6.8}$$

The calculation of the energy levels requires the use of a product of two basis sets.

In principle, the effective ground-state moments of inertia can be used to determine a structure. However, the molecules are usually too large to derive a complete set of α-constants and no experimental equilibrium geometry is available up to now. The published structures are either effective (r_0) or substitution (r_s) structures; see Chap. 5. On the other hand, many r_z structures have been determined by gas-phase electron diffraction. A typical example is cyclobutene, c-C_4H_8. A r_z structure was obtained by gas-phase electron diffraction [28] and to determine the ring puckering angle a two-dimensional potential was used because of the mixing of the ring-puckering and CH_2-rocking modes [29]. A semiexperimental structure was also obtained by the "extrapolation" method [30], see Sect. 5.6. The results are reported in Table 6.3. There is a good agreement for the angles including for the effective r_0 values [31], but there is a problem for the C–H bond lengths. In such a case, the CCSD(T)/cc-pVQZ values [32] are more reliable provided that the core correction is taken into account. The reasons of this discrepancy are multiple, and their explanation requires more work: for instance, the cc-pwCVDZ basis set used for the

Table 6.3 Structure of cyclobutane, *c*-C_4H_8 (distances in Å and angles in degree)

	r_0 [a]	r_z^b [b]	r_e^{se} [c]	r_e^{BO} [d]
CC	1.5549(5)		1.5474(1)	1.5468
CH_{ax}	1.0934(19)		1.0830(2)	1.0895
CH_{eq}	1.0910(9)		1.0810(2)	1.0875
θ [e]	28.32(23)	27.8(12) [f]	29.8	29.59
HCH	109.33(14)	106.4(13)	109.24	109.07
CCC	88.23(23)	88.3(2)	88.04	88.07

[a] Caminati W, Vogelsanger B, Meyer R, Grassi G, Bauder A (1988) J Mol Spectrosc 131: 172–184
[b] Egawa T, Fukuyama T, Yamamoto S, Takabayashi F, Kambara H, Ueda T, Kuchitsu K (1987) J Chem Phys 86: 6018–6026
[c] Kummli DS, Frey HM, Leutwyler S (2007) J Phys Chem A 111: 11936–11942
[d] CCSD(T)_fc/cc-pVQZ + core-valence correction from: Glendening ED, Halpern AM (2005) J Chem Phys 109: 635–642
[e] Ring-puckering dihedral angle
[f] Minimum of the ring-puckering potential, see: Egawa T, Yamamoto S, Ueda T, Kuchitsu K (1987) J Mol Spectrosc 126: 231–239

extrapolation might be too small or the calculation of the rovibrational corrections using the standard second-order perturbation might be inaccurate, see discussion of the structure of H_2O_2 in Sect. 6.4.2 and Table 6.6.

6.4 Internal Rotation

There are many reviews on this subject, see for instance Refs. [1, 33–38].

For molecules consisting of two groups connected by a single bond, the groups can rotate with respect to each other. In an asymmetric top, the internal rotor is called the top and the other group the frame. The rotation is not entirely free but is restricted by a potential barrier (for this reason, internal rotation is sometimes called hindered rotation). The mechanism permitting the observation of internal rotation is a coupling between the overall rotational angular momentum and the momentum associated with the internal rotation.

The top may be symmetric (for instance a methyl group) or asymmetric (for instance a NO_2 group). However, most of the studies are dedicated to the internal rotation of a methyl group. Furthermore, the moment of inertia of a methyl group is quite small, ~3.1 uÅ^2, whereas most other rotors have a much larger moment of inertia. Therefore, as mentioned in the introduction, the perturbation induced by the internal rotation of a heavy top will not be directly visible on the rotational spectra. A typical example is the internal rotation of a deuterated methyl group, CD_3. For instance, acetaldehyde, CH_3CHO, has large internal rotation splittings, 231.78 MHz for the $2_{11} \leftarrow 1_{10}$ ground state rotational transition [39], but this splitting is reduced to 29.39 MHz when CH_3 is replaced by CD_3 [40]. For these reasons we will limit ourselves to the internal rotation of a methyl group. Of course, the theory remains valid for similar tops as the silyl, germyl, and even trifluoromethyl groups, although the splitting exponentially decreases as the moment of inertia of the top increases. For a CH_3 group, $I_\alpha \approx 3.1$ uÅ^2, for a SiH_3 group, $I_\alpha \approx 5.9$ uÅ^2, for a GeH_3 group, $I_\alpha \approx 6.2$ uÅ^2, and for a CF_3 group, $I_\alpha \approx 89$ uÅ^2. For this reason, the splitting is difficult to be observed in the ground torsional state, in particular for the CF_3 top. However, there are some rare exceptions. A typical example is fluoral, CF_3CHO, where the splitting could be measured in the ground torsional state, thanks to a high coupling between internal rotation and overall rotation because the CF_3 top is heavier than the molecular frame, CHO [41].

6.4.1 *Internal Rotation of a Symmetric Top*

When the top has threefold (or higher) symmetry, which is the rule for a methyl group, the rotational constants of the molecule are independent on the internal rotation angle. The methyl group has rarely a C_{3v} local symmetry in an asymmetric molecule, but the assumption of C_{3v} symmetry may be used to derive the Hamiltonian [42]. The potential energy function is characterized by the barrier height and the internal rotation torsional angle α. The potential energy function is N-fold periodic, N representing the degree of symmetry of the rotor (i.e., 3 or 6). A Fourier expansion is generally used

$$V(\alpha) = \sum_i \frac{V_{Ni}}{2}(1 - \cos(iN\alpha) \tag{6.9}$$

This series converges rapidly. Furthermore, only the first term can be determined when only one torsional state has been analyzed.

A threefold barrier, the most frequent case, splits the torsional energy levels into a nondegenerate (A) level and a doubly degenerate (E) level. It generally produces a A-E doublet splitting of the rotational lines. The selection rules are: $A \leftrightarrow A$, $E \leftrightarrow E$.

The rigid frame-rigid top Hamiltonian is

$$\mathbf{H} = \mathbf{H}_\mathrm{R} + F(\mathbf{p}_\alpha - \mathbf{P})^2 + \frac{V_3}{2}(1 - \cos(3\alpha)) + \cdots \tag{6.10}$$

where $\mathbf{H}_\mathrm{R}$ is the rotational Hamiltonian, $\mathbf{p}_\alpha$ the internal rotation angular momentum conjugate to the torsional angle α and

$$\mathrm{P} = \sum_g \rho_g J_g \quad (g = a, b, c) \tag{6.11}$$

where J_g is a component of the rotational angular momentum and

$$\rho_g = \frac{\lambda_g I_\alpha}{I_g} \tag{6.12}$$

The λ_g are the direction cosines of the internal rotation axis i of the top in the principal axis system: $\lambda_g = \cos(\angle(i, g))$. I_g are the moments of inertia of the molecule in the principal axis system and I_α the moment of inertia of the top.

The inverse reduced moment of inertia F of the top defined by

$$F = \frac{\hbar^2}{2rI_\alpha} \tag{6.13}$$

with

$$r = 1 - \sum_g \lambda_g^2 \frac{I_\alpha}{I_g} \tag{6.14}$$

It is also useful to define the reduced barrier height, s

$$s = \frac{4V_3}{9F} \tag{6.15}$$

There are several ways to find the eigenvalues of the Hamiltonian. The first method, called principal axis method (PAM), uses the principal axis system which seems to be quite convenient because the rotational constants are the usual rotational constants. The perturbation term $-2F\mathbf{p}_\alpha\mathbf{P}$ is treated by successive van Vleck transformations.

With the PAM method, an effective Hamiltonian is obtained for each torsional state. It is [43]

$$\mathbf{H}_{v\sigma}^{\mathrm{PAM}} = \mathbf{H}_{\mathrm{R}} + F\sum_n W_{v\sigma}^{(n)}\mathbf{P}^n \tag{6.16}$$

The coefficients $W_{v\sigma}^n$ result from the van Vleck transformations. They are tabulated [33] and can be calculated easily [43, 44]. The quantum number of the torsional state is v, and $\sigma = A, E$ denotes the substate.

For the A-substates, $W_{vA}^{(2n+1)} = 0$. Thus, it is possible to fit the A-lines using a standard Watson Hamiltonian. For the E-substates, odd-order terms can be important and it is necessary to add to the Watson Hamiltonian at least the linear terms

$$\mathbf{H}_{\mathrm{in}} = \sum_g F\rho_g W_{1E}^{(1)} J_g \tag{6.17}$$

When the barrier is not too low, the rotational spectrum may be fitted using the IAMCALC program integrating Pickett's SPFIT/SPCALC programs [45]. An interesting alternative is to fit separately the A- and E-lines using one program such as SFLAMS (Separately Fitting of Large-Amplitude Motion Species) [46]. Then, it is easy to determine the unperturbed rotational constants and the internal rotation parameters [47].

The problem with the PAM Hamiltonian is that it converges slowly. Therefore, it is better to use another axis system which eliminate the $2F\rho_x pJ_x$ and the $2F\rho_y pJ_y$ terms leaving only the $2F\rho_z pJ_z$ term which gives rise to $\Delta K = 0$ matrix elements. In the case of a molecule with an (***a b***) symmetry plane (which is the most frequent case) and the I^r representation $[(a, b, c) = (z, x, y)]$, the Coriolis cross-term $2F\rho_y pJ_y$ does not exist. One coordinate axis is coincident with the **ρ** vector. This method is called reduced axis method (RAM).

In the particular case of an (***a b***) symmetry plane, this corresponds to a rotation about the c axis by an angle β, which is also called θ_{RAM} in the BELGI code [48], given by

$$\tan(\beta) = \frac{\rho_b}{\rho_a} \tag{6.18}$$

When there is no symmetry plane, a second angle γ has to be introduced. It is the angle between the principal axis ***c*** and the half line y, which is perpendicular to the plane defined by ***a*** and **ρ**, and points in the direction such that **ρ**, **a**, and y form a right-handed system.

The Hamiltonian may be written as

$$\mathbf{H}^{\mathrm{RAM}} = \mathbf{H}_{\mathrm{T}} + \mathbf{H}_{\mathrm{R}} + \mathbf{H}_{\mathrm{cd}} + \mathbf{H}_{\mathrm{INT}} \tag{6.19}$$

where $\mathbf{H}_{\mathrm{T}}$ is the torsional Hamiltonian, $\mathbf{H}_{\mathrm{R}}$ the rotational Hamiltonian, $\mathbf{H}_{\mathrm{cd}}$ the usual centrifugal distortion Hamiltonian, and $\mathbf{H}_{\mathrm{INT}}$ contains higher-order torsional-rotational interaction terms:

$$\mathbf{H}_{\mathrm{T}} = F(p_\alpha - \rho J_a)^2 + \frac{V_3}{2}(1 - \cos(3\alpha)) + \cdots \tag{6.20}$$

$$\mathbf{H}_{\mathrm{R}} = A_{\mathrm{RAM}} J_a^2 + B_{\mathrm{RAM}} J_b^2 + C_{\mathrm{RAM}} J_c^2 + D_{ab}(J_a J_b + J_b J_a) \tag{6.21}$$

It is easy to find the relationship between the rotational constants A, B, C in the principal axis system and the constants A_{RAM}, B_{RAM}, C_{RAM}, and D_{ab} in the rho-axis system using the definition of β or by diagonalizing the 3×3 matrix of the RAM rotational constants. In the particular case of an (***a b***) symmetry plane, it gives

$$\tan(2\beta) = \frac{2D_{ab}}{A_{\mathrm{RAM}} - B_{\mathrm{RAM}}} \tag{6.22}$$

The higher-order interaction terms of $\mathbf{H}_{\mathrm{INT}}$ can be obtained by multiplying the powers of the rotation angular momentum components J_g with the powers of the torsion operators p_α, $(1 - \cos 3\alpha)$, $(1 - \cos 6\alpha)$, etc.

This method is quite efficient to fit spectra, but it introduces off-diagonal rotational constants which are often heavily correlated with the diagonal rotational constants. Kleiner in her program BELGI [49] uses a global treatment including all the torsional levels for a given vibrational state, an approach which is particularly successful for low barriers or excited torsional states. Furthermore, it permits to take into account many higher-order interaction terms. For instance, it was necessary to determine 48 parameters to obtain a good fit of the rotational spectrum of acetaldehyde, CH_3CHO [39]. More recently, Ilyushin et al. [50] developed a similar but faster program called RAM36.

To avoid the difficulty of off-diagonal rotational constants, some authors transform the PAM rotational constants into RAM rotational constants. This is the combined axis method (CAM) used in the Effective Rotational-torsional Hamiltonian (ErHam) of Groner [51]. This method is also employed in the program XIAM of Hartwig and Dreizler [52].

There are several comparative studies of these methods, see for instance: [37, 53].

The CAM method permits to obtain unperturbed rotational constants (as the PAM method), and it has been used to determine the semiexperimental equilibrium structure of many molecules. By comparison of these structures with high-level ab initio structures, it was concluded that the torsion may indeed be treated as a small amplitude vibration. There are many typical examples such as methyl formate, $HCOOCH_3$ [37, 54], propene, $CH_2{=}CHCH_3$ [55], ethanol, CH_3CH_2OH [56], propanal, CH_3CH_2CHO [56], pyruvic acid, $CH_3C({=}O)COOH$ [54], dimethyl ether, $(CH_3)_2O$ [56], dimethyl sulfide $(CH_3)_2S$ [57], 2-acetylfuran, *c*-$C_6H_6O_2$ [58], etc.

In theory, the parameters derived from an internal rotation analysis can give useful information on the structure of the molecule. They are the moment of inertia of the top, I_α, and the angles between the axis of rotation, i, of the top and the principal axes of the molecule, $\angle(i, g)$ with $g = a, b, c$. Unfortunately, they are not very useful. I_α is highly correlated with the potential barrier and, thus, difficult to determine. The result depends of the method used and, furthermore, the range of I_α is small: 0.116 uÅ^2 for a mean value of 3.20 uÅ^2 [59, 60], see also Sect. 8.7. On the other hand, the angles $\angle(i, g)$ can be determined with a high accuracy, but they are vibrationally averaged parameters and it is not obvious that the internal rotation axis i coincides with the bond axis. In conclusion, in most cases, the values of the $\angle(i, g)$ angles are not compatible with the values of the equilibrium structure, see Table 6.4 for a few examples. This discrepancy is called methyl tilt [1].

The equilibrium structure of *syn* methyl formate is given in Table 6.5. The potential hindering the internal rotation is not high, $V_3 = 372.672(4)$ cm^{-1}; thus, it produces large splittings of the rotational transitions, even in the ground torsional state. Its structure has been determined several times, allowing us to make comparisons. The standard deviations of the semiexperimental structures are quite small, except for the distance $r(CH_s)$ and the angle $\angle(CCH_s)$. These two parameters are highly correlated, correlation coefficient $\rho = -0.9737$ with a rather large condition number, $\kappa = 511$. This correlation may be partly explained by the fact that the in-plane hydrogen of the methyl group is close to the principal b-axis, $b(H_s) = -0.246$ Å, see Sect. 5.5.

Table 6.4 Comparison of the internal rotation angle, $\angle(b, i)$ (in degree) from internal rotation analysis and equilibrium structure

Molecule	$\angle(b, i)$[a]	Ref	$\angle(b, CC)$[a]	Ref.
$HCOOCH_3$	37.011	b	31.93	b
Anti $C_4H_4OC(O)CH_3$	29.690	c	26.3	c
Syn $C_4H_4OC(O)CH_3$	36.263	c	32.5	c
$(CH_3)_2O$	59.00	d	55.63	e
$(CH_3)_2S$	51.701	f	49.29	g
$(CH_3)_2CO$	60.249	h	58.24	*i*

[a] *b* is the principal axis and *i* the internal rotation axis which is identical to the CC bond.
Demaison J, Margulès L, Kleiner I, Császár AG (2010) J Mol Spectrosc 259: 70–79
[c] Dindić C, Lüchow A, Vogt N, Demaison J, Nguyen HVL (2021) J Phys Chem A 125: 4986–4997
[d] Neustock W, Guarnieri A, Demaison J, Wlodarcak G (1990) Z Naturforsch A 45: 702–706
[e] Vogt N, Demaison J, Vogt J, Rudolph HD (2014) J Comput Chem 35: 2333–2342
[f] Vacherand JM, Wlodarczak G, Dubrulle A, Demaison J (1987) Can J Chem 65: 1159–1163
[g] Demaison J, Margulès L, Rudolph HD (2010) J Mol Struct 978: 229–233
[h] Vacherand JM, Van Eijck BP, Burie J, Demaison J (1986) J Mol Spectrosc 118: 355–362
[i] Lovas FJ, Groner P (2006) J Mol Spectrosc 236: 173–177

Obviously, the standard deviations do not seem to be not a good indicator of the accuracy as discussed in Sect. 5.12. Taking this observation into account, the agreement between the different determinations is satisfactory and a better estimate can be obtained with a median of the different determinations.

For lower potentials, line bands appear instead of a clear multiplet fine structure. It is still possible to use the RAM and CAM methods, see for instance [50, 53]. For the PAM method, the situation is different because of its slow convergence. In this case, instead of treating the top-frame coupling term as a perturbation, it is better to first solve the free internal rotation problem and to treat the potential barrier as a perturbation using the free rotor basis set [61].

When the frame has C_{2v} symmetry, the V_3 potential term vanishes, the first term in the Fourier expansion is V_6, and the symmetry axis of the methyl group coincides with one of the principal axes. For instance, it is the case for molecules such as toluene, c-$C_6H_5CH_3$ [50, 62], difluoromethylborane, CH_3BF_2 [63], and nitromethane, CH_3NO_2 [63]. The spectrum appears as line bands labeled with the free rotation quantum number, *m*, the selection rule being $\Delta m = 0$. For $m = 0$, the Hamiltonian is the same as a Watson Hamiltonian with the exception that the constant C_z is not the rotational constant of the entire molecule but of the frame part only

$$C_z = \frac{\hbar^2}{2(I_z - I_\alpha)} \tag{6.23}$$

where *z* is the symmetry axis of the top.

Table 6.5 Equilibrium structure of *syn* methyl formate, $HCOOCH_3$ (distances in Å and angles in degree)

Method[c]	r_e^{se}			r_e^{BO} [a]	Best
	MP2[d]	B3LYP[e]	B2PLYP[e]	[d]	Estimate[b]
Basis[c]	cc-pVTZ	SNSD	cc-pVTZ		
C_m–O	1.4341(5)	1.4358(16)	1.4347(13)	1.4355	1.43510(76)
O–C_c	1.3345(4)	1.3343(15)	1.3346(13)	1.3347	1.33455(18)
C_m–H_s	1.0793(10)	1.0845(30)	1.0848(25)	1.0837	1.0841(24)
C_m–H_a	1.0871(3)	1.0875(8)	1.0870(7)	1.0872	1.08715(22)
C_c–H	1.0930(5)	1.0925(18)	1.0923(15)	1.0934	1.09275(49)
C=O	1.2005(5)	1.2001(17)	1.2003(14)	1.2008	1.20040(31)
COC	114.32(4)	114.26(13)	114.30(11)	114.411	114.310(67)
CCH_s	106.05(16)	105.35(42)	105.40(35)	105.572	105.49(31)
CCH_a	110.19(2)	110.07(5)	110.15(5)	110.125	110.138(54)
OCH	109.60(5)	109.54(17)	109.58(14)	109.653	109.590(50)
O–C=O	125.50(5)	125.50(16)	125.47(14)	125.467	125.485(14)
H_aCOC	60.28(3)	60.36(8)	60.37(7)	60.248	60.280(50)

[a] CCSD(T)_fc/cc-pVQZ + MP2[cc-pV(Q ,5)Z + cc-pwCVQ Z(ae) – cc-pwCVQZ(fc) + aug-cc-pVQZ – 2× cc-pVQZ]
[b] Median of values, standard deviation derived from the range
[c] Used for the calculation of the anharmonic force field
[d] Demaison J, Margulès L, Kleiner I, Császár AG (2010) J Mol Spectrosc 259: 70–79
[e] Penocchio E, Piccardo M, Barone V (2015) J Chem Theory Comput 11: 4689–4707

Table 6.6 Equilibrium structure of H_2O_2 (distances in Å and angles in degree)

Method	r_e(O–H)	r_e(O–O)	∠(OOH)	τ(HOOH)	Ref.
Exp.[a]	0.967	1.4556	102.32	113.70	1
SE[b]	0.9617(2)	1.4524(3)	99.76(6)	113.6(3)	2
CCSD(T)_ae/cc-pCVQZ	0.9619	1.4497	99.99		3
AI [c]	0.9623	1.4515	100.07	113.17	2
AI [d]	0.9628	1.4509	100.09	112.81	4

[a] Minimum of the torsional potential
[b] Rovibrational corrections obtained from a variational calculation with a potential energy surface from[4]
[c] Extrapolation of optimized CCSD(T)_ae/cc-pCVTZ-5Z geometries supplemented by higher order perturbative quadruple correction calculated with the cc-pCVTZ basis set and the frozen core approximation
[d] CCSD(T)/aug-ccpV7Z with valence electrons correlated, corrections for core correlation, scalar relativistic correction and higher-order valence-electron correlation effects
[1] Pelz G, Yamada KMT, Winnewisser G (1993) J Mol Spectrosc 159: 507–520
[2] Baraban J, Changala PB, Stanton JF (2018) J Mol Spectrosc 343: 92–95
[3] Bak KL, Gauss J, Jørgensen P, Olsen J, Helgaker T, Stanton JF (2001) J Chem Phys 114: 6548–655
[4] Małyszek P, Koput J (2013) J Comput Chem 34: 337–345

Up to now, there is no experimental equilibrium structure published for such molecules, but there are empirical structures available, for instance for toluene, *c*-$C_6H_5CH_3$ [64].

6.4.2 *Internal Rotation of an Asymmetric Top*

In many cases, the moment of inertia of the top is too large to produce an observable splitting. Thus, a Watson Hamiltonian is appropriate to fit the spectrum.

However, when the moment of inertia of the top is small enough, internal rotation splittings may appear. A typical example is given by the rotational spectrum of monodeuterated propene, CH_2=$CHCH_2D$ [55, 65]. There are two conformers, one with the D atom in the symmetry plane (S), the other one with the D atom out-of-plane (A). For the S-conformer, a Watson Hamiltonian may be used. Furthermore, although some transitions are split into doublets for the A-conformer, it is still possible to use a Watson Hamiltonian to fit the mean frequency of the doublets. Different methods can be employed for a global fit of the spectrum: either using an Internal Axis Hamiltonian [55] as for water dimer [66] or by assuming that the two substates of the A-conformer are in Coriolis interaction [65]. A more detailed discussion of the theory may be found in [67].

The case of hydrogen peroxide, H_2O_2, is extremely interesting because it is the subject of a large-amplitude torsional motion [68], and its structure has been determined using different methods [69]. The structure was first determined by the variational calculation of the potential energy surface, then by the standard second-order vibrational perturbation theory (VPT2, i.e., assuming that the vibrations are of small amplitude), and, finally by the second-order vibrational Møller-Plesset theory (VMP2). In this method, a curvilinear reaction path coordinate system is used. It is necessary to properly treat a large-amplitude motion, multi-well system. Furthermore, a standard Eckart frame tied to the planar *trans* transition state geometry was used as well as a moving Eckart frame tied to the instantaneous principal axis system along the torsional isomerization coordinate [70]. The structures are compared in Table 6.6 and the calculated rovibrational corrections to the ground state rotational constants are given in Table 6.7. Table 6.7 confirms the superiority of the variational and the VMP2 methods which give results in good agreement. This is further confirmed by a high-level ab initio calculation of the equilibrium rotational constants [71]; see Table 6.7.

6.5 Quasi-linear Molecules, Bender, and MORBID

Molecules with only one configuration can also be nonrigid. Typical examples are the quasi-linear molecules HCNO and HNCO for which the HCN or HNC bending mode is a large-amplitude motion. In this case, the potential has a double minimum but there

Table 6.7 Rovibrational corrections to the ground state rotational constants for H_2O_2 (in MHz)

Method	$A_e - A_0$	$B_e - B_0$	$C_e - C_0$	Ref.
Variational	2346	281	507	1
VPT2	2731	320	445	2
VMP2	2342	260	529	1
AI[a]	2341	286	532	3

[a] Equilibrium rotational constants computed at the CCSD(T)/aug-ccpV7Z level of theory with valence electrons correlated, corrections for core correlation, scalar relativistic correction and higher-order valence-electron correlation effects
[1] Baraban J, Changala PB, Stanton JF (2018) J Mol Spectrosc 343: 92–95
[2] Bak KL, Gauss J, Jørgensen P, Olsen J, Helgaker T, Stanton JF (2001) J Chem Phys 114: 6548–655
[3] Małyszek P, Koput J (2013) J Comput Chem 34: 337–345

is no splitting of the energy levels. The problem was first observed during the analysis of the rotational spectra of alkali hydroxides [72]. The variation of the effective rotational constant with v_2 (bending mode) showed a strong nonlinear dependence and a drastic deuterium effect. This is a consequence of the extremely large amplitude of the bending motion. In the conventional theory of vibrations small-amplitude rectilinear displacements are assumed, whereas, in the present case, the atoms follow a curvilinear path. As a consequence, the cubic terms in the normal-coordinate expansion of the potential energy contain an important harmonic component. This large pseudo-anharmonic contribution almost cancels out the opposite signed anharmonic contribution to the α_2 bending vibration–rotation interaction constant so that the higher-order coefficient γ_{22} becomes dominant.

A general theory was developed by Hougen et al. [73]. A vibration–rotation Hamiltonian of a triatomic molecule was derived using a curvilinear bending coordinate and two rectilinear stretching coordinates. Then, this theory was extended by many authors, the model was called nonrigid bender. A simpler model, called semirigid bender, takes into account for both the relaxation of the molecular geometry as the molecule undergoes a large-amplitude vibration and for the coupling of the large-amplitude motion with the rotational motion. All other vibrations are neglected [74]. There is a generalization of the semirigid bender called generalized semirigid bender described in Ref. [75]. These models were reviewed by Jensen [76].

These bender models were used with success to study the spectra of many nonrigid molecules. However, the description of the small amplitude vibrations by a perturbation calculation is not always satisfactory. For this reason, a new model called Morse Oscillator Rigid Bender Internal Dynamics (MORBID) was developed by Jensen [77]. For a triatomic molecule, the rotation-vibration Hamiltonian is derived in terms of two bond-length displacements Δr_{12} and Δr_{32} and a bending coordinate ρ as in the bender model. The Hamiltonian is expanded as a power series in the variables $y_j = 1 - \exp(-a_j \Delta r_{j2})$, where the a_j are molecular constants obtained from the potential energy function. The matrix representation of the Hamiltonian is diagonalized in a truncated set of basis functions.

MORBID is based on the use of an approximate kinetic energy operator and experience has shown [77, 78] that this leads to small errors in potentials determined using it.

6.6 Structure of Weakly Bound Complexes

This is beyond the scope of this book to describe in details the methods used to determine the structure of a complex. Only the general principles will be presented. The experimental parameters which are useful for a structure elucidation will be listed, and the information they give will be discussed.

A large number of reviews have been published on this topic [79–93].

There are mainly two types of complexes. Those formed by a hydrogen bond which is an attractive interaction between a hydrogen atom of a molecule or a molecular fragment X–H, in which X is more electronegative than H, and an atom or a group of atoms in a different molecule [94]. There are also generally weaker complexes where atoms or molecules are held together by van der Waals forces (long-range electrostatic interactions and London dispersion forces). The binding energy is small and typically not greater than kT. Therefore, in a van der Waals molecule, there is very little redistribution of electronic charge. Thus, it is a good approximation to neglect it; however, see Sect. 6.6.6. When a molecular complex is formed, the number of translational and rotational degrees of freedom is reduced; they are replaced by low-frequency vibrational modes. It is a good approximation to neglect the coupling between the van der Waals modes and the modes of the monomers. Furthermore, it is often possible to treat separately the vibrational and rotational motions, so that the rotational spectra can be analyzed using the standard semirigid model (Watson Hamiltonian). The analysis of the spectra and the determination of the main molecular parameters (rotational constants, centrifugal distortion constants, nuclear quadrupole coupling constants) is generally not difficult.

The experimental parameters which are usually derived from spectroscopy are:

1. The ground state rotational constants whose major contribution comes from the moments of inertia of the complex. They are characteristic of the vibrationally averaged structure, which can be quite different from the equilibrium structure.
2. The centrifugal distortion constants which give information on the quadratic force constant of the intermolecular bond.
3. The nuclear quadrupole coupling constants which permit to determine the angle between a bond and a principal axis. The centrifugal distortion of the nuclear quadrupole coupling constants also furnishes useful information.
4. Spin-rotation and spin–spin coupling constants
5. The electric dipole moment furnishing the same information as the quadrupole coupling constants.
6. The vibrational frequencies measured in the far infrared or in the infrared.

An approximation which is generally made when determining the structure of a complex is to assume that the geometries of the monomers do not change upon complexation. This is a rough approximation for the strong hydrogen bonds because it is known that the length of the X–H bond in the complex X–H·Y increases on hydrogen bond formation [81, 95].

There are different ways to determine the structure of a complex.

6.6.1 From the Ground State Rotational Constants

The simplest method is a least-squares fit of the moments of inertia with bond lengths and angles of the monomers constrained to their free-molecule values, giving the so-called effective structure (r_0). In this case, the rovibrational corrections are neglected. The problem is that these corrections are large (see Sect. 5.2), typically one to two orders of magnitude larger than for a semirigid molecule. This is due to the fact that the vibration–rotation interaction constants, α_i are inversely proportional to the vibrational frequencies ω_i, some of them being are quite small for a complex, see Eqs. (5.3–5.5). For instance, In Ar·HCl, α_3 (corresponding to the stretching of the bond Ar···H whose vibrational frequency is 32.4 cm^{-1} [96]) is as large as about 100 MHz [97], whereas for a semirigid molecule with a rotational constant of the same order of magnitude, $B = 1678$ MHz, a value of a few MHz is expected, see Table 5.1. For this reason, the r_0 structure is not expected to be accurate. A typical example is the S···S bond length in $(CH_3)_2S.SO_2$: the r_0 value of this bond length is 3.008(3) Å whereas the semiexperimental equilibrium value is 2.947(3) Å [98]. In particular, when the van der Waals bond length can be determined using the rotational constant(s) from only one isotopologue, significant differences are observed. For instance, for Xe·HCl, r_0(Xe–Cl) = 4.2753 Å for ^{129}Xe.H^{35}Cl and 4.2875 Å for ^{129}Xe.D^{35}Cl, whereas the equilibrium value is 4.2767 Å [99]. The r_0 structure of a few complexes is compared to the r_e structure in Table 6.8. It confirms that the accuracy of the r_0 bond length is not better than 0.01 Å and can be as bad as 0.1 Å. For a semirigid molecule, the r_0 bond length is usually larger than the corresponding r_e value, see Sect. 5.8. This conclusion is not valid for a complex because the equilibrium structure can be extremely different from the vibrationally averaged structure; see Sect. 6.6.3.

Thus, it would be tempting to try to determine a full set of α-constants to obtain the equilibrium rotational constants. Unfortunately, there are several difficulties: (i) it is difficult to measure the rotational spectra in excited vibrational states because of the low-temperature of the molecular beam which is routinely used to record the spectra of complexes. For this reason, this was only done for a few molecules such as Ar·HCl; (ii) furthermore, the vibrational states are often perturbed by Coriolis interaction [96, 97] (iii) even assuming that the interactions could have been correctly analyzed, higher-order constants would probably be required because the α-constants are large (see Eq. (5.1)).

To overcome the first two difficulties, it seems simpler to calculate the rovibrational corrections from an anharmonic quantum-chemical force field. Up to now, it was only done for a few molecules. For instance, Kraus et al. [100] determined the semiexperimental structure of 16 complexes (45 bond lengths) using anharmonic force fields computed at the double-hybrid density functional level with an aug-cc-pVTZ basis set, see Sect. 2.3. They found a root-mean-square error of 0.055 Å for the bond lengths. A similar method was employed for several complexes, for instance $(CH_3)_2S{\cdot}SO_2$ [98] and $N_2O{\cdot}CO$ [101]. However, at least in the case of $N_2O.CO$ the results were not convincing. Indeed, the theory of the vibration–rotation interactions [102] assumes that the amplitude of the vibrations is small which is obviously not the case for the complexes. In this case, the second-order vibrational perturbation theory is probably inaccurate as it was shown on H_2O_2; see Sect. 6.4.2 and Ref. [69]. Actually, it is better to calculate the potential energy surface as explained in Sect. 6.6.8.

To avoid these difficulties, Kisiel [103] proposed to use the mass-dependent method (r_m, see Sect. 5.10.4) and obtained encouraging results for small molecules with many experimental rotational constants available such as Ar·HCl [104]. Table 6.8 compares the $r_m^{(1L)}$ bond length to the equilibrium value and confirms that the $r_m^{(1L)}$ value is significantly more accurate than the r_0 value.

Table 6.8 $r_m^{(1L)}$(Rg-X) distance (in Å) between a rare-gas atom and a halide atom in several weakly-bound Rg-hydrogen halide clusters

	Ar·HF	Ar·HCl	Ar·HBr	Kr·HBr
r_0[a]	3.5546(3)	4.0130(5)	4.1489(1)	4.2598(1)
$r_m^{(1L)}$	3.4416(6)[a]	4.0252(7)[b]	4.2440[a]	4.3194(7)[a]
r_e potential	3.485[c]	4.004[e]	4.243[g]	4.27[h]
r_e ab initio	3.482[d]	4.006[f]		

[a] Kisiel Z (2003) J Mol Spectrosc 218: 58–67
[b] Marshall FE, Neill JL, Muckle MT, Pate BH, Kisiel Z, Grubbs II GS (2018) J Mol Spectrosc 344: 34–38
[c] H6(4,3,2) potential: Chang HC, Tao F, Klemperer W, Healey C, Hutson JM (1993) J Chem Phys 99: 9337–9349
[d] CCSD(T)/CBS with counterpoise correction: Van Mourik T, Dunning TH (1997) J Chem Phys 107: 2451–2462
[e] H6(4,3,0) potential: Hutson JM (1992) J Phys Chem 96: 4237–4247
[f] CCSD(T)/CBS with counterpoise correction: Woon DE, Peterson KA, Dunning TH (1998) J Chem Phys 109: 2233–2241
[g] Scaled CCSD(T)/aug-cc-pVTZ potential: Willaert FF, McElmurry BA, Lucchese RR, Bevan, JW (2008) Chem Phys Lett 460: 525–530
[h] Scaled MP2/aug-cc-pVTZ potential: Wang Z, Lucchese RR, Bevan JW (2004) J Phys Chem A 108: 2884–2892

6.6.2 *Use of the Centrifugal Distortion Constant D_J*

When the stretching vibration of the intermolecular bond has a much lower frequency than the other vibrations, the centrifugal distortion constant D_J may be used to estimate the stretching force constant assuming that the vibration is harmonic. For a weakly bound linear dimer X.HA having a linear configuration at equilibrium, it is possible to use the pseudo-diatomic approximation

$$D_J = \frac{4\overline{B}^3}{\nu^2} \tag{6.24}$$

From ν, it is possible to derive the stretch force constant k_s. However, there is a more sophisticated approximation [105]

For a polyatomic dimer X·HA

$$D_J = \frac{16\pi^2 \mu_D B_0^3}{k_s}\left[1 - \frac{B_D}{B_X} - \frac{B_D}{B_{HA}}\right] \tag{6.25}$$

where B_D, B_X, and B_{HA} are the ground state rotational constants of the dimer, of X, and HA, respectively. This model has been generalized for different types of complexes [106]. Once k_s is known, it can be used in different ways depending of the approximations made.

The method used by Balle et al. [107] to determine an approximate equilibrium structure of Kr.HCl is described below.

In the pseudo-diatomic approximation, the stretching frequency of the intermolecular bond can be calculated

$$\nu_s = \frac{1}{2\pi}\sqrt{\frac{k_s}{\mu_D}} \tag{6.26}$$

where μ_D is the pseudo-diatomic reduced mass

$$\mu_D = \frac{m_1 m_2}{m_1 + m_2} \tag{6.27}$$

m_1 and m_2 being the masses of the two monomers. Then, the rotation-vibration interaction constant α is calculated

$$\alpha = \frac{36 B_e^2}{\nu_s} \tag{6.28}$$

Inserting α in the expression of the ground state rotational constant

$$B_0 = B_e - \alpha/2 \tag{6.29}$$

gives a quadratic equation in B_e whose solution is

$$B_e = \frac{1}{2}\left[\frac{\nu_s}{18} - \sqrt{\left(\frac{\nu_s}{18}\right)^2 - 4B_0\frac{\nu_s}{18}}\right] \tag{6.30}$$

The equilibrium bond length is derived from B_e

$$r_e = \sqrt{\frac{h}{8\pi^2 \mu_D B_e}} \tag{6.31}$$

Using a Lennard-Jones potential [107]

$$V(r) = \varepsilon\left[(r/r_e)^{12} - 2(r/r_e)^6\right] \tag{6.32}$$

where r_e is the equilibrium distance between the monomers in the complex and ε the energy depth of the well. One of the two parameters ε and r_e of this potential is determined by the force constant k_s of the stretching vibration

$$k_s = 72\frac{\varepsilon}{r_e^2} \tag{6.33}$$

The above methods are rather rudimentary, more sophisticated methods will be discussed in Sect. 6.6.8.

6.6.3 Use of the Quadrupole Coupling Constants [108]

The diagonal elements of the quadrupole coupling tensor, when available, are often used to get information on the structure of the complex. The experimental values of the quadrupole coupling constants, χ_{gg} ($g = a, b, c$) of the complex are the projections of the value of the quadrupole coupling constant of the monomer on the instantaneous g axes (Eckart axes, $g = a, b, c$) of the complex averaged over the zero-point motion

$$\chi_{gg} = \frac{3\langle\cos^2(\theta_g)\rangle - 1}{2}\chi_{\text{monomer}} \tag{6.34}$$

θ_g is the angle between the axis of the monomer and the Eckart axis g of the complex. This equation permits to deduce an average value of the angle θ_g. However, there are three limitations to the validity of this equation. First, it assumes that there is no redistribution of the electric field gradient upon complexation, see Sect. 6.6.6.

The second difficulty is due to the vibrational averaging. The value of θ will vary significantly from one isotopologue to the other one; see Table 6.9.

This variation can be explained with the help of the two-dimensional isotropic harmonic oscillator model [107]. The average value of the θ^2 is

$$\langle\theta^2\rangle = \frac{\hbar}{2\pi\mu_b\nu_b} \tag{6.35}$$

With the harmonic bending frequency ν_b

$$\nu_b = \frac{1}{2\pi}\sqrt{\frac{k_b}{\mu_b}} \tag{6.36}$$

μ_b is the reduced mass of the bending mode. For a complex X·YZ such as Ar·HCl, its value is

$$\frac{1}{\mu_b} = \frac{1}{M_X r_{X\text{-}Y}^2} + \frac{1}{M_Z r_{YZ}^2} + \frac{1}{M_Y}\left(\frac{1}{r_{X\text{-}Y}} - \frac{1}{r_{YZ}}\right)^2 \tag{6.37}$$

The third difficulty is that the change of geometry upon complexation is neglected. It is often a good approximation for weak van der Waals complexes such as N_2O.CO [101] but not for stronger complexes [81, 95] see also Sect. 6.6.7.

When an off-diagonal quadrupole coupling constant can be experimentally determined, it is possible to derive the value of the angle θ between an inertial axis and the bond axis. For instance, when the complex has a symmetry plane *ac* such as 2,5-dihydrofurane.HCl and χ_{ac} has been determined [109], the following Eq. (6.38) permits to obtain θ

$$\tan 2\theta = -2\frac{\chi_{ac}}{\chi_{aa} - \chi_{cc}} \tag{6.38}$$

Although the precision of the method is high (about 0.01°), the accuracy of the angle derived from this equation is limited because the bond axis does not always coincide with the principal axis of the quadrupole tensor, Kisiel [110] estimated that the difference is about 1°.

Finally, when the centrifugal distortion of the quadrupole coupling constants can be measured, it furnishes information on the coupling between radial and angular potentials [111].

When the complex rotates, the average intermolecular distance increases, so that the anisotropy felt by the quadrupole nucleus changes and the quadrupole coupling constant is a function of the rotational angular momentum J (in other words, it is a function of the stretching centrifugal distortion constant D_J).

Starting from a simple potential function

$$V(R,\theta) = \frac{1}{2}k_s(R - R_e)^2 + \frac{1}{2}k\theta^2 + \frac{1}{2}k_I\theta^2(R - R_e) \tag{6.39}$$

where the third term is the bend-stretch interaction term (k_I is the bend-stretch cubic force constant), they showed that the centrifugal distortion constant D_χ of the quadrupole coupling may be expressed as a function of the force constant k_I.

$$D_\chi = \frac{3\mu_b \chi_0 R_0 D_J k_\mathrm{I}}{8\hbar^2 B_0} \theta^5 \cos\theta \sin\theta \tag{6.40}$$

6.6.4 *Use of the Spin–Spin Coupling Constants [112]*

When the hyperfine structure due to the spin–spin interaction can be resolved (it is in particular the case for complexes with HF), the spin–spin coupling constants furnish in theory the same information as the quadrupole coupling constants, i.e., the average value of θ_g, the angle between the axis of the monomer and the Eckart axis g of the complex.

$$D_{gg} = \frac{3\langle\cos^2(\theta_g)\rangle - 1}{2} D_{\mathrm{monomer}} \tag{6.41}$$

However, there is a difference because the spin–spin coupling constant is a function of the distance between the interacting atoms, whereas the quadrupole coupling constant depends of the electric field gradient at the nucleus, see Sects. 6.6.3 and 6.6.6. If the electric field gradient is perturbed by the complexation, the quadrupole coupling constant and the spin–spin constant will deliver different results. This is the case for the planar, T-shaped, hydrogen-bonded complex DF·HC≡CH where the spin–spin coupling constant gives $\theta_a = 17.5(82)°$ whereas the quadrupole coupling constant gives $\theta_a = 21.5(11)°$ indicating that the electric field gradient at the deuterium site is probably perturbed [113].

6.6.5 *Use of the Electric Dipole Moment*

For a quasi-linear complex X·YZ, where X is a rare-gas and YZ a diatomic molecule, the component of the dipole moment along the *a*-principal axis is [114]

$$\mu_a = \mu\cos(\theta) + \alpha_\mathrm{X}\left[\frac{2\mu\cos(\theta)}{R^3} + \frac{3\Theta(3\cos^2\theta - 1)}{2R^4} + \frac{2\Omega(5\cos^3\theta - 3\cos\theta)}{R^5}\right]$$
$$\times\left[1 + \frac{2\alpha_\mathrm{YZ}}{R^3}\right]\left[1 + \frac{4\alpha_\mathrm{X}\alpha_\mathrm{YZ}}{R^6 - 4\alpha_\mathrm{X}\alpha_\mathrm{YZ}}\right] \tag{6.42}$$

Table 6.9 Comparison of the angle $\theta = \angle(a, XY)$ calculated with Eq. (6.34) and X = H, D and Y=Ar, Kr, Xe (in degree). Note that the equilibrium structure of these complexes is linear

Molecule	X = H	X = D	Ref.
$Ar.X^{35}Cl$	41.53	33.71	a
$Ar.X^{79}Br$	42.12	34.43	b
$Kr.X^{35}Cl$	37.96	30.93	c
$Kr.X^{79}Br$	38.10	31.11	b
$Xe{\cdot}X^{35}Cl$	34.69	28.30	d

[a] Novick SE, Davies P, Harris SJ, Klemperer W (1973) J Chem Phys 59: 2273–2279
[b] Keenan MR, Campbell EJ, Balle TJ, Buxton LW, Minton TK, Soper PD, Flygare WH (1980) J Chem Phys 72: 3070–3080
[c] Campbell EJ, Buxton LW, Keenan MR, Flygare WH (1981) Phys Rev A 24: 812–821
[d] Keenan MR, Buxton LW, Campbell EJ, Balle TJ, Flygare WH (1980) J Chem Phys 73: 3523–3529

μ, Θ, and Ω are respectively, the molecular dipole, quadrupole, and octupole moments of YZ. α_{YZ} and α_X are, respectively, the dipole polarizabilities of XY and X, R is the distance from the X nucleus to the center of mass of YZ. This equation yields a value of $\cos\theta$, the vibrational average of $\cos\theta$. The furnished information is similar to the quadrupole coupling constants but is less precise. The higher-order terms in Eq. (6.42) can be important. For instance, the value of the dipole moment of ClF is 0.8881 D, whereas the *a*-component of the dipole moment of Kr·ClF is $\mu_a = 1.136(3)$ D [114]. This is mainly due to the polarization of krypton by ClF. In the case of complexes, the simple method of vector sum of the free monomer dipole moments [115] can be extremely inaccurate, see Table 6.10 for a few examples.

6.6.6 *Redistribution of the Electric Field Gradient upon Complexation*

It is often assumed that there is no redistribution of the electric field gradient upon complexation. This assumption was investigated by several authors. See also Sect. 6.6.4.

A very simple way to check this assumption is to determine the angle θ_g, see Eq. (6.34), using two different quadrupole nuclei, for instance D and ^{35}Cl in Ar·DCl [79]. ^{35}Cl gives $\theta_a = 41.3°$ and D 40.88(35)°.

Campbell et al. [116] measured the rotational spectra of the Kr.HCl and Xe.DCl complexes and found a nuclear quadrupole hyperfine structure due to the nuclei ^{83}Kr and ^{131}Xe. Likewise, Jäger et al. [117] measured the rotational spectra of Ne.Xe, Ar.Xe, and Kr.Xe and could determine the quadrupole coupling constant due to ^{131}Xe. As the nuclear quadrupole hyperfine coupling constant of the free rare-gas

Table 6.10 Electric dipole moment and induced moment (D) for hydrogen-bonded dimers

Dimer	μ_a	μ_{ind} [a]	Ref	Dimer	μ_a	μ_{ind}	Ref.
$(HF)_2$	2.98865	0.60	b	$CO_2{\cdot}HF$	2.2465	0.60	b
$(H_2O)_2$	2.6429	0.46	b	SCO·HF	3.208	0.94	b
$H_2O{\cdot}HF$	4.073	0.68	b	OC·HCl	1.5178	0.388	c
HF·HCl	2.4095	0.14	b	OC·DCl	1.5623	0.418	c
$H_2S{\cdot}HF$	2.6239	0.78	b	$OC{\cdot}BF_3$	0.592	0.482	c
HCN·HF	5.612	0.80	b	$CO_2{\cdot}HCl$	1.4509	0.446	c
$(CH_2)_2O{\cdot}HF$	3.85	0.99	b				
Ar·ClF	1.053	0.18	d	Kr·ClF	1.136	0.26	d
Ne·Ar	0.0022		e	Ar·Kr	0.0034		e
Ne·Kr	0.011		e	Ar·Xe	0.014		f
Ne·Xe	0.012		f	Kr·Xe	0.007		f

[a] $\mu_{\text{ind}} = \mu_{\text{dimer}} - \mu_{\text{A}}\cos(\theta_{\text{A}}) - \mu_{\text{B}}\cos(\theta_{\text{B}})$, where θ_{X} is the angle between the *a*-principal axis and the symmetry axis of monomer X (X = A, B)
[b] Legon AC, Millen DJ (1986) Chem Rev 86: 635–657
[c] Altman RS, Marshall MD, Klemperer W (1983) J Chem Phys 79: 52–56
[d] Novick SE, Harris SJ, Janda KC, Klemperer W (1975) Can J Phys 53: 2007–2015
[e] Xu Y, Jäger, W, Djauhari J, Gerry MCL (1995) J Chem Phys 103: 2827–2833
[f] Jäger W, Xu Y, Gerry MCL (1993) J Chem Phys 99: 919–927

atom is zero, the effect must arise from a distortion of the spherical symmetry of the rare-gas atom upon complexation.

When the quadrupole nucleus is in a symmetry plane *xz* of the complex, the constant χ_{cc} of the complex should be identical to that of the monomer. This is sometimes far from true as can be seen in Table 6.11. Thus, it is important to check the validity of Eq. (6.34) when using it. Note that it is possible to estimate the variation of the electric field gradient upon complexation. This was done in particular for the Ar·ICl complex [118]. However, it is much easier to calculate ab initio this variation. Table 6.11 shows that, using the quadrupole coupling constant of the end nitrogen atom of the complex $X{\cdot}^{14}N_2{}^{16}O$ (with X=HC≡CH, Ar, or CO) is a better approximation.

6.6.7 *Bond Lengthening in the Monomers*

In the previous sections, it was assumed that the geometry of the monomers remains unchanged upon complexation and this assumption is made in most works. Two cases have to be distinguished: the weak van der Waals bond and the stronger hydrogen bond. Legon and Millen [95] have studied experimentally the bond lengthening δr of HF in hydrogen-bonded dimers B.HF. By combining the quadrupole coupling constant, χ(D) of B.DF and the spin–spin coupling constant *D*(HF) together with the electron charge distribution of B, they were able to determine δr for a series of complexes. They found that δr increases smoothly from zero for B = Ar to 0.016 Å for B = CH_3CN when plotted against the intermolecular stretching force constant k_σ. It is worth noting that a high-level ab initio calculation of the structure of H_2O.HF gives $\delta r = 0.0156$ Å [119] in perfect agreement with their experimental value, 0.015 Å.

For the weak van der Waals bonds, the lengthening is found almost negligible. For instance, for the dimer N_2O.CO it was found at the CCSD(T)_ae/cc-pwCVQZ level of theory that r(CO) = 1.1289 Å, r(NN) = 1.1262 Å, and r(NO) = 1.1847 Å, whereas for the monomers, the results are r(CO) = 1.1286 Å, r(NN) = 1.1267 Å, and r(NO) = 1.1848 Å [101].

6.6.8 Potential Energy Surface

The methods described in the previous sections give useful information on the structure of the complex. However, the results can be significantly different from the equilibrium structure. The best method to obtain a reliable equilibrium structure is to determine the potential surface of the complex. The equilibrium structure of the complex is determined by the absolute minimum of this surface. Generally, the Schrödinger equation is solved using a parametrized form of the potential and the experimental data. The first difficulty is that rotational spectroscopy furnishes information on the bottom of the potential and this is not enough to obtain a good potential. For that goal, infrared spectroscopy is quite useful, in particular far-infrared spectroscopy which permits to measure the low-frequency intermolecular vibrations: (i) the stretching vibrations give information on the dependence of the potential on the separation of the monomers; (ii) the bending vibrations give information on the anisotropy of the potential. Other sources are also useful: second-virial coefficients, viscosity, rotational line broadening, molecular beam scattering, etc. [86, 120].

The methodology will be described for the relatively simple case of a complex formed by an atom and a diatomic molecule, such as Ar·HCl. The interaction potential is a function of the intermolecular distance R, the length of the vector $\mathbf{R}$ from the atom to the center of mass of the diatomic molecule and the angle θ, the angle between $\mathbf{R}$ and the axis of the diatomic molecule.

Different forms of the potential are used. The simplest one is the Morse potential which was used to determine the structure of Ar.HF [121]. In the diatomic-like approximation, the Morse potential is written

$$V(R) = D\left(1 - e^{-a(R-R_e)}\right)^2 \tag{6.43}$$

R_e is the equilibrium value of R, D is the dissociation energy of the pseudo-diatomic molecule, and a is the Morse constant (see Sect. 3.5). Unfortunately, the Morse potential dies off too rapidly at large R. Actually, it is well established that the Morse potential is a rough approximation for diatomic molecules; see Sect. 3.6.

The 6–12 Lennard-Jones potential [122, 123] was often used

$$V(R) = \varepsilon\left[\left(\frac{R_{\mathrm{m}}}{R}\right)^{12} - 2\left(\frac{R_{\mathrm{m}}}{R}\right)^{6}\right] \tag{6.44}$$

where ε is the depth of the potential well and R_{m} the distance at which the potential reaches its minimum. This potential was used to study several complexes; see for instance [114].

However, as noted by Buckingham and Corner [124], the repulsive term varying as the inverse of the twelfth power of the distance of separation between the centers of the molecules is only an empirical approximation to the repulsion energy. It is more suitably described by an exponential form. This modified potential, known as Exp-Six potential [125], has the following form

$$V(R) = \frac{\varepsilon}{\alpha - 6}\left[6\exp\left[\alpha(1 - x) - \alpha x^{-6}\right]\right] \tag{6.45}$$

where $x = R/R_{\mathrm{m}}$, and α is a measure of the steepness of the repulsion energy.

A still more successful functional form has been proposed by Maitland and Smith [126], their potential is based on the Lennard–Jones function, but the exponent of the repulsive wall is allowed to vary with R

$$V(R) = \varepsilon(\theta)\left[\frac{6}{n - 6}x^{-n} - \frac{n}{n - 6}x^{-6}\right] \tag{6.46}$$

With

$$n = m(\theta) + \gamma(x - 1) \tag{6.47}$$

γ is an adjustable parameter and $m = 13$ for a rare gas.

Holmgren et al. [127] tried to determine the potential of Ar·HCl using spectroscopic results and found satisfactory results with the Exp-Six modified potential of Buckingham.

The choice of a good potential is crucial. Hutson made a thorough study of the different potential energy surfaces and published several seminal papers [86, 120, 128–131].

Finally, he proposed a potential called H6(4,3,2) which reproduces all available spectroscopic data for Ar·HCl as well as for Ar·HF [131]. The usual structural parameters are used for a van der Waals complex X.YZ: the vector **R** from the center of mass of the diatomic molecule YZ to the X atom, the vector **r** from the Z nucleus to the Y nucleus, and θ, the angle between **R** and **r**. The intermolecular potential is expanded

$$V(R, r, \theta) = \sum_{k=0}^{k_{max}} V_k(R, \theta)x^k \tag{6.48}$$

where x is the fractional extension of the diatom bond length from its equilibrium value

$$x = (r - r_e)/r_e \tag{6.49}$$

The expectation value of this potential for a vibrational state v is a smooth function of the mass-reduced quantum number η

$$\eta = (v + 1/2)/\mu_{YZ}^{1/2} \tag{6.50}$$

where μ_{YZ} is the reduced mass of the diatom.

The analytical form of this potential is

$$V(R, \eta, \theta) = A(\eta, \theta)\exp[-\beta(\eta, \theta)R] + V_{ind}(R, \eta, \theta) - \sum_{n=6}^{10} C_n(\eta, \theta)D_n(R, \eta, \theta)R^{-n} \tag{6.51}$$

The first term corresponds to the repulsion interaction which decreases exponentially with R, it is due to the intermolecular Coulomb and exchange terms involving electrons. V_{ind} is the effect of the polarization of the charge cloud of each monomer by the electrostatic field of the other one, it is calculated theoretically. The last term corresponds to the dispersion. It is an attractive interaction due to the correlation of the fluctuating instantaneous multipoles. The well depth ε, the radial position of the minimum, and the potential hardness parameter β are expanded as a double series in powers of η and Legendre polynomials of $\cos\theta$. Some parameters such as C_6 and C_7 are taken from theory. D_n is a damping factor

$$D_n(R) = 1 - \exp(-\beta R) \sum_{m=0}^{n} \frac{(\beta R)^m}{m!} \tag{6.52}$$

The number of adjustable parameters is 19 for H(4,3,0) and 22 for H(4,3,2).

Hutson, in particular, redetermined the potential many times, each time improving it. A comparison of the different potentials is shown in Table 6.12, where they are compared to two recent high-level ab initio structures.

It is also possible to use ab initio calculations to determine the potential but, even using large basis sets and high-level correlation treatment, they still do not always reproduce acceptably experimental data. However, the ab initio potential energy surface may be "morphed" in order to obtain a better agreement with the experimental data. Scaling functions for both the energy and the molecular distance are used. They are parametrized and determined by least-squares fitting to the experimental data.

Table 6.11 Values of the quadrupole coupling constant $\chi_{cc}(^{14}N)$ of N_2O in several complexes $X{\cdot}^{14}N_2{}^{16}O$ (in MHz)

X	Central ^{14}N	End ^{14}N	Ref.
Monomer	133.79(19)	386.88(14)	a
HC≡CH	162.51(71)	395.63(45)	b
Ar	140.66(64)	387.41(42)	c
CO	155.74(77)	397.52(48)	d

[a] Reinartz JMLJ, Meerts WL, Dymanus A (1978) Chem Phys 31: 19–29
[b] Leung HO (1997) J Chem Phys 107/ 2232–2241. https://doi.org/10.1062/1.474620
[c] Leung HO, Gangwani D, Grabow J-U (1997) J Mol Spectrosc 184: 106–112
[d] Ngari MS, Xu Y, Jäger W (1999) J Mol Spectrosc 244–253

First, the potential energy surface is interpolated between the ab initio points at each value of R using Legendre polynomials

$$V(R,\theta) = \sum_{\lambda=0} V_\lambda(R) P_\lambda(\cos\theta) \tag{6.53}$$

The morphing procedure can be expressed in the following way

$$V_{\text{morph}}(R,\theta) = v(R,\theta) V_{\text{orig}}(\rho(\theta)\cdot R,\theta) \tag{6.54}$$

The scaling functions v and ρ may be written

$$\rho(\theta) = \sum_{\lambda} \rho_\lambda P_\lambda(\cos\theta), \tag{6.55}$$

$$v(R,\theta) = \sum_{\lambda k} v_{\lambda k} P_\lambda(\cos\theta) f_k(R) \tag{6.56}$$

The number of terms in the sum depends of the quantity and variety of experimental data. The coefficients of the scaling functions are determined by a least-squares fit to the experimental data. In these equations, $\rho(\theta)$ is a distance scaling function.

This morphing method was used to determine the potential energy surface of several van der Waals complexes such as Ne·HF [132], Kr·BrH [133], or Ar·HI [134].

Table 6.12 Distance from center of mass of HCl to Ar atom for Ar···HCl

Potential	R_m (Å)	Ref.
B1[a]	3.8672(84)	1
M1[b]	3.8744(85)	1
M2[b]	3.9701(47)	1
M3[b]	3.9997(81)	1
M5[b]	4.0093	2
H6(2)	3.9925	2
H6(3)	3.9898	2
H6(4,3,0)	4.004	3
AI[c]	4.02	4
AI[d]	4.006	5

[a] Buckingham alpha-6 potential
[b] Maitland-Smith potential
[c] From a force field calculated at the CCSD(T)/aug-cc-PVQZ-33211 level of theory
[d] CCSD(T)/CBS
[1] Hutson JM, Howard BJ (1981) Mol Phys 43: 493–516
[2] Hutson JM (1988) J Chem Phys 89: 4550–4557
[3] Hutson JM (1992) J Phys Chem 96: 4237–4247
[4] Jouypazadeh H, Solimannejad M, Farrokhpour H (2016) Comput Theo Chem 1083: 64–71
[5] Woon DE, Peterson, KA, Dunning TH (1998) J Chem Phys 109: 2233–2241

6.6.9 Accuracy

Ar·HCl is a benchmark model of van der Waals complex, and its structure has been determined many times; see Table 6.13. First, it is obvious that the substitution structure, r_s, is quite inaccurate. It is a general behavior for weak complexes. Therefore, this method is not to be recommended for weak complexes. If the effective structure, r_0, is excluded, the largest value of R_m (distance of the Ar atom to the center of mass of HCl) is 4.061 Å for the $r_m^{(1L)}$ value and the smallest 4.006 Å for the ab initio CCSD(T)/CBS value, i.e., a range of 0.055 Å corresponding to a standard deviation of 0.025 Å ($s = R/(1.40n^{1/3}$, see Ref. [135]). The $r_m^{(1L)}$ value seems to be too large compared to the other values. If it is excluded, the standard deviation decreases to 0.014 Å.

The structure of $N_2O{\cdot}CO$ was recently redetermined using different methods, ab initio, experimental, and semiexperimental [101]. From the range of the different results, the value of R_m was found 3.825(13) Å. Results of similar accuracy were found for the complex $(CH_3)_2S{\cdot}SO_2$ for which the S···S semiexperimental bond length was found to be 2.947(3) Å, whereas an ab initio value was 2.9575 [98].

Table 6.13 Structure of Ar·HCl

Method	$r(\text{Ar}\cdots\text{Cl})$	R_m [a]	θ	r (HCl)	Ref.
r_0 [b]	4.0131[c]	4.0491		1.2839	1
r_0^{av} [b]	4.00654[d]	4.0425	41.53[e]		1
r_s	3.9843(11)				
$r_m^{(1L)}$	4.02521(72)	4.0612		1.2746	3
r_e [f]	4.0061	4.0421		1.28388	4
M3		3.9997(81)			5
M5		4.0093(293)			6
H6(3)		3.990			6
H6(4, 3, 0)[g]		4.004			7
V* [h]		[3.98 4.01]			8
AI[i]		4.02		1.2904	9
AI[j]		4.006		1.2776	10

[a] Distance from center of mass of HCl to Ar atom
[b] For $^{40}\text{Ar}\cdots\text{H}^{35}\text{Cl}$
[c] Evaluated from B_0 by assuming $r_0(\text{HCl})_{\text{dimer}} = r_0(\text{HCl})_{\text{free}}$ 1.2839 Å
[d] Evaluated from B_0 by assuming that the moment of inertia of the complexed HCl is given by $(1/2)I_b(\text{HCl})_{\text{free}}\left(1+\langle\cos^2\theta_{\text{av}}\rangle\right)$
[e] Defined by $\chi(\text{dimer}) = (1/2)\chi(\text{HCl})\left(3\cos^2\theta_{\text{av}}-1\right)$
[f] Using a Lennard-Jones 6–12 potential
[g] Potential obtained by fitting to results from high-resolution microwave, far-infrared, and infrared spectroscopy
[h] A quantum inversion procedure is used to obtain the interaction potential (called V*) based on the rotationally resolved state-to-state inelastic cross sections as well as rovibrational spectral data
[I] From a force field calculated at the CCSD(T)/aug-cc-PVQZ-33211 level of theory.
[j] CP-CCSD(T)/CBS

1 Kisiel Z, Pszczółkowski L (1998) Chem Phys Lett 291: 190–196.
2 Kisiel Z, Pszczółkowski L (1998) Chem Phys Lett 291: 190–196.
3 Marshall FE, Neill JL, Muckle MT, Pate BH, Kisiel Z, Grubbs II GS (2018) J Mol Spectrosc 344: 34–38
4 Keenan MR, Campbell EJ, Balle TJ, Buxton LW, Minton TK, Soper PD Flygare WH (1980) J Chem Phys 72: 3070–3080
5 Hutson JM, Howard BJ (1981) Mol Phys 43: 493–516
6 Hutson JM, Howard BJ (2002) Mol Phys 100: 151–164
7 Hutson JM (1992) J Phys Chem 96: 4237–4247
8 Geremia JM, Rabitz H (2001) J Chem Phys 115, 8899–8912
9 Jouypazadeh H, Solimannejad M, Farrokhpour H (2016) Comput Theo Chem 1083: 64–71
10 Woon, D. E., Peterson, K. A., & Dunning, T. H. (1998) J Chem Phys 109: 2233–2241

6.7 Conclusion

In conclusion, it is important to note that the rovibrational spectrum of a molecule with large-amplitude motion can be analyzed using different methods. For instance, the spectrum of H_2O_2 can be analyzed with the variational method, Sects. 4.8 and 6.4.2 [136], the semirigid bender model [137], or a Watson Hamiltonian with Fermi resonances and Coriolis interactions [138].

However, for many molecules with a large-amplitude motion, such as ring-puckering or quasi-linear molecules, the best method is to determine the minimum of the potential energy surface. Ab initio, this is easy. Experimentally, it is more difficult. The first problem is to parametrize correctly the potential. The second problem is that the difficulty increases rapidly with the number of degrees of freedom. For this reason, the use of this method is limited to rather small molecules.

References

1. Gordy W, Cook RL (1984) Microwave molecular spectra. Wiley, New York
2. Wilson EB Jr, Decius JC, Cross PC (1955) Molecular vibrations: the theory of infrared and Raman vibrational spectra. McGraw Hill, New York
3. Cleeton CE, Williams NH (1934) Electromagnetic waves of 1.1 cm wavelength and the absorption spectrum of ammonia. Phys Rev 45:234–237
4. Schnabel E, Törring T, Wilke W (1965) Zum Inversionsspektrum des NH_3. Z Phys 188:167–171
5. Helminger P, Gordy W (1969) Submillimeter-wave spectra of ammonia and phosphine. Phys Rev 188:100–108
6. Papoušek D, Stone JMR, Špirko V (1973) Vibration-inversion-rotation spectra of ammonia. A vibration-inversion-rotation Hamiltonian for NH_3. J Mol Spectrosc 48:17–37
7. Urban Š (1988) Effective rotational Hamiltonians of pyramidal XY_3 molecules with the inversion splittings of energy levels. J Mol Spectrosc 131:133–153
8. Chen P, Pearson JC, Pickett HM, Matsuura S, Blake GA (2006) Measurements of $^{14}NH_3$ in the $v_2 = 1$ state by a solid-state, photomixing, THz spectrometer, and a simultaneous analysis of the microwave, terahertz, and infrared transitions between the ground and ν_2 inversion–rotation levels. J Mol Spectrosc 236:116–126
9. Nguyen HVL, Gulaczyk I, Kręglewski M, Kleiner I (2021) Large amplitude inversion tunneling motion in ammonia, methylamine, hydrazine, and secondary amines: From structure determination to coordination chemistry. Coord Chem Rev 436:213797/1-21
10. Benedict WS, Plyler K (1957) Vibration-rotation bands of ammonia: II. The molecular dimensions and harmonic frequencies of ammonia and deuterated ammonia. Can J Phys 35:1235–1241
11. Puzzarini C, Heckert M, Gauss J (2008) The accuracy of rotational constants predicted by high-level quantum chemical calculations. I. Molecules containing first-row atoms. J Chem Phys 128:194108/1-9
12. Pawlowski F, Jørgensen P, Olsen J, Hegelund F, Helgaker T, Gauss J, Bak KL, Stanton JF (2002) Molecular equilibrium structures from experimental rotational constants and calculated vibration-rotation interaction constants. J Chem Phys 116:6482–6496
13. Sears TJ, Bunker PR, Davies PB, Johnson SA, Špirko V (1985) Diode laser absorption spectroscopy of D_3O^+: determination of the equilibrium structure and potential function of the oxonium ion. J Chem Phys 83:2676–2685

14. Araki M, Ozeki H, Saito S (1998) Experimental determination of the ground-state inversion splitting in D_3O^+ by microwave spectroscopy. J Chem Phys 109:5707–5709
15. Araki M, Ozeki H, Saito S (1999) Microwave spectrum of the inversion—rotation transitions of the D_3O^+ ion: $\Delta k = \pm 3n$ interaction and equilibrium structure. Mol Phys 97:177–183
16. Rajamäki T, Miani A, Halonen L (2003) Six-dimensional ab initio potential energy surfaces for H_3O^+ and NH_3: approaching the sub-wave number accuracy for the inversion splittings. J Chem Phys 118:10929–10938
17. Huang X, Carter S, Bowman J (2003) Ab initio potential energy surface and rovibrational energies of H_3O^+ and its isotopomers. J Chem Phys 118:5431–5441
18. Davies PB, Smith DM (1994) Diode laser spectroscopy and coupled analysis of the ν_2 and ν_4 fundamental bands of SiH_3^+. J Chem Phys 100:6166–6174
19. Yang J, Li J, Hao Y, Zhou C, Mo Y (2006) Inversion vibration of PH_3^+ (X^2A_2'') studied by zero kinetic energy photoelectron spectroscopy. J Chem Phys 125:054311/1–054311/12
20. Legon AC (1980) Equilibrium conformations of four- and five-membered cyclic molecules in the gas phase: determination and classification. Chem Rev 80:231–262
21. Laane J (1993) Vibrational potential energy surfaces of non-rigid molecules in ground and excited electronic states. In: Laane J et al (eds) Structures and conformations of non-rigid molecules. Kluwer Academic Publishers, pp 65–98
22. Laane J (1987) Determination of vibrational potential energy surfaces from Raman and infrared spectra. Pure Appl Chem 59:1307–1326
23. Harris DO, Harrington HW, Luntz AC, Gwinn WD (1966) Microwave spectrum, vibration-rotation interaction, and potential function for the ring-puckering vibration of trimethylene sulfide. J Chem Phys 44:3467–3480
24. Harris DO, Engerholm GG, Tolman CA, Luntz AC, Keller RA, Kim H, Gwinn WD (1969) Ring puckering in five-membered rings I. General theory. J Chem Phys 50:2438–2445
25. Chan SI, Stelman D (1963) Some energy levels and matrix elements of the quartic oscillator. J Mol Spectrosc 10:278–299
26. Perrin A, Demaison J, Flaud J-M, Lafferty W, Sarka K (2011) Spectroscopy of polyatomic molecules: determination of the rotational constants. In: Demaison J, Boggs JE, Császár AG (eds) Equilibrium molecular structures. From spectroscopy to quantum chemistry. CRC Press, Boca Raton, pp 89–124
27. Lesarri A, López JC, Alonso JL, Wlodarczak G, Demaison J (1999) Rotational spectrum, ring-puckering vibration and ab initio calculations on 1,1-difluorocyclobutane. J Chem Phys 111:6375–6384
28. Egawa T, Fukuyama T, Yamamoto S, Takabayashi F, Kambara H, Ueda T, Kuchitsu K (1987) Molecular structure and puckering potential function of cyclobutane studied by gas electron diffraction and infrared spectroscopy. J Chem Phys 86:6018–6026
29. Egawa T, Yamamoto S, Ueda T, Kuchitsu K (1987) Two-dimensional analysis of the ring-puckering and methylene-rocking modes of cyclobutane. J Mol Spectrosc 126:231–239
30. Kummli DS, Frey HM, Leutwyler S (2007) High-accuracy structure of cyclobutane by femtosecond rotational Raman four-wave mixing. J Phys Chem A 111:11936–11942
31. Caminati W, Vogelsanger B, Meyer R, Grassi G, Bauder A (1988) Rotational spectrum, dipole moment, and ring-puckering potential of cyclobutane-1, 1-d_2. J Mol Spectrosc 131:172–184
32. Glendening ED, Halpern AM (2005) Ab initio study of cyclobutane: molecular structure, ring-puckering potential, and origin of the inversion barrier. J Phys Chem A 109:635–642
33. Wollrab JE (1967) Rotational spectra and molecular structure. Academic Press, New York
34. Lin CC, Swalen JD (1959) Internal rotation and microwave spectroscopy. Rev Mod Phys 31:841–892
35. Dreizler H (1968) Mikrowellenspectroskopische Bestimmung von Rotationsbarrieren freier Moleküle. Fortschr Chem Forsch 10:59–155
36. Kleiner I (2010) Asymmetric-top molecules containing one methyl-like internal rotor: methods and codes for fitting and predicting spectra. J Mol Spectrosc 260:1–18
37. Demaison J, Margulès L, Kleiner I, Császár AG (2010) Equilibrium structure in the presence of internal rotation: a case study of *cis*-methyl formate. J Mol Spectrosc 259(2010):70–79

38. Nguyen HVL, Kleiner I (2020) Understanding (coupled) large amplitude motions: the interplay of microwave spectroscopy, spectral modeling, and quantum chemistry. Phys Sci Rev. https://doi.org/10.1515/psr-2020-0037, online
39. Kleiner I, Lovas FJ, Godefroid M (1996) Microwave spectra of molecules of astrophysical interest. XXIII. Acetaldehyde. J Phys Chem Ref Data 25:1113–1210
40. Kleiner I, Lopez JC, Blanco S, McKellar ARW, Moazzen-Ahmadi N (1999) The ground and first torsional states of CD_3CHO. J Mol Spectrosc 197:275–288
41. Bermudez C, Motiyenko RA, Cabezas C, Ilyushin VV, Margulès L, Endo Y, Guillemin JC (2022) Internal rotation analysis of the microwave and millimeter wave spectra of fluoral (CF_3CHO). Spectrochim Acta A 274:121071
42. Szalay V, Császár AG, Senent ML (2002) Symmetry analysis of internal rotation. J Chem Phys 117:6489–6492
43. Herschbach DR (1959) Calculation of energy levels for internal torsion and over-all rotation. III J Chem Phys 31:91–108
44. Van Eijck BP, van Opheusden J, van Schaik MMM, van Zoeren E (1981) Acetic acid: microwave spectra, internal rotation and substitution structure J Mol Spectrosc 86:465–479
45. Pickett H (1991) The fitting and prediction of vibration-rotation spectra with spin interactions. J Mol Spectrosc 148:371–377
46. Herbers S, Fritz SM, Mishra P, Nguyen HVL, Zwier TS (2020) Local and global approaches to treat the torsional barriers of 4-methylacetophenone using microwave spectroscopy. J Chem Phys 152: 074301/1-8
47. Demaison J, Vogt N, Nguyen HVL (2022) Determination of rotational constants in the presence of methyl internal rotation: 2-acetylfuran and 2-acetylthiophene as examples. Chem Phys Lett 794:139488
48. Hougen JT, Kleiner I, Godefroid M (1994) Selection rules and intensity calculations for a C_s asymmetric top molecule containing a methyl group internal rotor. J Mol Spectrosc 163:559–586
49. Kleiner I, Hougen JT (2003) Rho-axis-method Hamiltonian for molecules having one methyl rotor and C_1 point-group symmetry at equilibrium. J Chem Phys 119:5505–5509
50. Ilyushin VV, Kisiel Z, Pszczółkowski L, Mäder H, Hougen JT (2010) A new torsion-rotation fitting program for molecules with a sixfold barrier: application to the microwave spectrum of toluene. J Mol Spectrosc 259:26–38
51. Groner P (1997) Effective rotational Hamiltonian for molecules with two periodic large-amplitude motions. J Chem Phys 107:4483–4498
52. Hartwig H, Dreizler H (1996) The microwave spectrum of trans-2,3-dimethyloxirane in torsional excited states. Z Naturforsch A 51:923–932
53. Herbers S, Zingsheim O, Nguyen HVL, Bonah L, Heyne B, Wehres N, Schlemmer S (2021) Internal rotation arena: program performances on the low barrier problem of 4-methylacetophenone. J Chem Phys 155:224302
54. Penocchio E, Piccardo M, Barone V (2015) Semiexperimental equilibrium structures for building blocks of organic and biological molecules: the B2PLYP route. J Chem Theo Comput 11:4689–4707
55. Demaison J, Craig NC, Gurusinghe R, Tubergen MJ, Rudolph HD, Coudert LH, Szalay PG, Császár AG (2017) Fourier transform microwave spectrum of propene-$3d_1$ (CH_2=$CHCH_2D$), quadrupole coupling constants of deuterium, and a semiexperimental equilibrium structure of propene. J Phys Chem A 121:3155–3166
56. Vogt N, Demaison J, Vogt J, Rudolph HD (2014) Why it is sometimes difficult to determine the accurate position of a hydrogen atom by the semiexperimental method: structure of molecules containing the OH or the CH_3 group. J Comput Chem 35:2333–2342
57. Demaison J, Margulès L, Rudolph HD (2010) Accurate determination of an equilibrium structure in the presence of a small coordinate: the case of dimethyl sulfide. J Mol Struct 978:229–233
58. Dindić C, Lüchow A, Vogt N, Demaison J, Nguyen HVL (2021) Equilibrium structure in the presence of methyl internal rotation: Microwave spectroscopy and quantum chemistry study of the two conformers of 2-acetylfuran. J Phys Chem A 125:4986–4997

59. Demaison J, Maes H, Van Eijck BP, Wlodarczak G, Lasne MC (1987) Determination of the moment of inertia of methyl groups: analysis of the millimeter-wave spectra of cis-propanal and methylthioethyne. J Mol Spectrosc 125:214–224
60. Demaison J, Wlodarczak G, Siam K, Ewbank JD, Schäfer L (1988) Comparison of *ab initio* calculated and experimental methyl-top moments of inertia. Chem Phys 120:421–428
61. Wilson EB, Lin CC, Lide DR (1955) Calculation of energy levels for internal torsion and over-all rotation. I. CH_3BF_2-type molecules. J Chem Phys 23:136–142
62. Rudolph HD, Dreizler H, Jaeschke A, Wendling P (1967) Mikrowellenspektrum, Hinderungspotential der internen Rotation und Dipolmoment des Toluols. Z Naturforsch A 22:940–944
63. Ilyushin V (2018) Millimeter wave spectrum of nitromethane. J Mol Spectrosc 345:64–69
64. Amir-Ebrahimi V, Choplin V, Demaison J, Roussy G (1981) Microwave spectrum of the ^{13}C-ring-monosubstitued toluenes and structure of toluene. J Mol Spectrosc 89:42–52
65. Hirota E, Hirooka T, Morino Y (1968) Microwave spectrum of propylene. Direct determination of the internal rotation splitting for the $CH_2DCH{=}CH_2$ species. J Mol Spectrosc 26:351–367
66. Coudert LH, Hougen JT (1988) Tunneling splittings in the water dimer: further development of the theory. J Mol Spectrosc 130:86–119
67. Schwoch D, Rudolph HD (1975) The microwave spectra of ortho-fluorotoluene with the asymmetric internal rotors CH_2D and CHD_2. J Mol Spectrosc 57:47–74
68. Petkie DT, Goyette TM, De Lucia FC, Helminger P, Belov SP, Winnewisser G (1998) Millimeter and submillimeter-wave spectrum of hydrogen peroxide in the ground and $\nu_3 = 1$ vibrational states. J Mol Spectrosc 192:25–31
69. Baraban J, Changala PB, Stanton JF (2018) The equilibrium structure of hydrogen peroxide. J Mol Spectrosc 343:92–95
70. Changala PB, Baraban JH (2016) Ab initio effective rotational and rovibrational Hamiltonians for non-rigid systems via curvilinear second order vibrational Møller-Plesset perturbation theory. J Chem Phys 145:174106
71. Małyszek P, Koput J (2013) Accurate ab initio potential energy surface and vibration-rotation energy levels of hydrogen peroxide. J Comput Chem 34:337–345
72. Lide DR, Matsumura C (1969) Structure of the alkali hydroxides. IV. Interpretation of vibration-rotation interactions in CsOH and RbOH and refinement of structures. J Chem Phys 50:3080–3086
73. Hougen JT, Bunker PR, Jones JWC (1970) The vibration-rotation problem in triatomic molecules allowing for a large-amplitude bending vibration. J Mol Spectrosc 34:136–172
74. Bunker PR, Landsberg BM (1977) The rigid bender and semirigid bender models for the rotation-vibration Hamiltonian. J Mol Spectrosc 67:374–385
75. Winnewisser BP, Winnewisser M, Medvedev IR, De Lucia FC, Ross SC, Koput J (2010) Analysis of the FASSST rotational spectrum of NCNCS in view of quantum monodromy. Phys Chem Chem Phys 12:8158–8189
76. Jensen P (1983) The nonrigid bender Hamiltonian for calculating the rotation-vibration energy levels of a triatomic molecule. Comp Phys Rep 1:1–55
77. Jensen P (1988) A new Morse oscillator-rigid bender internal dynamics (MORBID) Hamiltonian for triatomic molecules. J Mol Spectrosc 128:478–501
78. Polyansky OL, Jensen P, Tennyson J (1996) The potential energy surface of $H_2{}^{16}O$. J Chem Phys 105:6490–6497
79. Klemperer W (1977) Rotational spectroscopy of van der Waals molecules. Farad Discuss Chem Soc 62:179–184
80. Le Roy RJ, Carley JC (1980) Spectroscopy and potential energy surfaces of van der Waals molecules. In: Lawley KP (ed) Potential energy surfaces. Wiley, New York, pp 354–420
81. Legon AC, Millen DJ (1986) Gas-phase spectroscopy and the properties of hydrogen-bonded dimers: HCN-HF as the spectroscopic prototype. Chem Rev 86:635–657
82. Buckingham AD, Fowler PW, Hutson JM (1988) Studies of van der Waals molecules and intermolecular forces. Chem Rev 88:963–988

83. Nesbitt DJ (1988) High-resolution infrared spectroscopy of weakly bound molecular complexes. Chem Rev 88:843–870
84. Saykally RJ (1989) Far-infrared laser spectroscopy of van der Waals bonds: a powerful new probe of intermolecular forces. Acc Chem Res 22:295–300
85. Cohen RC, Saykally RJ (1991) Multidimensional intermolecular potential surfaces from vibration-rotation-tunneling (VRT) spectra of van der Waals complexes. Annu Rev Phys Chem 42:369–392
86. Hutson JM (1990) Intermolecular forces from the spectroscopy of van der Waals molecules. Annu Rev Phys Chem 41:123–154
87. Leopold KR, Fraser GT, Novick SE, Klemperer W (1994) Current themes in microwave and infrared spectroscopy of weakly bound complexes. Chem Rev 94:1807–1827
88. Bacic Z, Miller RE (1996) Molecular clusters: structure and dynamics of weakly bound systems. J Phys Chem 100:12945–12959
89. Wormer PES, van der Avoird A (2000) Intermolecular potentials, internal motions, and spectra of van der Waals and hydrogen-bonded complexes. Chem Rev 100:4109–4144
90. Tao F-M (2001) Bond functions, basis set superposition errors and other practical issues with ab initio calculations of intermolecular potentials. Int Rev Phys Chem 20:617–643
91. Klemperer W, Chuang CC, Higgens KJ, Miller AS, Fu HC (2001) Spectroscopy of van der Waals molecules: isomers and vibrational predissociation. Can J Phys 79:101–108
92. Caminati W, Grabow J-U (2009) Microwave spectroscopy, molecular systems. In: Laane J (ed) Frontiers of molecular spectroscopy. Elsevier, Amsterdam, pp 589–598
93. Legon AC (2017) Tetrel, pnictogen and chalcogen bonds identified in the gas phase before they had names: a systematic look at non-covalent interactions. Phys Chem Chem Phys 19:14884
94. Arunan E, Desiraju GR, Klein RA et al (2011) Defining the hydrogen bond: an account (IUPAC technical report). Pure Appl Chem 83:1619–1636
95. Legon AC, Millen DJ (1986) The lengthening of the HF bond on formation of heterodimers B···HF: determination and dependence on the strength of the hydrogen bond. Proc B Soc Lond 404:89–99
96. Lovejoy CM, Nesbitt DJ (1988) Infrared-active combination bands in ArHCl. Chem Phys Lett 146:582–588
97. Chuang C, Gutowsky HS (1991) Rotational spectra of the $\sum$ bend states of Ar–H/DCl and the $\sum$ stretch of Ar–HCl. J Chem Phys 94:86–91
98. Obenchain DA, Spada L, Alessandrini S, Rampino S, Herbers S, Tasinato N, Mendolicchio M, Kraus P, Gauss J, Puzzarini C, Grabow J-U, Barone V (2018) Unveiling the sulfur–sulfur bridge: Accurate structural and energetic characterization of a homochalcogen intermolecular bond. Angew Chem Int Ed 57:15822–15826
99. Keenan MR, Buxton LW, Campbell EJ, Balle TJ, Flygare WH (1980) ^{131}Xe nuclear quadrupole coupling and the rotational spectra of XeHCl. J Chem Phys 73:3523–3529
100. Kraus P, Obenchain DA, Frank I (2018) Benchmark-quality semiexperimental structural parameters of van der Waals complexes. J Phys Chem A 122:1077–1087
101. Demaison J, Vogt N, Jin Y, Saragi RT, Juanes M, Lesarri A (2021) How accurate is the determination of equilibrium structures for van der Waals complexes? The dimer $N_2O\cdots CO$ as an example. J Chem Phys 154:194302
102. Mills IM (1972) Vibration-rotation structure in asymmetric and symmetric molecules. In: Rao KN, Mathews CW (eds) Molecular spectroscopy: modern research, vol 1. Academic Press, New York
103. Kisiel Z (2003) Least-squares mass-dependence molecular structures for selected weakly bound intermolecular clusters. J Mol Spectrosc 218:58–67
104. Marshall FE, Neill JL, Muckle MT, Pate BH, Kisiel Z, Grubbs GS II (2018) Observation of $^{36}ArH^{37}Cl$, $^{38}ArH^{35}Cl$ and $^{38}ArH^{37}Cl$ in natural abundance using CP-FTMW spectroscopy. J Mol Spectrosc 344:34–38
105. Novick SE (1977) Centrifugal distortion in weakly bound linear triatomic molecules. J Mol Spectrosc 68:77–80

106. Millen DJ (1985) Determination of stretching force constants of weakly bound dimers from centrifugal distortion constants. Can J Phys 63:1477–1479
107. Balle TJ, Campbell EJ, Keenan MR, Flygare WH (1980) A new method for observing the rotational spectra of weak molecular complexes: KrHCl. J Chem Phys 72:922–932
108. Legon AC (1994) Non-linear hydrogen bonds and rotational spectroscopy: measurement and rationalisation of the deviation from linearity. Faraday Discuss 97:19–33
109. Legon AC, Thorn JC (1994) Can the nonlinearity of hydrogen bonds B···HX be measured precisely? A method from rotational spectroscopy and its application to 2,5-dihydrofuran···HCl. Chem Phys Lett 227:472–479
110. Kisiel Z (1998) Nuclear quadrupole coupling in Cl_2C=CHCl and $Cl_2C=CH_2$: evidence for systematic differences in orientations between internuclear and field gradient axes for terminal quadrupolar nuclei. J Chem Phys 109:10263–10272
111. Keenan MR, Campbell EJ, Balle TJ, Buxton LW, Minton TK, Soper PD, Flygare WH (1980) Rotational spectra and molecular structures of ArHBr and KrHBr. J Chem Phys 72:3070–3080
112. Keenan MR, Buxton LW, Campbell EJ, Legon AC, Flygare WH (1981) Molecular structure of ArDF: an analysis of the bending mode in the rare gas–hydrogen halides. J Chem Phys 74:2133–2137
113. Read WG, Flygare WH (1982) The microwave spectrum and molecular structure of the acetylene-HF complex. J Chem Phys 76:2238–2246
114. Novick SE, Harris SJ, Janda KC, Klemperer W (1975) Structure and bonding of KrClF: intermolecular force fields in van der Waals molecules. Can J Phys 53:2007–2015
115. Smyth CP (1955) Dielectric behavior and structure. MacGraw Hill, New York
116. Campbell EJ, Buxton LW, Keenan MR, Flygare WH (1981) ^{83}Kr and ^{131}Xe nuclear quadrupole coupling and quadrupolar shielding in KrHCl and XeDCl. Phys Rev A 24:812–821
117. Jäger W, Xu Y, Gerry MCL (1993) Pure rotational spectra of the mixed rare gas van der Waals complexes Ne–Xe, Ar–Xe, and Kr–Xe. J Chem Phys 99:919–927
118. Davey JB, Legon AC, Waclawick ER (1999) Iodine and chlorine nuclear quadrupole coupling in the rotational spectra of Ar···ICl and ICl: intramolecular charge transfer induced in ICl by Ar. Chem Phys Lett 306:133–144
119. Demaison J, Liévin J (2008) Equilibrium structure of hydrogen-bonded dimer H_2O···HF. Mol Phys 106:1249–1256
120. Hutson JM, Howard BJ (1981) The intermolecular potential energy surface of Ar · HCl. Mol Phys 43:493–516
121. Cousins BL, O'Brien S̀C, Lisy JM (1984) Argon isotope effect in the microwave spectra of ArHF. J Phys Chem 88:5142–5143
122. (Lennard-)Jones JE (1924) On the determination of molecular fields. I. From the variation of the viscosity of a gas with temperature. Proc R Soc (London) A206:441–462
123. (Lennard-)Jones JE (1924) On the determination of molecular fields. II. From the equation of state of a gas. Proc R Soc (London) A206: 463–477
124. Buckingham RA, Corner J (1947) Tables of second virial and low-pressure Joule-Thomson coefficients for intermolecular potentials with exponential repulsion. Proc R Soc (London) A189:118–129
125. Mason EA (1954) Transport properties of gases obeying a modified Buckingham (ExpSix) potential. J Chem Phys 22:169–186
126. Maitland GC, Smith EB (1973) A simplified representation of intermolecular potential energy. Chem Phys Lett 22:443–446
127. Holmgren SL, Waldman M, Klemperer W (1978) Internal dynamics of van der Waals complexes. II. Determination of a potential energy surface for ArHCl. J Chem Phys 69:1661–1669
128. Hutson JM, Howard BJ (1982) Anisotropic intermolecular forces. I. Rare gas–hydrogen chloride systems. Mol Phys 45:769–790
129. Hutson JM (1988) The intermolecular potential of Ar–HCl: determination from high resolution spectroscopy. J Chem Phys 89:4550–4557

130. Hutson JM (1992) Vibrational dependence of the anisotropic intermolecular potential of Ar–HF. J Chem Phys 96:6752–6767
131. Hutson JM (1992) Vibrational dependence of the anisotropic intermolecular potential of Ar-HCl. J Phys Chem 96:4237–4247
132. Meuwly M, Hutson JM (1999) Morphing ab initio potentials: a systematic study of Ne–HF. J Chem Phys 110:8338–8347
133. Wang Z, Lucchese RR, Bevan JW (2004) A Kr-BrH global minimum structure determined on the basis of potential morphing. J Chem Phys 108:2884–2892
134. McIntosh AL, Wang Z, Castillo-Chara J, Lucchese RR, Bevan JW, Suenram RD, Legon AC (1999) The structure and ground state dynamics of Ar–IH. J Chem Phys 111:5764–5770
135. Kyker GC (1983) The range of data set: a quick estimator of the standard deviation. Am J Phys 51:852–852
136. Koput J, Carter S, Handy NC (1998) Potential energy surface and vibrational-rotational energy levels of hydrogen peroxide. J Phys Chem A 102:6325–6330
137. Pelz G, Yamada KMT, Winnewisser G (1993) Torsional dependence of the effective rotational constants of H_2O_2 and H_2S_2. J Mol Spectrosc 159:507–520
138. Flaud J-M, Johns JWC, Lu Z, Winnewisser G, Klein H (2001) The torsion-rotation spectrum of D_2O_2. Can J Phys 79:367–374

Chapter 7
Equilibrium Molecular Structure as Determined by Gas-Phase Electron Diffraction

Abstract Gas-phase electron diffraction is a very productive method in the structural chemistry of isolated molecules particularly for large-size ones. Besides purely experimental thermal-average internuclear distances (r_g/r_a), semiexperimental equilibrium structure (r_e^{se}) can be precisely derived from electron diffraction data accounting for vibrational corrections calculated with anharmonic force field from high-level quantum-chemical calculations. The equilibrium structure, being universal for different experimental and theoretical methods, is the basis for benchmarking and joint use of these methods. Procedures of structure determinations by conventional gas-phase electron diffraction are reviewed in this chapter.

7.1 Introduction

Our fundamental knowledge about structures of free neutral molecules, radicals, and ions as well as weakly bound complexes has been obtained mainly by gas-phase electron diffraction (GED) and high-resolution (HR) spectroscopy (microwave (MW), infrared (IR), Raman (Ra), etc.). A considerable amount of all known molecular structures has been determined by GED (see Fig. 7.1 with the histogram of structure determinations over the years according to the MOGADOC database [1, 2]). Molecules with different properties, namely both stable and unstable, both polar and non-polar, both small-size and relatively large (with several dozens of atoms) can be studied by this method. The existence of different isotopic species is not relevant for a structure determination by GED. Thus, this method has obvious advantages in comparison with MW and HR IR spectroscopy.

Before going to the main topic, the experimental and theoretical basis of conventional (real-time) electron diffraction will be briefly reviewed in the Sect. 7.2 and Sect. 7.3. The time-resolved (ultrafast) electron diffraction (see papers [3–5] and reviews [6, 7]), being applied only to a few structure determinations, is not a topic of this chapter.

N. Vogt and J. Demaison, *Equilibrium Structure of Free Molecules*, Lecture Notes in Chemistry 111, https://doi.org/10.1007/978-3-031-36045-9_7

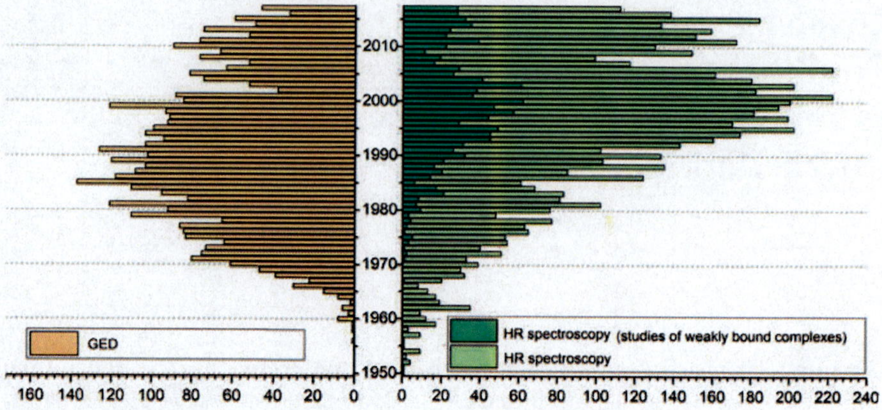

Fig. 7.1 Histogram of structure determinations by GED and HR spectroscopy over the years [1]

7.2 Conventional Gas-Phase Electron Diffraction Experiment

First experiments with electron scattering on randomly oriented molecules in a gas were carried out by the German scientists Mark and Wierl in 1930 [8]. They observed strong interference effects in the intensity of electron scattering by free molecules. These effects depend on internuclear distances in molecules and, thus, can be used for the determination of molecular structure.

In the 1950s, the electron diffraction method was established due to development of electron scattering theory accounting for molecular vibrations (mainly by Karle and Karle [9–11]) and construction of the first precision diffractometers [12, 13]. Since the 1960s, this method has been actively used for structure determinations (see Fig. 7.1). Several years later, the GED structural analysis was accelerated by the advent of fast computers.

The main components of gas-phase diffractometers are the following:

(1) an electron gun (1),
(2) a focusing system (2),
(3) a sample inlet system (3–5), and
(4) a detector (9).

The scheme of a conventional GED experiment is shown in Fig. 7.2.

The electrons thermally emitted from the cathode of an electron gun are accelerated by a stable potential (in the range from 40 to 100 kV) between the cathode and the anode. The monochromatic electron beam (7) focused by means of diaphragms and electromagnetic lenses (2) has a diameter of about 0.1 mm.

A gas-phase sample is introduced into a diffraction chamber (10) through the nozzle (3) of an effusion cell (4) initially created by Frost et al. [14]. A specially

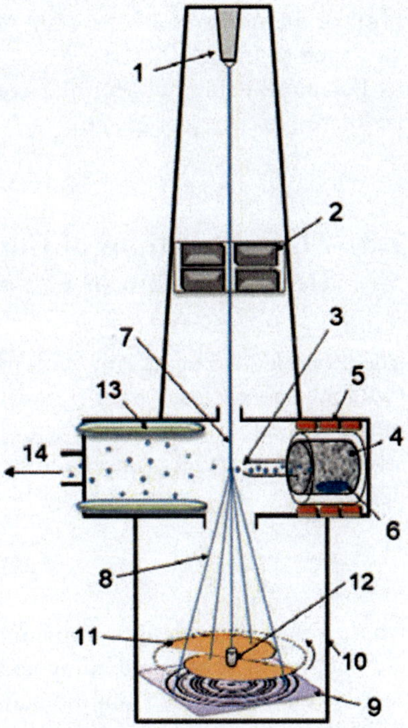

Fig. 7.2 Scheme of a gas-phase electron diffraction experiment: (1) electron gun, (2) focusing lenses, (3) nozzle, (4) effusion cell, (5) evaporator, (6) sample, (7) electron beam, (8) diffracted electrons, (9) detector, (10) diffraction chamber, (11) rotating sector, (12) beam stop, (13) cold trap, (14) connection to pump

constructed evaporator is used to heat compounds with a low volatility up to temperatures of about 2300 K [15]. The required sample pressure at the nozzle is usually from a few tenths to a few mm Hg. The pressure in the diffraction chamber of about $10^{-6}-10^{-5}$ mm Hg is maintained by pump.

The diffraction patterns arising from electron scattering by molecules are registered by photo films, image plates, or other detectors. Recording the scattering intensity by the image plates with radiation-sensitive layers was introduced by Iijima et al. [16]. In comparison to photo films, the image plates are characterized by a very large dynamic range, and they can be used instead of photo films without modification of the diffractometers [17–19].

The rotating sector (11), a metallic disk of special shape, is placed in front of the detector. First proposed by Finbak [20], it is used to compensate the rapid decrease of the electron scattering intensity which is inversely proportional to the fourth power of the scattering coordinate.

For the first time, to study unstable species and vapors with complex composition, Hargittai et al. suggested the use of electron diffraction in combination with mass spectrometry [21]. Later, the synchronous electron diffraction/mass-spectrometric experiments were regularly carried out by Girichev et al. [22, 23].

The currently used gas-phase electron diffraction technique is either an improved commercial gas-phase electron diffractometer, such as the Balzers Eldigraph KD-G2

[18], or a commercial solid-state electron diffractometer adapted to the study of a gas, such as EMR-100 [24].

For more details concerning conventional GED experiment, the reader can be addressed to reviews [25, 26].

7.3 Main Equations of Conventional GED Method. Determination of Experimental Molecular Structures

The total electron scattering with intensity $I_T(s)$ includes elastic electron scattering (without energy loss) both by each pair of nuclei in the molecule, the so-called molecular electron scattering with intensity $I_M(s)$, and by each single nucleus, the so-called atomic electron scattering with intensity $I_A(s)$. The scattering variable s is defined as:

$$s = (4\pi/\lambda)\sin(\theta/2), \tag{7.1}$$

where λ is the wavelength of incoming electrons, θ is the scattering angle between momentum vectors of incoming and scattered electrons.

Also taking into account the inelastic electron scattering (with energy loss) with $I_{\text{inel}}(s)$:

$$\begin{aligned} I_T(s) &= I_M(s) + I_A(s) + I_{\text{inel}}(s), \\ I_T(s) &= I_M(s) + I_B(s), \end{aligned} \tag{7.2}$$

where the electron scattering by the background

$$I_B(s) = I_A(s) + I_{\text{inel}}(s).$$

For a rigid N-atomic molecule, $I_M(s)$ is expressed as:

$$I_M(s) = \frac{K^2}{L^2} I_0 \sum_{i=1}^{N}\sum_{j=1}^{N} |f_i(s)||f_j(s)| \cos[\eta_i(s) - \eta_j(s)] \frac{\sin sr_{ij}}{sr_{ij}}, \tag{7.3}$$

where $i \neq j$; K is a constant, $K = 8\pi^2 me^2/h^2$ with the mass of electron m and the charge of electron e; L is the distance between the diffraction center and the recording point; I_0 is the primary intensity; $f(s)$ and $\eta(s)$ are the atomic electron scattering amplitudes and phases, respectively.

It is worth noting that the interference occurs only at scattering by pairs of nuclei ij, i.e., by molecules. Being dependent on the magnitudes of internuclear distances r_{ij}, the $I_M(s)$ component of the total electron scattering intensity contains information about molecular structure.

For single atoms i, the $I_B(s)$ function is defined as:

$$I_B(s) = \frac{K^2}{L^2} I_0 \sum_i (|f_i(s)|^2 + 4S_i(s)/s^4), \tag{7.4}$$

where $S_i(s)$ is a function of incoherent scattering.

For a semirigid molecule, vibrational effects are taken into account by employing the probability distribution function $P(r)$ for the internuclear distance r in the interval between r and $r + dr$:

$$I_M(s) = \frac{K^2}{L^2} I_0 \sum_{i=1}^{N} \sum_{j=1}^{N} |f_i(s)||f_j(s)| \cos[\eta_i(s) - \eta_j(s)] \int_0^\infty P_{ij}(r) \frac{\sin sr_{ij}}{sr_{ij}} dr. \tag{7.5}$$

The dependence of the $P(r)$ function on the population of vibrational states at the temperature T can be expressed using the Boltzmann distribution:

$$P(r) = \sum_\upsilon P_\upsilon(r) \exp(-\frac{E_\upsilon}{kT}) / \sum_\upsilon \exp(-\frac{E_\upsilon}{kT}), \tag{7.6}$$

where υ is the vibrational quantum number, $P_\upsilon(r) = |\psi_\upsilon(r)|^2$, $\psi_\upsilon(r)$ is the wave function, E_υ is the vibrational energy, and k is the Boltzmann constant.

For an anharmonic oscillator, the $P(r)$ function can be approximated using a distorted Gaussian function:

$$P(r) = \frac{1}{u\sqrt{2\pi}} \exp(-(r - r_e)^2/2u^2)[1 + \sum_n c_n(r - r_e)^n], \tag{7.7}$$

where r_e is the distance between the equilibrium positions of nuclei i and j; u is the root-mean-square amplitude of vibration of these nuclei at the temperature T, it corresponds to a half-width at a half-maximum of the Gaussian function. The coefficients c_n depending on the anharmonicity of potential energy function (PEF) can be estimated due to the links to the Morse anharmonicity constant a [27]:

$$\begin{aligned} c_1 &= a, \\ c_2 &= a^2/2, \\ c_3 &= a/6u^2 + a^3/6. \end{aligned} \tag{7.8}$$

With the expression for $P(r)$ one finally gets the $I_M(s)$ function in the following form:

$$I_M(s) = \frac{K^2}{L^2} I_0 \sum_{i=1}^{N} \sum_{j=1}^{N} g_{ij}(s) \exp(-\frac{1}{2}u_{ij}^2 s^2) \frac{\sin[s(r_{a,ij} - \kappa_{ij} s^2)]}{s r_{e,ij}}, \quad (7.9)$$

where $g_{ij}(s)$ is the function of atomic scattering, $g_{ij}(s) = |f_i(s)||f_j(s)|cos[\eta_i(s) - \eta_j(s)]$, and κ_{ij} is the asymmetry constant. The effective internuclear distance r_a corresponds to the center of gravity of the *P*(*r*)/*r* distribution [27, 28]:

$$r_a = \int_0^\infty P(r)dr / \int_0^\infty (P(r)/r)dr, \quad (7.10)$$

and, thus, it has not a clear physical meaning. However, the r_a distances as being the arguments of the $I_M(s)$ function have an operational significance and can be directly determined from experimental data. The r_a parameters are related to the thermal-average internuclear distances r_g as follows [27, 29]:

$$r_a = r_g - u^2/r_e, \quad (7.11)$$

where

$$r_g = \int_0^\infty r P(r)dr / \int_0^\infty P(r)dr \quad (7.12)$$

The r_g distance corresponding to the center of gravity of the *P*(*r*) distribution has a clear physical meaning and a status of experimental value. For the given molecule, the thermal-average internuclear distances are not constant because vibrational effects depend on the temperature of experiment, and they are trigonometrically inconsistent. For instance for linear triatomic XY_2-type molecules, twice the bonded distance r_g(X–Y) is not equal to the non-bonded distance r_g(Y…Y), namely $\delta_g = 2r_g$(X–Y)$-r_g$(Y…Y) > 0. Correspondingly, for planar XY_3-type molecules, $\delta_g = \sqrt{3}\, r_g$(X–Y)$-r_g$(Y…Y) > 0 (see [30] for other types of molecules). This phenomenon first observed by Bastiansen and Morino [31, 32] is called shrinkage effect. Therefore, the r_g structures cannot be directly compared with experimental structures determined by rotational spectroscopy (r_0, r_s, r_e) or calculated by quantum-chemical methods (r_e).

For polyatomic molecules, the relation between the κ and *a* values was obtained by Kuchitsu [28] in the following form:

$$\kappa = \left(a u_T^4/6\right)\left(3 - 2u_0^4/u_T^4\right), \quad (7.13)$$

where u_T^4 and u_0^4 are the root-mean-square vibrational amplitudes at the temperatures T and 0 K, respectively. For application of this equation to study vibrational anharmonicity, see Ref. [33].

The equilibrium distances r_e can be roughly estimated from the r_g values [29]:

$$r_e = r_g - (3/2)au^2 - \delta r_{cent}, \quad (7.14)$$

where δr_{cent} is a small displacement due to centrifugal force [34].

For many diatomic molecules, the a constants were calculated by Kuchitsu et al. [35] from experimental values of equilibrium rotational constants B_e, harmonic vibrational frequencies ω and rovibrational constants α taken from the handbook by Huber and Herzberg [36]. The calculated values range between 1.3 and 2.6 $Å^{-1}$, and their median of about 2 $Å^{-1}$ was often used for rough estimation of equilibrium distances.

In comparison to $I_B(s)^{theor}$, the experimental background $I_B(s)^{exp}$ contains also the extraneous scattering, $I_{extraneous}(s)$, depending on some experimental conditions and, thus, cannot be reliably determined. The experimental background is approximated by a smooth line through the oscillations of the total intensity $I_T(s)^{exp}$ (as an example, see experimental curves for 1-methylthymine (1,5-dimethyl-2,4(1*H*,3*H*)-pyrimidinedione) [37] in Fig. 7.3). To offset the steep decrease of both $I_M(s)$ and $I_B(s)$ with increase of scattering coordinate s, the ratio of these functions (the so-called reduced molecular electron scattering intensity, $sM(s)$) is used in structural analysis:

$$sM(s)^{exp} = \frac{I_T(s)^{exp} - I_B(s)^{exp}}{I_B(s)^{exp}} s. \quad (7.15)$$

The $sM(s)$ function is a superposition of damped sinusoids with the periods and the sinusoid damping defined by the r_a distances and the mean vibrational amplitudes, respectively.

Due to the sine Fourier transformation, this function can be converted to the so-called radial distribution function $f(r)$ which is related to the probability distribution of the r_a distances:

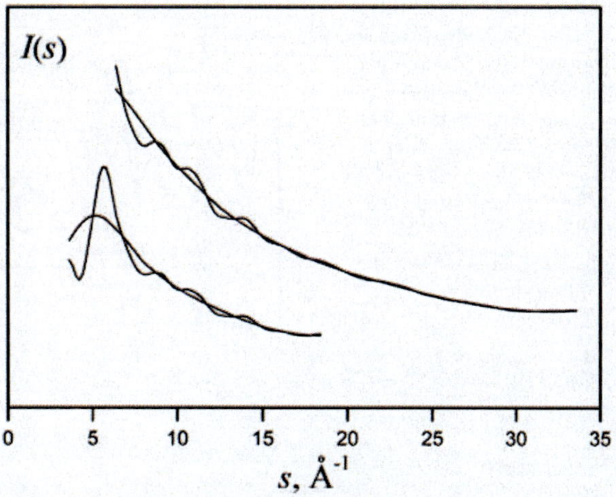

Fig. 7.3 Total electron scattering intensities $I_T(s)^{exp}$ for 1-methylthymine measured at the long (above) and short (below) nozzle-to-plate distances of $L = 362.28$ mm and $L = 193.94$ mm, respectively, with empirical background $I_B(s)^{exp}$ through the oscillations of the total intensity $I_T(s)^{exp}$. Reprinted with permission from: [37]. Copyright 2014 American Chemical Society

$$f(r)^{\exp} = \int_0^{\infty} sM(s)^{\exp} \exp(-bs^2) \sin(sr) ds, \tag{7.16}$$

where additional damping factor $\exp(-bs^2)$ is chosen so that the contribution of the missing range of $sM(s)^{\exp}$ after $s_{\max}$ is negligible [9–11].

The determination of the internuclear distances $r_{a,ij}$, the root-mean-square vibrational amplitudes u_{ij} and sometimes the asymmetry parameters κ_{ij} is carried out by fitting the theoretical function $sM(s)^{\text{theor}}$ calculated for the given molecular model to the experimental curves $sM(s)^{\exp}$ using the method of least squares. The quality of the fitting is characterized by the goodness (disagreement) factor R_f calculated as:

$$R_f = \left[\sum_j w_j (sM(s)_j^{\exp} - sM(s)_j^{\text{theor}})^2 / \sum_j w_j (sM(s)_j^{\exp})^2 \right]^{1/2} \times 100\%, \tag{7.17}$$

where $sM(s)^{\text{theor}}$ is the function calculated for the final molecular model, w_j is the weight of the point j.

An example of the $sM(s)^{\exp}$ intensities and adjusted theoretical counterparts $sM(s)^{\text{theor}}$ is shown in Fig. 7.4, whereas the corresponding radial distribution curves $f(r)$ are presented in Fig. 7.5.

For more details concerning the theory of the conventional electron diffraction, the reader can be addressed to reviews [38, 39].

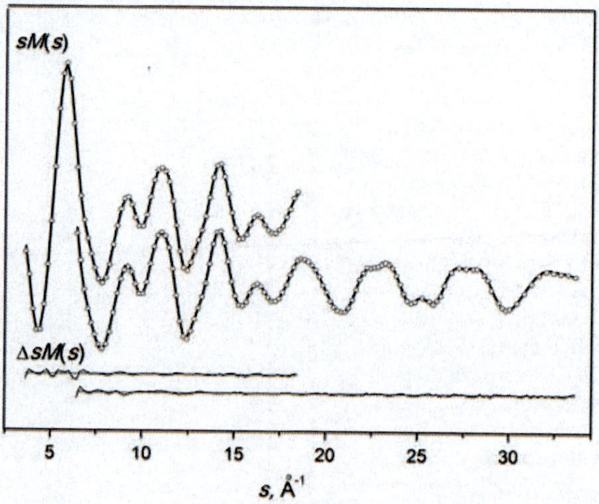

Fig. 7.4 Reduced molecular electron scattering intensities $sM(s)^{\exp}$ (open circles) for 1-methylthymine for the long (above) and short (below) nozzle-to-plate distances and their theoretical counterparts $sM(s)^{\text{theor}}$ (solid lines) for the final molecular model ($R_f = 3.7\%$). Difference curves $\Delta sM(s) = sM(s)^{\exp} - sM(s)^{\text{theor}}$. Reprinted with permission from: [37] Copyright 2014 American Chemical Society

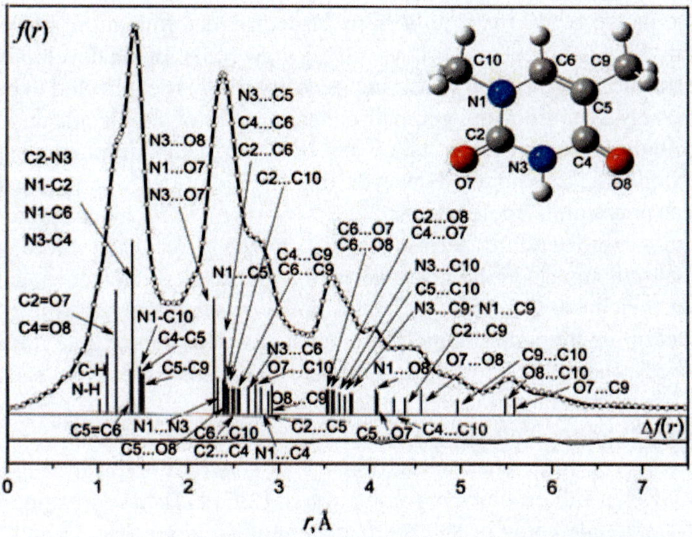

Fig. 7.5 Experimental (open circles) and theoretical (solid line) radial distribution curves $f(r)$ for 1-methylthymine with vertical bars of the terms corresponding to final molecular model (damping factor $b = 0.0015$; the difference curve $\Delta f(r) = f(r)^{\text{exp}} - f(r)^{\text{theor}}$). Reprinted with permission from: [37] Copyright 2014 American Chemical Society

7.4 Equilibrium Molecular Structure from GED Data

7.4.1 General Remarks

The anharmonic part of potential energy function (PEF) or surface (PES) significantly improves the description of molecular dynamics. Therefore, this part has to be taken into account for the precise determination of equilibrium structure (r_e) from experimental data.

The r_e structure can be determined from GED data by two different ways. According to the first one suggested by Bartell, Morino, and Kuchitsu [27–29, 40, 41] several years ago, the r_e internuclear distances can be simply derived from the experimental thermal-average r_g values by subtracting the estimated vibrational corrections (see Eq. (7.14)).

Later, due to the availability of the accurate quantum-chemical force field calculations, this method became very important. Taking into account vibrational corrections calculated from harmonic and anharmonic force fields, the interconversion between the thermal-average and equilibrium distances can be carried out on each step of the $sM(s)$ fitting, and, thus, the so-called semiexperimental equilibrium structure, r_e^{se}, can be refined. One of the methods of vibrational correction calculations from quantum-chemical force fields was suggested by Sipachev [42, 43]. It is predominantly used in the modern electron diffraction structural analysis.

The alternative way for equilibrium structure determination by GED was suggested by Spiridonov et al. [44] (see also review [45]). In the developed method, the reduced molecular electron scattering intensity $sM(s)$ is connected to parameters of PEFs, namely to equilibrium geometrical parameters of a molecule and harmonic and anharmonic force constants, which can be directly and simultaneously refined by the method of least squares. However, this excellent idea was realized only for the simplest polyatomic molecules (of the XY_n-type) using anharmonic potential functions with cubic or quartic terms (see Sect. 7.4.2 and Sect. 7.5). Moreover, it was revealed that in many cases the low quality of experimental intensities, namely low accuracy of their measurements and small range of scattering variable s, does not allow to determine the anharmonic force constants with appropriate uncertainties. In other words, the application of this method is limited by the low sensitivity of electron diffraction to anharmonicity effects.

Several equilibrium structures determined by the method of PEFs are reviewed by Spiridonov et al. [46]. The r_e(Be–F) of 1.374(4) Å for the linear BeF_2 molecule studied by GED in anharmonic approximation of PEF [47] can be mentioned here as an exclusive example being in an excellent agreement with r_e(Be–F) of 1.37297(1) Å determined by high-resolution infrared spectroscopy [48].

7.4.2 Structure Determination in Terms of Potential Energy Functions

This method of equilibrium structure determinations was first realized in harmonic approximation of PEF using rectilinear coordinates [44]. The equation for sM(s) was expressed through the equilibrium internuclear distances, r_e^h, and the generalized vibrational amplitudes, which are linked to the harmonic force constants using the standard normal coordinate analysis [30]. However, as it was soon revealed, the use of rectilinear coordinates leads to the noticeable overestimation of vibrational effects (see review [46]). For example, for the dichlorides MCl_2 (M = Mn, Fe, Co, and Ni), the differences $\Delta r = r_g$(M–Cl) $- r_e^h$ (M–Cl) are equal to 0.049, 0.046, 0.049, and 0.048 (in Å) [33] with uncertainties of about 0.007 Å.

The expression for $sM(s)$ linked to the parameters of the PEF expanded up to cubic terms was obtained by Gershikov and Spiridonov [49, 50] at the level of the first-order perturbation theory. The following vibrational Hamiltonian was used:

$$H = \frac{1}{2}\sum_{r}\omega_r(p_r^2 + q_r^2) + \sum_{r,s,t} k_{rst}q_r q_s q_t, \quad (7.18)$$

where q_r are the dimensionless normal coordinates and p_r are the corresponding momenta, ω_r are the harmonic frequencies, and k_{rst} are the anharmonic force constants in wavenumber units.

The obtained equation for sM(s) has the form:

$$sM(s) = \sum_{i \neq j}^{N} \frac{g_{ij}(s)}{r_{e,ij}} \exp\left[-s^2 \langle \Delta z^2 \rangle_{ij}/2\right] \times \left[R_{ij}(s) \sin sr_{e,ij} + L_{ij}(s) \cos sr_{e,ij}\right], \tag{7.19}$$

with (omitting the subscripts ij)

$$R(s) = 1 + \langle \Delta z^2 \rangle / r_e^2 - U/2r_e^2 - s^2\left[\langle \Delta z^2 \rangle^2 / r_e^2 - W/2r_e^2\right],$$

$$L(s) = s\left[\langle \Delta z \rangle - \langle \Delta z^2 \rangle / r_e + U/2r_e\right] - s^3[W/2r_e - \chi],$$

$$\langle \Delta z^2 \rangle = \sum_{k=1}^{n} a_k^2 T_k,$$

$$U = \sum_{k=1}^{n} b_{kk}^2 T_k,$$

$$W = \sum_{k,l=1}^{n} a_k a_l b_{kl} T_k T_l,$$

$$\langle \Delta z \rangle = \frac{1}{2} \sum_{i,k=1}^{n} a_k T_i \left[S^{iik}(1 + 2\delta_{ik}) + S_{ii}^k\right],$$

$$\chi = \frac{1}{4} \sum_{i \leq j \leq k}^{n} S^{ijk} a_i a_j a_k - \frac{1}{4} \sum_{i \leq j,k}^{p} S_{ij}^k a_i a_j a_k T_i T_j,$$

$$T_k = \coth(hc\omega_k/2k\mathrm{T}),$$

where r_e is the equilibrium internuclear distance corresponding to the minimum of the anharmonic PEF, n is the number of vibrational degrees of freedom, δ_{ik} is the Kronecker symbol, and a_k and b_{kl} are the transformation coefficients in the expansion of the instantaneous internuclear distance in normal coordinates,

$$r = r_e + \sum_{k=1}^{n} a_k q_k + \frac{1}{2r_e} \sum_{k,l=1}^{n} b_{kl} q_k q_l \tag{7.20}$$

The terms S^{ijk} and S_{ij}^k are defined as follows:

$$S^{ijk} = 2k_{ijk}\omega_i\omega_j\omega_k D$$

and

$$S_{ij}^k = -k_{ijk}\omega_k(\omega_k^2 - \omega_i^2 - \omega_j^2)(1 + \delta_{ik} + \delta_{jk})D \tag{7.21}$$

with

$$D = \left[(\omega_i + \omega_j + \omega_k)(\omega_i - \omega_j - \omega_k)(\omega_i - \omega_j + \omega_k)(\omega_i + \omega_j - \omega_k)\right]^{-1}$$

Linking the $\langle \Delta z \rangle$, $\langle \Delta z^2 \rangle$, U, W, and χ terms to force constants, the last ones together with the r_e distance can be directly refined by the least-squares method. For instance for linear XY_2-type molecules, the r_e(X–Y) internuclear distance and the stretching f_r, bending f_α, stretch-stretch interaction f_{rr}, and cubic f_{rrr} force constants were determined for metal dihalides [33].

In comparison to vibrational effects evaluated in harmonic approximation, the anharmonic vibrational corrections, $\Delta r = r_g$(M–Cl)$-r_e$(M–Cl), are noticeably smaller, being 0.018, 0.023, 0.023, and 0.020 (in Å) for $MnCl_2$, $FeCl_2$, $CoCl_2$, and $NiCl_2$, respectively [33].

The topic of this section is considered in detail in the review papers [45, 46].

7.4.3 *Cumulant-Moment Method*

In this method suggested by Kuchitsu [28, 29] for treating the effects of vibrational anharmonicity on electron scattering, the determination of the $P(r)/r$ and $sM(s)$ functions is reduced to the calculations of the thermally averaged moments of displacements in any appropriate coordinates. It is effectively realized in terms of the cumulant averages (cumulants).

In the cumulant-moment method [51, 52], the expression for molecular electron scattering intensity restricted to the third-order terms has a form similar to the conventional equation for $sM(s)$:

$$sM(s) = \sum_{i \neq j}^{N} \frac{g_{ij}(s)}{\langle r_{ij} \rangle_c} \exp\left[-\frac{s^2}{2}\langle r_{ij}^2 \rangle_c\right] \times \sin\left[s\langle r_{ij} \rangle_c - \frac{s^3}{6}\langle r_{ij}^3 \rangle_c\right] \tag{7.22}$$

where $\langle r_{ij}^n \rangle_c$ are the cumulants defined with respect to the $P_{ij}(r)/r$ function.

The cumulants are equal to zero in points of reference, and they are expressed as follows (omitting the subscripts ij):

$$\begin{aligned} \langle r \rangle_c &= r_e + \langle \Delta r \rangle_c, \\ \langle r^2 \rangle_c &= \langle \Delta r^2 \rangle_c, \\ \langle r^3 \rangle_c &= \langle \Delta r^3 \rangle_c, \end{aligned} \tag{7.23}$$

where $\Delta r = r - r_e$.

The cumulants, except for the first one, do not depend on the point of reference. They can be expressed in terms of the more common moments:

$$\langle \Delta r \rangle_c = \langle \Delta r \rangle,$$

$$\langle \Delta r^2 \rangle_c = \langle \Delta r^2 \rangle - \langle \Delta r \rangle^2,$$
$$\langle \Delta r^3 \rangle_c = \langle \Delta r^3 \rangle - 3\langle \Delta r^2 \rangle \langle \Delta r \rangle + 2\langle \Delta r \rangle^3, \tag{7.24}$$

which can be related to the thermal average normal coordinate moments on the basis of the potential energy functions expanded up to cubic (or quartic) terms using conventional perturbation theory [52].

It is worth noting that

$$\langle r \rangle_c = r_a,$$
$$\langle \Delta r^2 \rangle_c = l^2 + \delta_1,$$
$$\frac{1}{6}\langle \Delta r^3 \rangle_c = \kappa + \delta_2, \tag{7.25}$$

where δ_1 and δ_2 are minor correction terms.

For the further development and application of the cumulant-moment method, see for instance [53–55].

7.4.4 Vibrational Corrections to Thermal-Average Internuclear Distances

In the GED structural analysis, the force field is required for the calculations of vibrational corrections to experimental internuclear distances, Δr_{vib}, and of mean vibrational amplitudes, u.

Within the well-known Wilson ***GF*** matrix formalism [56], the harmonic (quadratic) force field calculations are based on the definition of n linearized internal displacement coordinates (for nonlinear and linear molecules with N atoms, $n = 3N$–6 and $n = 3N$–5 respectively). At this point, ***F*** is the matrix of the force constants, and ***G*** is the matrix of kinetic energy in the momentum representation depending on the nuclear masses and equilibrium geometrical parameters.

Later, it became evident that molecular dynamics cannot be satisfactorily described in the linearized coordinates, especially in a wide range of displacements from equilibrium, and the natural curvilinear internal coordinates are preferable for this purpose. The advantages of the use of curvilinear coordinates were first considered by Morino et al. [57–59]. Later, several electron diffraction models were formulated using nonlinear transformations (see, for instance [49, 50, 60, 61]). In these cases, vibrational corrections to thermal-average internuclear distances can be presented as follows:

$$\Delta r_{\text{vib}} = \langle \Delta r_1 \rangle + \langle \Delta r_2 \rangle + \langle \Delta r_3 \rangle \ldots \tag{7.26}$$

where $\langle\Delta r_1\rangle$ and $\langle\Delta r_3\rangle$ are the harmonic and anharmonic components, respectively. The additional anharmonic term $\langle\Delta r_2\rangle$ appears due to nonlinear coordinate transformation being taken into account via the kinetic part of Hamiltonian (the so-called kinetic anharmonicity). It is interesting to note that the $\langle\Delta r_1\rangle$ and $\langle\Delta r_2\rangle$ terms usually compensate each other being of the same order of magnitude but of opposite signs. Thus, the total vibrational corrections to the experimental internuclear distances are mainly defined by the $\langle\Delta r_3\rangle$ terms, i.e., by dynamic anharmonicity. As an example, the data of Table 7.1 demonstrate relations between the mentioned components of Δr_{vib} for picolinonitrile (2-cyanopyridine) at $T_{\text{nozzle}} = 327$ K [62].

The formalism of vibrational correction calculations in curvilinear internal coordinates was realized by Sipachev in 1985 [42]. In the suggested method [63], frequently used in the modern GED structural analysis, the vector of mean atomic displacements from equilibrium positions is written as:

$$\langle\mathbf{r}\rangle = -(1/2)\mathbf{B}_0^{-1}\sum_n \sigma_n \Delta\mathbf{B}_n \xi_n \tag{7.27}$$

where n is the number of vibrational modes, $\boldsymbol{\xi}_n$ is the vector of atomic displacements in Cartesian coordinates, σ_n is the frequency factor taking into account vibrational state populations, $\mathbf{B}_0$ is the transformation matrix between the internal and Cartesian coordinates calculated at equilibrium geometry, and $\Delta\mathbf{B}_n$ is the increment corresponding to a shift of the system by $\boldsymbol{\xi}_n$ from the equilibrium configuration.

Table. 7.1 Total corrections Δr to experimental internuclear distances between the heavy atoms for picolinonitrile (all values in Å)

Bonds [a]	r_e^{se} [b,c]	δ_{cent} [d]	$\langle\Delta r_1\rangle$ [d]	$\langle\Delta r_2\rangle$ [d]	$\langle\Delta r_3\rangle$ [d]	Δr [e]
N1–C6	1.332(2)[1]	0.000_0	0.005_7	-0.004_8	0.006_8	0.007_7
N1–C2	1.337(2)[1]	0.000_0	0.003_4	-0.002_4	0.006_5	0.007_5
C4–C5	1.384(2)[2]	0.000_0	0.004_6	-0.003_7	0.007_5	0.008_4
C3–C4	1.387(2)[2]	0.000_0	0.006_4	-0.005_3	0.007_5	0.008_6
C5–C6	1.391(2)[2]	0.000_0	0.004_8	-0.003_9	0.007_6	0.008_5
C2–C3	1.393(2)[2]	0.000_2	0.003_4	-0.002_4	0.007_4	0.008_5
C2–C7	1.445(3)	0.000_0	0.004_5	-0.003_1	0.006_9	0.008_3
C7–N8	1.158(5)	0.000_0	0.018_6	-0.018_1	0.005_7	0.006_2

[a]See Fig. 7.6 for atom numbering

[b]Estimated errors (3σ) are given in parentheses in units of the last significant digit

[c]Parameters noted by equal superscripts were refined in one group; differences between parameter values in each group were fixed at the computed values of CCSD(T)_ae/cc-pwCVQZ quality

[d]See text for definition

[e]$\Delta r = \Delta r_{\text{vib}} + \delta_{\text{cent}}$. Vibrational corrections were calculated from harmonic and anharmonic (cubic) force constants computed at the MP2/cc-pVTZ level of theory

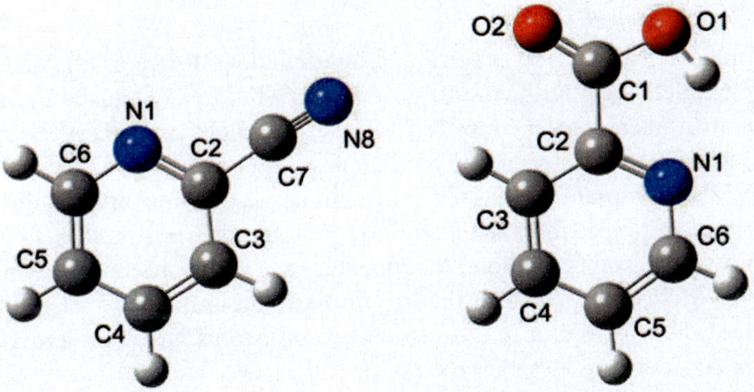

Fig. 7.6 Molecular models of picolinonitrile (left) and picolinic acid (right)

It is worth noting that the two components of Eq. 7.26, $\langle \Delta r_1 \rangle$ and $\langle \Delta r_2 \rangle$, appear here as one, $\langle \Delta r_{1+2} \rangle$ (first-order curvilinear correction), which is expected to be small. Thus, the magnitude of Δr_{vib} is defined mainly by the anharmonic term $\langle \Delta r_3 \rangle$ (as an example, see Table 7.2 with data for picolinic acid (2-pyridinecarboxylic acid) studied at T_{nozzle} = 375(3) K [64]).

The $\langle \Delta r_3 \rangle$ values are usually calculated from an anharmonic (cubic) force field computed by high-level quantum-chemical methods. The nonlinear transformation of the cubic force constants in Cartesian coordinates, $f^{(x)}_{ijk}$, to those in internal coordinates, f_{ijk}, is performed in the following way [43]:

$$f_{ijk} = f^{(x)}_{ijk} b^i_j b^j_k b^k_i, \tag{7.28}$$

Table. 7.2 Vibrational corrections to experimental internuclear distances and centrifugal distortion effects, δ_{cent}, for picolinic acid (all values in Å)[a]

Bonds	r_a [64] [b]	δ_{cent} [c]	$\langle \Delta r_{1+2} \rangle$ [c]	$\langle \Delta r_3 \rangle$ [c]
C1–O2	1.208(3)	0.000_7	0.000_0	0.004_2
C6–N1	1.344(4)	0.001_5	-0.001_0	0.007_2
C1–O1	1.345(4)	0.001_5	0.000_0	0.008_2
C2–N1	1.346(4)	0.001_6	-0.001_2	0.007_2
C3–C4	1.396(3)	0.000_8	0.001_0	0.006_5
C4–C5	1.396(3)	0.000_8	0.001_2	0.006_2
C2–C3	1.398(3)	0.000_6	-0.000_3	0.008_3
C5–C6	1.398(3)	0.000_9	0.000_2	0.007_3
C1–C2	1.513(4)	0.001_2	0.000_0	0.009_1

[a]See Fig. 7.6 for molecular model with atom numbering
[b]3σ values are given in parentheses in units of the last significant digit
[c]See text for definition, the values are calculated with MP2/cc-pVTZ quadratic and cubic force constants

where b is defined as $x^i = b^i_j q^j$.

To complete corrections to experimental internuclear distances, the distortions of molecular geometry at rotation of the molecule as whole, δ_{cent}, caused by centrifugal forces are also taken into account [65]. The δ_{cent} values are relatively small (see Tables 7.1 and 7.2), but they can be nevertheless significant (see Table 7.2).

Usually, the complete vibrational corrections, Δr_{vib}, are noticeably larger in magnitude than the experimental uncertainties of internuclear distances. For instance, for the single C–C and C–N bond lengths, they are of the order of 0.01 Å, whereas the estimated experimental errors (3σ) are only a few thousandths of Å (see Table 7.1). It means that there is a good reason to take into account the complete vibrational corrections Δr_{vib}.

However, because quantum-chemical calculations of anharmonic force fields are still very expensive and time-consuming, the $\langle \Delta r_3 \rangle$ terms are often neglected, particularly for large-size molecules with low symmetry. In these cases, the derived structure, r_{h1}, corresponds to a minimum of the harmonic potential energy function (surface) in curvilinear coordinates (see several examples in the handbook [66]). The usual relation $\langle \Delta r_{1+2} \rangle \approx 0$ (or $\langle \Delta r_1 \rangle \approx -\langle \Delta r_2 \rangle$) leads to a disadvantageous equality of the r_{h1} and r_g values having different physical meaning. A more dramatic situation arises when vibrational corrections include only the $\langle \Delta r_1 \rangle$ terms (zeroth-order rectilinear corrections). The $\langle \Delta r_1 \rangle$ corrections alone as being over-estimated for the bonded distances and under-estimated for the non-bonded distances lead to strong deformation of the refined structure (in literature denoted as r_α, r_e^h or r_{h0}). It is a reason why such kinds of structures are outside of current interest in the field of GED structure determinations [1].

Although significance of the r_e^h/r_α structures (equilibrium structures in harmonic approximation) derived from GED data is similar to that of the spectroscopic r_z one (see Sect. 5.7, Eq. (5.10)), their accuracy is noticeably lower because contributions of vibrational anharmonicity to r_g values is too large to be neglected, particularly for the high temperatures of GED experiment.

It is clear that the equilibrium structures, derived from experimental data and computed anharmonic vibrational corrections, are not completely experimental. In spectroscopy, such structures are specified as semiexperimental, r_e^{se}. In electron diffraction literature, this point is very often not emphasized, and semiexperimental structures are denoted simply as r_e.

The equilibrium molecular structures determined by GED with accounting for calculated anharmonic vibrational corrections are presented in Chap. 9 of this book being systematically denoted as r_e^{se}.

7.5 Large Amplitude Motions

The methods of structure determinations presented above fail very often for molecules with large amplitude internal motions, such as inversion, hindered rotation, ring puckering, etc. Adiabatic approximation is the simplest basis for treating the large amplitude motions. In this approximation, small amplitude rigid-frame vibrations and large amplitude motion(s) as having noticeably different periods and energies are considered separately. After first paper by Bastiansen et al. [67], various types of low frequency motions were described using the so-called dynamic models, *i.e.*, the method of pseudo-conformers (see review paper by Novikov and Vilkov [68]). In this approximation, the non-rigid (floppy) molecular model is presented by a set of rigid molecular configurations (frames) differing in the magnitude of the large amplitude coordinate ρ. The probability distribution function is written as follows:

$$P(r) = \int P_r(r, \rho) \cdot P(\rho) d\rho, \tag{7.29}$$

where $P_r(r, \rho)$ is the function related to the rigid frame vibrations at the fixed value of coordinate ρ. The $P(\rho)$ function can be simply defined by the classical Boltzmann distribution:

$$P(\rho) = N_A \exp(-V(\rho)/k\mathrm{T}), \tag{7.30}$$

where N_A is a normalization constant.

The method of pseudo-conformers was essentially improved by Samdal [69] due to accounting for the relaxation of structural parameters considered as functions of a large amplitude coordinate. These structural dependences are usually assumed according to quantum-chemical predictions, and the distribution of pseudo-conformers is defined according to calculated or analytically approximated potential energy function.

Later, the equilibrium geometry linked to the thermal-average internuclear distances via vibrational corrections (see Sect. 7.4.4) could also be refined for the floppy molecules (see for instance [70, 71]).

In the more advanced large amplitude model based on the theory developed by Hougen et al. for spectroscopy [72], the rotation-bending-stretching Hamiltonian is separated into a rotation-bending part as a function of the curvilinear coordinate ρ and a stretching part in rectilinear coordinates taking into account dependence from ρ [60]. At first, it was used to derive expression for $sM(s)$ of XY_2-type molecules in terms of the bending quadratic-quartic potential function [60]:

$$V(\rho) = \frac{1}{2} k_2 \rho^2 + k_4 \rho^4 \tag{7.31}$$

Besides the internuclear distance X–Y and the force constants describing the rigid vibrations, the quadratic and quartic force constants could be refined for metal dihalides in the GED analysis [33, 61] by a least-squares method. If quadratic force constant $k_2 \geq 0$, the molecule is linear ($\rho_e = 0$, where ρ_e is the equilibrium value of ρ); whereas at $k_2 < 0$, the molecule is nonlinear with $\rho_e = (-k_2/4k_4)^{1/2}$.

In the more universal treatment of large amplitude motions suggested by Kochikov et al. [73], the small amplitude vibrations are described in the approximation of harmonic oscillator with anharmonic corrections derived from the higher terms of potential energy functions, whereas a numerical solution is used for the description of large amplitude motion. Along with vibrational terms, rotational effects as well as interactions between overall rotation and intramolecular motion were taken into account in this approach, which was used for the studies of 1,4-cyclohexadiene, 1,4-dioxin [73], and 1,4-disilacyclohexa-2,5-diene [55] with large amplitude ring-puckering motions, tetrafluorodiborane [74] and nitroethane [75] with one-dimensional internal rotation, etc. For general descriptions of different types of large amplitude motions, see Chap. 6.

Finally, it was revealed that the large amplitude approach is not necessarily required if the large amplitude motion can be described by approximately quadratic potential function [73].

The method of pseudo-conformers was initially established for molecules with single non-rigid coordinate. However, there are no additional problems to use this formalism also for molecules with multiple large amplitude motions, except for a lot of technical efforts due to increase of the number of pseudo-conformers and provided that the adiabatic approximation is still appropriate. One of the first studies of molecules with two different large amplitude motions was realized for 3-nitrostyrene (1-ethenyl-3-nitrobenzene) with large amplitude internal rotations of the ethenyl and nitro groups around the C–N and C–C bonds, respectively [76].

7.6 Joint Analysis of Data from Different Methods

In comparison with experimental structures determined by GED (r_g and r_a) and HR spectroscopy (r_0, r_s, etc.), the equilibrium structure r_e is universal for both these experimental methods and quantum-chemical calculations. In other words, the r_e structure is a proper basis for joint analysis of data from different methods.

The structure determination by GED being the solution of inverse problem may be ambiguous. To stabilize the inverse problem solution, regularizing algorithms are required. In the GED analysis, the regularizing effects are usually reached due to support by data from other methods. The joint analysis can be carried out by minimization of the subsequent functional:

$$G = \sum_k p_k \left[y_k^{\text{exp}} - y_k^{\text{theor}}(Q_1, \ldots, Q_m) \right]^2 = \min \tag{7.32}$$

where y_k^{exp} are the experimental observables (for instance, electron diffraction intensities and spectroscopic rotational constants), y_k^{theor} are their theoretical counterparts, k is the number of experimental measurements, m is the number of the fitted parameters Q, p_k are the weights of observables which can be defined as inversely proportional to their estimated uncertainties.

The regularizing algorithm suggested by Kochikov et al. [53] combines the data of different experimental methods (namely electron scattering intensities, $sM(s)^{\exp}$, spectroscopic equilibrium rotational constants, A_e, B_e, and C_e, as well as vibrational frequencies, ω) and quantum-chemical calculations (force field) on the basis of equilibrium geometry. The minimized functional has the form:

$$\begin{aligned} \boldsymbol{M}^{\alpha} &= \left\|\omega(\boldsymbol{F}, \boldsymbol{R}) - \omega^{\exp}\right\|^2 + \left\|s\boldsymbol{M}(s)[\boldsymbol{F}, \boldsymbol{R}] - s\boldsymbol{M}(s)^{\exp}\right\|^2 \\ &+ \left\|[A_e, B_e, C_e](\boldsymbol{F}, \boldsymbol{R}) - [A_e, B_e, C_e]^{\exp}\right\|^2 \\ &+ \alpha_1\left\|\boldsymbol{F} - \boldsymbol{F}^0\right\|^2 + \alpha_2\left\|\boldsymbol{R} - \boldsymbol{R}^0\right\|, \end{aligned} \tag{7.33}$$

where $\boldsymbol{F}$ and $\boldsymbol{R}$ are the force field and equilibrium geometry matrices, respectively, and α are the regularization parameters, which ensure the stability of the solution with respect to the experimental uncertainties. Results of high-level quantum-chemical computations ($\boldsymbol{F}^0$ and $\boldsymbol{R}^0$) can be used advantageously as stabilizers. For basic principles of the regularization method, see the book by Yagola et al. [77].

The general problem of the GED structural analysis is that the similar structural parameters of a molecule, being close to each other in the magnitude and/or strongly correlated, cannot be separately determined by this method alone. Due to the low resolution of the GED method, the structure studies of large and overcrowded molecules, conformational and tautomeric mixtures, etc. are not possible without support by quantum-chemical methods. In such cases, the small differences between values of related structural parameters and the values of some parameters themselves are constrained to the values derived from the computed structure. For the first time, this routine (Molecular Orbital Constrained Electron Diffraction abbreviated as MOCED) was suggested by Schäfer et al. [78, 79], see also reviews [80, 81].

There are several modern techniques developed for the analysis of GED data in combination with data from other experimental methods and/or restraints from quantum–chemical calculations, namely

SARACEN (acronym for structure analysis restrained by ab initio calculations for electron diffraction),
SEMTEX (structure enhancement methods for theory and experiment),
DYNAMITE (dynamic interaction of theory and experiment),
STRADIVARIUS (structural analysis using diffraction and various other data), etc. (see review by Masters [82] and references therein).

7.7 Error Estimation and Accuracy of Molecular Structure Determined by GED

The total errors, σ_{tot}, of the experimental internuclear distances include random, σ_{rand}, and systematic, σ_{syst}, errors:

$$\sigma_{\mathrm{tot}} = \left(\sigma_{\mathrm{rand}}^2 + \sigma_{\mathrm{syst}}^2\right)^{1/2} \tag{7.34}$$

The random errors of geometrical parameters, refined in the fitting of the theoretical intensity $sM(s)$ to the experimental one by the least-squares method, are estimated by the corresponding root-mean-square deviation, σ, usually multiplied by a factor 2, 2.5 or 3 depending on the confidence level. For bonded distances, the σ values are usually of the order of 0.1% of the distance values themselves and can even be reduced to about 0.01% [83]. In principle, the random errors reliably indicate only the relative errors of the geometrical parameters in the molecule.

The systematic errors of internuclear distances arise due to the errors of electron scattering intensity measurements, including measurements of electron wavelength, camera distance, etc. The σ_{syst} values are somewhat different from laboratory to laboratory being estimated to be 0.1% or 0.2% of the internuclear distance value. Further systematic errors are propagated from theoretical models, which can be more or less adequate for the description of molecular structure and dynamics. The estimation of systematic errors is very difficult and in some cases not possible. Therefore, the total systematic errors are rather underestimated than real.

Uncertainties of the $r_{\mathrm{e}}^{\mathrm{se}}$ structure depend also on the accuracy of vibrational corrections. A comprehensive analysis revealed that the highest accuracy of calculated vibrational corrections is a few percent of the correction values themselves [84]. Even the less accurate force field calculations increase uncertainties of corrections to 10–20%, which consequently increase the total errors, for example, of the C—C and C—N bond lengths by about 0.001 Å only.

Comparison of structural datasets determined by different methods (electron diffraction, rotational spectroscopy, and high-level ab initio calculations) is a reliable way to estimate real accuracy providing that the compared structures have the same physical meaning. Several benchmark studies (see for instance [85–88]) demonstrate remarkable agreement between the equilibrium structures derived from both electron diffraction and microwave spectroscopy data as well as computed at the level of coupled-cluster theory and, thus, high accuracy of the $r_{\mathrm{e}}^{\mathrm{se}}$ structures (see Table 7.3 and Fig. 7.7 with data for uracil (2,4(1*H*,3*H*)-pyrimidinedione) [88]). Differences between the r_{g} and $r_{\mathrm{e}}^{\mathrm{se}}$(GED) values correspond to calculated corrections. Very large discrepancies between the r_{s} and $r_{\mathrm{e}}^{\mathrm{se}}$(MW) structures point out that the isotopic substitution method of Kraitchman [89] fails in this case.

Thus, the estimated total errors of the $r_{\mathrm{e}}^{\mathrm{se}}$ bond lengths and bond angles usually being of a few thousandths of Å and a few tenths of degree, respectively, except for parameters containing hydrogen atom(s), essentially agree with accuracy of the GED method revealed by comparison with spectroscopic and computed molecular structures.

Table. 7.3 Bond lengths (in Å) between heavy atoms of uracil determined by GED in comparison with data from other methods (see Fig. 7.7 for atom numbering)

Distance	r_e^{se}(GED)[a,b][88]	r_e^{se}(MW)[90]	r_e^{BO} [c][88]	r_g(GED)[a][88]	r_s(MW)[91]	r(UED)[d][92]
N1–C2	1.381(2)[1]	1.38175(53)	1.379_6	1.395(2)	1.386(5)	1.406
C2–N3	1.379(2)[1]	1.3763	1.377_5	1.390(2)	–	1.392
N3–C4	1.402(2)[1]	1.39793(40)	1.400_5	1.413(2)	1.38(2)	1.426
C4–C5	1.454(8)[e]	1.45500(57)	1.457_8	1.465(8)	1.451(4)	1.487
C5–C6	1.339(18)	1.34496(59)	1.344_6	1.348(18)	1.379(4)	1.348
N1–C6	1.374(2)[1]	1.37196(55)	1.372_9	1.388(2)	1.352(14)	1.376
C2–O	1.210(1)[2]	1.21025(21)	1.211_1	1.215(1)	1.219(4)	1.219
C4–O	1.212(1)[2]	1.21278(24)	1.213_1	1.217(1)	1.22(2)	1.233

[a]Parenthesized uncertainties are 3σ values in units of the last significant digits; parameters with equal superscripts were refined in one group

[b]Vibrational corrections to experimental thermal-average distances were calculated from MP2_ae/cc-pVTZ quadratic and cubic force constants

[c]Additive structure of CCSD(T)_ae/aug-cc-pwCVQZ quality estimated as $r_e^{BO} = r_e$(CCSD(T)_ae/cc-pwCVTZ) + [r_e(MP2_ae/aug-cc-pwCVQZ) – r_e(MP2_ae/cc-pwCVTZ)]

[d]Type of this structure was not specified

[e]Dependent parameter

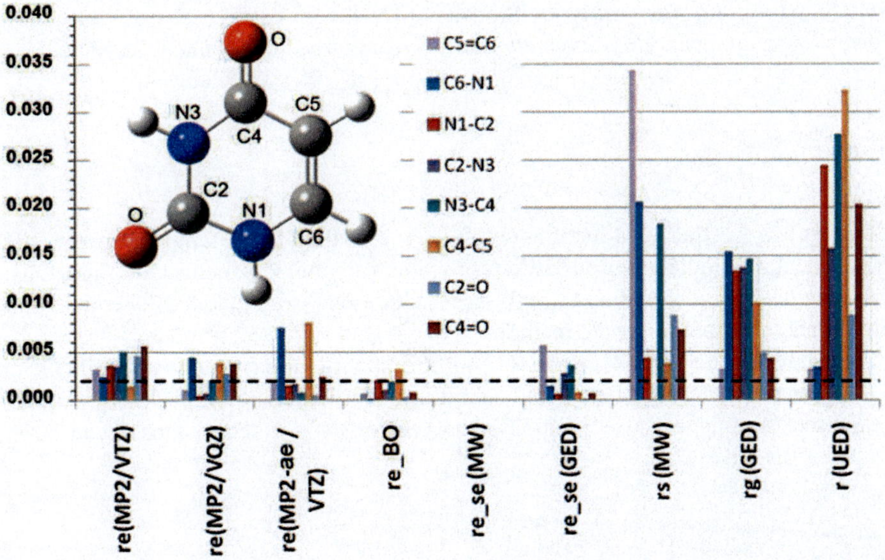

Fig. 7.7 Histogram of the bond length deviations (in Å) relative to the r_e^{se} (MW) values for uracil (see Table 7.3 for details; VTZ and VQZ are abbreviations for cc-pVTZ and cc-pVQZ basis sets, respectively) [88]: Copyright 2013, with permission from Elsevier

7.8 Software

Several computer programs are currently used for the primary processing of experimental GED data and for the performance of the structural analysis by the method of least squares. The oldest of them is the KCED program package presented by Andersen et al. in 1969 [93]. Later, Gundersen et al. supplemented it by the programs for structural analysis of floppy molecules [94] and for processing of experimental intensities recorded by means of image plates (PIMAG) [17]. Due to free access to open-source code of the Norwegian software package, there are many its adapted and extended versions, for example, the KCED version modified by Belyakov [95], the program PIMAG-VIEW extended due to coupling with a graphical user interface by Vogt et al. [19], etc. Furthermore, there are some alternative programs created for modern electron diffraction analysis, namely the PLATE program for the primary processing of experimental data (by Kochikov et al. [96]), the ed@ed program package for the GED structural analysis (by Hinchley et al. [97]), the UNEX program package for complete treatment of electron diffraction data (by Vishnevskiy [98]), the software package Symm/Disp/Eldiff/Large for combined structural analysis including calculations of vibrational amplitudes, vibrational and vibration-rotational contributions to experimental internuclear distances, and ground-state rotational constants, respectively, etc. (by Kochikov et al. [53, 72, 73]). The very popular SHRINK program was written by Sipachev [42, 43] for calculations of vibrational corrections and mean vibrational amplitudes from quantum-chemical force fields.

7.9 Conclusion

Contrary to experimental structures (r_g and r_a) depending on the temperature of GED experiment, the equilibrium structural parameters are molecular constants, and, thus, they can be benchmarked and jointly used with those from quantum-chemical and high-resolution spectroscopic methods.

Accuracy of the semiexperimental equilibrium structure (r_e^{se}) derived from GED data taking into account calculated vibrational corrections can reach a few thousandths of Å unit and a few tenths of degree for the bonded distances and bond angles, respectively.

References

1. MOGADOC database, section of chemical information systems, University of Ulm, update 2017–2018
2. Vogt J, Mez-Starck B, Vogt N, Hutter W (1999) MOGADOC—a database for gasphase molecular spectroscopy and structure. J Mol Struct 485–486:249–254

3. Ischenko AA, Golubkov VV, Spiridonov VP, Zgurskii AV, Akhmanov AS, Vabischevich MG (1983) A stroboscopical gas-electron diffraction method for the investigation of short-lived molecular species. Appl Phys B 32:161–163
4. Ben-Nun M, Martínez TJ, Weber PM, Wilson KR (1996) Direct imaging of excited electronic states using diffraction techniques: theoretical consideration. Chem Phys Lett 262:405–414
5. Srinivasan R, Lobastov VA, Ruan C-Y, Zewail AH (2003) Ultrafast electron diffraction (UED). A new development for the 4D determination of transient molecular structures. Helv Chim Acta 86:1763–1838
6. Carley RE, Heesel E, Fielding HH (2005) Femtosecond lasers in gas phase chemistry. Chem Soc Rev 34:949–969
7. Ischenko AA, Weber PM, Miller RJD (2017) Capturing chemistry in action with electrons: realization of atomically resolved reaction dynamics. Chem Rev 117:11066–11124
8. Mark H, Wierl R (1930) The determination of molecular structure by the diffraction of electrons by a stream of vapor (in German). Z Elektrochem 36:675–676
9. Karle IL, Karle J (1949) Internal motion and molecular structure studies by electron diffraction. J Chem Phys 17:1052–1057
10. Karle J, Karle IL (1950) Internal motion and molecular structure studies by electron diffraction. II Interpretation and method. J Chem Phys 18:957–962
11. Karle IL, Karle J (1950) Internal motion and molecular structure studies by electron diffraction. III. Structure of CH_2CF_2 and CF_2CF_2. J Chem Phys 18:963–971
12. Brockway LO, Bartell LS (1954) A high precision electron-diffraction unit for gases. Rev Sci Instrum 25:569–575
13. Bastiansen O, Hassel O, Risberg E (1955) The Oslo electron diffraction units for gas work. Acta Chem Scand 9:232–238
14. Frost AV, Akishin PA, Gurvich LV, Kurkchi GA, Konstantinov AA (1953) Electron diffraction investigation of molecular structure. Vestn Mosk Univ, Ser Khim 8(12):85–95
15. Ivanov AA, Spiridonov VP, Demidov AV, Zasorin EZ (1974) A reconstructed EG-100 electron diffraction camera for investigating the structure of nonvolatile compounds. Prib Tekh Eksp/ Instrum Exp Tech (Engl Transl) 2:270–272/622–625
16. Iijima T, Suzuki W, Yano YF (1998) Use of imaging plates in gas-phase electron diffraction. Jpn J Appl Phys 37:5064–5065
17. Gundersen S, Samdal S, Strand TG, Volden HV (2007) Benzene; high level quantum chemical calculations, gas electron diffraction pattern recorded on Fuji imaging plates and a method to explore systematic discrepancies which was used to determine an improved sector correction. J Mol Struct 832:164–171
18. Berger RJF, Hoffmann M, Hayes SA, Mitzel NW (2009) An improved gas electron diffractometer—the instrument, data collection, reduction and structure refinement procedures. Z Naturforsch (B) 64(11, 12):1259–1268
19. Vogt N, Rudert R, Rykov AN, Karasev NM, Shishkov IF, Vogt J (2011) Use of imaging plates (IPs) in the gas-phase electron diffraction (GED) experiments on the EG-100 M apparatus. The tetrachloromethane molecule as a test object. Struct Chem 22 (2):287–291
20. Finbak C (1937) Electron diffraction by gases. Avh Norsk Vidensk-Akad Oslo 13:1–17
21. Tremmel J, Bohatka S, Berecz I, Hargittai I (1978) Attachment of a quadrupole mass spectrometer to the EG-100A electronograph. Prib Thekh Eksp/Instrum Exp Tech (Engl Transl) 4:251–252/1118–1120
22. Girichev GV, Shlykov SA, Revichev YF (1986) Apparatus for study of molecular structure of valence-unsaturated compounds. Prib Thekh Eksp/Instrum Exp Tech (Engl Transl) 4:167–169/ 939–942
23. Girichev GV, Shlykov SA, Petrova VN, Subbotina NY, Lapshina SB, Danilova TG (1988) Devices and technique of combined electron diffraction and mass-spectroscopic experiment and their aplication in study of titanium trihalide molecules (in Russian). Izv Vyssh Uchebn Zaved, Khim Khim Tekhnol 31:46–51
24. Girichev GV, Utkin AN, Revichev YF (1984) Upgrading the EMR-100 electron-diffraction camera for use with gases. Prib Thekh Eksp/Instrum Exp Tech (Engl Transl) (2/2):187–190/ 457–461

25. Tremmel J, Hargittai I (1988) Gas electron diffraction experiment. In: Hargittai I, Hargittai M (eds) Stereochemical applications of gas-phase electron diffraction. Part A. The electron diffraction experiment. VCH Publishers, Inc., New York, pp 191–225
26. Demaison J, Vogt N (2020) Accurate structure determination of free molecules. Springer Nature, Switzerland, chapter 7, pp 167–204
27. Kuchitsu K, Bartell LS (1961) Effects of anharmonicity of molecular vibrations on the diffraction of electrons. II. Interpretation of experimental structural parameters. J Chem Phys 35:1945–1949
28. Kuchitsu K (1967) Effect of molecular vibrations on gas electron diffraction. II. Probability distribution function and phase parameter for polyatomic molecules. Bull Chem Soc Jpn 40:505–510
29. Kuchitsu K (1967) Effect of molecular vibrations on gas electron diffraction. I. Probability distribution function and molecular intensity for diatomic molecules. Bull Chem Soc Jpn 40:498–504
30. Cyvin SJ (1968) Molecular vibrations and mean square amplitudes. Elsevier, Amsterdam, p 424
31. Breed H, Bastiansen O, Almenningen A (1960) Structure anomalies and molecular vibrations in carbon suboxide. Acta Cryst 13:1108–1108
32. Morino Y (1960) The effect of thermal vibration of gaseous molecules in electron diffraction. Acta Cryst 13:1107–1107
33. Hargittai M, Subbotina NY, Kolonits M, Gershikov AG (1991) Molecular structure of first-row transition metal dihalides from combined electron diffraction and vibrational spectroscopic analysis. J Chem Phys 94:7278–7291
34. Iwasaki M, Hedberg K (1962) Centrifugal distortion of bond distances and bond angles. J Chem Phys 36:2961–2963
35. Kuchitsu K, Nakata M, Yamamoto S (1988) Joint use of electron diffraction and high-resolution spectroscopic data for accurate determination of molecular structure. In: Hargittai I, Hargittai M (eds) Stereochemical applications of gas-phase electron diffraction. Part A. the electron diffraction experiment. VCH Publishers, Inc., New York, pp 227–263
36. Huber KP, Herzberg G (1979) Constants of diatomic molecules. Molecular spectra and molecular structure. Springer, Boston, MA, pp 8–689
37. Vogt N, Marochkin II, Rykov AN (2015) From the determination of the accurate equilibrium structure of 1-methylthymine by gas electron diffraction and coupled cluster computations to the observation of methylation and flexibility effects in pyrimidine nucleobases. J Phys Chem A 119:152–159
38. Hargittai I (1988) A survey: the gas-phase electron diffraction technique of molecular structure determination. In: Hargittai I, Hargittai M (eds) Stereochemical applications of gas-phase electron diffraction. Part A. The electron diffraction technique. VCH Publishers, Inc., New York, pp 1–54
39. Fink M, Kohl DA (1988) Temperature dependence of electron diffraction structural parameters: theory and experiment. In: Hargittai I, Hargittai M (eds) Stereochemical applications of gas-phase electron diffraction. Part A. The electron diffraction technique. VCH Publishers, Inc., New York, pp 139–190
40. Bartell LS (1955) Effects of anharmonicity of vibration on the diffraction of electrons by free molecules. J Chem Phys 23:1219–1222
41. Morino Y, Nakamura Y, Iijima T (1960) Mean square amplitudes and force constants of tetrahedral molecules. I. Carbon tetrachloride and germanium tetrachloride. J Chem Phys 32:643–652
42. Sipachev VA (1985) Calculation of shrinkage corrections in harmonic approximation. J Mol Struct (Theochem) 121:143–151
43. Sipachev VA (2000) Anharmonic corrections to structural experiment data. Struct Chem 11:167–172
44. Spiridonov VP, Gershikov AG, Zasorin EZ, Butayev BS (1981). The determination of harmonic potential functions from diffraction information. In: Hargittai I, Orville-Thomas WJ (eds) Diffraction studies of non-crystalline substances. Akadémiai Kiadó, Budapest, pp 159–195

45. Spiridonov VP (1988) Spectroscopic information from electron diffraction. In: Hargittai I, Hargittai M (eds) Stereochemical applications of gas-phase electron diffraction. Part A. The electron diffraction technique. VCH Publisher, Inc., New York, pp 265–299
46. Spiridonov VP, Vogt N, Vogt J (2001) Determination of molecular structure in terms of potential energy functions from gas-phase electron diffraction supplemented by other experimental and computational data. Struct Chem 12:349–376
47. Vogt N, Girichev GV, Vogt J, Girichev AG (1995) Equilibrium structure of beryllium difluoride from least-squares analysis of gas-phase electron diffraction data. J Mol Struct 352–353:175–179
48. Frum CI, Engelman R, Bernath PF (1991) Fourier transform emission spectroscopy of BeF_2 at 6.5 μm. J Chem Phys 95:1435–1440
49. Gershikov AG (1982) Anharmonic theory of electron scattering by polyatomic molecules. Khim Fiz 5:587–593 (in Russian)
50. Gershikov AG, Spiridonov VP (1982) Anharmonic force field of CO_2 as determined by gas-phase electron diffraction study. J Mol Struct 96:141–149
51. Spiridonov VP, Ischenko AA, Ivashkevich LS (1981) A new intensity equation for electron diffraction analysis: a barrier to pseudorotation in PF_5 from diffraction data. J Mol Struct 72:153–164
52. Ischenko AA, Spiridonov VP, Tarasov YI, Stuchebryukhov AA (1988) The cumulant method in difraction analysis of polyatomic molecules. J Mol Struct 172:255–273
53. Kochikov IV, Tarasov YI, Spiridonov VP, Kuramshina GM, Yagola AG, Saakjan AS, Popik MV, Samdal S (1999) Extension of a regularizing algorithm for the determination of equilibrium geometry and force field of free molecules from joint use of electron diffraction, molecular spectroscopy and ab initio data on systems with large-amplitude oscillatory motion. J Mol Struct 485–486:421–443
54. Kochikov IV, Tarasov YI, Spiridonov VP, Kuramshina GM, Rankin DWH, Saakjan AS, Yagola AG (2001) The equilibrium structure of thiophene by the combined use of electron diffraction, vibrational spectroscopy and microwave spectroscopy guided by theoretical calculations. J Mol Struct 567–568:29–40
55. Dakkouri M, Kochikov IV, Tarasov YI, Vogt N, Vogt J, Bitschenauer R (2002) Equilibrium structure and large amplitude motion investigation of 1,4-disilacyclohexa-2,5-diene by means of electron diffraction, vibrational spectroscopic data, and ab initio calculations. J Mol Struct 607:195–206
56. Wilson EB, Decius JC, Cross PC (1955) Molecular vibrations. The Theory of Infrared and Raman Vibrational Spectra. McGraw-Hill, New York, p 388
57. Morino Y, Nakamura J, Moore PW (1962) Shrinkage effect by thermal vibrations in linear-skeleton molecules. J Chem Phys 36:1050–1056
58. Morino Y, Iijima T (1963) The distribution function of internal displacement coordinates in linear XY_2 molecules. Bull Chem Soc Jpn 36:412–419
59. Bartell LS (1963) Calculation of mean atomic positions in vibrating polyatomic molecules. J Chem Phys 38:1827–1833
60. Gershikov AG, Spiridonov VP (1986) Semirigid model of the bending-rotational Hamiltonian in electron diffraction analysis of triatomic molecules. I. Theory. Zh Strukt Khim 27:30–35/699–704 (Engl Transl)
61. Gershikov AG, Subbotina NY, Girichev GV (1986) Semirigid model of the bending-rotational Hamiltonian in electron diffraction analysis of triatomic molecules. II. Difluorides of manganese, iron, cobalt, and nickel. Zh Strukt Khim 27:36–41/704–709 (Engl Transl)
62. Vogt N, Khaikin LS, Rykov AN, Grikina OE, Batiukov AA, Vogt J, Kochikov IV, Shishkov IF (2019) The quilibrium molecular structure of 2-cyanopyridine from combined analysis of gas-phase electron diffraction and microwave data and results of ab initio calculations. Struct Chem 30:1699–1706
63. Sipachev VA (1999) Vibrational effects in diffraction and microwave experiments: a start on the problem. Adv Mol Struct Res 5:263–311

64. Vogt N, Marochkin II, Rykov AN (2018) Experiment and theory at the convergence limit: accurate equilibrium structure of picolinic acid by gas-phase electron diffraction and coupled-cluster computations. Phys Chem Chem Phys 20:9787–9795
65. Sipachev VA (2001) Local centrifugal distortions caused by internal motions of molecules. J Mol Struct 567–568:67–72
66. Vogt N, Vogt J (2019) Structure data of free polyatomic molecules. Springer Nature, Switzerland, p 926
67. Bastiansen O, Kveseth K, Møllendal H (1979) Structure of molecules with large amplitude motion as determined from electron-diffraction studies in the gas phase. Top Curr Chem 81:99–172
68. Novikov VP, Vilkov LV (2000) Development of dynamic model in gas electron diffraction. Adv Mol Struct Res 6:299–340
69. Samdal S (1994) The effect of large amplitude motion on the comparison of bond distances from ab initio calculations and experimentally determined bond distances, and on root-mean-square amplitudes of vibration, shrinkage, asymmetry constants, symmetry constraints, and inclusion of rotational constants using the electron diffraction method. J Mol Struct 318:133–141
70. Dorofeeva OV, Vogt N, Vogt J, Popik MV, Rykov AN, Vilkov LV (2007) Molecular structure and conformational composition of 1,3-dihydroxyacetone studied by combined analysis of gas-phase electron diffraction data, rotational constants, and results of theoretical calculations. Ideal gas thermodynamic properties of 1,3-dihydroxyacetone. J Phys Chem A 111:6434–6442
71. Vogt N, Khaikin LS, Grikina OE, Karasev NM, Vogt J, Vilkov LV (2009) Flexibility of the saturated five-membered ring in 2,5-pyrrolidinedione (succinimide): electron diffraction and quantum-chemical studies with use of vibrational spectroscopy data. J Phys Chem A 113:931–937
72. Hougen JT, Bunker PR, Johns JWC (1970) The vibration-rotation problem in triatomic molecules allowing for a large-amplitude bending vibration. J Mol Spectrosc 34:136–172
73. Kochikov IV, Tarasov YI, Vogt N, Spiridonov VP (2002) Large-amplitude motion in 1,4-cyclohexadiene and 1,4-dioxin: Theoretical background for joint treatment of spectroscopic, electron diffraction and ab initio data. J Mol Struct 607:163–174
74. Kochikov IV, Tarasov YI (2003) Equilibrium structure and internal rotation in B_2F_4 from electron diffraction and spectroscopic data and quantum chemical calculations. Struct Chem 14:227–238
75. Tarasov YI, Kochikov IV, Vogt N, Stepanova AV, Kovtun DM, Ivanov AA, Rykov AN, Deyanov RZ, Novosadov BK, Vogt J (2008) Electron diffraction and quantum chemical study of the structure and internal rotation in nitroethane. J Mol Struct 872:150–165
76. Kovtun DM, Kochikov IV, Tarasov YI (2015) Electron diffraction analysis for the molecules with multiple large-amplitude motions. 3-Nitrostyrene—a molecule with two internal rotors. J Phys Chem A 119:1657–1665
77. Yagola AG, Kochikov IV, Kuramshina GM, Pentin YA (1999) Inverse problems of vibrational spectroscopy. VSP, Utrecht, The Netherlands, p 297
78. Chiu NS, Ewbank JD, Askari M, Schäfer L (1979) Molecular orbital constrained electron diffraction studies. J Mol Struct 54:185–195
79. Chiu NS, Sellers HL, Schäfer L, Kohata K (1979) Molecular orbital constrained electron diffraction studies. Conformational behavior of 1,2-dimethylhydrazine. J Amer Chem Soc 101:5883–5889
80. Schäfer L, Ewbank JD, Siam K, Chiu NS, Sellers HL (1988) Molecular orbital constrained electron diffraction (MOCED) studies. In: Hargittai I, Hargittai M (eds) Stereochemical applications of gas-phase electron diffraction. Part A. The electron diffraction technique. VCH Publisher, Inc., New York, pp 301–319
81. Geise HJ, Pyckhout W (1988) Self-consistent molecular models from a combination of electron diffraction, microwave, and infrared data together with high-quality theoretical calculations. In: Hargittai I, Hargittai M (eds) Stereochemical applications of gas-phase electron diffraction. Part A. The electron diffraction technique. VCH Publisher, Inc., New York, pp 321–346
82. Masters SL (2013) Gas phase structure of small molecules. Compr Inorg Chem II 9:89–109

83. Vishnevskiy YV (2007) The initial processing of the gas electron diffraction data: an improved method for obtaining intensity curves from diffraction patterns. J Mol Struct 833:30–41
84. Vogt N, Vogt J, Demaison J (2011) Accuracy of the rotational constants. J Mol Struct 988:119–127
85. Vogt N, Khaikin LS, Grikina OE, Rykov AN, Vogt J (2008) Study of thymine molecule: equilibrium structure from joint analysis of gas-phase electron diffraction and microwave data and assignment of vibrational spectra using results of ab initio calculations. J Phys Chem A 112:7662–7670
86. Vogt N, Demaison J, Ksenafontov DN, Rudolph HD (2014) A benchmark study of molecular structure by experimental and theoretical methods: equilibrium structure of thymine from microwave rotational constants and coupled-cluster computations. J Mol Struct 1076:483–489
87. Vogt N, Altova EP, Ksenafontov DN, Rykov AN (2015) Benchmark study of molecules with large-amplitude ring-twisting motion: accurate equilibrium structure of succinic anhydride from gas electron diffraction data and coupled-cluster computations. Struct Chem 26:1481–1488
88. Vogt N, Khaikin LS, Grikina OE, Rykov AN (2013) A benchmark study of molecular structure by experimental and theoretical methods: Equilibrium structure of uracil from gas-phase electron diffraction data and coupled-cluster calculations. J Mol Struct 1050:114–121
89. Kraitchman J (1953) Determination of molecular structure from microwave spectroscopic data. Am J Phys 21:17–24
90. Puzzarini C, Barone V (2011) Extending the molecular size in accurate quantum-chemical calculations: the equilibrium structure and spectroscopic properties of uracil. Phys Chem Chem Phys 13:7189–7197
91. Vaquero V, Sanz ME, López JC, Alonso JL (2007) The structure of uracil: A laser ablation rotational study. J Phys Chem A 111:3443–3445
92. Gahlmann A, Park TS, Zewail AH (2008) Structure of isolated biomolecules by electron diffraction-laser desorption: uracil and guanine. J Am Chem Soc 131:2806–2808
93. Andersen B, Seip HM, Strand TG, Stølevik R (1969) Computer programs for the structure determination of gaseous molecules from electron diffraction data. Acta Chem Scand 23:3224–3234
94. Gundersen G, Samdal S, Seip HM (1981) Least Squares Refinement Program Based on Gas Electron Diffraction Data. Internal report, University of Oslo, Oslo, p 116
95. Belyakov AV, Baskakov AA, Berger RJF, Mitzel NW, Oberhammer H, Arnason I, Wallevik SÒ (2012) Molecular structure and conformational preferences of gaseous 1-iodo-1-silacyclohexane. J Mol Struct 1012:126–130
96. Kochikov IV, Kovtun DM, Tarasov, YI (2008) Vychislitel'nye metody i programmirovanie 8:44 (in Russian)
97. Hinchley SL, Robertson HE, Borisenko KB, Turner AR, Johnston BF, Rankin DWH, Ahmadian M, Jones JN, Cowley AH (2004) The molecular structure of tetra-tert-butyldiphosphine: an extremely distorted, sterically crowded molecule. Dalton Trans 2469–2476
98. Vishnevskiy YV (2019) UNEX program package, http://unexprog.org

Chapter 8
Other Methods, Mainly for the X–H Bond (X = C, N, O)

Abstract This chapter briefly describes other methods than rotational spectroscopy and gas-phase electron diffraction. Particular emphasis is put on the determination of the X–H bond length (X = C, O, N, Si, etc.)

8.1 Introduction

The hydrogen atom is ubiquitous. However, it is sometimes difficult to determine its position in a molecule with appropriate accuracy.

The semiexperimental method (Sect. 5.5), although powerful, gives values which are less accurate for the position of the hydrogen atom than for many other atoms. For instance, for glycidol, $C_3H_6O_2$, it gives r_e^{se} (O–H) = 0.993(4) Å, whereas the true value is close to 0.963 Å [1]. Unfortunately, this pathological behavior is not rare and has been investigated in Ref. [2]. The causes are multiple. Because of its small mass, the hydrogen atom makes a small contribution to the moment of inertia. Furthermore, when the errors of the rovibrational corrections of the parent species and of the isotopologues are nearly constant (which is often the case for a heavy atom), they almost cancel each other out when the fit is performed. But when a hydrogen atom is isotopically substituted by deuterium, the rovibrational correction may change a great deal and, therefore, also its error. Gas-phase electron diffraction (GED) is not better because the charge of the hydrogen nucleus predominantly contributing to electron scattering is quite small. A typical example is the r_z structure of methyl cyanide, CH_3CN, in Table 5.11. The accuracy of the C–C and C≡N bond lengths is better than 0.001 Å, whereas that of C–H is as worse as 0.0024 Å.

Fortunately, there are several ways to improve the situation.

N. Vogt and J. Demaison, *Equilibrium Structure of Free Molecules*, Lecture Notes in Chemistry 111, https://doi.org/10.1007/978-3-031-36045-9_8

8.2 Other Diffraction Methods

Although X-ray diffraction is mainly used to determine interatomic distances in crystals, it can also be employed in gas-phase. The first application is due to Debye [3] who obtained the structure of CCl_4 in gas-phase. Unfortunately, X-ray diffraction has also a weak point because the photons interact with the electron shell of the atoms and is, thus, not appropriate for hydrogen atoms. Furthermore, there is a lack of sensitivity. However, the advent of free-electron lasers improved the situation. It is now possible to study oriented molecules (by a near-infrared laser for instance) in gas phase. A typical example is the determination of the distance between the two iodine atoms in 2,5-diodothiophene [4].

Neutron diffraction of thermal neutrons ($\lambda \approx 0.5-10$ Å) has a big advantage because neutrons are scattered by nuclei and hydrogen (and deuterium) has a high scattering cross section. The experiment is usually performed in liquid or solid phase. There are very few gas-phase measurements [5].

The main disadvantage of neutron diffraction is the requirement for a source (nuclear reactor). The technique requires a large sample. The advantages of the technique are many - sensitivity to light atoms, ability to distinguish isotopes, absence of radiation damage.

8.3 Nuclear Magnetic Resonance (NMR) Spectroscopy

8.3.1 Liquid Crystal NMR

Direct spin–spin coupling can be observed in nuclear magnetic resonance (NMR) spectra of solutions in oriented crystal solvents. The liquid–crystal solvent molecules in the presence of a magnetic field partially orientate the molecules under investigation. In this anisotropic fluid, the intramolecular direct dipolar couplings are no longer averaged to zero, and so become measurable. This coupling is due to the interaction between two magnetic dipoles, and it depends on the inverse cube of the internuclear distance [6]. This method may provide complementary information on ratios of average distances. It is particularly useful for hydrogen atoms which have a spin $I = ½$. The principal difficulty is the number of adjustable parameters due to the introduction of the so-called interaction parameters, which are used to describe the external torques experienced by the bonds of the solute molecule [7]. Tables 8.1 and 8.2 compare the r_z structure of some molecules by liquid–crystal NMR and gas-phase electron diffraction. Table 8.2 shows that the combined use of GED and NMR significantly increases the accuracy of the molecular parameters.

Table 8.1 Comparison of the r_z structures from liquid crystal NMR, LCNMR, and gas-phase electron diffraction, GED, (angles in degree)

Molecule		r_z(CH)/r_z(CC)		$\angle_z$(CCH)		Ref	
		LCNMR	GED	LCNMR	GED	NMR	GED
Allene	C_3H_4	0.8315(14)	0.8298(15)	118.12(5)	118.3(3)	a	b
Cyclopropane	C_3H_6	0.7154(16)	0.7166(14)	115.17(13)	114.5(9)	c	d
Benzene	C_6H_6	0.77592(24)	0.7767(15)			e	f
Methyl cyanide	CH_3CN	0.7514(18)	0.7489(16)	109.22(6)	109.85(10)	g	h
Methyl isocyanide	CH_3NC	0.7681(4)	0.7654(9)	109.67(1)	109.6	i	h

[a]Diehl P, Baraldi C, Kellerhals M, Wasser R (1987) J Mol Struct 162: 333–339
[b]Ohshima Y, Yamamoto S, Nakata M, Kuchitsu K (1987) J Phys Chem 91: 4696–4700
[c]Kellerhals M, Diehl P, Lounila J, Wasser R (1987) J Mol Struct 156: 255–260
[d]Yamamoto S, Nakata M, Fukuyama T, Kuchitsu K (1985) J Phys Chem 89: 3298–3302
[e]Wasser R, Kellerhals M, Diehl P (1989). Magn Reson Chem 27: 335–339
[f]Kjeldseth Moe OE, Strand TG (1985) J Mol Struct 128: 13–19
[g]Hiltunen Y, Jokisaari J, Lounila J, Pulkkinen A, Dombi G (1988) Chem Phys Lett148: 353–357
[h]Halonen L, Mills IM (1978) J Mol Spectrosc 73: 494–502
[i]Lounila J, Ala-Korpela M, Jokisaari J (1990) J Chem Phys 93: 8514–8523

Table 8.2 Average structure, r_z, of some molecules by gas-phase electron diffraction, GED, and in combination with liquid crystal NMR, LCNMR (distances in Å, angles in degree)

			GED	GED + LCNMR	Ref
Pyrazine	$C_4H_4N_2$	CH	1.096(6)	1.0831(37)	a
		CCH	117(7)	119.96(8)	
Pyrimidine	$C_4H_4N_2$	C2H7	1.088(4)	1.082(4)	
		C4H8	1.088(4)	1.079(2)	
		C5H9	1.088(4)	1.087(3)	
		C5C4H8	116(10)	120.9(3)	
Thiophene	C_4H_4S	C2H6	1.116(5)	1.0688(6)	b
		C3H7		1.0812(11)	
		H6C2S	119.3(13)	120.24(8)	
		H7C3C4	126.3(21)	124.06(6)	

[a] Cradock S, Liescheski PB, Rankin DWH, Robertson HE (1988) J Am Chem Şoc 110: 2758–2763
[b] Liescheski PB, Rankin DWH (1988) J Mol Struct 178: 227–241

8.3.2 $^1J(^{13}C–H)$ Spin–Spin Coupling Constant

Muller and Pritchard [8], using empirical bond lengths, found a linear correlation between the ^{1} J(^{13}C–H) spin–spin coupling constant and the C–H internuclear distance. This correlation was later confirmed with equilibrium bond lengths [9], the correlation coefficient being $\rho = 0.94$. However, the spin–spin coupling constants are sensitive to solvent effects and, even when they are determined in gas-phase, they show a density dependence [10]. Furthermore, the rovibrational effects are not negligible [11]. For instance, for $H^{13}C{\equiv}^{13}CH$, the equilibrium value of $^1J(^{13}C\text{–}H)$ is 242.69 kHz, whereas the ground state value is 247.56 kHz [10]. In conclusion, this correlation with a standard deviation of 0.004 Å is not accurate but may be used to compare similar molecules.

8.3.3 *Spin-Relaxation*

The nuclear Overhauser effect (NOE) can be used to determine internuclear distances [12–14]. This effect arises from dipole–dipole interactions between nearby active nuclei. It does not depend upon through-bond J coupling but only on the spatial proximity of the nuclei. The NOE may be observed between protons that are up to 4 Å apart, while the upper limit for large molecules is about 5 Å. It is particularly useful for large biomolecules, especially proteins.

8.4 Spin–spin Interaction in Rotational Spectra

The spin–spin interaction between nuclei can also lead to a hyperfine splitting of the rotational lines, and the determination of the spin–spin coupling tensor **D** permits to derive the distance between two atoms as already discussed in Sect. 6.6.4 [15].

In particular, the distance H···H has been determined for some molecules [16]. However, what is obtained is an effective distance r_{eff}

$$r_{\text{eff}} = \left[\left\langle v \left| \frac{1}{r^3} \right| v \right\rangle\right]^{-1/3} \tag{8.1}$$

where v denotes the vibrational state. A series expansion of r_{eff} measured in the ground vibrational state $v = 0$ gives

$$r_{\text{eff}} = r_e + \langle 0|\delta r|0\rangle + \ldots \approx r_z \tag{8.2}$$

where r_z is the average distance between nuclear positions, see Sect. 5.7 which explains how to calculate $r_z - r_e$. See Table 8.3 which shows the good agreement between $r_z - r_e$ and $r_{\text{eff}} - r_e$. Unfortunately, there are two difficulties inherent in this method: (i) the hyperfine splitting is quite small and (ii) the spin-rotation interaction has also to be taken into account. For these reasons, it is necessary to use molecular beam spectroscopies (electronic resonance, magnetic resonance [17], or beam maser [18]) as for instance in Ref. [16, 19] or Fourier transform microwave spectroscopy

Table 8.3 Distance $d(H \cdots H)$ (Å) from spin–spin interaction

Molecule	r_{eff} [a]	r_e	$r_{eff}-r_e$	r_z-r_e [e]
$H_2C{=}O$	1.894(4)	1.8736[b]	0.021(4)	0.022
$H_2C{=}C{=}O$	1.890(2)	1.8810	0.009(2)	0.008
$H_2C{=}S$	1.860(2)	1.8456[d]	0.014(2)	0.016

[a]Fabricant B, Krieger D, Muenter JS (1977) J Chem Phys 1576–1586
[b]Morgan WJ, Matthews DA, Ringholm M, Agarwal J, Gong JZ, Ruud K, Allen WD, Stanton JF, Schaefer HF (2018) J Chem Theory Comput 14: 1333–1350
[c]Guarnieri A, Demaison J, Rudolph HD (2010) J Mol Struct 969: 1–8
[d]Müller HSP, Maeda A, Thorwirth S, Lewen F, Schlemmer S, Medvedev IR, Winnewisser M, DeLucia FC, Herbst E (2019) Astron Astrophys 621: A143
[e]Calculated from a MP2/cc-pVTZ cubic force field

[20, 21]). Unfortunately, as the splittings are quite small, the spin–spin coupling constants cannot be determined with accuracy and the derived interatomic distances are not better than an empirical effective structure, as confirmed by Table 8.3.

8.5 Coulomb Explosion Imaging (CEI) [22]

When valence electrons responsible for the chemical bonding are expelled, either by laser radiation or by bombardment with ions, the remaining nuclei fly apart in a Coulomb explosion under the forces of electrostatic repulsion. The measurement of the velocities of all the fragments provides us with the information necessary to determine the structure of the original molecule. The reconstruction relies on the knowledge of the potential that describes the interaction of the charged particles during the explosion, generally a Coulomb potential is used. This method furnishes an image of the entire, isolated molecule, including all the hydrogen atoms, see for instance the example of the iodopyridine, C_5H_4IN, and iodopyrazine, $C_4H_3IN_2$, molecules [23]. The weak point is that it is not accurate. For instance, for D_2O, the accuracy of the bond length is only 0.3 Å and that of the bond angle 15° [24]. On the other hand, it is well adapted to the study of floppy molecules such as van der Waals complexes [25].

8.6 Infrared Spectroscopy

High-resolution infrared spectroscopy is useful to determine rotational constants, in particular in vibrationally excited states, see Sect. 4.3. Furthermore, infrared spectroscopy delivers another useful parameter: the frequency of the stretching vibration of the X–H bond. When the upper level of the transition is isolated (i.e., without interaction with nearby levels), there is a relationship between the vibrational frequency

ν_{is} and the equilibrium bond length r_e(X–H). It was first noticed by Bernstein [26] and developed by McKean [27, 28]. When several hydrogen atoms are present, to obtain an isolated C–H stretching vibration, McKean used selective deuteration, all hydrogens but one being replaced by deuterium. This method is sometimes not easy. Furthermore, anharmonic resonances are possible and may affect the frequency. An alternative way is to use the local mode method where overtone bands of the stretching vibration are measured. The motion of the individual bonds becomes more localized upon increasing excitation [29], and the unperturbed fundamental frequency may be obtained by a Birge-Sponer plot [30]. For the C–H bond length, a least-squares fit of 60 equilibrium bond lengths gave the following linear relationship [31]

$$r_e(\mathrm{C-H})[\mathring{A}] = 1.3047(29) - 7.311(97) \times 10^{-5}\nu_{is}(\mathrm{C-H})\left[\mathrm{cm}^{-1}\right] \quad (8.3)$$

with a correlation coefficient $\rho = 0.990$ and a standard deviation $\sigma = 0.001$ Å. For the O–H bond length, a similar relationship exists [32]. A fit of 28 equilibrium bond lengths gave the following linear expression

$$r_e(\mathrm{O-H})[\mathring{A}] = 1.2261(76) - 7.29(21) \times 10^{-5}\nu_{is}(\mathrm{O-H})\left[\mathrm{cm}^{-1}\right] \quad (8.4)$$

with a correlation coefficient $\rho = 0.979$ and a standard deviation $\sigma = 0.0015$ Å. There is also a good correlation between r_e(N–H) and ν_{is}(N–H), but it does not seem to be linear [33, 34]. In this case, it would be interesting to check the presence of anharmonic resonances.

McKean extended this method to other bonds, in particular the Si–H bond [35, 36]. However, to establish the correlation, he used empirical bond lengths. Equilibrium Si–H are not very numerous; nevertheless, it is possible to confirm the correlation; see Table 8.4. A least-squares fit of 10 data gives the following linear expression

$$r_e(\mathrm{Si-H})[\mathring{A}] = 1.8804(54) - 1.862(24) \times 10^{-4}\nu_{is}(\mathrm{Si-H})\left[\mathrm{cm}^{-1}\right] \quad (8.5)$$

with a correlation coefficient $\rho = 0.999$ and a standard deviation $\sigma = 0.0009$ Å. Inspection of Table 8.4 confirms the excellent accuracy of Eq. (8.5).

8.7 Planar Moments of Inertia

According to the ligand close packing model [37], the distance between two ligands attached to the same atom is almost constant, i.e., for the H…H distance 1.8 Å. However, the ligand radius depends on the central atom and its charge. Therefore, significant differences may be observed.

For molecules with a symmetry plane and with hydrogen atoms in symmetrically equivalent pairs, the planar moment equation may be used to determine the H…H

Table 8.4 ν_{is}(Si–H) values (cm^{-1}) and r_e(Si–H) bond lengths (Å)

	ν_{is}(Si–H)	Ref	r_e(Si–H)	Ref	r_{calc}	exp−calc
SiHCl	1968.7	a	1.5140	b	1.5138	0.0002
SiH_2	1973.3	c	1.513	d	1.5129	0.0001
$SiHD_3$	2182	e	1.4742	d	1.4743	-0.0001
SiH_3Br	2205.61	f	1.4696	g	1.4699	-0.0003
SiH_3Cl	2206.7	f	1.46878	h	1.4697	-0.0009
SiH_3F	2207.4	f	1.46948	i	1.4696	0.0000
SiH_2Cl_2	2231.1	j	1.4671	k	1.4652	0.0020
SiH_2F_2	2248	l	1.4615	m	1.4621	-0.0004
$SiHF_3$	2316.9	n	1.4487	n	1.4493	-0.0003
SiH_3CN	2217.4	o	1.4680	p	1.4677	0.0003

[a]Harper WW, Clouthier DJ (1997) J Chem Phys 106: 9461–9473
[b]Vázquez J, Stanton JF (2002) J Phys Chem A 106: 4429–4434
[c]Fredin L, Hauge RH, Kafafi ZH, Margrave JL (1985) J Chem Phys 82: 3542–3545
[d]Coriani S, Marchesan D, Gauss J, Hättig C, Helgaker T, Jørgensen P (2005) J Chem Phys 123: 184,107
[e]Shimanouchi T (1972) Tables of Molecular Vibrational Frequencies Consolidated Vol I, National Bureau of Standards, pp 1–160
[f]McKean DC (1981) J Mol Struct 113: 251–266
[g]Demaison J, Cosléou J, Bürger H, Mkadmi EB (1997) J Mol Spectrosc 185: 384–391
[h]Cazzoli G, Puzzarini C, Gauss J (2012) Mol Phys 110: 2359–2369
[i]Puzzarini C, Cazzoli G, Gauss J (2010) J Mol Spectrosc 262: 37–41
[j]Christensen DH, Nielsen OF (1968) J Mol Specrosc 27: 489–498
[k]Davis RW, Robiette AG, Gerry, MCL (1981) J Mol Spectrosc 85: 399–415
[l]Bunnell J, Ford TA (1983) J Mol Spectrosc 100: 215–233
[m]D'Eu J-F, Demaison J, Bürger H (2003) J Mol Spectrosc 218: 12–21
[n]Demaison J, Margulès L, Breidung J, Thiel W, Bürger H (1999) Mol Phys 97: 1053–1067
[o]McKean DC (1992) Spectrochim Acta A 48: 1335–1345
[p]Priem D, Cosléou J, Demaison J, Merke I, Stahl W, Jerzembeck W, Bürger H (1998) J Mol Spectrosc 191: 183–198

distance. For a symmetric pair of hydrogen atoms

$$2P_{\gamma}^{0} = I_{\alpha}^{0} + I_{\beta}^{0} - I_{\gamma}^{0} = 2\sum_{i} m_{\mathrm{H}}\gamma_i^2 + \Delta = m_{\mathrm{H}}d_{\mathrm{HH}}^2 + \Delta \qquad (8.6)$$

where $\alpha, \beta, \gamma = a, b, c$, Δ is the vibrational contribution and m_{H} is the mass of the hydrogen atom.

A good correlation was found between the ground state planar moment of inertia and its equilibrium value with a correlation coefficient $\rho = 0.97$ [38]. This correlation may be used to estimate the d_{HH} distance.

However, there is perhaps a better method. It was found that the vibrational contribution Δ, although large, has the same order of magnitude with $\Delta(XH_2)$ being slightly larger than $\Delta(XD_2)$. For this reason, Laurie [39] proposed to calculate d_{HH} by using

Table 8.5 $d(\mathrm{H}\cdots\mathrm{H})$ distance (in Å) and rovibrational contribution, Δ, (in uÅ^2) to the planar moment of inertia of a few molecules

Molecule	$d(\mathrm{H}\cdots\mathrm{H})$			Δ from	Δ from	Ref
	d_0	d_e	Equation (8.7)	Force field	r_e	
CH_2F_2	1.810	1.818	1.829	0.0139	0.0134	a
CD_2F_2	1.820			-0.0078	-0.0076	a
CH_2Cl_2	1.761	1.790	1.800	0.0520	0.0519	a
CD_2Cl_2	1.781			0.0370	0.0342	a
SiH_2F_2	2.439	2.459	2.470	0.0582	0.0504	b
SiD_2F_2	2.454			0.0223	0.0244	b

[a] Vogt N, Demaison J, Rudolph HD (2014) Mol Phys 112: 2873–2883
[b] D'Eu, J-F, Demaison J, Bürger H (2003) J Mol Spectrosc 218: 12–21

the difference between $P^0_\gamma[\mathrm{H_2}]$ and $P^0_\gamma[\mathrm{D_2}]$

$$P^0_\gamma[\mathrm{XD_2}] - P^0_\gamma[\mathrm{XH_2}] = \frac{1}{2}(m_\mathrm{D} - m_\mathrm{H})d^2_\mathrm{HH} \tag{8.7}$$

Unfortunately, the results are not satisfactory because the difference between $\Delta(\mathrm{XH_2})$ and $\Delta(\mathrm{XD_2})$ is far from negligible; see Table 8.5.

8.8 Ab Initio Calculations

In Sect. 2.8, it was indicated that the MP2_fc/cc-pVTZ level of theory can give satisfactory results. It is particularly true for the X–H bond (X = C, N, O), see Table 8.6, which show that it is possible to predict these X–H bonds with an accuracy of about 0.001 Å. As indicated in Sect. 2.8, it is possible to extend this method to other bonds.

Table 8.6 Typical offset values $\Delta r = r_e - r[\mathrm{MP2}]$ for the X–H bond length (in Å)

Bond	Basis	Mean	Median	STD	Max	Ref
C–H	cc-pVTZ	0.0002	0.0003	0.0007	0.0018	a
	cc-pVQZ	0.0011	0.0011	0.0008	0.0029	
O–H	cc-pVTZ	−0.0031	−0.0031	0.0010	0.0054	b
	cc-pVQZ	−0.0015	−0.0015	0.0011	0.0038	
N–H	cc-pVTZ	−0.0013	−0.0013	0.0010	0.0032	c
	cc-pVQZ	0.0000	-0.0001	0.0009	0.0023	

[a] Demaison J, Craig NC (2011) J Phys Chem A 115: 8049–8054
[b] Demaison J, Herman M, Liévin J (2007) Intern Rev Phys Chem 26: 391–420
[c] Demaison J, Margulès L, Boggs JE (2000) Chem Phys 260: 65–81

8.9 Other Methods

The most useful methods have been described in the previous sections. However, there are many correlations that may also be used. Some of them are reviewed [40]. For instance, it was observed that the C–H bond length in CH_3X molecules increases with the electronegativity of the X atom, whereas it decreases in HC≡CX molecules [41].

For the structure determination of a molecule of n atoms without symmetry, it is generally assumed that there are $3n-6$ independent parameters. Actually, some parameters are not independent as confirmed by the presence of cross-terms in the force fields using internal coordinates. In particular, the adjacent structural parameters in a molecule are not independent and, in some cases, it is possible to find a correlation [42]. For instance, there is a linear correlation between the C–H and the C≡C bond lengths in HC≡CX molecules [41]. In many cases, there is a relationship between a bond length and the adjacent bond angle [42]. Particularly interesting is the correlation between C–H bond lengths and ∠(HCH) bond angles in methyl groups [43]. It was extended with success to CH_2X_2 molecules [44]. However, it has to be noted that it fails when X = F. A fit of 24 data from methyl groups, CH_3X, gives

$$r_e(\mathrm{C-H})[\text{Å}] = 1.305(32) - 0.00199(29) \times \angle(\mathrm{HCH})[\mathrm{degree}] \tag{8.8}$$

with a standard deviation of 0.002 Å; see Table 8.7. There are four outliers: dimethylether, dimethylsulfide, methanol, and propanal. A new fit without these four data gives

$$r_e(\mathrm{C-H})[\text{Å}] = 1.314(21) - 0.00208(19) \times \angle(\mathrm{HCH})[\mathrm{degree}] \tag{8.9}$$

with a standard deviation of 0.001 Å. To try to explain the outliers, the ligand close packing model [37] may be used: there are small variations depending on the charge of the carbon atom, i.e., on the electronegativity of the substituent X. This should affect the $d(\mathrm{H}\cdots\mathrm{H})$ non-bonded distance. However, as shown in Table 8.7, this is not the correct explanation because the outliers have a regular value. A more likely explanation is that this correlation does not work well when the methyl group is highly asymmetric with large value of $r(\mathrm{CH_A})-r(\mathrm{CH_S})$ and $\angle(\mathrm{XCH_A})-\angle(\mathrm{XCH_S})$ as shown in Table 8.7. It is worth noting that in these cases, the hydrogen atom is close to an electronegative atom (O, N, S).

A correlation between the C–H bond length and the deuterium quadrupole coupling constant has also been observed [9, 45], the main difficulty being to obtain accurate constants in the bond axis system.

Table 8.7 Structure of the methyl group (distances in Å, angles in degree)

				Exp−Calc				
	r_e(CH)	$\angle_e$(HCH)	Ref.[a]	All data	Equation (8.9)[b]	d(H...H)	CH_A–CH_S	$\Delta\angle$[c]
CH_3I	1.0819	111.40	1	−0.0011	−0.0004	1.7875	0.0000	0.0
CH_3Br	1.0825	111.17	1	−0.0010	−0.0003	1.7861	0.0000	0.0
CH_3Cl	1.0842	110.21	1	−0.0012	−0.0006	1.7785	0.0000	0.0
CH_4	1.0862	109.47	2	−0.0007	−0.0002	1.7737	0.0000	0.0
CH_3NC	1.0862	109.50	3	−0.0006	−0.0001	1.7741	0.0000	0.0
CH_3CN	1.0865	109.10	4	−0.0011	−0.0006	1.7702	0.0000	0.0
CH_3F	1.0870	110.26	5	0.0017	0.0023	1.7837	0.0000	0.0
CH_3CCl_3	1.0870	109.8	6	0.0009	0.0014	1.7790	0.0000	0.0
$HCOOCH_3$	1.0871	108.81	7	−0.0011	−0.0006	1.7679	0.0034	4.1
CH_3CH_2CN	1.0874	108.16	8	−0.0021	−0.0017	1.7612	−0.0002	0.9
CH_3CH_2CHO	1.0878	107.37	9	**−0.0032**	−0.0029	1.7530	−0.0002	0.0
CH_3CCH	1.0885	108.33	10	−0.0006	−0.0002	1.7649	0.0000	0.0
CH_3CH_2OH	1.0886	108.40	9	−0.0004	0.0000	1.7659	−0.0009	−0.1
CH_3NO	1.0887	107.69	11	−0.0017	−0.0014	1.7581	0.0011	−4.0
CH_3CH_3	1.0887	107.67	12	−0.0017	−0.0014	1.7578	0.0000	0.0
CH_3C_4H	1.0890	108.42	13	0.0001	0.0005	1.7668	0.0000	0.0
$(CH_3)_2S$	1.0897	109.73	14	**0.0034**	0.0039	1.7823	0.0011	3.3
CH_3CHO	1.0906	107.05	11	−0.0011	−0.0008	1.7539	0.0050	−1.1
$(CH_3)_2CO$	1.0912	107.04	15	−0.0005	−0.0002	1.7548	–.–	−0.1
CH_3OH	1.0912	108.93	16	**0.0033**	0.0037	1.7760	0.0056	5.2
$CH_2(CH_3)_2$	1.0908	107.58	17	0.0002	0.0005	1.7603	0.0014	−0.9
CH_3CHCH_2	1.0920	108.02	18	0.0022	0.0026	1.7670	0.0033	−0.1

(continued)

Table 8.7 (continued)

Exp−Calc								
	r_e(CH)	$\angle_e$(HCH)	Ref.[a]	All data	Equation (8.9)[b]	d(H...H)	CH_A–CH_S	$\Delta\angle$[c]
$(CH_3)_2O$	1.0951	107.61	9	**0.0046**	0.0049	1.7675	0.0086	3.7
CH_3Li	1.0958	105.69	24	0.0014	0.0016	1.7467	0.0000	0.0

[a]Reference for equilibrium structure
[b]See text
[c]$\Delta\angle = \angle(XCH_A) - \angle(XCH_S)$
References
1. Demaison J, Margulès L, Boggs JE (2003) Struct Chem 14: 159–174
2. Puzzarini C, Heckert M, Gauss J (2008) J Chem Phys 128: 194108
3. Margulès L, Demaison J, Rudolph HD (2001) J Mol Struct 599: 23–30
4. Puzzarini C, Cazzoli G (2006) J Mol Spectrosc 240: 260–264
5. Demaison J, Breidung J, Thiel W, Papousek D (1999) Struct Chem 10: 129–133
6. Margulès L, Carvajal M, Demaison J (2008) J Mol Spectrosc 247: 160–166
7. Demaison J, Margulès L, Kleiner I, Császár AG (2010) J Mol Spectrosc 259: 70–79
8. Demaison J, Margulès L, Mäder H, Sheng M, Rudolph HD (2008) J Mol Spectrosc 169–175
9. Vogt N, Demaison J, Vogt J, Rudolph HD (2014) J Comput Chem 35: 2333–2342
10. Müller HSP, Thorwirth S, Lewen F (2020) J Mol Spectrosc 1207: 127769
11. Present work, CCSD(T)_ae/wCVQZ structure
12. Puzzarini C, Taylor PR (2004) J Chem Phys 122: 054315
13. Cazzoli G, Cludi L, Contento M, Puzzarini C (2008) J Mol Spectrosc 251: 229–234
14. Demaison J, Margulès L, Rudolph HD (2010) J Mol Struct 978: 229-233
15. Lovas, FJ, Groner P (2006) J Mol Spectrosc 236: 173–177
16. Demaison J, Herman M, Liévin J (2007) Int Rev Phys Chem 26: 391–420
17. Present work, semiexperimental structure
18. Demaison J, Craig NC, Gurusinghe R, Tubergen MJ, Rudolph HD, Coudert LH, Szalay P, Császár AG (2017) J Phys Chem A 121:3155–3166
19. Breidung J, Thiel W (2001) J Mol Struct 599: 239–254

References

1. Demaison J, Craig NC, Conrad AR, Tubergen MJ, Rudolph HD (2012) Semiexperimental equilibrium structure of the lower energy conformer of glycidol by the mixed estimation method. J Phys Chem A 116:9116–9122
2. Vogt N, Demaison J, Vogt J, Rudolph HD (2014) Why it is sometimes difficult to determine the accurate position of a hydrogen atom by the semiexperimental method: structure of molecules containing the OH or the CH_3 group. J Comput Chem 35:2333–2342
3. Debye P, Bewilogua L, Ehrhardt P (1929) Zerstreuung von Röntgenstrahlen an einzelnen Molekeln. Phys Z 30:84–87
4. Kierspel T, Morgan A, Wiese J, Mullins T, Aquila A, Barty A, Küpper J (2020) X-ray diffractive imaging of controlled gas-phase molecules: Toward imaging of dynamics in the molecular frame. J Chem Phys 152:084307
5. Zemach AC, Glauber RJ (1956) Neutron diffraction by gases. Phys Rev 101:129–136
6. Diehl P (1992) Nuclear magnetic resonance spectroscopy and precise molecular geometries. In: Domenicano A, Hargittai I (eds) Accurate molecular structures. Oxford University Press, Oxford, p 299
7. Lounila J, Ala-Korpela M, Jokisaari J (1990) A special method for analyzing anisotropic nuclear magnetic resonance parameters: acetonitrile in liquid crystals. J Chem Phys 93:8514–8523
8. Muller N, Pritchard DE (1959) ^{13}C Splittings in proton magnetic resonance spectra. II. Bonding in substituted methanes. J Chem Phys 31:1471–1476

9. Demaison J, Wlodarzcak G (1994) The equilibrium C-H bond length. Struct Chem 5:57–66
10. Jackowski K (2003) Gas-phase study of spin-spin coupling constants. Int J Mol Sci 4:135–142
11. Ruden TA, Lutnaes OB, Helgaker T (2003) Vibrational corrections to indirect nuclear spin–spin coupling constants calculated by density-functional theory. J Chem Phys 118:9572–9581
12. Neuhaus D, Williamson MP (2000) The nuclear Overhauser effect in structural and conformational analysis. Wiley-VCH, New York
13. Canet D, Bouguet-Bonnet S, Leclerc S, Yemloul M (2011) Carbon-13 heteronuclear longitudinal spin relaxation for geometrical (and stereochemical) determinations in small or medium size molecules. Ann Rep NMR Spectrosc 74:89–123
14. Khodov IA, Efimov SV, Klochkov VV, Batista de Carvalho LAE, Kiselev MG (2016) The importance of suppressing spin diffusion effects in the accurate determination of the spatial structure of a flexible molecule by nuclear Overhauser effect spectroscopy. J Mol Struct 1106:373–381
15. Legon AC, Demaison J (2011) Other spectroscopic sources of molecular properties: intermolecular complexes as examples. In: Demaison J, Boggs JE, Császár AG (eds) Equilibrium molecular structures. From Spectroscopy to Quantum Chemistry. CRC Press, Boca Raton, pp 205–231
16. Fabricant B, Krieger D, Muenter JS (1977) Molecular beam electric resonance study of formaldehyde, thioformaldehyde, and ketene. J Chem Phys 1576–1586.
17. Dyke TR, Muenter JS (1975) The properties of molecules from molecular beam spectroscopy. In: Buckingham AD (ed) Molecular structure and properties. MTP International review of science, physical chemistry, series two, vol 2. Butterworths, London, Chapt. 2, p 27
18. Dymanus A (1992) Beam-maser spectroscopy. At Mol Beam Methods 2:58–102
19. Ozier I, Crapo LM, Lee SS (1968) Nuclear radio-frequency spectra of a series of tetrahedral molecules. Phys Rev 172:63–82
20. Stahl W, Dreizler H (1987) Determination of the spin-rotation and spin-spin interaction constants of germane by microwave Fourier transform measurements. Z Naturforsch 42a:1402–1404
21. Wolf P, Stahl W (1989) The determination of spin-rotation and spin-spin interaction constants from the microwave spectrum of methane. Z Naturforsch 44a:651–654
22. Vager Z, Naaman R, Kanter EP (1989) Coulomb explosion imaging of small molecules. Science 244:426–431
23. Boll R, Schäfer JM, Richard B et al (2022) X-ray multiphoton-induced Coulomb explosion images complex single molecules. Nat Phys 18:423–428
24. Légaré F, Lee KF, Litvinyuk IV, Dooley PW, Wesolowski SS, Bunker PR, Dombi P, Krausz F, Bandrauk AD, Villeneuve DM, Corkum PB (2005) Laser Coulomb-explosion imaging of small molecules. Phys Rev A 71:013415
25. Wu J, Kunitski M, Schmidt LPH, Jahnke T, Dörner R (2012) Structures of N_2Ar, O_2Ar, and O_2Xe dimers studied by Coulomb explosion imaging. J Chem Phys 137:104308
26. Bernstein HJ (1962) The average XH stretching frequency as a measure of XH bond properties. Spectrochim Acta 18:161–170
27. McKean DC (1978) Individual CH bond strengths in simple organic compounds: effects of conformation and substitution. Chem Soc Rev 7:399–422
28. McKean DC (1989) CH bond dissociation energies, isolated stretching frequencies, and radical stabilization energy. Int J Chem Kinet 21:445–464
29. Henry BR (1987) The local mode model and overtone spectra: a probe of molecular structure and conformation. Acc Chem Res 20:429–435
30. Birge RT, Spooner H (1926) The heat of dissociation of non-polar molecules. Phys Rev 28:259–283
31. Demaison J, Rudolph HD (2008) Ab initio anharmonic force field and equilibrium structure of propene. J Mol Spectrosc 248:66–76
32. Demaison J, Herman M, Liévin J (2007) The equilibrium OH bond length. Intern Rev Phys Chem 26:391–420

33. Evans DF, Missen PH, Upton MW (1982) Isolated N-H stretching frequencies in compounds containing NH_2 groups. J Mol Struct 82:147–149
34. Demaison J, Margulès L, Boggs JE (2000) The equilibrium N-H bond length. Chem Phys 260:65–81
35. McKean DC (1981) New light on the stretching vibrations, lengths and strengths of CH, SiH and GeH bonds. J Mol Struct 113:251–266
36. McKean DC, Torto I, Boggs JE, Fan K (1992) Infrared and ab initio studies of molecules containing SiH bonds. J Mol Struct Theochem 260:27–46
37. Gillespie RJ, Popelier PLA (2001) Chemical bonding and molecular geometry. Oxford University Press, New York
38. Demaison J, Wlodarczak G, Siam K, Ewbank JD, Schäfer, (1988) Comparison of *ab initio* calculated and experimental methyl-top moments of inertia. Chem Phys 120:421–428
39. Laurie VW (1958) Note on the determination of molecular structure from spectroscopic data. J Chem Phys 28:704–706
40. Mastryukov VS, Simonsen SH (1996) Empirical correlations in structural chemistry. In: Hargittai M, Hargittai I (eds) Advances in molecular structure research, vol 2. JAI Press, Stanford, pp 163–189
41. Demaison J, Margulès L, Boggs JE (2003) The equilibrium C-Cl, C–Br, and C–I bond lengths from ab initio calculations, microwave and infrared spectroscopies, and empirical correlations. Struct Chem 14:159–174
42. Mastryukov VS, Schaefer HF III, Boggs JE (1994) Self-consistent changes of geometrical parameters: experiment and theory. Acc Chem Res 27:242–249
43. McKean DC (1976) A correlation between isolated CH stretching frequencies and HCH bond angles in methyl groups. J Mol Struct 34:181–185
44. Demaison J, Wlodarczak G, Rück H, Wiedenmann KH, Rudolph HD (1996) Accurate structures of simple dicyanides. J Mol Struct 376:399–411
45. Huber H (1985) Deuterium quadrupole coupling constants. A theoretical investigation. J Chem Phys 83:4591–4598

Chapter 9
Database with Equilibrium Structures of Free Molecules

Abstract Equilibrium structures of free molecules determined by experimental methods and published after 2017 are systematized. The presentation of structural datasets is similar to that in the handbooks "Structure Data of Free Polyatomic Molecules" previously published by Springer, and this chapter is recommended as a current supplement to these handbooks.

9.1 Introduction

The equilibrium structures of free molecules derived from experimental data (in most cases in combination with quantum-chemical (QC) anharmonic force fields, *i.e.*, semiexperimental equilibrium structures, r_e^{se}) and published after 2017 are systematized in this chapter. For the methods of equilibrium structure determination, see Sects. 5.5 and 7.4. A few r_m structures being close in magnitude to the equilibrium ones are also presented. Most of these structures were obtained by an analysis of gas-phase electron diffraction (GED) intensities or ground state rotational (in a few cases also vibrational) constants determined from microwave (MW) and high-resolution infrared (IR) spectra.

The data presentation in this chapter is similar to that in the handbooks "Structure Data of Free Polyatomic Molecules" (Landolt-Börnstein volumes II/25A-D [1–4], II/28A-D [5–8] and II/30A,B [9, 10] edited by M. Tanimoto, N. Vogt, and/or K. Kuchitsu and their recent update [11]). These handbooks include both effective and equilibrium structures of ca. 7000 polyatomic molecules and cover the literature published between 1960 and 2017. Thus, this chapter with 100 datasets can be used as the following update for the comprehensive collection of experimental structural data [1–11].

The bibliographic information was taken from MOGADOC database (acronym for MOlecular GAs-phase DOCumentation) established by B. Mez-Starck, J. Vogt, and N. Vogt [12–16] (see also Chap. 10 in Ref. [17]).

N. Vogt and J. Demaison, *Equilibrium Structure of Free Molecules*, Lecture Notes in Chemistry 111, https://doi.org/10.1007/978-3-031-36045-9_9

9.2 Data Presentation

The molecular structure(s) for each chemical compound is presented in a separate file (document). Each file begins with the identification data, namely:

No	document number;
SF	sum formula of molecule in Hill system [18];
CN	chemical name(s) of compound;
Symmetry	point-group symmetry;
Methods	methods of structure determination;
Ref.	cross-reference to the original publication with the structural data.

The document numbers (from 1 to 98) follow in alphabetical order of the sum formulae in Hill system [18]. According to this system, each formula starts with carbon and hydrogen atoms (if available) followed by the atoms of remaining chemical elements in alphabetical order of their symbols.

The chemical names are taken mostly from the original paper following the rules of IUPAC (International Union of Pure and Applied Chemistry) and Chemical Abstracts dialect. Trivial chemical names are additionally given in some cases.

Complexes consisting of different weakly bound components in a ratio n: m... are given with the notation of a ratio, *e.g.*, argon – hydrogen bromide (1/1), where $n = 1$ and $m = 1$.

The point-group symmetry describes the symmetry of the molecular structure at a stable equilibrium corresponding to a minimum of the potential energy function (PEF) or surface (PES). This minimum is used as the reference point for the description of internal motions (vibrations) of the atoms in a molecule. In linear and nonlinear molecules containing N atoms, there are different numbers of vibrational degrees of freedom ($3N$–5 and $3N$–6, respectively). The whole potential surface can have several minima corresponding to different stable conformations (conformers). For non-rigid molecules, the potential barriers between minima can be very low and do not even rise above the zero-point level, for instance for weakly bound complexes. In such cases, the symmetry of the molecular structure corresponding to a PES maximum between equivalent shallow minima can be specified as effective symmetry.

The symmetry of a molecule can be deduced from experimental data, but very often it is assumed according to results of quantum-chemical (QC) computations.

The experimental **method** used for the structure determination is given in abbreviated form as shown below in the List of abbreviations. The source yielding supplemental information for the given structure determination is also indicated, for instance QC (density functional theory (DFT) and ab initio) calculations.

The cross-reference (Ref.) to original publication is given for each dataset.

The tables with structural data (internuclear distances, bond angles and dihedral angles) follow after the identification data. The distances between directly bonded atoms are indicated by solid lines and between non-bonded atoms by dotted lines. The bond orders (single, aromatic, double and triple) are not specified in these tables. The values of distances and angles taken from the original publications are presented in Å (1 Å = 100 pm) and degrees, respectively.

It is worth to note that the internuclear distances being arguments of intensity function $sM(s)$ are primary structural data determined by GED, whereas the angles or nuclear coordinates are derived from these distances. Contrary to that, nuclear positions in the principal axis system are directly derived from the spectroscopic rotational constants, and bond distances and angles are then derived from the determined coordinates.

It is clear that the equilibrium structures determined from experimental data with use of QC force field are not totally experimental. In spectroscopy, such structures are called as semiexperimental (or semiempirical) ones, r_e^{se}. However, in the electron diffraction literature, the discrepancy between experimental and semiexperimental structures was very often not emphasized, and both experimental and semiexperimental equilibrium structures were denoted as r_e. In the following tables, the semiexperimental equilibrium structures are consistently indicated as r_e^{se}. The procedure of equilibrium structure determinations by electron diffraction and spectroscopic methods are described in Chaps. 5, 6, 7.

Uncertainties in structural parameters given in parenthesis are taken from original paper.

The total errors, σ_{tot}, in the internuclear distances determined by GED can be estimated as:

$$\sigma_{tot} = \left(\sigma_{rand}^2 + \sigma_{syst}^2\right)^{1/2}, \tag{9.1}$$

where σ_{rand} and σ_{syst} are the random and systematic errors, respectively. The experimental σ_{syst} is a result of error propagation of electron scattering intensity I_M(s) measurements including measurements of electron wave lengths and camera distance, sector calibration, etc. In many studies, they were estimated to be 0.1 or 0.2% of the values of refined internuclear distances. The estimation of analytical systematic errors arising due to application of some theoretical approximations is very difficult or even impossible. Therefore, the estimated systematic errors are rather underestimated than real.

The random errors are estimated in a least-squares analysis from differences between experimental values and their theoretical counterparts. The obtained least-squares deviations, σ_{LS}, are usually multiplied by a factor two or three according to the chosen significance level. Random errors indicate relative errors in the structural parameters of a molecule. In many GED studies, σ_{rand} were obtained to be about 0.1% of the distance values. Currently, the total errors are simply estimated by three times the root-mean-square deviations approximately covering both random and systematic errors. For more details concerning estimations of total errors and accuracy of r_e^{se} structure determined by GED, see Sect. 7.7.

The random errors in the structural parameters determined by spectroscopic methods are propagated from uncertainties in rotational constants. Since rotational constants are usually determined with six to eight significant digits, the σ_{LS} values for the structural parameters are very small. For more details concerning the error estimations of equilibrium structure determined by spectroscopic methods, see Sects. 4.11 and 5.12.

The systematic errors arising by conversion of B_0 to B_e^{se} and r_a to r_e^{se} are estimated to be from a few to about 10–20% of the calculated values of vibrational corrections themselves.

The Figure represents the ball-stick model(s) of the molecule. All figures in this chapter are created in the same style by means of the GaussView program [19]. The atoms in the model are labeled by symbols of chemical elements, except for hydrogens in many cases. The individual atoms are discriminated using letters, numbers, and other characters. The atom numbering in the table is adjusted to that in the respective figure and can be different from that in the original paper. The single, double, or triple line (stick) connecting two atoms corresponds to the bond order, which cannot necessarily be unambiguous. Aromatic bonds are shown in two different manners, namely by combination of the solid and dotted lines or by alternating single and double bonds.

The comments given after the footnotes to the tables contain important information, namely about the following items:

assumptions made in the structure determination;

effective temperature of the sample molecules in the GED experiment which is close to the measured temperature of the effusion cell, $T_{\text{effusion cell}}$, or its nozzle, T_{nozzle} (it defines both conformational composition and magnitudes of vibrational corrections to experimental internuclear distances);

ratio of conformers;

barrier to internal rotation given in units used in original paper, *i.e.*, in kcal mol^{-1}, kJ mol^{-1} or cm^{-1} (interconversion factors between these units are the following: 1 kcal mol^{-1}≈4.184 kJ mol^{-1}≈349.76 cm^{-1} [20]);

QC methods used for calculations of supporting data about molecular structure, conformation composition, force fields, barriers to internal rotation, etc. For abbreviations of DFT and ab initio methods as well as of basis sets, see Chap. 2 (except for frozen core approximation, fc, which is by default in this chapter) and cited original papers.

List of abbreviations

CEI	Coulomb explosion imaging
DFT	Density functional theory (calculations)
GED	Gas-phase electron diffraction
HR	High-resolution (spectroscopy)
FT	Fourier transform (spectroscopy)
IR	Infrared (spectroscopy)
MB	Molecular beam
MW	Microwave, millimeter-wave and/or submillimeter-wave (spectroscopy)
PES	Potential energy surface

PEF Potential energy function
QC Quantum-chemical (*ab initio* and DFT calculations)

9.3 Equilibrium Structures of Free Molecules

No	**1**
SF	**ArF_6S**
CN	**Sulfur hexafluoride – argon (1/1)**
Symmetry	C_{3v} (effective symmetry)
Method	**IR**
Ref	[21]

Distance	r_e [Å] [a]
S…Ar	4.3021(44)

[a] Parenthesized uncertainty given in units of the last significant digit was not specified, probably 1σ

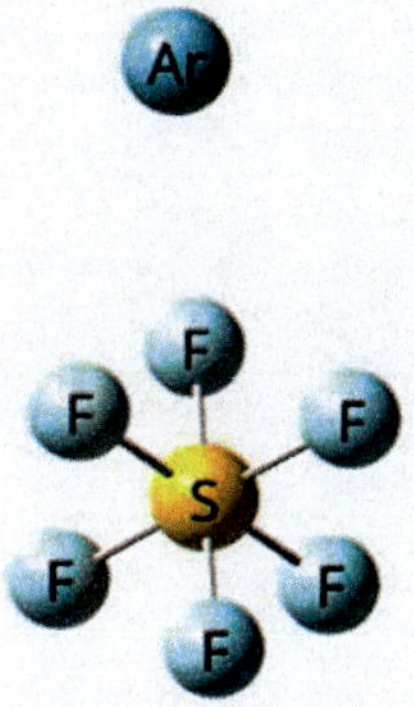

The vibration–rotation spectrum of the title van der Waals heterocluster was recorded in the ν_3 region of SF_6 using a high-resolution infrared spectrometer in combination with a cavity quantum cascade laser coupled to a pulsed supersonic jet.

No	**2**
SF	**CCl_2NO_2P**
CN	**Phosphorisocyanatidic dichloride** Dichlorophosphoryl isocyanate

(continued)

(continued)

Symmetry	C_s (*syn*) C_s (*anti*)
Methods	**GED** supplemented by **IR** and **QC**
Ref	[22]

Distances	r_e^{se} [Å] [a]	
	syn	*anti*
P–O	1.433(2)	1.428(2)
P–Cl	1.985(1)	1.994(1)
P–N	1.637(2)	1.632(2)
N–C	1.203(2) [b]	1.200(2) [b]
C–O	1.146(2) [b]	1.147(2) [b]

Bond angles	θ_e^{se} [deg] [a]	
	syn	*anti*
O–P–Cl	114.6(2)	114.4(2)
Cl–P–Cl	103.0(2)	102.9(2)
O–P–N	117.7(6)	115.0(6)
Cl–P–N	102.4(4)	104.4(4)
P–N–C	135.8(7)	141.1(7)
N–C–O	175.3(7)	175.8(7)

Dihedral angles	τ_e [deg]	
	syn	*anti*
O–P–N–C	0.0 [c]	180.0 [c]
P–N–C–O	180.0 [d]	180.0 [d]

[a] Parenthesized uncertainties in units of the last significant digit are 3σ

[b] Difference between the N=C and C=O bond lengths was assumed at the value of MP2/aug-cc-pVTZ structure

[c] According to assumed C_s point-group symmetry

[d] Assumed

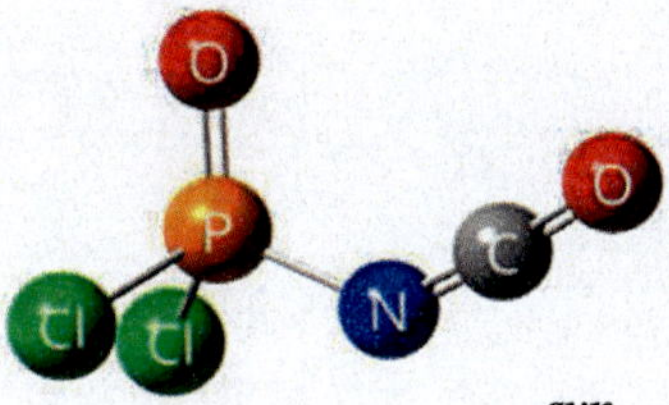

syn

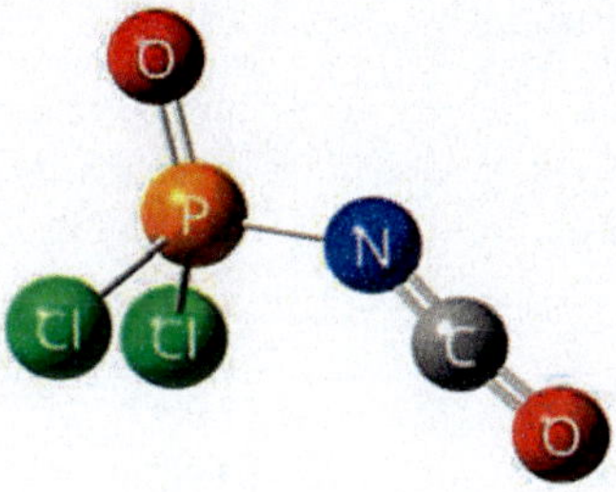

anti

Two conformers with the P=O bond being in *syn* or *anti* position with respect to the N=C bond were predicted to be close in energy by several ab initio and DFT calculations. The *syn* conformer is lower in energy than the *anti* conformer (MP2/cc-pVTZ) by about 2 kJ mol^{-1}.

In the GED analysis, the model of pseudo-conformers was used to describe large-amplitude motions around the P–N bonds in both conformers. Structural parameter differences between pseudo-conformers were assumed at the values from MP2/aug-cc-pVTZ computations. Vibrational corrections to the experimental internuclear distances (T_{nozzle}≈296 K) were calculated with quadratic and cubic force constants from computation at the same level of theory.

No	**3**
SF	**CCl_2O_4S**
CN	**Chlorosulfuric acid chlorocarbonyl ester** Chlorosulfuryl chloroformate
Symmetry	**C_1**
Methods	**GED** supplemented by **QC**
Ref	[23]

Distances	r_e^{se} [Å] [a]

(continued)

(continued)

C–Cl	1.711(2)
C–O8	1.192(3)
C–O5	1.379(4)
S–O2	1.400(1)
S–O3	1.399(1)
S–O5	1.635(2)
S–Cl	1.967(1)

Bond angles	θ_e^{se} [deg] [a]
S–O–C	113.9(6)
Cl7–C–O8	108.1(5)
Cl4–S–O5	104.1(5)
O8–C–O5	124.0(5)
O2–S–O3	122.0(12)
Cl4–S–O3	107.9(3)

Dihedral angles	τ_e^{se} [deg] [a]
Cl4–S–O5–C	68.3(8)
O2–S–O5–C	−47.6(8)
Cl7–C–O5–S	172.2(12)
O8–C–O5–S	−8.7(15)

The table data are reprinted with permission from: [23], Copyright 2018, American Chemical Society

[a] Parenthesized uncertainties in units of the last significant digit are 3σ

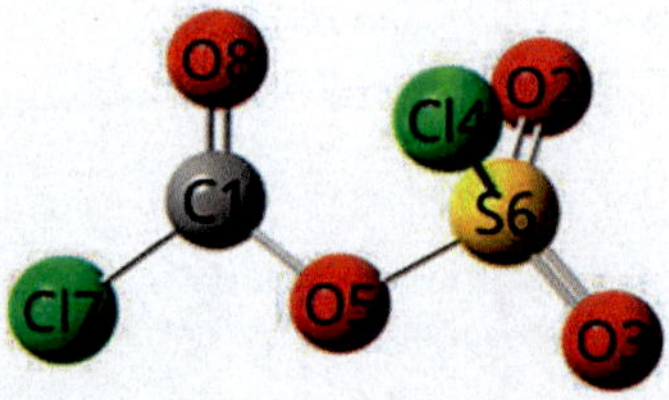

According to predictions of QC calculations (MP2, B3LYP, and B2PLYP with cc-pVTZ basis set), the title molecules exist as a mixture of *sp-sc* (up to 85%) and *sp-ap* conformers characterized by the *synperiplanar* O8–C–O5–S fragment and differing by the value of the Cl4–S–O5–C torsional angle (*synclinal* or *anticlinal*). The existence of the *sp-ap* conformer was neither confirmed nor excluded by GED.

In the GED analysis, differences between structural parameters refined in one group were assumed at the values of the calculated structure (MP2/cc-pVTZ). Vibrational corrections to the experimental internuclear distances (T_{nozzle}≈293 K) were calculated with quadratic and cubic force constants from MP2/cc-pVTZ computation.

No	**4**
SF	**CF_2NOPS**
CN	**Phosphorisocyanatidothioic difluoride** Difluorothiophosphoryl isocyanate
Symmetry	**C_s**
Methods	**GED** supplemented by **QC**
Ref	[24]

Distances	r_e^{se} [Å] [a,b]
P–S	1.871(1)
P–F	1.532(1) [1]
P–N	1.626(1) [1]
N–C	1.216(1) [2]
C–O	1.161(1) [2]

Bond angles	θ_e^{se} [deg] [a]
S–P–F	116.4(1)
S–P–N	119.2(3)
F–P–F	98.3(2)
F–P–N	101.6(2)
P–N–C	136.0(4)
N–C–O	174.0(4)

Dihedral angles	τ_e^{se} [deg] [a]
F–P–N–C	± 129.5(1)
S–P–N–C	0.0 [c]

[a] Parenthesized uncertainties in units of the last significant digit are 3σ

[b] Parameters denoted by the same numerical superscript were refined in one group. Difference between parameter values in each group was assumed at the values of M06-2X/cc-pVTZ structure

[c] According to symmetry

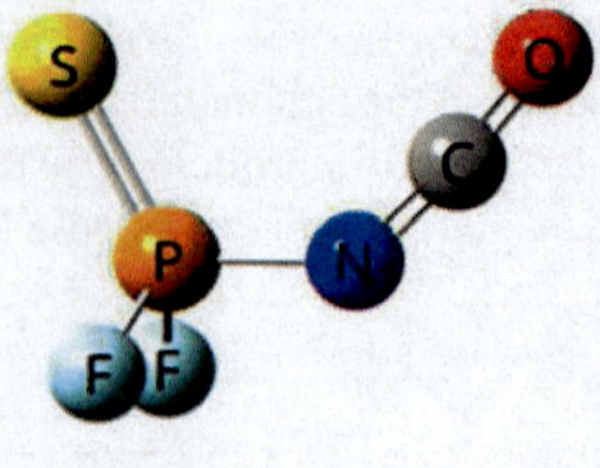

According to prediction of MP2/cc-pVTZ calculations, the title molecules exist as the *syn* and *anti* conformers characterized by different values of the S=P–N=C torsional angle, the *syn* conformer is being more stable. The barrier to internal rotation around the C–N bond was predicted to be about 9 kJ mol^{-1}.

Only one conformer, *syn*, was found in the gas phase by GED.

Vibrational corrections to the experimental internuclear distances at the temperature of experiment (T_{nozzle}≈296 K) were calculated with quadratic and cubic force constants from M06-2X/cc-pVTZ computation.

No	**5**
SF	**CHI_3**
CN	**Triiodomethane** Iodoform
Symmetry	**C_{3v}**
Methods	**GED** supplemented by **ab initio**
Ref	[25]

Distances	r_e^{se} [Å] [a]
C–I	2.123(3)
C–H	1.081 [b]

Bond angles	θ_e^{se} [deg] [a,b]
I–C–I	112.4(4)
H–C–I	106.3(5)

The table data are reprinted from: [25], with permission of AIP Publishing

[a] Parenthesized uncertainties in units of the last significant digit are 1σ

[b] Assumed at the value of CCSD(T)_ae/CBS-PP structure

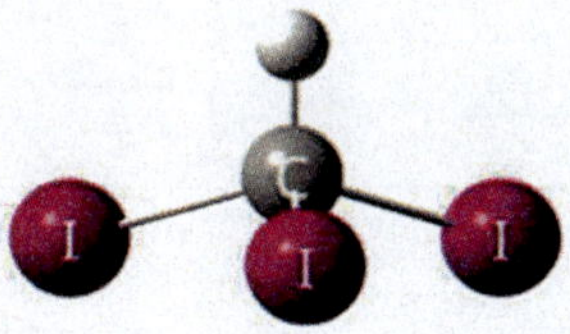

Vibrational corrections to the experimental internuclear distances at the temperature of low-pressure experiment of 289 K were calculated with quadratic and cubic force constants from MP2/SDB-cc-pVTZ computations.

No	**6a**
SF	**CHO**
CN	**Methylidyneoxonium** Oxomethylium
Symmetry	$C_{\infty v}$
Methods	**MW** supplemented by **ab initio**
Ref	[26]

Distances	r_e^{se} [Å] [a,b]	r_e^{se} [Å] [a,c]
H–C	1.091981(7)	1.09180(3)
C–O	1.105615(2)	1.105874(9)

[a] Parenthesized uncertainties given in units of the last significant digit are 1σ
[b] Derived using atomic masses
[c] Derived using nuclear masses

Experimental ground-state rotational constants of the HCO^+ ion and of the singly ($HC^{17}O^+$, $HC^{18}O^+$, $H^{13}CO^+$ and DCO^+) and doubly ($H^{13}C^{18}O^+$, $DC^{18}O^+$ and $D^{13}CO^+$) substituted isotopologues were taken from literature.

Rovibrational corrections, ΔB_{vib}, to experimental ground-state rotational constants were calculated with quadratic and cubic force constants from CCSD(T)/cc-pVQZ computations.

No	**6b**
Symmetry	$C_{\infty v}$
Methods	**CEI** supplemented by **ab initio**
Ref	[27]

Distances	r_e^{se} [Å] [a]
H–C	1.09780(35)
C–O	1.107035(61)

[a] Parenthesized uncertainties given in units of the last significant digit are 1σ

Potential energy surface was refined in a simultaneous fitting of experimentally derived and ab initio (CCSD(T)_ae/aug-cc-pCVQZ) data.

No	**7**
SF	**CHO**
CN	**Oxoniomethylene** Hydroxymethyliumylidene
Symmetry	$C_{\infty v}$
Methods	**MW** supplemented by **ab initio**
Ref	[26]

Distances	r_e^{se} [Å] [a,b]	r_e^{se} [Å] [a,c]
H–C	0.990482(7)	0.99041(3)
C–O	1.154468(2)	1.154692(7)

[a] Parenthesized uncertainties given in units of the last significant digit are 1σ

[b] Derived using atomic masses

[c] Derived using nuclear masses

Experimental ground-state rotational constants of the HO^+C ion and for all singly substituted isotopic species ($H^{18}O^+C$, $HO^{+13}C$, and DO^+C) were taken from literature.

Rovibrational corrections, ΔB_{vib}, to experimental ground-state rotational constants were calculated with quadratic and cubic force constants from CCSD(T)/cc-pVQZ computations.

No	**8**
SF	CH_2BrNO_2
CN	**Bromonitromethane**
Symmetry	C_s
Methods	**GED** supplemented by **QC**
Ref	[28]

Distances	r_e^{se} [Å] [a]
C–H	1.082 [b]
N–O	1.217(5)
C–N	1.484(18)
C–Br	1.919(5)

Bond angles	θ_e^{se} [deg] [c]
N–C–Br	109.6(9)
O–N–O	125.9(9)
C–N–O	117.0(5)
N–C–H	107.7(3)
H–C–H	113.9 [b]

Dihedral angle	τ_e [deg]
χ [d]	0.4 [b]

[a] Parenthesized uncertainties in units of the last significant digit include 2.5σ and systematic errors estimated as 0.2% of r_e^{se}

[b] Assumed at the values from MP2/cc-pVTZ calculations

[c] Parenthesized uncertainties in units of the last significant digit are 3σ

[d] Angle between the C–N bond and the ONO plane

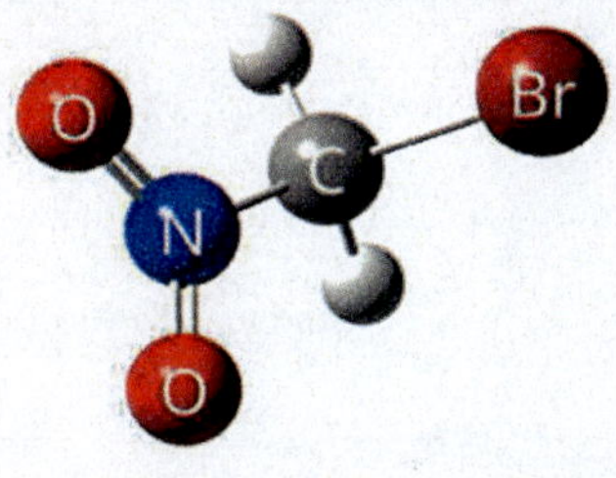

Structural analysis was carried out using the model of pseudo-conformers describing the large-amplitude motion of the nitro group. It was found that the equilibrium configuration with the C–Br and N=O bonds in orthogonal position with respect to each other is preferable, and the barrier height to internal rotation of the NO_2 group does not exceed 280 cm^{-1}.

Vibrational corrections to experimental internuclear distances at the temperature of experiment (T_{nozzle}≈308 K) were calculated using scaled quadratic and cubic force fields from B3LYP/cc-pVTZ computations.

No	**9**
SF	**CH_2FNO_3**
CN	**Nitric acid fluoromethyl ester** Fluoromethyl nitrate
Symmetry	**C_1**
Methods	**GED** supplemented by **QC**
Ref	[29]

Distances	r_e^{se} [Å] [a]
C–O2	1.385(3)
O2–N	1.454(2)
N–O5	1.190(2) [b]
N–O4	1.185(1) [b]
C–F	1.336(2)
C–H	1.083 [c]

Bond angles	θ_e^{se} [deg] [a]
C–O2–N	115.3(2)
O2–N–O5	115.1(3)
O2–N–O4	111.9(11)
O4–N–O5	133.0(13)

Dihedral angle	τ_e^{se} [deg] [a]
F–C–O2–N	74.7(8)

The table data are reprinted from: [29], Copyright 2019, with permission from Wiley–VCH Verlag GmbH & Co. KGaA, Weinheim

[a] Parenthesized uncertainties in units of the last significant digit are total errors estimated by Monte Carlo simulations

[b] Difference between the N = O4 and N = O5 bond lengths was assumed at the value from MP2_ae/cc-pwCVTZ calculation

[c] Average value

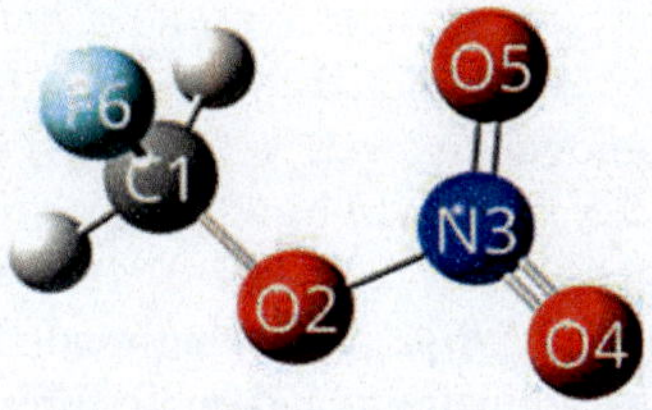

According to prediction of quantum-chemical calculations, the title molecule exists as a single conformer (*gauche*) at the temperature of the GED experiment (T_{nozzle} = 298 K).

Vibrational corrections to experimental internuclear distances were calculated with TPSSh/def2-TZVP quadratic and cubic force constants.

No	**10**
SF	**CH_2OS**
CN	**Methanethione *S*-oxide** Sulfine
Symmetry	**C_s**
Methods	**MW** supplemented by **QC**
Ref	[30]

Distances	r_e^{se} [Å] [a]
C–S	1.60838(6)
S–O	1.46504(5)
C–H4	1.0797(2)
C–H5	1.0817(4)

Bond angles	θ_e^{se} [deg] [a]
C–S–O	114.801(1)
S–C–H4	123.047(8)
S–C–H5	115.19(4)

The table data are reprinted from: [30], Copyright 2020, with permission from Elsevier

[a] Parenthesized uncertainties in units of the last significant digit are 1σ

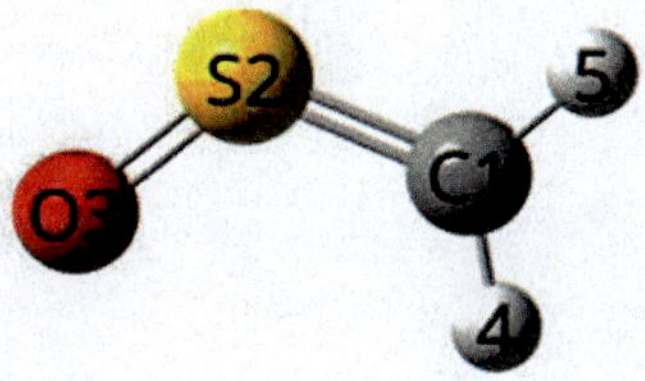

MW spectra and ground-state rotational constants for the parent species and all of its singly substituted isotopologues as well as for the D_2 and $D_2{}^{34}S$ isotopic species were taken from literature. Experimental rotational constants for the parent species were refitted.

Rovibrational corrections, ΔB_{vib}, to experimental ground-state rotational constants were calculated with quadratic and cubic force constants from B2PLYP-D3/aug-cc-pV(T + d)Z computations.

The composite structure of CCSD(T)_ae/cc-pwCVQZ quality and the r_e^{se} structure were found to be in excellent agreement.

No	**11**
SF	**CH_2S**
CN	**Methanethione** Thioformaldehyde
Symmetry	**C_{2v}**
Methods	**MW** supplemented by **QC**
Ref	[31]

Distances	r_e^{se} [Å] [a,b]	r_e^{se} [Å] [a,c]	r_e^{se} [Å] [a,d]
C–S	1.60975(2)	1.60934(6)	1.60909(1)
C–H	1.08526(6)	1.08556(15)	1.08531(2)

Bond angle	θ_e^{se} [deg] [a,b]	θ_e^{se} [deg] [a,c]	θ_e^{se} [deg] [a,d]
H–C–S	121.706(5)	121.759(13)	121.758(2)

The table data are reproduced from: [31], with permission © 2019 ESO

[a] Parenthesized uncertainties in units of the last significant digit are 1σ

[b] Vibrational corrections to experimental rotational constants were calculated with B3LYP/aug-cc-pwCVQZ quadratic and cubic force field

[c] Vibrational corrections to experimental rotational constants were calculated with MP2/aug-cc-pwCVQZ quadratic and cubic force fields

[d] Vibrational corrections to experimental rotational constants were calculated with CCSD(T)_ae/cc-pwCVQZ quadratic and cubic force fields

Rotational spectra were measured by a fast scan submillimeter-wave spectroscopic technique and a solid-state-based spectrometer system. Ground-state rotational constants were determined for the parent, five singly substituted (^{33}S, ^{34}S, ^{36}S, ^{13}C, and D) and two doubly substituted ($^{13}C^{34}S$ and D_2) isotopic species.

The r_e^{se} structure was found to be in very good agreement with the structure calculated at the CCSD(T)_ae/cc-pwCVQZ level of theory.

No	**12**
SF	**CH_3NO_3**
CN	**Nitric acid methyl ester** Methyl nitrate
Symmetry	**C_s**
Methods	**GED** supplemented by **QC** and **MW**
Ref	[29]

Distances	r_e^{se} [Å] [a]
C–O2	1.425(3)
O2–N	1.403(2)
N–O5	1.205(1) [b]
N–O4	1.198(1) [b]
C–H	1.080(5) [c]

Bond angles	θ_e^{se} [deg] [a]
C–O2–N	113.6(3)
O2–N–O5	116.3(3)
O2–N–O4	112.3(2)
O4–N–O5	131.4(4)

[a] Parenthesized uncertainties in units of the last significant digit are total errors estimated by Monte Carlo simulations

[b] Difference between the N=O4 and N=O5 bond lengths was assumed at the value from MP2/cc-pVTZ calculation

[c] Average value

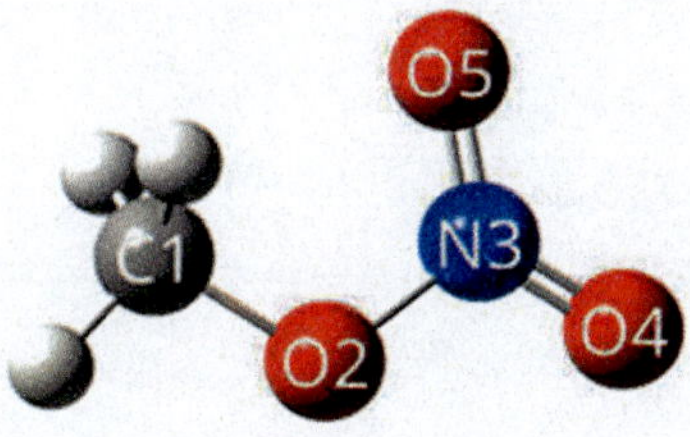

Vibrational corrections to experimental internuclear distances at the temperature of experiment (T_{nozzle}≈298 K) and to experimental ground-state rotational constants taken from literature, $\Delta B = B_0 - B_e$, were calculated with TPSSh/def2-TZVP quadratic and cubic force constants.

No	**13**
SF	$\mathbf{CI_4}$
CN	**Tetraiodomethane** Carbon tetraiodide
Symmetry	$\mathbf{T_d}$
Methods	**GED** supplemented by **ab initio**
Ref	[25]

Distance	r_e^{se} [Å] [a]
C–I	2.133(7)

The table data are reprinted from: [25], with permission of AIP Publishing
[a] Parenthesized uncertainty in units of the last significant digit is 1σ

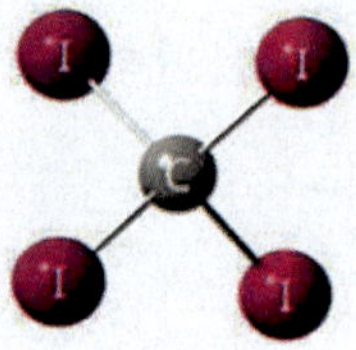

Vibrational corrections to the experimental internuclear distances at the temperature of low-pressure experiment of 290 K were calculated with quadratic and cubic force constants from MP2/SDB-cc-pVTZ computation.

No	**14**
SF	$\mathbf{CN_2O_2}$
CN	**Carbon monoxide – dinitrogen monoxide (1/1)**
Symmetry	$\mathbf{C_s}$
Methods	**MW** supplemented by **QC**
Ref	[32]

Distances	r_e^{se} [Å] [a]
N1–N2	1.1260(16)

(continued)

(continued)

N2–O3	1.1856(12)
C5…O3	3.291(5)
C5–O4	1.1267(27)
R_{cm} [b]	3.811

Angles	θ_e^{se} [deg] [a]
C5…O3–N2	74.6(2)
O4–C5…O3	147.0(3)
θ_1 [c]	81.9
θ_2 [d]	9.4
θ_3 [e]	107.6

The table data are reprinted from: [32], with permission of AIP Publishing

[a] Parenthesized uncertainties in units of the last significant digit are 1σ

[b] Distance between the centers of masses of CO and N_2O monomers (cm1 and cm2, respectively)

[c] $\theta_1 = \angle$(cm1…cm2–O3)

[d] $\theta_2 = \angle$(cm2…cm1–C5)

[e] θ_3 is angle between the symmetry axes of the monomers

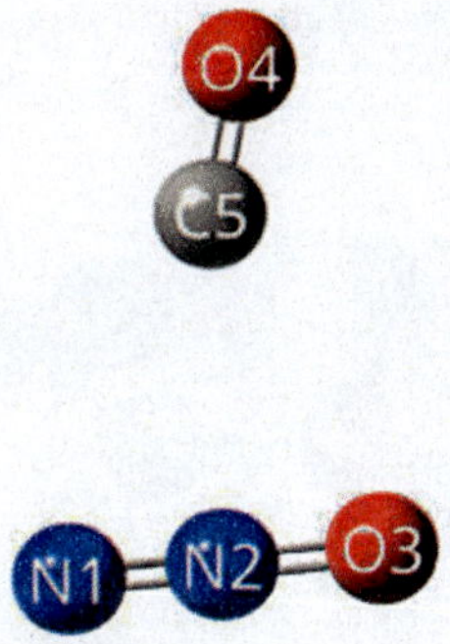

Microwave spectra of the parent species of the van der Waals dimer $N_2O \cdot CO$ and two of its isotopic species ($^{15}N_2O \cdot {}^{12}CO$ and $^{15}N_2O \cdot {}^{13}CO$) were measured, expanding previous measurements.

Rovibrational corrections, ΔB_{vib}, to experimental ground-state rotational constants were calculated with quadratic and cubic force constants from CP-B2PLYP-D3/6–311+G(3df) computations.

The semiexperimental and CCSD(T)_ae/cc-pwCVQZ equilibrium structures were found to be in fair agreement.

No	**15**
SF	$\mathbf{C_2Cl_3F_3}$
CN	**1,1,1-Trichloro-2,2,2-trifluoroethane**
Symmetry	$\mathbf{C_{3v}}$
Methods	**MW** supplemented by **ab initio**
Ref	[33]

Distances	r_e^{se} [Å] [a]
C–Cl	1.7586(20)
C–C	1.5493(12) [b]
C–F	1.3251(21)
Cl…Cl	2.8855(5)

Bond angles	θ_e^{se} [deg] [a]
C–C–Cl	108.68(21)
C–C–F	110.19(31)

The table data are reprinted from: [33], Copyright 2018, with permission from Elsevier Inc
[a] Parenthesized uncertainties given in units of the last significant digit are 1σ
[b] Estimated by comparison of experimental and computed results for related molecules

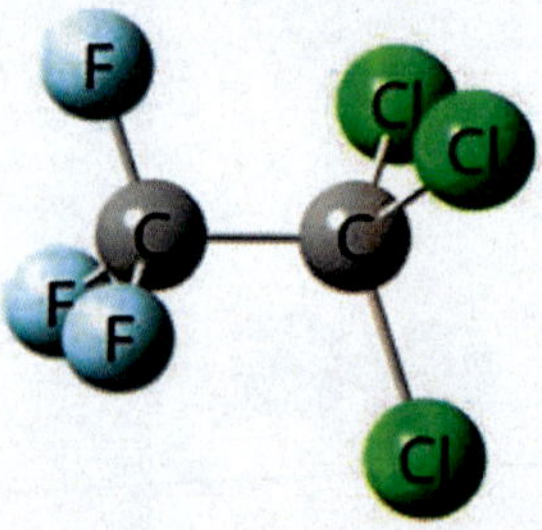

Rotational spectra were studied in supersonic expansion with chirped pulse and cavity FT microwave spectrometers and with a broadband millimeter-wave spectrometer.

Ground-state rotational constants were determined for the parent species and three possible ^{37}Cl isotopologues (singly, doubly, and triply substituted).

Rovibrational corrections, ΔB_{vib}, to experimental ground-state rotational constants were calculated with quadratic and cubic force constants from MP2/aug-cc-pVDZ computations.

No	**16**
SF	**C_2H_2ClF**
CN	**1-Chloro-1-fluoroethene**
Symmetry	**C_s**
Methods	**MW** supplemented by **QC**
Ref	[34]

Distances	r_e^{se} [Å] [a,b]	r_e^{se} [Å] [a,c]	r_e^{se} [Å] [a,d]
C–F	1.3287(52)	1.3294(8)	1.3282(1)
C–C	1.3233(37)	1.3220(6)	1.3224(1)
C–H [e]	1.0780(13)	1.0780(2)	1.0777(4)
C–Cl	1.7081(31)	1.7089(5)	1.7099(1)
C–H [f]	1.0750(16)	1.0765(2)	1.0762(5)

Bond angles	θ_e^{se} [deg] [a,b]	θ_e^{se} [deg] [a,c]	θ_e^{se} [deg] [a,d]
C–C–F	125.44(30)	122.48(4)	122.56(10)
C–C–Cl	125.52(32)	125.61(5)	125.54(1)
C–C–H [f]	120.02(23)	119.85(3)	119.90(8)
C–C–H [e]	119.30(20)	119.33(3)	119.30(7)

The table data are reproduced from: [34], with permission from the PCCP Owner Societies, copyright 2019

[a] Parenthesized uncertainties in units of the last significant digit are 1σ

[b] Rovibrational corrections, ΔB_{vib}, to experimental ground-state rotational constants were calculated with quadratic and cubic force constants from CCSD(T)_ae computations (with cc-pCVTZ basis set for C and Cl and aug-cc-pCVTZ basis set for F)

[c] Rovibrational corrections were calculated with B2PLYP/cc-pVTZ quadratic and cubic force fields

[d] Rovibrational corrections were calculated with B3LYP/SNSD quadratic and cubic force fields

[e] *Trans* to the Cl atom

[f] *Cis* to the Cl atom

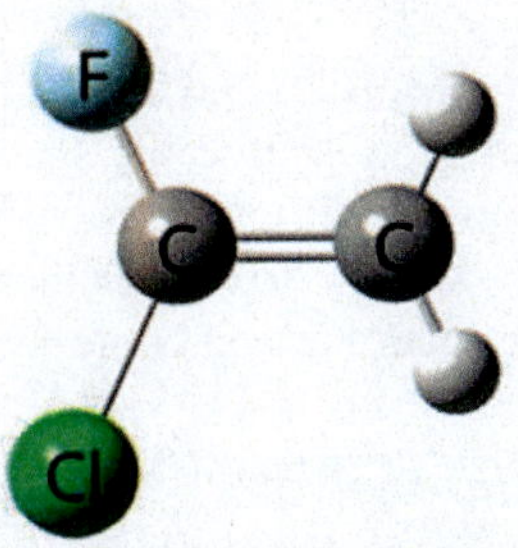

Experimental ground-state rotational constants for the parent species and five singly substituted (^{37}Cl, two ^{13}C, and two D) and two doubly substituted (^{13}C ^{37}Cl) isotopic species were taken from literature.

No	**17**
SF	C_2H_2S
CN	**Ethynethiol**
Symmetry	C_s
Methods	**MW** supplemented by **ab initio**
Ref	[35]

Distances	r_e^{se} [Å] [a]
C–H	1.0627(1)
C–C	1.2082(3)
C–S	1.6892(2)
S–H	1.3403(4)

Bond angles	θ_e^{se} [deg] [a]
C–C–S [b]	174.93(18)
C–S–H	96.04(5)

[a] Parenthesized uncertainties in units of the last significant digit are 1σ

[b] Bent away from the H(–S) atom

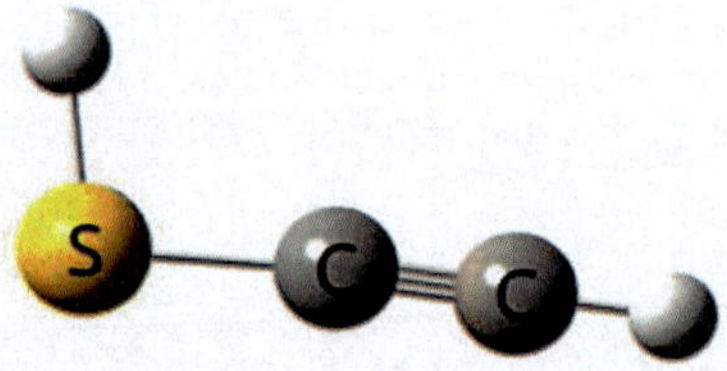

Rotational spectra of the parent species and all singly substituted (^{13}C, D and ^{34}S) as well as one doubly substituted ^{13}CD isotopic species were measured using FT microwave and direct absorption submillimeter-wave spectrometers.

Rovibrational corrections, ΔB_{vib}, to experimental ground-state rotational constants were calculated with quadratic and cubic force constants from CCSD(T)/ANO0 computations.

The r_e^{se} and calculated CCSD(T)_ae/cc-pCVQZ structures were found to be in very good agreement.

No	**18**
SF	**C_2H_3P**
CN	**Ethylidynephosphine** 1-Phosphapropyne
Symmetry	**C_{3v}**
Methods	**MW** supplemented by **QC**
Ref	[36]

Distances	r_e^{se} [Å] [a]
C–C	1.461951(44)
C–H	1.090214(33)
C–P	1.545609(34)

Bond angle	θ_e^{se} [deg] [a]
C–C–H	110.5613(14)

[a] Parenthesized uncertainties given in units of the last significant digit are 1σ

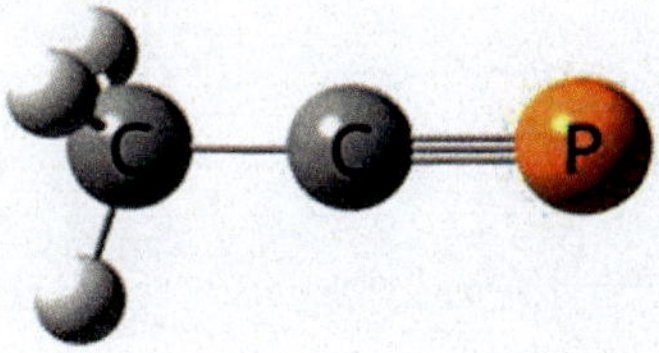

Rotational spectra of the D_3 isotopic species in the ground state and two excited low-lying vibrational states, ν_8 (CCP bending) and $2\nu_8$, were recorded in the millimeter-wave region. Rotational spectra of the parent species in the excited vibrational states ($\nu_8 = 1,2$ and $\nu_4 = 1$) were also recorded.

Rovibrational corrections, ΔB_{vib}, to experimental ground-state rotational constants were calculated with quadratic and cubic force constants from MP2/cc-pVTZ computations.

No	**19**
SF	**$C_2H_6O_2S_2$**
CN	**1,1'-Thiobismethane – sulfur dioxide (1/1)** Dimethyl sulfide – sulfur dioxide (1/1)
Symmetry	**C_s** (effective symmetry)

(continued)

(continued)

Methods	**MW** supplemented by **QC**
Ref	[37]

Distances	r_e^{se} [Å] [a]
S1…S10	2.947(3)
O–S10	1.446(6)
C2–S1	1.790(5)
H4–C2	1.0883 [b]
H5–C2	1.0864 [b]
H6–C2	1.0870 [b]

Bond angles	θ_e^{se} [deg] [a]
O–S10…S1	95.0(2)
C2–S1…S10	91.7(2)
H4–C2–S1	110.67 [b]
H5–C2–S1	107.14 [b]
H6–C2–S1	109.97 [b]

Dihedral angles	τ_e^{se} [deg] [a]
O11–S10…S1…O12	−118.2(9)
C2–S1…S10–O12	9.7(7)
H4–C2–S1…S10	26.06 [b]
H5–C2–S1…S10	−92.56 [b]
H6–C2–S1…S10	148.63 [b]

[a] Parenthesized uncertainties in units of the last significant digit are 1σ

[b] Fixed to the value of B2PLYP-D3/maug-cc-pVTZ-*d*H structure

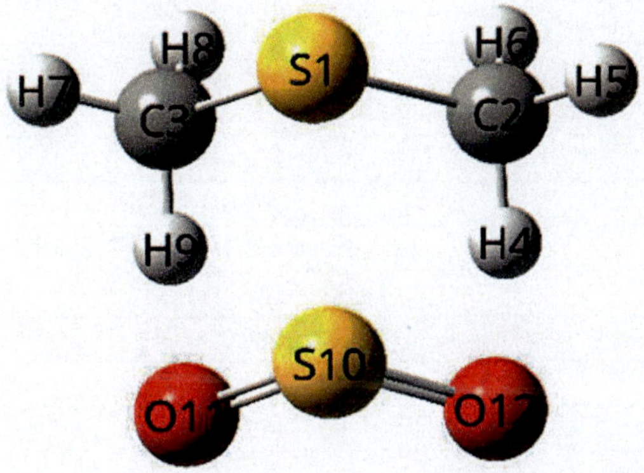

Rotational spectra were measured for the parent and six singly substituted isotopic (^{16}O, ^{13}C, two ^{34}S, and two ^{33}S) species by a MB FTMW spectrometer.

Rovibrational corrections, ΔB_{vib}, to experimental ground-state rotational constants were calculated with quadratic and cubic force constants from computations at the level of theory as indicated above.

No	**20**
SF	**$C_2H_{11}B_{10}Br$**
CN	**9-Bromo-1,2-dicarbadodecaborane(12)**
Symmetry	**C_s**
Methods	**GED** supplemented by **QC**
Ref	[38]

Distances	r_e^{se} [Å] [a,b]
C–C	1.619(2)
C–B	1.701(2)
B–B	1.775(2)
B–Br	1.936(1)
C–H	1.084(2)
B–H	1.185(2)

Bond angles	θ_e^{se} [deg] [a,b]
C–C–B (narrow)	61.9(1)
C–C–B (wide)	111.6(1)
C–B–C	56.3(1)
C–B–B (narrow)	58.5(1)

(continued)

(continued)

C–B–B (wide)	103.9(1)
B–C–B (narrow)	63.0(1)
B–C–B (wide)	116.0(1)
B–B–B (narrow)	60.0(1)
B–B–B (wide)	108.1(1)
B–B–Br	121.7(1)
C–C–H	116.0(2)
B–B–H	123.4(1)
C–B–H	117.8(1)
B–C–H	118.4(1)

The table data are reprinted from: [38], Copyright 2020, with permission from Wiley–VCH Verlag GmbH

[a] Parenthesized uncertainties in units of the last significant digit are 1σ

[b] Distances and bond angles (refined in Cartesian coordinates) were constrained to those of the MP2/SDB-cc-pVTZ structure. Average values are given, except for unique parameters

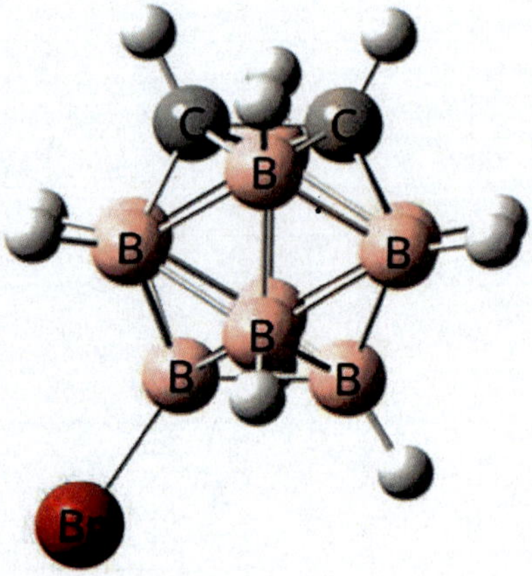

Vibrational corrections to the experimental internuclear distances at the temperature of experiment (T_{nozzle}≈443 K) were calculated with quadratic and cubic force constants from PBE0/SDB-cc-pVTZ computation.

No	**21**
SF	$\mathbf{C_2H_{12}B_{10}S_2}$
CN	**1,2-Dicarbadodecaborane(12)-1,2-dithiol**
Symmetry	$\mathbf{C_2}$ **(I)** $\mathbf{C_s}$ **(II)**

(continued)

(continued)

Methods	**GED** supplemented by **QC**
Ref	[39]

Distances	r_e^{se} [Å] [a,b]
B–H	1.186(5)
S–H	1.341(5)
C–B	1.702(4)
C…C	1.755(7)
B–B	1.777(5)
C–S	1.755(4)

Bond angles	θ_e^{se} [deg] [a,b]
C…C–B (narrow)	59.2(2)
C…C–B (wide)	109.5(2)[c]
C–B–C	61.6(2)
C–B–B (narrow)	58.4(2)
C–B–B (wide)	105.7(2)[c]
B–C–B (narrow)	63.3(2)
B–C–B (wide)	113.4(2)
B–B–B (narrow)	60.0(2)
B–B–B (wide)	107.9(2)
S–C…C	118.0(1)
S–C–B	119.6(2)
C–S–H	95.9(2)
C–B–H	116.7(3)
B–B–H	123.1(3)

Dihedral angles	τ_e^{se} [deg] [a]
C…C–S–H	−89.9(4) [c]
C…C–S–H	± 95.2(2) [d]

[a] Parenthesized uncertainties in units of the last significant digit are standard deviations

[b] Average values for parameters of the same type in both I and II conformers

[c] C_2 conformer

[d] C_s conformer

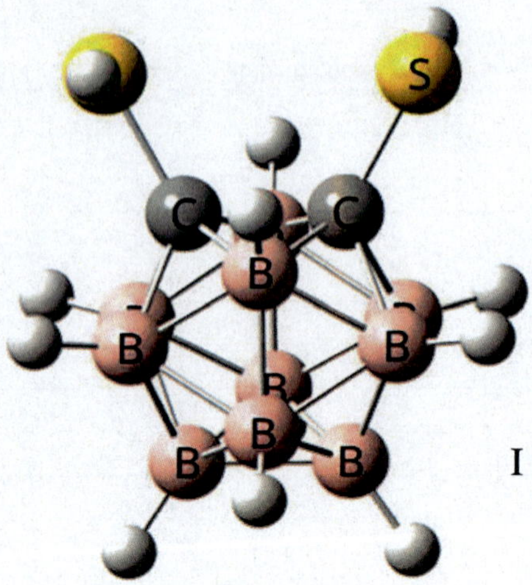

I

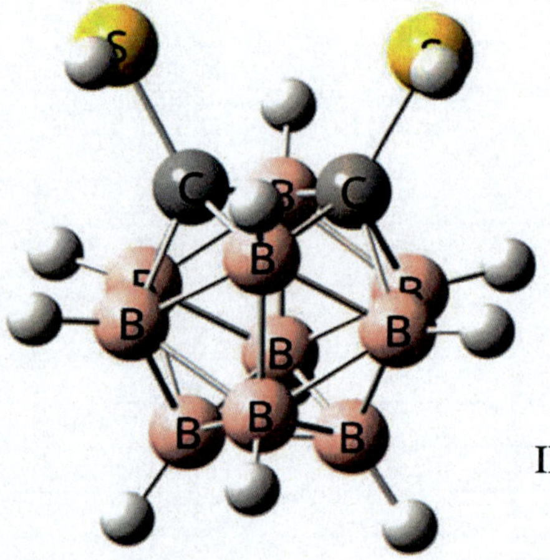

II

According to results of ab initio calculations, the title compound exist as a mixture of two conformers, I and II, with C_2 and C_s symmetry, respectively, differing by magnitudes of the C…C–S–H torsional angles. A ratio of conformers was predicted to be I : II = 55 : 45 (MP2/cc-pVTZ, for the temperature of experiment T_{nozzle} = 410 K) and fixed in the GED analysis.

Vibrational corrections to the experimental internuclear distances were calculated with quadratic and cubic force constants from B3LYP/cc-pVTZ computation.

No	**22**
SF	$C_2H_{12}B_{10}Se_2$
CN	**1,2-Dicarbadodecaborane(12)-1,2-diselenol**
Symmetry	C_2 (I) C_s (II)
Methods	**GED** supplemented by **QC**
Ref	[39]

Distances	r_e^{se} [Å] [a,b]
B–H	1.188(4)
Se–H	1.457(4)
C–B	1.699(3)
C…C	1.726(5)
B–B	1.775(3)
C–Se	1.904(3)

Angles	θ_e^{se} [deg] [a,b]
C…C–B (narrow)	59.7(1)
C…C–B (wide)	109.9(1)
C–B–C	60.6(2)
C–B–B (narrow)	58.4(1)
C–B–B (wide)	105.3(2)
B–C–B (narrow)	63.3(1)
B–C–B (wide)	114.0(2)
B–B–B (narrow)	60.0(1)
B–B–B (wide)	107.9(2)
Se–C…C	119.3(1)
Se–C–B	118.9(2)
C–Se–H	94.0(2)
C–B–H	117.1(2)
B–B–H	123.1(2)

Dihedral angles	τ_e^{se} [deg] [a]
C…C–Se–H	−84.6(3) [c]
C…C–Se–H	± 92.7(1) [d]

[a] Parenthesized uncertainties in units of the last significant digit are standard deviations

[b] Average values for parameters of the same type in both I and II conformers

[c] C_2 conformer

[d] C_s conformer

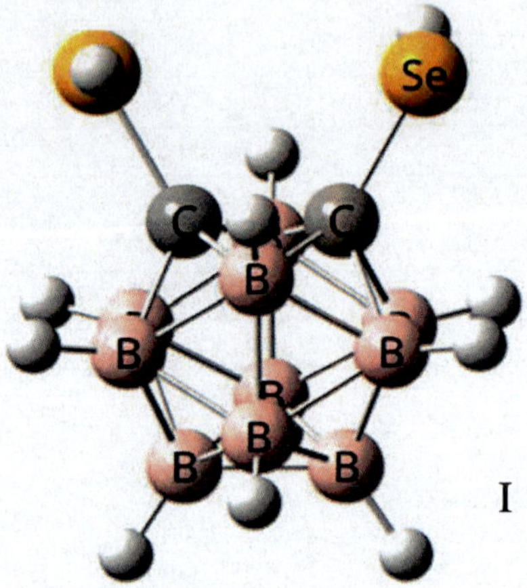

I

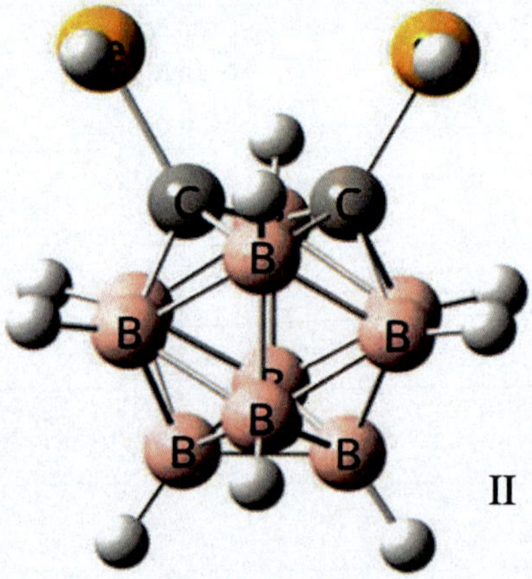

II

According to results of ab initio calculations, the title compound exists as a mixture of two conformers, I and II, with C_2 and C_s symmetry, respectively, differing by magnitudes of the C…C–Se–H torsional angles. A ratio of conformers was predicted to be I : II = 42 : 58 (MP2/cc-pVTZ, for the temperature of experiment T_{nozzle} = 407 K) and fixed in the GED analysis.

Vibrational corrections to the experimental internuclear distances were calculated with quadratic and cubic force constants from B3LYP/cc-pVTZ computation.

No	**23**
SF	$C_2N_4O_5$
CN	**3,4-Dinitrofurazan**
Symmetry	C_2
Methods	**GED** supplemented by **QC**
Ref	[40]

Distances	r_e^{se} [Å] [a,b]
C3–N6	1.449(17) [1]
C3–C4	1.416(16) [2]
C4–N7	1.449(17) [1]
C3–N2	1.300(16) [2]
O1–N2	1.367(16) [2]
O1–N5	1.367(16) [2]
O8–N6	1.220(5) [3]
O10–N6	1.214(5) [3]

Bond angles	θ_e^{se} [deg] [a,b]
C4–C3–N6	128.7(11)
N2–C3–C4	109.1(8) [4]
O1–C2–N3	104.9(8) [4]
N2–O1–N5	112.1(6)
O8–N6–C3	114.5(19) [5]
O10–N6–C3	117.2(19) [5]

Dihedral angles	θ_e [deg]
N6–C3–C4–N7	8.7 [c]
N2–C3–C4–N7	−176.1 [c]
O1–N2–C3–C4	0.6 [c]
N5–O1–N2–C3	−0.3 [c]
O8–N6–C3–C4	37.0 [d]
O10–N6–O8…C3	179.2 [c]

The table data are reprinted from: [40], Copyright 2021, with permission from Elsevier

[a] Parenthesized uncertainties in units of the last significant digit are 3σ

[b] Parameters denoted by the same numerical superscript were refined in one group. Differences between parameter values in each group were constrained to the calculated values of CCSD(T)_ae/aug-cc-pwCVQZ quality

[c] Assumed at the value calculated as indicated above

[d] Constrained to the value calculated as indicated above

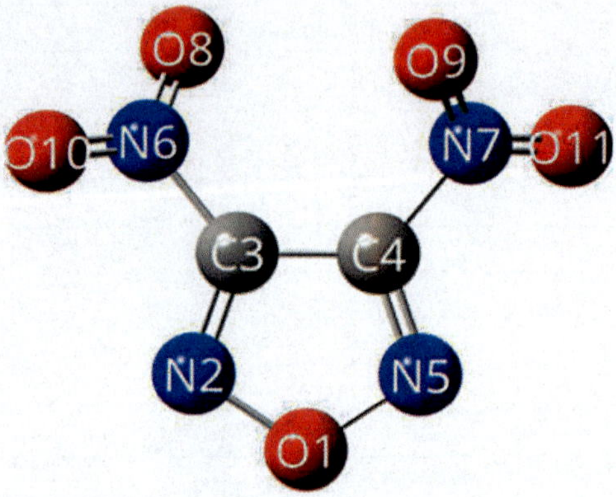

Large-amplitude motions of the nitro groups around the C–N bonds were described by a model of pseudo-conformers.

Vibrational corrections to the experimental internuclear distances (T_{nozzle} = 306 K) were calculated with quadratic and cubic force constants from B3LYP/SNST computations.

The r_e^{se} and computed CCSD(T) structures were found to be in excellent agreement.

No	**24**
SF	C_3
CN	**1,2-Propadiene-1,3-diylidene** Tricarbon
Symmetry	$D_{\infty h}$
Method	**MW**
Ref	[41]

Distance	r_m [Å] [a]
C–C	1.29397(10)

[a] Parenthesized uncertainty in units of the last significant digit was not specified, probably 1σ

Rotational constants of the parent species and all of its ^{13}C substituted isotopologues were taken from literature.

Improved bond length determination technique for linear molecules with a large-amplitude bending vibration (Bent-Structure Model) was suggested.

The determined r_m structure was found to be in excellent agreement with the ab initio value of $r_e = 1.29397(10)$ Å published previously.

No	**25**
SF	**C_3HNS**
CN	**Isothiocyanatoethyne** Ethynyl isothiocyanate
Symmetry	**$C_{\infty v}$**
Methods	**MW** supplemented by **ab initio**
Ref	[42]

Distances	r_e^{se} [Å] [a]
H–C1	1.0621 [b]
C1–C2	1.207(1)
C2–N	1.304(3)
N–C4	1.192(5)
C4–S	1.573(3)

The table data are reprinted from: [42], with permission of AIP Publishing

[a] Parenthesized uncertainties given in units of the last significant digit were not specified, probably 1σ

[b] Fixed to the value from CCSD(T)_ae/cc-pwCVTZ computation

According to predictions of CCSD(T) calculations with different basis sets, the title molecule is linear (cc-pV(T+d)Z and cc-pwCVTZ) with extremely flat bending C=N=C potential or quasi-linear (aug-cc-pwCVTZ).

Rotational spectra of the parent species and all of its singly substituted ^{13}C and ^{34}S isotopologues were recorded using chirped pulse and Balle–Flygare FTMW spectrometers.

Rovibrational corrections, ΔB_{vib}, to experimental ground-state rotational constants were calculated with quadratic and cubic force constants from CCSD(T)_ae/cc-pwCVTZ computations.

The r_e^{se} and calculated CCSD(T)_ae/cc-pwCVTZ structures were found to be in good agreement.

No	**26**
SF	$C_3H_2F_4$
CN	**(*E*)-1,3,3,3-Tetrafluoro-1-propene**
Symmetry	C_s
Methods	**GED** supplemented by **QC**
Ref	[43]

Distances	r_e^{se} [Å] [a,b]
C1–C2	1.321(1) [1]
C1–F	1.326(1) [1]
C3–F4	1.336(1) [1]
C3–F5	1.337(1) [1]
C2–C3	1.480(4)
C1–H	1.102(17) [2]
C2–H	1.101(17) [2]

Bond angles	θ_e^{se} [deg] [a,b]
C1–C2–C3	120.8(7)
C1–C2–H	121.6 [c]
F–C1–H	116.1(7)
C3–C2–H	117.6(7)
F4–C3–C2	113.1(1) [3]
F5–C3–C2	111.7(1) [3]
F5–C3–F6	106.2(1)

Dihedral angle	τ_e [deg]
F6–C3–C2–C1	120.6 [c]

[a] Parenthesized uncertainties in units of the last significant digit are 3σ
[b] Parameters denoted by the same numerical superscript were refined in one group. Differences between parameter values in each group were assumed at the values of MP2/cc-pVTZ structure
[c] Assumed at the value calculated at the level of theory as indicated above

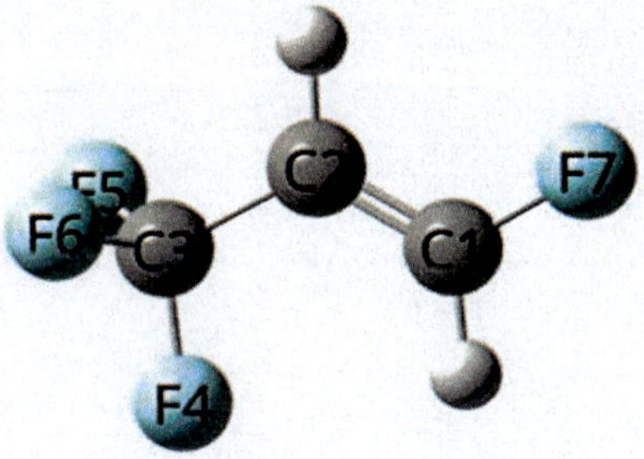

Vibrational corrections to the experimental internuclear distances at the temperature of experiment (T_{nozzle}≈293 K) were calculated with quadratic and cubic force constants from PBE0/cc-pVTZ computation. The barrier to internal rotation of the CF_3 group around the C2–C3 bond was predicted to be 8 kJ mol^{-1} at the same level of theory.

No	**27**
SF	$\mathbf{C_3H_2F_4}$
CN	**2,3,3,3-Tetrafluoro-1-propene**
Symmetry	$\mathbf{C_s}$
Methods	**GED** supplemented by **MW** and **QC**
Ref	[43]

Distances	r_e^{se} [Å] [a,b]
C1–C2	1.323(4)
C2–C3	1.498(2)
C2–F	1.333(1)
C3–F4	1.332(1)
C3–F5	1.334(1)
C1–H [c]	1.083(11)
C1–H [d]	1.085(11)

Bond angles	θ_e^{se} [deg] [a,b]
C1–C2–C3	125.8(2)
C1–C2–F	122.8(1)
H–C1–H	120.4 [e]
C3–C2–F	111.4(1)
F4–C3–C2	111.3(1)
F5–C3–C2	111.0(1)
F4–C3–F5	108.0(1)
F5–C3–F6	107.3(2)
C2–C1–H	119.4 [e]
C2–C1–H	120.2 [e]

Dihedral angle	τ_e^{se} [deg] [a]
F6–C3–C2–C1	120.4(1)

The table data are adapted from: [43], with permission from Wiley–VCH GmbH

[a] Parenthesized uncertainties in units of the last significant digit are 3σ

[b] Parameters denoted by the same numerical superscript were refined in one group. Differences between parameter values in each group were assumed at the values of MP2/cc-pVTZ structure

[c] H atom in the *cis* position with respect to the C2–C3 bond

[d] H atom in the *trans* position with respect to the C2–C3 bond

[e] Assumed at the value calculated at the level of theory as indicated above

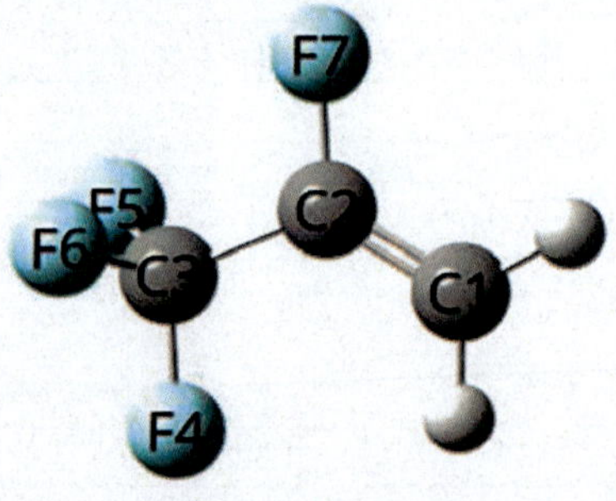

Vibrational corrections to the experimental internuclear distances at the temperature of experiment (T_{nozzle}≈293 K) and rovibrational corrections, ΔB_{vib}, to experimental ground-state rotational constants taken from literature were calculated with quadratic and cubic force constants from PBE0/cc-pVTZ computations.

The barrier to internal rotation of the CF_3 group around the C2–C3 bond was predicted to be 10 kJ mol^{-1} at the same level of theory.

No	**28**
SF	$\mathbf{C_3H_2O}$
CN	**2-Cyclopropen-1-one**
Symmetry	$\mathbf{C_{2v}}$
Methods	**MW** supplemented by **ab initio**
Ref	[44]

Distances	r_e^{se} [Å] [a]
C–O	1.2005(10)
C1–C2	1.4288(8)[b]
C2–C3	1.34480(4)
C–H	1.0784(2)

Bond angles	θ_e^{se} [deg] [a]
O–C–C	151.93(2)
C2–C1–C3	56.15(3) [b]
C1–C2–H	153.73(3)

The table data are reproduced from: [44], with permission © 2021 ESO

[a] Parenthesized uncertainties given in units of the last significant digit are 1σ

[b] Derived, *i.e.*, dependent parameter

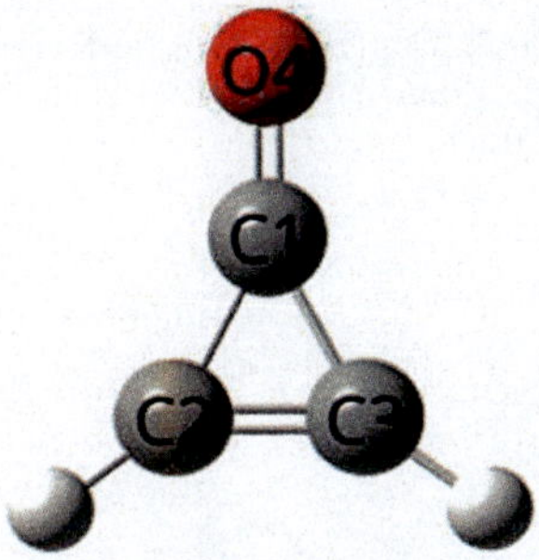

Rotational spectra for the parent species and all singly (two ^{13}C, ^{18}O, and D) and doubly substituted (D_2) isotopologues were recorded in the microwave and millimeter-wave regions.

Rovibrational corrections, ΔB_{vib}, to experimental ground-state rotational constants were calculated with quadratic and cubic force constants from MP2/cc-pVTZ computations.

No	**29**
SF	C_3H_3N
CN	**2-Propenenitrile** Vinyl cyanide
Symmetry	C_s
Methods	**MW** supplemented by **ab initio**
Ref	[45]

Distances	r_e^{se} [Å] [a]
C1–C2	1.3351
C2–C3	1.4315
C3–N4	1.1584
C1–H5	1.0797
C1–H6	1.0785
C2–H7	1.0800

Bond angles	θ_e^{se} [deg] [b]
C1–C2–C3	122.03
C2–C3–N4	179.14 [c]
C2–C1–H5	121.31
C2–C1–H6	120.48
C1–C2–H7	121.66

[a] Estimated uncertainties (1σ) are less than 0.0004 Å

[b] Estimated uncertainties (1σ) are less than 0.07°

[c] Toward H7

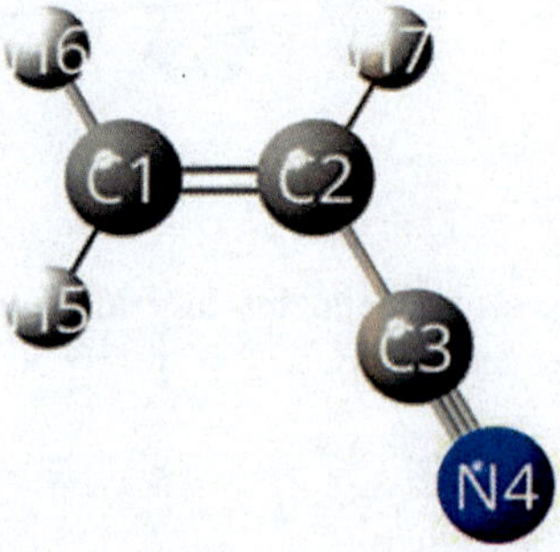

Experimental rotational constants for fourteen isotopologues were available from literature.

Some spectra were refitted in this study.

Rovibrational corrections, ΔB_{vib}, to experimental ground-state rotational constants were calculated with quadratic and cubic force constants from CCSD(T)/cc-pVTZ computations.

The r_e^{se} structure was found to be in very good agreement with the structure calculated at the CCSD(T)_ae/cc-pwCV5Z level of theory.

No	**30**
SF	**C_3H_3N**
CN	**(1Z)-2-Propyn-1-imine** (Z)-Propargylimine
Symmetry	**C_s**
Methods	**MW** supplemented by **DFT**
Ref	[46]

Distances	r_e^{se} [Å] [a]
N–C	1.272(3)
C3–C2	1.436(2)
C2–C1	1.207(3)
N–H	1.021(2)
C3–H	1.0840 [b]
C1–H	1.059(1)

Bond angles	θ_e^{se} [deg] [a]
H–N–C	109.99 [b]
N–C–H	118.03 [b]
H–C–C	116.3(4)
C–C–H	180.1 [c]

[a] Parenthesized uncertainties given in units of the last significant digit were not specified, probably 1σ

[b] Kept fixed at the value for similar molecules

[c] Toward to nitrogen atom

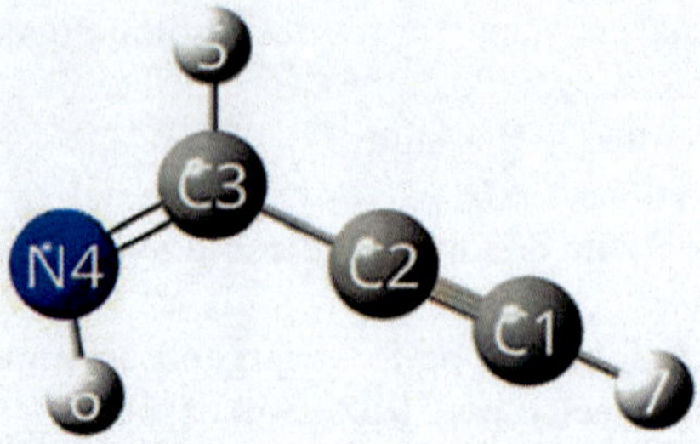

Rotational spectra for the parent species and for all of its singly substituted ^{13}C and two deuterated (D7 and D6) isotopologues were taken from literature.

Rovibrational corrections, ΔB_{vib}, to experimental ground-state rotational constants were calculated with quadratic and cubic force constants from B3LYP/jun-cc-pVDZ (partially augmented) computations.

No	**31**
SF	C_3H_3NS
CN	**Thiazole**
Symmetry	C_s
Methods	**MW** supplemented by **ab initio**
Ref	[47]

Distances	r_e^{se} [Å] [a]
C4–C5	1.36182(48)
S1–C5	1.70977(28)
C5–H	1.07679(29)
C4–H	1.07778(20)
N3–C4	1.37334(31)
S1–C2	1.71963(36)
C2–H	1.07909(41)
C2–N3	1.30062(90)

Bond angles	θ_e^{se} [deg] [a]
S1–C5–C4	109.713(17)
C4–C5–H	128.793(57)
C5–C4–H	124.971(37)
N3–C4–C5	115.754(19)
C2–S1–C5	89.219(14)
S1–C2–H	120.65(12)

(continued)

(continued)

S1–C2–N3	115.365(32)
C2–N3–C4	109.941(36)
N3–C4–H	119.263(38)

The table data are reprinted from: [47], with permission of AIP Publishing

[a] Parenthesized uncertainties given in units of the last significant digit are possibly 2σ

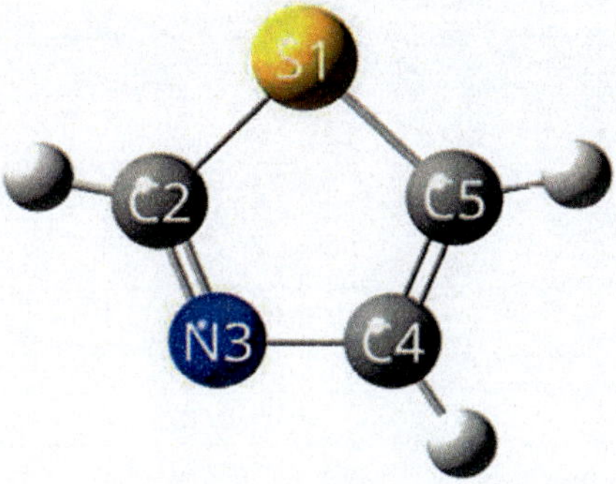

Pure rotational spectra of the title molecule were recorded in the millimeter-wave region.

Rotational constants of the parent species and 23 isotopologues including ^{13}C, ^{15}N, ^{33}S, ^{34}S and several D isotopically substituted species were used to determine the molecular geometry.

Rovibrational corrections, ΔB_{vib}, to experimental ground-state rotational constants were calculated with quadratic and cubic force constants from CCSD(T)_ae/cc-pCVTZ computations.

No	**32**
SF	$\mathbf{C_3H_3N_3O_4}$
CN	**3-Methyl-4-nitro-1,2,5-oxadiazole 2-oxide** 3-Methyl-4-nitrofuroxan
Symmetry	$\mathbf{C_s}$
Methods	**GED** supplemented by **QC**
Ref	[48]

Distances	r_e^{se} [Å] [a,b]
O1–N2	1.462(9)
N2–C3	1.333(9)
C3–C4	1.414(9)

(continued)

(continued)

C4–N5	1.304(9)
O1–N5	1.354(9)
N2–O6	1.215(9)
C3–C7	1.488(9)
C4–N8	1.456(9)
N8–O9	1.228(9)
N8–O10	1.216(9)
C7–Hs [c]	1.091(9)
C7–Ha [d]	1.091(9)

Bond angles	θ_e^{se} [deg] [a,b]
C3–N2–O1	107.2(5)
C4–C3–N2	104.6(5)
N5–C4–C3	113.9(5)
O1–N5–C4	106.1(10)
N2–O1–N5	108.1(10)
O6–N2–O1	118.1 [e]
C7–C3–N2	121.2 [e]
N8–C4–C3	125.9 [e]
O9–N8–C4	115.4(15)
O10–N8–C4	118.2(15)
Hs–C7–C3 [c]	108.5 [e]
Ha–C7–C3 [d]	110.5 [e]

Dihedral angles	τ_e [deg]
Ha–C7–C3…Hs [c,d]	± 120.4 [e]

[a] Parenthesized uncertainties in units of the last significant digit are 3σ

[b] Small differences between values of similar geometric parameters were constrained to the values of CCSD(T)/cc-pVTZ structure

[c] Hs is hydrogen atom in symmetry plane

[d] Ha is hydrogen atom out of symmetry plane

[e] Constrained to the value from calculation at the level of theory as indicated above

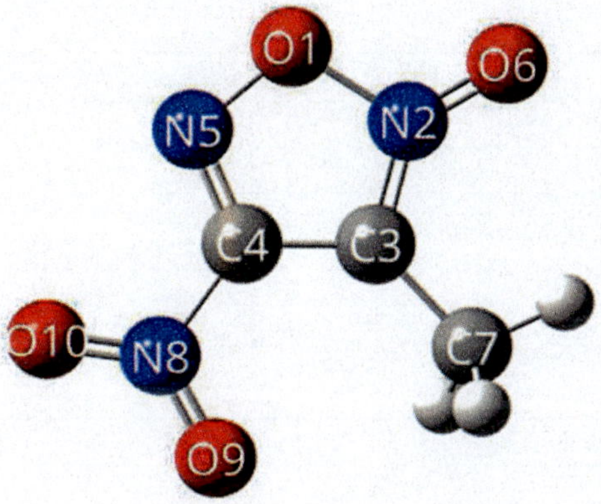

Large-amplitude torsional vibration of the nitro group around the C–N bond was described by the model of pseudo-conformers. The barrier to internal rotation was determined to be $V_2 = 1.3(6)$ kcal mol^{-1}.

Vibrational corrections to the experimental internuclear distances at the temperature of experiment (T_{nozzle}≈331 K) were calculated with quadratic and cubic force constants from B3LYP/SNDS computation.

No	**33**
SF	$\mathbf{C_3H_3N_3O_4}$
CN	**4-Methyl-3-nitro-1,2,5-oxadiazole 2-oxide** 4-Methyl-3-nitrofuroxan
Symmetry	$\mathbf{C_s}$
Methods	**GED** supplemented by **QC**
Ref	[48]

Distances	r_e^{se} [Å] [a,b]
O1–N2	1.382(6)
N2–C3	1.307(6)
C3–C4	1.422(6)
C4–N5	1.340(6)
O1–N5	1.429(6)
N5–O6	1.205(6)
C3–C7	1.488(6)
C4–N8	1.434(6)

(continued)

(continued)

N8–O9	1.229(6)
N8–O10	1.219(6)
C7–Hs [c]	1.090(6)
C7–Ha [d]	1.090(6)

Bond angles	θ_e^{se} [deg] [a,b]
C3–N2–O1	107.5(3)
C4–C3–N2	109.2(3)
N5–C4–C3	109.2(3)
O1–N5–C4	104.7(4)
N2–O1–N5	109.4(5)
O6–N5–O1	118.6 [e]
C7–C3–N2	121.1 [e]
N8–C4–C3	129.3 [e]
O9–N8–C4	114.9 [e]
O10–N8–C4	118.4 [e]
Hs–C7–C3 [c]	108.6 [e]
Ha–C7–C3 [d]	110.3 [e]

Dihedral angles	τ_e [deg]
Ha–C7–C3…Hs [c,d]	± 120.6 [e]

The table data are reprinted from: [48], Copyright 2020, with permission from Elsevier

[a] Parenthesized uncertainties in units of the last significant digit are 3σ

[b] Small differences between values of similar geometric parameters were constrained to the values of CCSD(T)/cc-pVTZ structure

[c] Hs is hydrogen atom in symmetry plane

[d] Ha is hydrogen atom out of symmetry plane

[e] Constrained to the value from calculation at the level of theory as indicated above

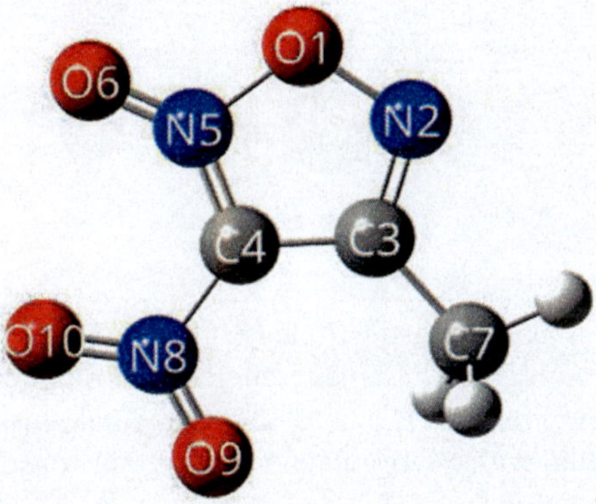

Large-amplitude torsional vibration of the nitro group around the C–N bond was described by the model of pseudo-conformers. The barrier to internal rotation was determined to be $V_2 = 2.9(9)$ kcal mol^{-1}.

Vibrational corrections to the experimental internuclear distances at the temperature of experiment ($T_{nozzle} \approx 332$ K) were calculated with quadratic and cubic force constants from B3LYP/SNDS computation.

No	**34**
SF	$\mathbf{C_3H_4}$
CN	**1-Propyne**
Symmetry	$\mathbf{C_{3v}}$
Methods	**MW** supplemented by **ab initio**
Ref	[49]

Distances	r_e^{se} [Å] [a]
C1–C2	1.20460(7)
C1–H	1.06146(3)
C2–C3	1.45884(6)
C3–H	1.08852(4)

Bond angle	θ_e^{se} [deg] [a]
C–C–H	110.591(2)

The table data are reprinted from: [49], Copyright 2020, with permission from Elsevier

[a] Parenthesized uncertainties in units of the last significant digit are 1σ

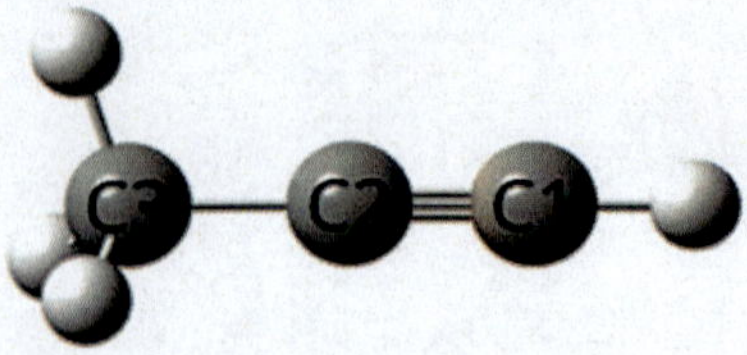

Rotational spectra of all singly substituted ^{13}C isotopic species were recorded in the submillimeter region by the Cologne terahertz spectrometer.

Rovibrational corrections, ΔB_{vib}, to experimental ground-state rotational constants were calculated with quadratic and cubic force constants from CCSD(T)_ae/cc-pwCVTZ computations.

The structure calculated at the CCSD(T)_ae/cc-pwCV5Z level of theory was found to be very close to the r_e^{se} structure.

No	**35**
SF	$C_3H_6Cl_3N$
CN	**1-Chloro-*N*,*N*-bis(chloromethyl)methanamine** Tris(chloromethyl)amine
Symmetry	C_s
Methods	**GED** supplemented by **DFT**
Ref	[50]

Distances	r_e^{se} [Å] [a]
N–C2	1.391(2)
N–C4	1.400(2)
C2–Cl	1.841(1)
C4–Cl	1.830(2)

Bond angles	θ_e^{se} [deg] [a]
N–C2–Cl	114.14(17)
N–C4–Cl	111.98(21)
C2–N–C3	119.40(20)
C2–N–C4	117.13(17)

The table data are reprinted from: [50], by permission from Springer Nature, Springer Science + Business Media, LLC. Copyright 2018

[a] Estimated uncertainties given in parentheses in units of the last significant digit were not specified, possibly estimated total errors

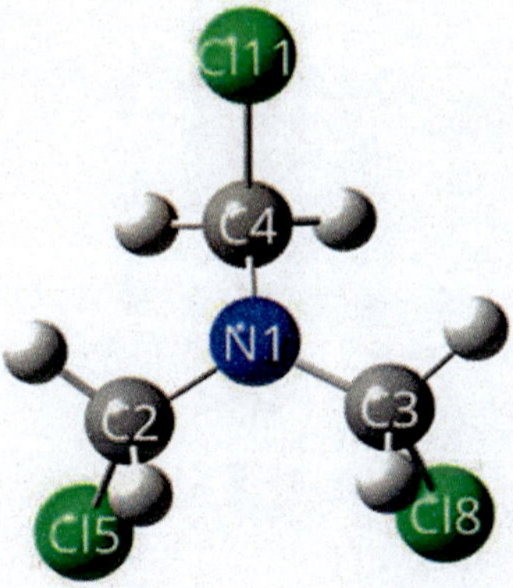

Vibrational corrections to the experimental internuclear distances (T_{nozzle} = 403 K) were computed using semiempirical molecular dynamics simulations in order to account for large-amplitude vibrational modes of the C4H$_2$Cl group.

No	**36**
SF	C_3O_2
CN	**1,2-Propadiene-1,3-dione** Carbon suboxide
Symmetry	$D_{\infty h}$
Methods	**GED** supplemented by **IR** and **QC**
Ref	[51]

Distances	r_e^{se} [Å] [a]
C–C	1.273(2) [b]
C–O	1.161(2) [b]

[a] Parenthesized uncertainties in units of the last significant digit are 3σ

[b] Differences between values of linear reference configuration and pseudo-conformers were assumed according to computations at the level of theory as indicated below

This floppy linear molecule characterized by the extremely large CCC bending was reinvestigated using the pseudo-conformer approach. Vibrational corrections to the experimental internuclear distances ($T_{exp} = 293$ K) were calculated with quadratic and cubic force constants from MP2/6–311+G(d) computation.

No	**37**
SF	$C_4H_3FN_2O_2$
CN	**5-Fluoro-2,4(1*H*,3*H*)-pyrimidinedione** 5-Fluorouracil
Symmetry	C_s
Methods	**GED** supplemented by **ab initio**
Ref	[52]

Distances	r_e^{se} [Å] [a,b]
N1–C2	1.378(2) [1]
C2–N3	1.382(2) [1]
N3–C4	1.393(2) [1]
N1–C6	1.379(2) [1]
C4–C5	1.458(7) [c]
C5–C6	1.342(5) [2]
C2–O	1.212(1) [3]
C4–O	1.212(1) [3]
C5–F	1.329(5) [2]
N1–H	1.005 [d]
N3–H	1.009 [d]
C6–H	1.079 [d]

Bond angles	θ_e^{se} [deg] [b,c]
N1–C2–N3	113.1(1) [4]
C2–N3–C4	128.7(1) [4]
N3–C4–C5	112.6(2) [c]
C4–C5–C6	121.2(2) [c]
C4–C5–F	117.3 [d]
N3–C4–O	121.9 [d]
N3–C2–O	123.5 [d]
C2–N1–C6	123.8(1) [4]
N1–C6–C5	120.7(1) [4]

The table data are reprinted from: [52], with permission from Elsevier
[a] Parenthesized uncertainties in units of the last significant digit are 3σ
[b] Parameters denoted by the same numerical superscript were refined in one group. Differences between parameter values in each group were assumed at the values of CCSD(T)_ae/aug-cc-pCVQZ quality
[c] Dependent parameter
[d] Assumed at the value calculated at the same level of theory as indicated above

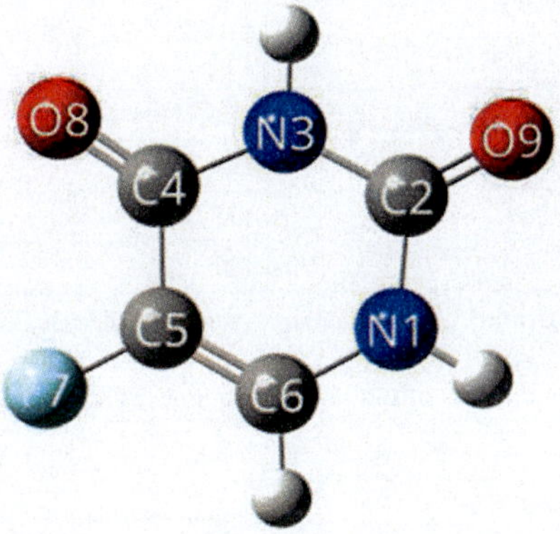

Vibrational corrections to the experimental internuclear distances at the temperature of experiment ($T_{nozzle} = 484$ K) were calculated with quadratic and cubic force constants from MP2/aug-cc-pVTZ computation.

The r_e^{se} and calculated CCSD(T)_ae/cc-pwCVTZ structures were found to be in excellent agreement.

No	**38**
SF	$\mathbf{C_4H_3NO_2}$

(continued)

(continued)

CN	**1*H*-Pyrrol-2,5-dione** Maleimide
Symmetry	C_{2v}
Methods	**MW** supplemented by **QC**
Ref	[53]

Distances	r_e^{se} [Å] [a]
N1–C2	1.384(2)
C2–C3	1.491(2)
C3–C4	1.347(5)
C2–O6	1.209(2)
C3–H	1.077(3)
N1–H	1.009(2)

Bond angles	θ_e^{se} [deg] [a]
N1–C2–C3	105.7(2)
C2–C3–C4	108.5(2)
C2–N1–C5	111.7(1)
N1–C2–O6	126.1(1)
C2–N1–H	124.1(1)
C4–C3–H	129.3(2)
C2–C3–H	122.2(3)

[a] Parenthesized uncertainties in units of the last significant digit were not specified, probably 1σ

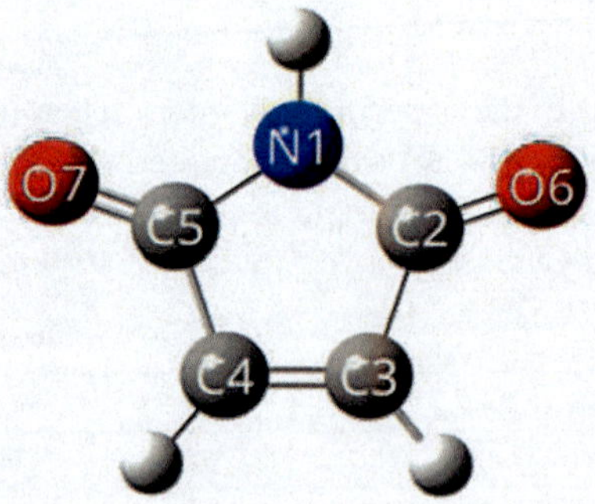

Ground-state rotational constants for the parent and three isotopically substituted (^{13}C2&C5, ^{13}C3&C4 and ND) species were taken from literature.

Rovibrational corrections, ΔB_{vib}, to experimental ground-state rotational constants were calculated with quadratic and cubic force constants from B2PLYP/cc-pVTZ computations.

No	**39**
SF	C_4H_4
CN	**1-Buten-3-yne** Vinylacetylene
Symmetry	C_s
Methods	**MW** supplemented by **ab initio**
Ref	[45]

Distances	r_e^{se} [Å] [a]
C1–C2	1.3381
C2–C3	1.4267
C3–C4	1.2072
C1–H5	1.0798
C1–H6	1.0798
C2–H7	1.0819
C4–H8	1.0617

Bond angles	θ_e^{se} [deg] [b]
C1–C2–C3	123.33
C2–C3–C4	178.48 [c]
C3–C4–H8	180.79 [d]
C2–C1–H5	121.21
C2–C1–H6	120.51
C1–C2–H7	120.07

The table data are reprinted from: [45], Copyright 2018, with permission from Elsevier Inc

[a] Estimated uncertainties (1σ) are less than 0.0004 Å

[b] Estimated uncertainties (1σ) are less than 0.20°

[c] Toward H7

[d] Away from H7

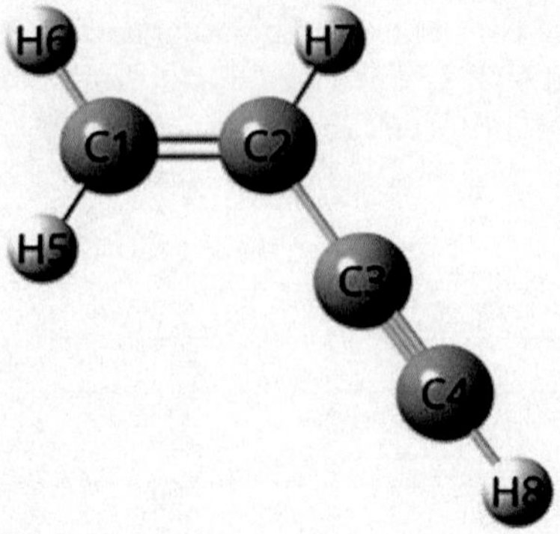

The microwave spectra were measured for twelve isotopic species (all singly substituted ^{13}C and D, doubly substituted ^{13}C3^{13}C4 as well as 2,4-D and two 1,4-D) using a pulsed supersonic jet FTMW spectrometer.

Rovibrational corrections, ΔB_{vib}, to experimental ground-state rotational constants were calculated with quadratic and cubic force constants from CCSD(T)/cc-pVTZ computations.

The r_e^{se} structure was found to be in very good agreement with the structure calculated at the CCSD(T)_ae/cc-pwCV5Z level of theory.

No	**40**
SF	$C_4H_4N_2$
CN	**Pyrimidine**
Symmetry	C_s
Methods	**MW** supplemented by **ab initio**
Ref	[54]

Distances	r_e^{se} [Å] [a]
N1–C2	1.33333(38)
C4–C5	1.38675(45)
C2…C5	2.65395(52)
C2–H	1.08165(39)
C4–H	1.08224(30)
C5–H	1.07929(36)

Bond angles	θ_e^{se} [deg] [a]
N1–C2–H	116.332(24)
C4–C5–H	121.665(22)
C5–C4–H	121.261(38)

The table data are reprinted from: [54], with permission of AIP Publishing
[a] Parenthesized uncertainties given in units of the last significant digit are 2σ

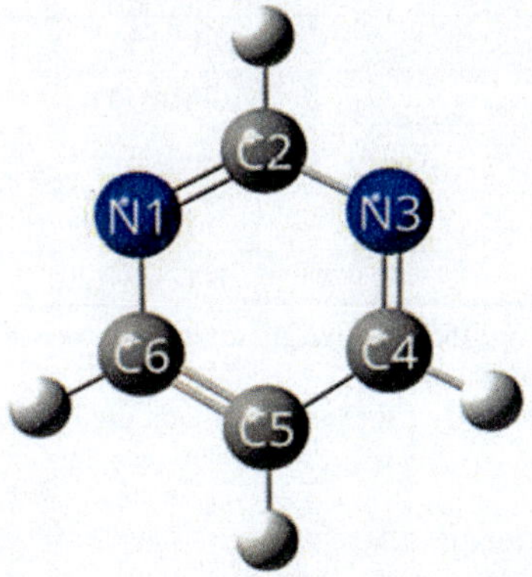

Rotational spectrum was recorded in millimeter-wave region. Transitions for the parent species, as well as for all four singly substituted ^{13}C and ^{15}N and all eleven possible deuterium-substituted isotopic species, were assigned.

Rovibrational corrections, ΔB_{vib}, to experimental ground-state rotational constants were calculated with quadratic and cubic force constants from CCSD(T)/ANO1 computations.

The determined r_e^{se} structure and the equilibrium structure calculated at the CCSD(T)_ae/cc-pCV5Z level of theory were found to be in excellent agreement (within the estimated experimental uncertainties).

No	**41**
SF	$\mathbf{C_4H_4O_3}$
CN	**Dihydro-2,5-furandione** Succinic anhydride
Symmetry	$\mathbf{C_{2v}}$
Methods	**MW** supplemented by **QC**
Ref	[55]

Distances	r_e^{se} [Å] [a]

(continued)

(continued)

C–O1	1.3824(5)
C2–O6	1.1878(5)
C2–C3	1.5116(3)
C–H	1.090(2)

Bond angles	θ_e^{se} [deg] [a]
C–O1–C	111.32(3)
O6–C2–O1	121.52(3)
O1–C2–C3	109.77(5)
C2–C3–H	108.34(11)

Dihedral angles	τ_e^{se} [deg] [a]
O1–C–C–H	± 121.84(20)

[a] Parenthesized uncertainties in units of the last significant digit are 1σ

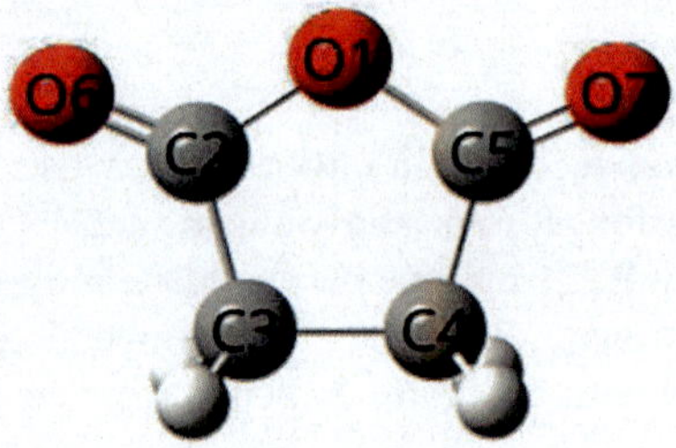

The title molecule is predicted to possess C_{2v} point-group symmetry at the B3LYP/cc-pVTZ level of theory, whereas it has non-planar skeleton (C_2 symmetry) according to MP2/cc-pVTZ prediction. At the CCSD(T)_ae level of theory, the small non-planarity of molecular skeleton decreases with enlarging of basis set and becomes very close to zero from cc-pwCVQZ basis set to CBS.

Rotational spectra of the parent species and all singly substituted heavy-atom isotopic species were recorded by Stark-modulated free-jet and pulsed supersonic jet-expansion FTMW spectrometers.

Rovibrational corrections, ΔB_{vib}, to experimental ground-state rotational constants were calculated with quadratic and cubic force constants from B3LYP/cc-pVTZ computations.

The r_e^{se} and CCSD(T)_ae/cc-pwCVQZ structures were found to be in excellent agreement.

No	**42**
SF	$\mathbf{C_4H_4S}$
CN	**Thiophene**
Symmetry	$\mathbf{C_{2v}}$
Methods	**MW** supplemented by **QC**
Ref	[56]

Distances	r_e^{se} [Å] [a]
C–S	1.71023(11)
C2–C3	1.36625(17)
C3–C4	1.4223(7) [b]
C2–H	1.07667(50)
C3–H	1.07838(21)

Bond angles	θ_e^{se} [deg] [a]
S–C–H	120.203(97)
S–C2–C3	111.5951(58)
C2–C3–H	123.459(15)
C–S–C	92.0787(82)

The table data are reprinted from: [56], Copyright 2022, with permission from the Royal Society of Chemistry

[a] Parenthesized uncertainties in units of the last significant digit are 1σ

[b] Derived parameter

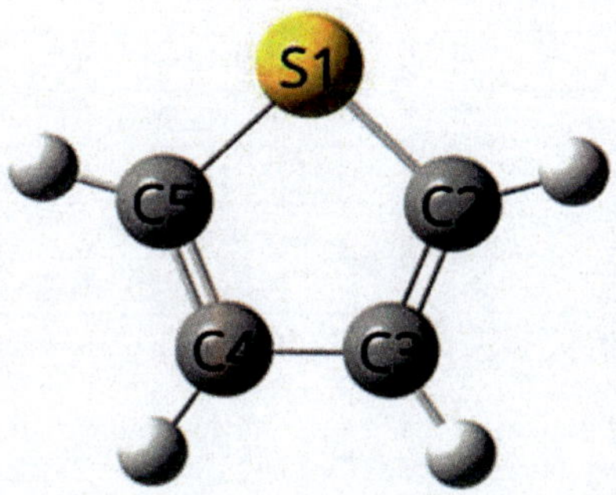

Microwave spectra of the parent species and its singly substituted ^{13}C2, ^{13}C3, and ^{34}S isotopic species were remeasured.

Rovibrational corrections, ΔB_{vib}, to experimental ground-state rotational constants were calculated with quadratic and cubic force constants from B2PLYP-D3/6–311+G(3df,2pd) computations.

The r_e^{se} and calculated CCSD(T)_ae/cc-pwCVQZ structures were found to be in excellent agreement.

No	**43**
SF	C_4H_6
CN	**1,3-Butadiene**
Symmetry	C_2
Methods	**MW** supplemented by **ab initio**
Ref	[57]

Distances	r_e^{se} [Å] [a]
C2–C3	1.48(2)
C1–C2	1.33(1)
C1–H6	1.08(1)
C1–H5	1.07983 [b]
C2–H7	1.083(5)

Bond angles	θ_e^{se} [deg] [a]
C1–C2–C3	124.1(5)
C2–C1–H6	123(2)
C2–C1–H5	118(1)
C3–C2–H7	116.2(6)

Dihedral angles	τ_e^{se} [deg] [a]
C1–C2–C3–C4	33.8(13)
C3–C2–C1–H6	2(1)
C4–C3–C2–H7	−146.4(7)
H5–C1–C2–H7	2.410 [b]

[a] Parenthesized uncertainties in units of the last significant digit are 1σ

[b] Fixed to the value of geometry extrapolated as CCSD(T)_ae/cc-pCVTZ(T- > 5)Z and supplemented by a perturbative quadruples (Q) correction calculated in fc approximation and with ANO0 basis set

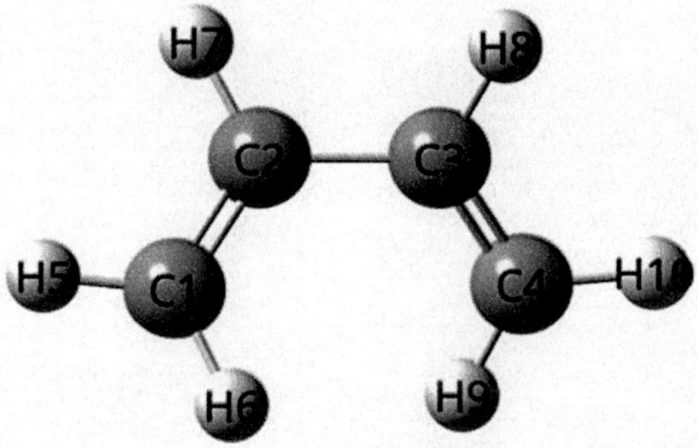

The microwave spectrum of the second stable conformer of butadiene, *gauche*, was measured for six isotopically substituted species ($^{13}CH_2CHCHCH_2$, $CH_2{}^{13}CHCHCH_2$, $CH_2CDCDCH_2$, $CD_2CHCHCD_2$, $CH_2CHCHCH_2$, and $CD_2CDCDCD_2$) using a cavity-enhanced FTMW spectrometer in combination with a supersonic expansion and chirped-pulse FTMW technique.

Rovibrational corrections, ΔB_{vib}, to experimental ground-state rotational constants were calculated with quadratic and cubic force constants from CCSD(T)/ANO1 computations.

No	**44**
SF	**$C_4H_6O_2$**
CN	**2-Methyl-2-propenoic acid** Methacrylic acid
Symmetry	**C$_s$** (*s-trans*) **C$_s$** (*s-cis*)
Methods	**MW** supplemented by **DFT**
Ref	[58]

Distances	r_e^{se} [Å] [a]	
	s-trans	*s-cis*
C1–C5	1.4931(44)	1.4962(45)
C5–C6	1.3394(46)	1.3368(44)
C5–C9	1.4803(47)	1.4857(43)
C9–O10	1.3502(39)	1.3546(41)
C9–O11	1.2076(36)	1.2035(30)
H2–C1	1.0931(6)	1.0960(6)
H4–C1	1.0910(6)	1.0941(6)

(continued)

(continued)

H7–C6	1.0826(6)	1.0866(6)
H8–C6	1.0850(6)	1.0875(6)
H12–O10	1.0095(375)	0.9775(350)

Bond angles	θ_e^{se} [deg] [a]	
	s-trans	*s-cis*
C6–C5–C1	124.08(19)	123.99(17)
C9–C5–C1	115.38(33)	119.15(34)
O10–C9–C5	114.21(31)	111.84(28)
O11–C9–C5	123.81(36)	126.27(41)
C5–C1–H2	110.77 [b]	111.12 [b]
C5–C1–H4	110.65 [b]	110.40 [b]
H7–C6–C5	121.49 [b]	120.46 [b]
H8–C6–C5	120.75 [b]	121.15 [b]
H12–O1O–C9	106.58 [b]	106.75 [b]

Dihedral angles	τ_e [deg]	
	s-trans	*s-cis*
H2–C1–C5–C6	121.03 [b]	120.67 [b]
H3–C1–C5–C6	−121.03 [b]	−120.67 [b]

The table data are reprinted from: [58], with permission of AIP Publishing

[a] Parenthesized uncertainties in units of the last significant digit are 3σ

[b] Fixed at calculated value

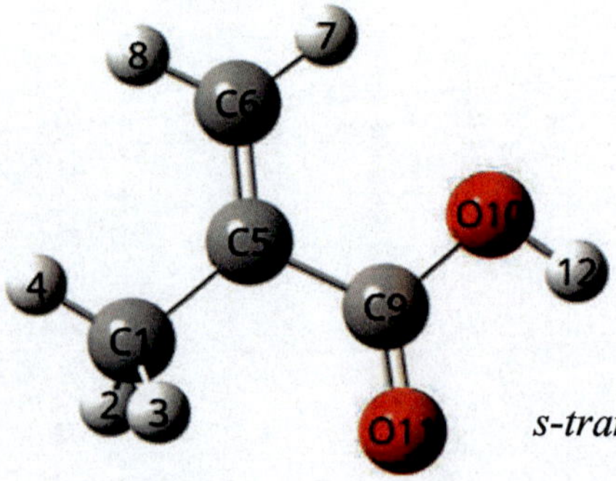

s-trans

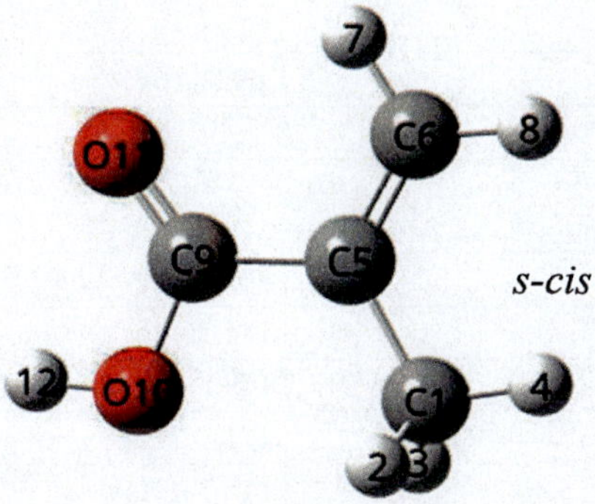

s-cis

Rotational spectra for the parent and singly substituted isotopic (all ^{16}O and all ^{13}C) species were measured by a MB FTMW spectrometer.

Rovibrational corrections, ΔB_{vib}, to experimental ground-state rotational constants were calculated with quadratic and cubic force constants from CAM-B3LYP-D3BJ/aug-cc-pVTZ computations.

No	**45**
SF	$\mathbf{C_4H_8O_4}$
CN	**(*R*)-1,3,4-Trihydroxybutan-2-one** Erythrulose
Symmetry	$\mathbf{C_1}$
Methods	**MW** supplemented by **QC**
Ref	[59]

Distances	r_e^{se} [Å] [a]
O5–C1	1.3985(16)
C1–C2	1.5052(15)
C2–C3	1.5154(15)
C3–C4	1.5187(15)
C4–O8	1.4111(15)
C2–O6	1.2149(16)
C3–O7	1.4053(16)
O5–H	0.9649(16)
C1–H	1.0926(16)
C1–H′	1.0958(16)
C3–H	1.0978(16)
O7–H	0.9670(16)
C4–H′	1.0884(16)

(continued)

(continued)

C4–H	1.0923(16)
O8–H	0.9614(16)

Bond angles	θ_e^{se} [deg] [a]
O5–C1–C2	111.36(21)
C1–C2–C3	119.57(21)
C2–C3–C4	111.76(17)
C3–C4–O8	110.74(21)
C1–C2–O6	120.29(23)
C2–C3–O7	109.66(23)
C1–O5–H	105.74(41)
O5–C1–H	110.07(24)
O5–C1–H′	111.47(24)
C2–C3–H	107.20(24)
C3–O7–H	105.61(49)
C3–C4–H	108.26(24)
C3–C4–H′	110.77(24)
C4–O8–H	106.17(24)

Dihedral angles	τ_e^{se} [deg] [a]
O5–C1–C2–C3	169.26(39)
C1–C2–C3–C4	50.33(27)
C2–C3–C4–O8	56.61(39)
O5–C1–C2–O6	−9.87(44)
C1–C2–C3–O7	170.90(36)
C3–C4–O8–H	55.39(41)
C1–C2…O5–H	−167.92(49)
C2…O5–C1–H	121.74(49)
C2…O5–C1–H′	−119.23(41)
C1–C2–C3–H	−68.72(49)
C2–C3–O7–H	12.95(49)
C2–C3–C4–H′	−62.14(41)
C2–C3–C4–H	179.07(41)

[a] Parenthesized uncertainties in units of the last significant digit are 1σ

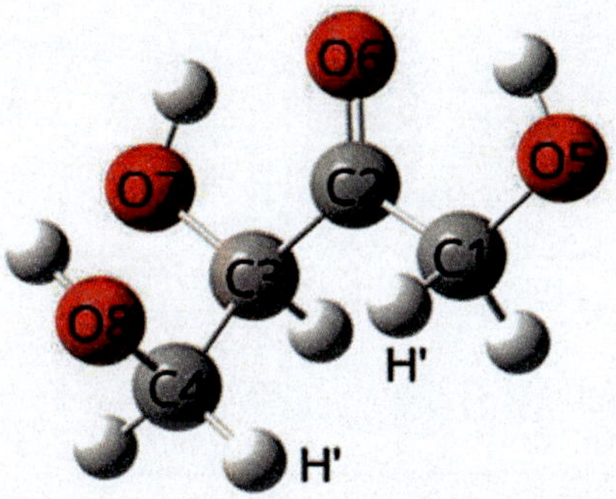

Rotational spectra of the parent and all monosubstituted ^{13}C isotopic species were recorded by a high-sensitive MB-FTMW spectrometer. As the number of determined ground-state rotational constants was not large enough to determine a complete structure, the predicate-regression mixed-estimation method was used.

Rovibrational corrections, ΔB_{vib}, to experimental ground-state rotational constants were calculated with quadratic and cubic force constants from MP2/cc-pVTZ computations.

Semiexperimental equilibrium structure was determined for the lowest energy conformer predicted at the MP2 and B3LYP levels of theory (with 6–311++G(d,p) basis set). This conformer (*sc*,*sp*,*sc*$^{-}$,*sc*$^{+}$) is characterized by *synclinal* (*sc*), *synperiplanar* (*sp*), *synclinal* (*sc*$^{-}$), and *synclinal* (*sc*$^{+}$) conformations of the C1–C2–C3–C4, O5–C1–C2=O6, O8–C4–C3–O7, and C3–C4–O8–H fragments, respectively.

The calculated structure of CCSD(T)_ae/cc-pwCVQZ quality agrees well with the determined r_e^{se} structure.

No	**46**
SF	**C_5BrF_4N**
CN	**4-Bromo-2,3,5,6-tetrafluoropyridine**
Symmetry	**C_{2v}**
Methods	**GED** supplemented by **DFT**
Ref	[60]

Distances	r_e^{se} [Å] [a,b]
C4–Br	1.840(4)
C3–C4	1.387(2) [1]
C2–C3	1.385(2) [1]
C3–F	1.321(1) [2]
N1–C2	1.299(1) [2]
C2–F	1.322(1) [2]

Bond angles	θ_e^{se} [deg] [a,b]
Br–C4–C3	121.5(2)
C3–C4–C5	117.0(5)
C2–C3–C4	118.7(2)
C4–C3–F	120.6(3) [3]
C2–C3–F	120.8(2)
N1–C2–C3	123.8(2)
C3–C2–F	118.3(3) [3]
N1–C2–F	117.8(3)
C2–N1–C6	118.0(5)

The table data are reprinted from: [60], Copyright 2019, with permission from Wiley–VCH Verlag GmbH & Co. KGaA, Weinheim

[a] Parenthesized uncertainties in units of the last significant digit are 3σ

[b] Parameters denoted by the same numerical superscript were refined in one group. Differences between parameter values in each group were assumed at the values of PBE0/cc-pVTZ structure

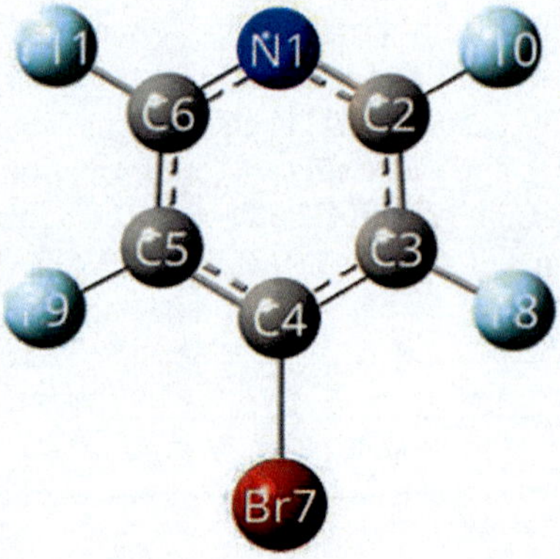

Vibrational corrections to experimental internuclear distances (T_{nozzle} = 303 K) were calculated with quadratic and cubic force fields from computations at the level of theory as indicated above.

No	**47**
SF	**C_5ClF_4N**
CN	**4-Chloro-2,3,5,6-tetrafluoropyridine**
Symmetry	**C_{2v}**
Methods	**GED** supplemented by **DFT**
Ref	[60]

Distances	r_e^{se} [Å] [a,b]
C4–Cl	1.675(3)
C3–C4	1.394(1) [1]
C2–C3	1.390(1) [1]
C3–F	1.318(1) [2]
N1–C2	1.297(1) [2]
C2–F	1.319(1) [2]

Bond angles	θ_e^{se} [deg] [a,b]
Cl–C4–C3	121.7(2)
C3–C4–C5	116.6(4)
C2–C3–C4	118.8(2)
C4–C3–F	120.2(2) [3]
C2–C3–F	121.0(1)
N1–C2–C3	123.5(2)
C3–C2–F	118.2(2) [3]
N1–C2–F	118.3(2)
C2–N1–C6	118.8(3)

The table data are reprinted from: [60], Copyright 2019, with permission from Wiley–VCH Verlag GmbH & Co. KGaA, Weinheim

[a] Parenthesized uncertainties in units of the last significant digit are 3σ

[b] Parameters denoted by the same numerical superscript were refined in one group. Differences between parameter values in each group were assumed at the values of PBE0/cc-pVTZ structure

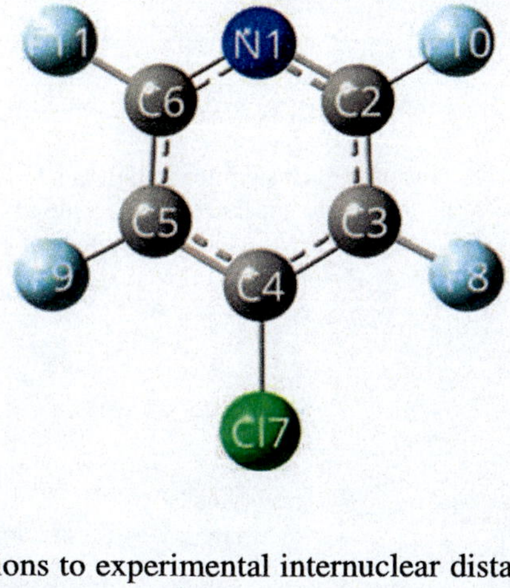

Vibrational corrections to experimental internuclear distances ($T_{nozzle} = 300$ K) were calculated with quadratic and cubic force fields from computations at the level of theory as indicated above.

No	**48**
SF	C_5F_4IN
CN	**4-Iodo-2,3,5,6-tetrafluoropyridine**
Symmetry	C_{2v}
Methods	**GED** supplemented by **DFT**
Ref	[60]

Distances	r_e^{se} [Å] [a,b]
C4–I	2.038(5)
C3–C4	1.374(1) [1]
C2–C3	1.374(1) [1]
C3–F	1.317(1) [2]
N1–C2	1.293(1) [2]
C2–F	1.316(1) [2]

Bond angles	θ_e^{se} [deg] [a,b]
I–C4–C3	121.5(1)
C3–C4–C5	117.0(2)
C2–C3–C4	118.8(1)
C4–C3–F	121.8(1) [3]
C2–C3–F	119.4(2)
N1–C2–C3	123.7(1)
C3–C2–F	119.4(1) [3]
N1–C2–F	116.9(2)
C2–N1–C6	117.9(2)

[a] Parenthesized uncertainties in units of the last significant digit are 3σ

[b] Parameters denoted by the same numerical superscript were refined in one group. Differences between parameter values in each group were assumed at the values of PBE0/cc-pVTZ structure

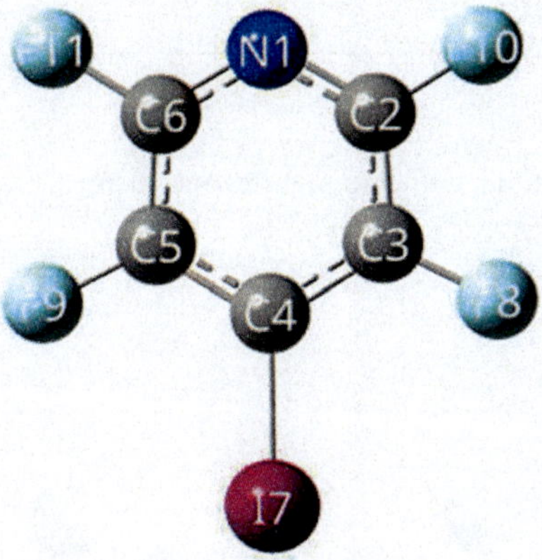

Vibrational corrections to experimental internuclear distances (T_{nozzle}≈318 K) were calculated with quadratic and cubic force fields from computations at the level of theory as indicated above.

No	**49**
SF	**C_5F_5N**
CN	**Pentafluoropyridine**
Symmetry	**C_{2v}**
Methods	**GED** supplemented by **DFT**
Ref	[60]

Distances	r_e^{se} [Å] [a]
C–N	1.304(2)
C2–C3	1.3814(8)
C3–C4	1.3838(8)
C2–F	1.321(2)
C3–F	1.323(2)
C4–F	1.316(2)

Bond angles	θ_e^{se} [deg] [a]
C2–N–C6	122.7(19)
N–C1–C2	120.9(13)
C2–C3–C4	117.7(2)
C3–C4–C5	120.0(2)

(continued)

(continued)

C3–C2–F	120.3(6)
C4–C3–F	120.0(5)

[a] Parenthesized uncertainties in units of the last significant digit are 1σ estimated by Monte Carlo method

[b] Average value

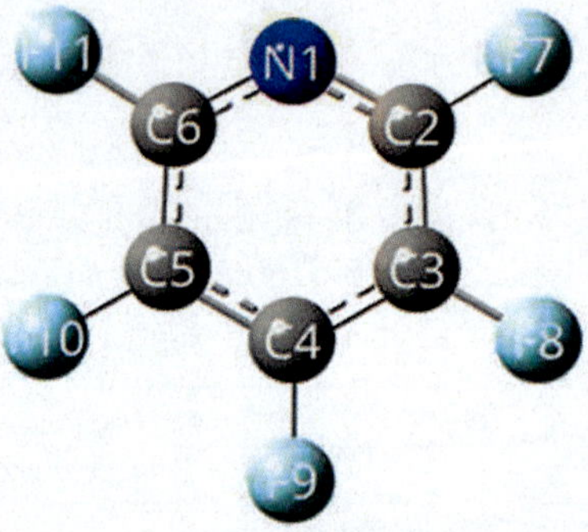

GED intensities were analyzed in combination with ground-state rotational constants taken from literature. Differences between values of similar parameters were fixed at the values of B3PW91/cc-pVTZ structure.

Vibrational corrections to experimental internuclear distances (T_{nozzle} = 299 K) and to experimental ground-state rotational constants, $\Delta B = B_e - B_0$, were calculated with quadratic and cubic force fields from computations at the DFT level as indicated above.

No	**50**
SF	$C_5H_4N_2O_4$
CN	**1,2,3,6-Tetrahydro-2,6-dioxo-4-pyrimidinecarboxylic acid** Orotic acid
Symmetry	C_s (*anti*) C_s (*syn*)
Methods	**GED** supplemented by **ab initio**
Ref	[61]

Distances	r_e^{se} [Å] [a,b]	
	anti	*syn*
N1–C2	1.380(5) [1]	1.378(5) [1]
C2–N3	1.377(5) [1]	1.379(5) [1]
C6–N1	1.394(5) [1]	1.395(5) [1]
C5–C6	1.463(3) [2]	1.462(3) [2]
C4–N3	1.371(5) [1]	1.375(5) [1]
C4–C5	1.348(13) [3]	1.348(13) [3]
O7–C2	1.209(1) [4]	1.209(1) [4]
O8–C6	1.212(1) [4]	1.211(1) [4]
C9–C4	1.490(3) [2]	1.493(3) [2]
O10–C9	1.205(1) [4]	1.199(1) [4]
O11–C9	1.337(13) [3]	1.349(13) [3]
O11–H	0.971(9) [5]	0.971(9) [5]
N3–H	1.013(9) [5]	1.011(9) [5]
N1–H	1.014(9) [5]	1.014(9) [5]
C5–H	1.081(9) [5]	1.082(9) [5]

Bond angles	θ_e^{se} [deg] [a,b]	
	anti	*syn*
N1–C2–N3	114.0(8) [6]	114.2(8) [6]
C2–N1–C6	127.7(4) [7]	127.7(4) [7]
C5–C6–N1	113.7(6) [8]	113.6(6) [8]
O7–C2–N3	123.4(11) [9]	123.0(11) [9]
O8–C6–C5	125.4(11) [9]	125.6(11) [9]
C9–C4–N3	110.8(7) [10]	115.1(7) [10]
O10–C9–C4	122.5(4) [11]	124.5(4) [11]
O11–C9–C4	113.1(4) [11]	111.4(4) [11]
C9–O11–H	106.7 [c]	106.9 [c]
C2–N3–H	118.1 [c]	116.7 [c]
C2–N1–H	115.1 [c]	115.2 [c]
C6–C5–H	119.0 [c]	119.6 [c]

[a] Parenthesized uncertainties in units of the last significant digit are 2.5σ

[b] Parameters denoted by the same numerical superscript were refined in one group. Differences between parameter values in each group were assumed at the calculated values of CCSD(T)_ae/cc-pwCVQZ quality

[c] Assumed at the calculated value as indicated above

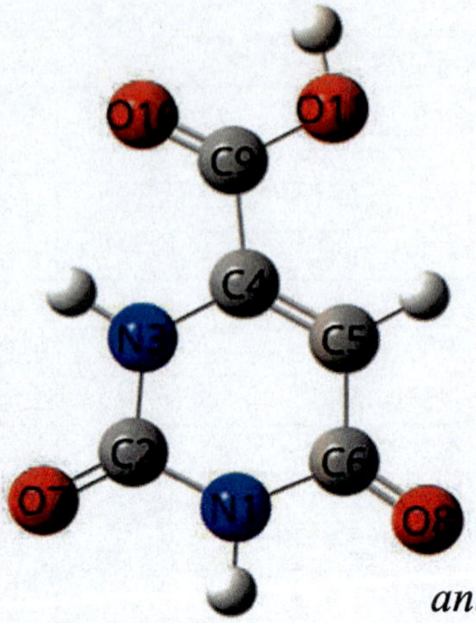

anti

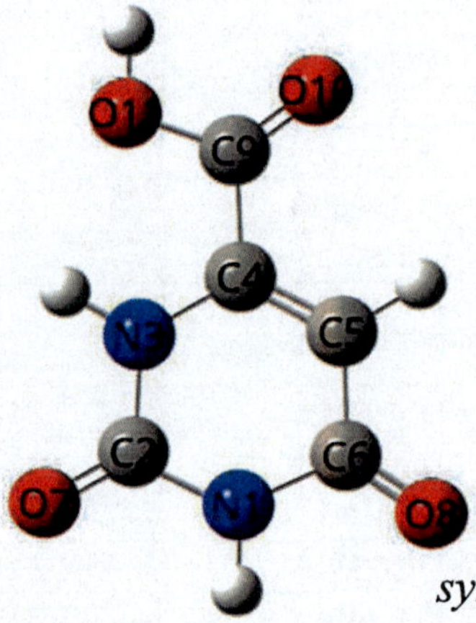

syn

Two lowest energy conformers of the title molecule, *anti* and *syn*, were predicted by MP2 and B3LYP calculations with cc-pVTZ basis set. They differ by conformation of the C5=C4–C9=O10 fragment (*anti* versus *syn*). In both conformers, the carboxylic group has *synperiplanar* conformation.

In the GED analysis, large-amplitude motion of the carboxylic group around the C4–C9 bond was described by the model of pseudo-conformers weighted according to MP2 potential energy function.

Vibrational corrections to the experimental internuclear distances at the temperature of experiment ($T_{nozzle} = 520$ K) were calculated with MP2 quadratic and cubic force constants.

No	**51**
SF	C_5H_4OS
CN	**2-Thiophenecarboxaldehyde**
Symmetry	C_s (*syn*)
Methods	**MW** supplemented by **QC**
Ref	[62]

Distances	r_e^{se} [Å] [a]
C2–C3	1.3738(43)
C2–S1	1.7173(42)
C2–C6	1.4567(43)
C3–C4	1.4173(25)
C4–C5	1.3718(36)
C5–S1	1.7069(45)
C6–O7	1.2139(51)

Bond angles	θ_e^{se} [deg] [a]
C3–C2–S1	111.73(31)
S1–C2–C6	121.78(29)
C2–C3–C4	112.48(19)
C3–C4–C5	111.79(10)
C4–C5–S1	112.55(14)
C2–S1–C5	91.45(20)
C2–C6–O7	124.24(30)

[a] Parenthesized uncertainties given in units of the last significant digit are 1σ

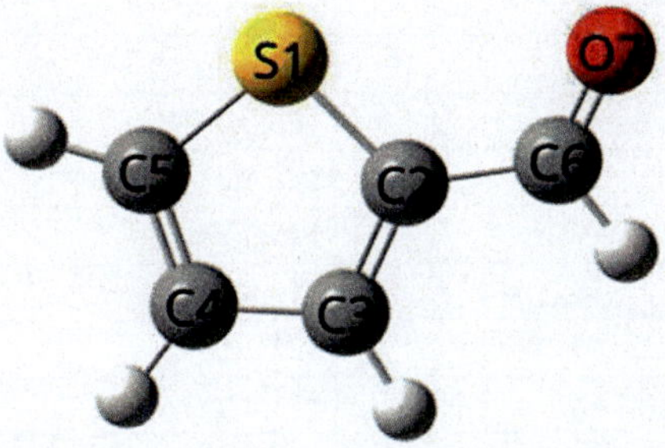

The existence of two conformers, *syn* and *anti*, differing in the magnitude of the S–C2–C6=O torsional angle (0 and 180°, respectively) was predicted by ab initio calculations. The *syn* conformer is lower in energy in comparison with the *anti* one by about 6 kJ mol^{-1} (MP2/6–311++G(d,p)). Both conformers were detected by molecular jet FTMW spectroscopy. The spectra of the parent species and all singly substituted heavy-atom (^{13}C, ^{18}O and ^{34}S) isotopic species were measured for the most abundant conformer.

Rovibrational corrections, ΔB_{vib}, to experimental ground-state rotational constants were calculated with quadratic and cubic force constants from CAM-B3LYP-D3BJ/aug-cc-pVTZ computations.

No	**52**
SF	$\mathbf{C_5H_5N_3O}$
CN	**2-Pyrazinecarboxamide** Pyrazinamide
Symmetry	$\mathbf{C_s}$
Methods	**GED** supplemented by **QC**
Ref	[63]

Distances	r_e^{se} [Å] [a,b]
C1–N2	1.339(1)
C1–C6	1.398(1)
C1–C7	1.506(1)
N2–C3	1.337(1)
C3–C4	1.397(1)
C–H	1.081(2) [c]
C4–N5	1.338(1)
N5–C6	1.338(1)
C7–N8	1.352(1)
C7–O9	1.221(1)
N–H	1.003(2) [c]

Bond angles	θ_e^{se} [deg] [a,b]
N2–C1–C6	122.0(1)
N2–C1–C7	118.5(1)
C1–N2–C3	115.9(1)
C6–C1–C7	119.6(2)
C1–C6–N5	122.3(2)
C1–C7–N8	113.8(1)

(continued)

(continued)

C1–C7–O9	121.3(1)
N2–C3–C4	121.9(1)
C3–C4–N5	122.5(1)
C4–N5–C6	115.5(1)
N8–C7–O9	125.0(1)
H–N–H	121.6(1)

[a] Parenthesized uncertainties in units of the last significant digit are 1σ
[b] Refined parameters were flexibly restrained to the values of CCSD(T)_ae/cc-pwCVTZ structure
[c] Average value

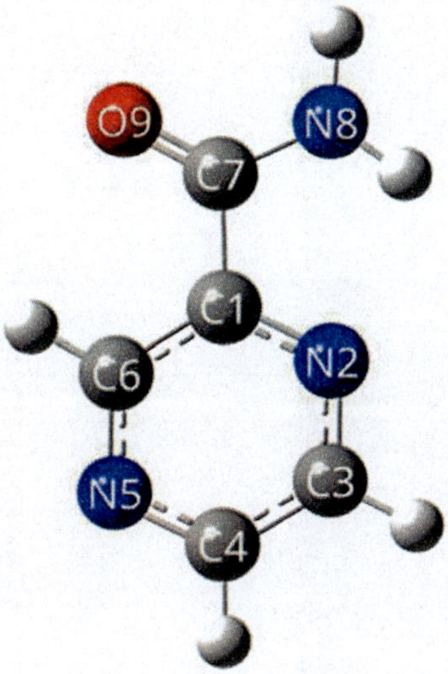

Diffraction patterns of gaseous pyrazinamide measured in three laboratories were analyzed in the same manner. The average experimental precision of these experiments was estimated to be 0.004 Å and 0.2° for distances and angles, respectively.

The structure corresponding to the best convergence criteria, R_f factor, was determined from experimental data obtained in the Moscow laboratory, see table above.

Vibrational corrections to the experimental internuclear distances (T_{nozzle} = 402 K) were calculated with quadratic and cubic force constants from B3LYP/def2-TZVP computations.

No	**53**
SF	$C_5H_8O_2$
CN	**2-Methyl-2-propenoic acid methyl ester** Methyl methacrylate
Symmetry	C_s (*s-trans*) C_s (*s-cis*)
Methods	**MW** supplemented by **DFT**
Ref	[58]

Distances	r_e^{se} [Å] [a]	
	s-trans	*s-cis*
C1–C5	1.4950(42)	1.4988(15)
C5–C6	1.3390(50)	1.3332(12)
C5–C9	1.4823(46)	1.4858(11)
C9–O10	1.3429(54)	1.3480(12)
C9–O15	1.2124(57)	1.2102(12)
C11–O10	1.4309(20)	1.4321(5)
H2–C1	1.0919(3)	1.0931(1)
H4–C1	1.0899(3)	1.0913(1)
H12–C11	1.0872(3)	1.0884(1)
H13–C11	1.0902(3)	1.0913(1)
H7–C6	1.0813(3)	1.0837(1)
H8–C6	1.0839(3)	1.0846(1)

Bond angles	θ_e^{se} [deg] [a]	
	s-trans	*s-cis*
C6–C5–C1	123.88(16)	123.92(5)
C9–C5–C1	115.24(41)	118.88(10)
C11–O10–C9	114.32(22)	114.35(6)
O10–C9–C5	114.06(46)	111.67(10)
O15–C9–C5	123.28(53)	125.90(13)
C5–C1–H2	110.77 [b]	111.13 [b]
C5–C1–H4	110.67 [b]	110.39 [b]
H7–C6–C5	121.43 [b]	120.42 [b]
H8–C6–C5	120.75 [b]	121.17 [b]
H12–C11–O10	105.77 [b]	105.83 [b]
H13–C11–O10	110.48 [b]	110.40 [b]

Dihedral angles	τ_e [deg]	
	s-trans	*s-cis*
H13–C11–O10–C9	−60.28 [b]	−60.21 [b]
H14–C11–O10–C9	60.29 [b]	60.21 [b]
H2–C1–C5–C6	121.06 [b]	−120.69 [b]
H3–C1–C5–C6	−121.06 [b]	120.69 [b]

The table data are reprinted from: [58], with permission of AIP Publishing

[a] Parenthesized uncertainties in units of the last significant digit are 3σ

[b] Fixed at calculated value

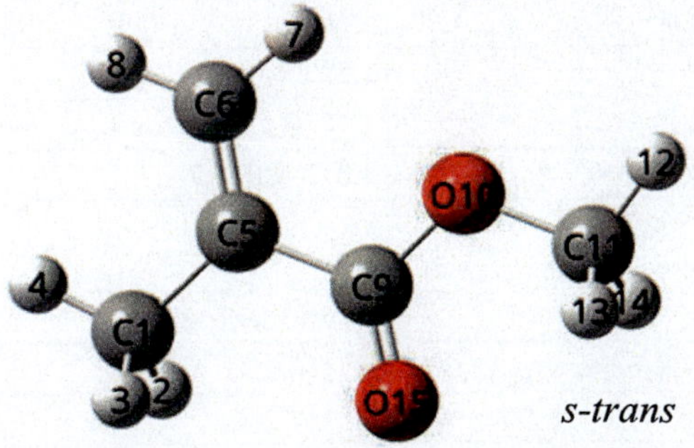

s-trans

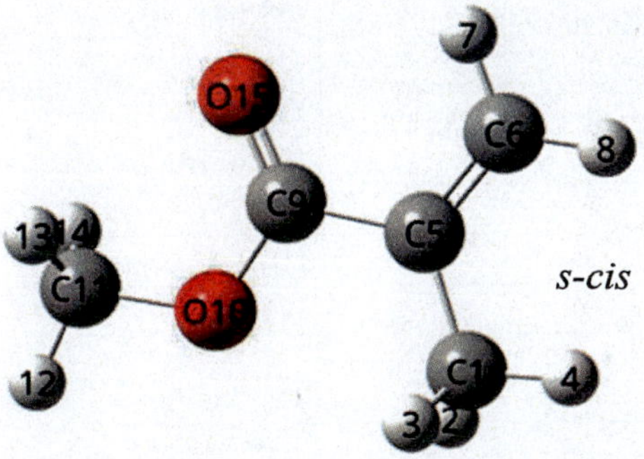

s-cis

Rotational spectra for the parent and singly substituted isotopic (all ^{16}O and all ^{13}C) species were measured by a MB FTMW spectrometer.

Rovibrational corrections, ΔB_{vib}, to experimental ground-state rotational constants were calculated with quadratic and cubic force constants from CAM-B3LYP-D3BJ/aug-cc-pVTZ computations.

No	**54**
SF	$C_5H_{10}N_2$
CN	**6-Methyl-1,5-diazabicyclo[3.1.0]hexane**
Symmetry	C_s (*boat*)
Methods	**GED** supplemented by **ab initio**
Ref	[64]

Distances	r_e^{se} [Å] [a,b]
C6–N1	1.440(2) [1]
N1–N5	1.512(2) [1]
C2–N1	1.466(2) [1]
C2–C3	1.523(2) [1]
C6–C7	1.487(2) [1]
C–H	1.073–1.077(7)

Bond angles	θ_e^{se} [deg] [b,c]
N1–C2–C3	110.2(4)
C7–C6–H	117.2(3) [2]
N1–C6–H	117.8(3) [2]
C2–C3–C4	99.9(4)
N1–N5–C4	106.1(2)
N1–N5–C6	58.3(1)
N1–C6–N5	63.3(1)
N1–C6–C7	114.9(6)
N1–C2–H(ax) [d]	105.2(6) [3]
N1–C2–H(eq) [e]	107.2(6) [3]
C–C–H [f]	110.9–113.2(4)
C6–N1–C2	111.8(1)

Dihedral angles	τ_e^{se} [deg] [c]
N5–N1–C2–C3	17.3(1)
N1–C2–C3–C4	−26.8(2)

(continued)

(continued)

$\varphi 1$ [g]	−26.2(3)
$\varphi 2$ [h]	74.0(1)

The table data are reproduced from: [64], Copyright 2020, with permission from the PCCP Owner Societies

[a] Parenthesized uncertainties in units of the last significant digit are 3σ

[b] Parameters denoted by the same numerical superscript were refined in one group. Differences between parameter values in each group were assumed at the values of MP2/aug-cc-pVTZ structure

[c] Parenthesized uncertainties in units of the last significant digit are 2σ

[d] H atom in axial position

[e] H atom in equatorial position

[f] In the five-membered ring

[g] Acute angle between the C2C3C4 and C2N1N5C4 planes

[h] Acute angle between the N1C6N5 and C2N1N5C4 planes

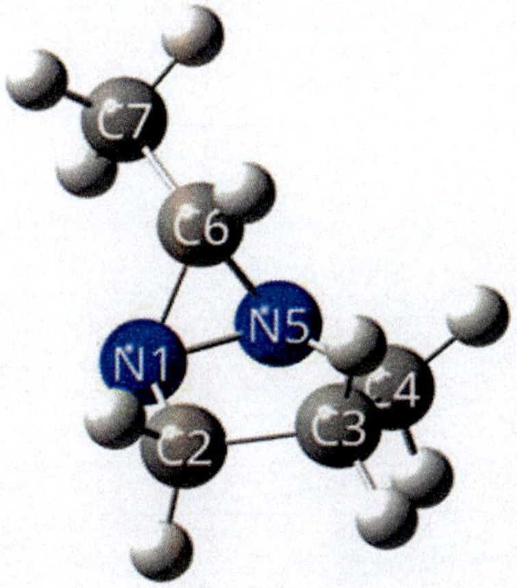

According to predictions of MP2/aug-cc-pVTZ calculations, the title molecule can exist as the *boat* and *chair* conformers. The *boat* conformer is lower in energy by 3.5 kcal mol^{-1} in comparison to the chair one.

Only one conformer (*boat*) was found to be present at the temperature of the GED experiment ($T_{nozzle} = 293$ K). Ring-puckering motion as a function of coordinate φ1 was well described by both semirigid and large-amplitude motion models.

Vibrational corrections to the experimental internuclear distances were calculated using scaled quadratic and cubic force fields from computation at the level of theory as indicated above.

No	**55**
SF	**$C_6H_2F_{10}O_2$**

(continued)

(continued)

CN	**2,2,3,3,4,4,5,5,6,6-Decafluorocyclohexane-1,1-diol**
Symmetry	C_1
Methods	**GED** supplemented by **DFT**
Ref	[65]

Distances	r_e^{se} [Å] [a,b]
C1–O2	1.378(5) [1]
C1–O3	1.358(5) [1]
C1–C5	1.547(1) [2]
C1–C6	1.553(1) [2]
O2–H	0.967(13) [3]
O3–H	0.967(13) [3]
C5–C7	1.540(1) [2]
C5–F10	1.326(1) [1]
C5–F11	1.348(1) [1]
C6–C8	1.538(1) [2]
C6–F12	1.343(1) [1]
C6–F13	1.342(1) [1]
C7–C9	1.542(1) [2]
C7–F14	1.333(1) [1]
C7–F15	1.332(1) [1]
C8–C9	1.540(1) [2]
C8–F16	1.333(1) [1]
C8–F17	1.334(1) [1]
C9–F18	1.330(1) [1]
C9–F19	1.335(1) [1]

Bond angles	θ_e^{se} [deg] [a,b]
O2–C1–O3	111.2(11)
O2–C1–C5	109.7(2) [4]
O2–C1–C6	108.6(2) [4]
C1–O2–H	109.3(19) [5]
O3–C1–C5	106.0(2) [4]
O3–C1–C6	109.7(2) [4]
C1–O3–H	108.0(19) [5]
C5–C1–C6	111.7(11) [c]
C1–C5–C7	115.1(8) [6]

(continued)

(continued)

C1–C5–F10	109.4(2) [7]
C1–C5–F11	107.1(2) [7]
C1–C6–C8	115.2(8) [6]
C1–C6–F12	107.6(2) [7]
C1–C6–F13	109.3(2) [7]
C5–C7–C9	108.0(9) [c]
C5–C7–F14	108.7(2) [7]
C5–C7–F15	108.7(2) [7]
C6–C8–C9	112.9(10) [c]
C6–C8–F16	108.9(2) [7]
C6–C8–F17	108.4(2) [7]
C7–C9–C8	113.7(9) [c]
C7–C9–F18	108.7(2) [7]
C7–C9–F19	108.6(2) [7]
C8–C9–F18	108.4(2) [7]
C8–C9–F19	108.1(2) [7]

Dihedral angles	τ_e^{se} [deg] [a,b]
H–O2–C1–O3	178.0(33)
O2–C1–O3–H	65.2(33)
O2–C1–C5–C7	−170.6(10) [8]
O2–C1–C5–F10	72.7(12) [9]
O2–C1–C6–C8	164.1(10) [8]
O2–C1–C6–F12	−71.6(12) [9]
H–O3–C1–C5	−175.6(36) [c]
O3–C1–C5–F11	−163.9(14) [10]
O3–C1–C6–F13	170.1(14) [10]
C7–C5–C1–C6	−50.2(10) [c]
C5–C1–C6–C8	43.1(14) [c]
C1–C5–C7–C9	56.3(16) [c]
C1–C6–C8–C9	−44.6(19) [c]
F14–C7–C5–F10	−65.5(12) [9]
F15–C7–C5–F11	172.2(14) [10]
F16–C8–C6–F12	70.4(12) [9]
F17–C8–C6–F13	−163.3(14) [10]
C6–C8–C9–C7	52.1(8) [c]

[a] Distance and angle values are rounded here to thousandths of Å and tenths of degree, respectively. Parenthesized uncertainties in units of the last significant digit are 3σ
[b] Parameters with equal superscript were refined in one group. Differences between parameters in each group were assumed at the values derived from the PBE0/cc-pVTZ structure
[c] Dependent parameter

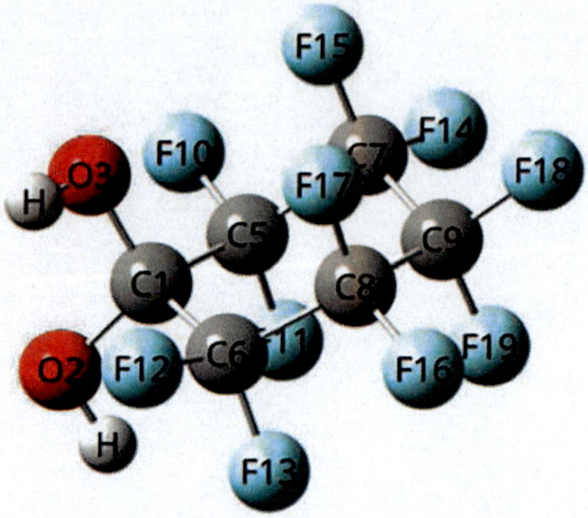

Vibrational corrections to the experimental internuclear distances (T_{nozzle} = 298 K) were calculated with quadratic and cubic force constants from computation at the same level of theory as indicated above.

No	**56**
SF	C_6H_4
CN	**1-Hexene-3,5-diyne** Vinyldiacetylene
Symmetry	C_s
Methods	**MW** supplemented by **ab initio**
Ref	[45]

Distances	r_e^{se} [Å] [a]
C1–C2	1.338(2)
C2–C3	1.422(5)
C3–C4	1.214(7)
C4–C5	1.370(5)
C5–C6	1.210(2)
C6–H	1.061(1)

(continued)

(continued)

C1–H7	1.081(6)
C1–H8	1.079(6)
C2–H	1.084(5)

Bond angles	θ_e^{se} [deg] [a]
C1–C2–C3	123.37(10)
C2–C1–H7	121.18(15)
C2–C1–H8	120.7(13)
C1–C2–H	120.23(23)

The table data are reprinted from: [45], Copyright 2018, with permission from Elsevier Inc

[a] Parenthesized uncertainties in units of the last significant digit are 1σ

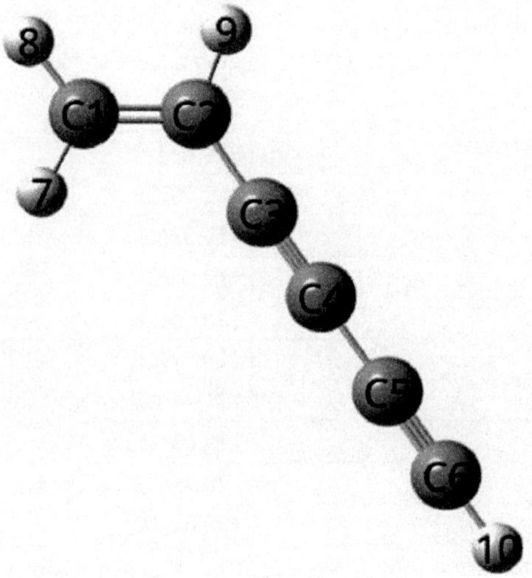

The microwave spectra for the parent species and all its singly substituted ^{13}C and D isotopologues as well as three doubly deuterated species (2,4-D and two 1,4-D) were measured by a pulsed supersonic jet FTMW spectrometer.

Rovibrational corrections, ΔB_{vib}, to experimental ground-state rotational constants were calculated with quadratic and cubic force constants from CCSD(T)/cc-pVTZ computations.

The r_e^{se} structure was found to be in good agreement with the structure calculated at the CCSD(T)_ae/cc-pwCV5Z level of theory.

No	**57**
SF	$C_6H_4Cl_2$
CN	**1,2-Dichlorobenzene**
Symmetry	C_{2v}
Methods	**MW** supplemented by **ab initio**
Ref	[66]

Distances	r_e^{se} [Å] [a]
Cl–C	1.7228(3)
C1–C2	1.3921(6)
C2–C3	1.3921(4)
C3–C4	1.3878(4)
C4–C5	1.3929(7)
C3–H	1.0787(2)
C4–H	1.0801(2)

Bond angles	θ_e^{se} [deg] [a]
Cl9–C1–C2	121.06(5)
Cl10–C2–C3	119.01(2)
C1–C2–C3	119.94(2)
C2–C3–C4	120.05(3)
H–C3–C2	118.82(2)
H–C3–C4	121.14(2)
H–C4–C3	119.63(3)
H–C4–C5	120.36(2)
C3–C4–C5	120.02(1)

[a] Parenthesized uncertainties given in units of the last significant digit are 1σ

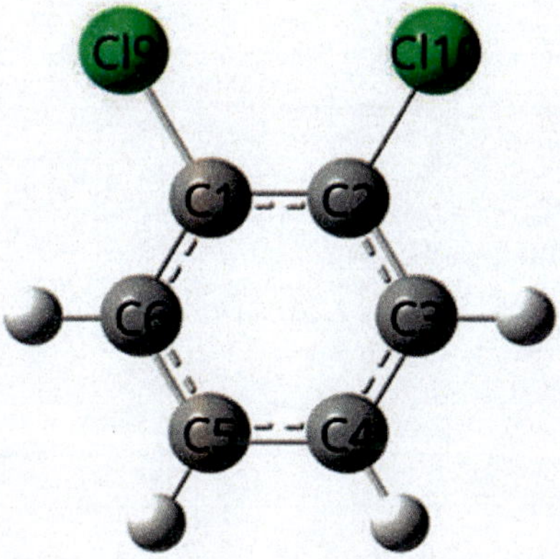

Supersonic expansion rotational spectroscopy with rectangular pulse and chirp excitation was used to study spectra of the parent species and its isotopologues with the single (^{37}Cl, three ^{13}C and four D) and double (^{37}Cl^{37}Cl and four ^{37}ClD) substitution.

Rovibrational corrections, ΔB_{vib}, to experimental ground-state rotational constants were calculated with quadratic and cubic force constants from MP2/6–311+G(2d,p) computations.

No	**58**
SF	$C_6H_4N_2$
CN	**2-Cyanopyridine**
Symmetry	C_s
Methods	**GED** supplemented by **MW** and **ab initio**
Ref	[67]

Distances	r_e^{se} [Å] [a,b]
N–C2	1.337(2) [1]
C2–C3	1.393(2) [2]
C3–C4	1.387(2) [2]
C4–C5	1.384(2) [2]
C5–C6	1.391(2) [2]
N1–C6	1.332(2) [1]
C2–C7	1.445(3)
C7–N8	1.158(5)

(continued)

(continued)

C3–H	1.079(6) [3]
C4–H	1.080(6) [3]
C5–H	1.080(6) [3]
C6–H	1.081(6) [3]

Bond angles	θ_e^{se} [deg] [a,b]
C2–N1–C6	117.1(2) [c]
N1–C2–C3	124.3(2) [c]
N1–C6–C5	123.2(2) [c]
C4–C5–C6	118.9(3) [4]
C3–C4–C5	118.9(3) [4]
C2–C3–C4	117.6(3) [4]
C3–C2–C7	119.3(2) [5]
N1–C2–C7	116.5(2) [5]
C2–C3–H	120.2 [d]
C4–C3–H	122.2 [d]
C3–C4–H	120.3 [d]
C5–C4–H	120.7 [d]
C4–C5–H	121.1 [d]
C6–C5–H	120.0 [d]
N1–C6–H	116.1(3) [c]
C5–C6–H	120.7 [d]
C2–C7–N8	177.6(5) [c,e]

[a] Parenthesized uncertainties in units of the last significant digit are 3σ and 2σ for the distances and angles, respectively

[b] Parameters with the same numerical superscripts were refined in one group; differences between parameters in each group were assumed at the calculated values of CCSD(T)_ae/cc-pwCVQZ quality

[c] Dependent parameter

[d] Assumed at the value from calculations as indicated above

[e] Cyano group is bent away from the N1 atom

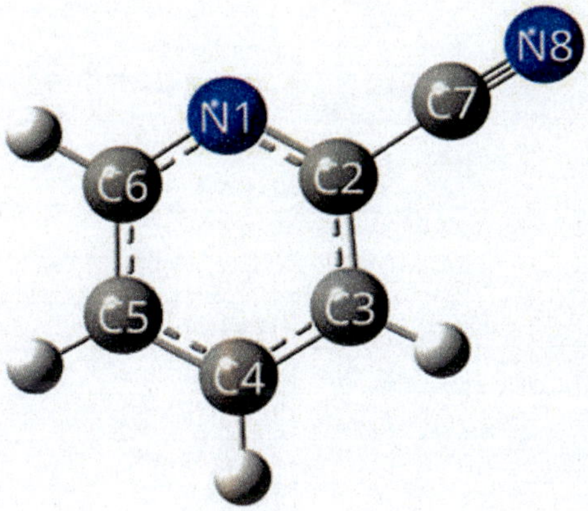

Vibrational corrections to experimental internuclear distances at the temperature of experiment (T_{nozzle} = 327 K) were calculated using scaled quadratic and cubic force fields (MP2/cc-pVTZ); rovibrational corrections, ΔB_{vib}, to experimental ground-state rotational constants taken from literature were calculated with the same force constants.

No	**59**
SF	**$C_6H_4N_2$**
CN	**3-Cyanopyridine**
Symmetry	**C_s**
Methods	**GED** supplemented by **MW** and **QC**
Ref	[68]

Distances	r_e^{se} [Å] [a,b]
N1–C2	1.330(1) [1]
C2–C3	1.397(1) [1]
C3–C4	1.391(1) [1]
C4–C5	1.381(1) [1]
C5–C6	1.389(1) [1]
C6–N1	1.336(1) [1]
C3–C7	1.432(2) [2]
C7–N8	1.157(2) [2]
C2–H	1.088(17) [3]
C4–H	1.086(17) [3]
C5–H	1.086(17) [3]
C6–H	1.088(17) [3]

Bond angles	θ_e^{se} [deg] [a,b]
C6–N1–C2	116.5(3) [c]
N1–C2–C3	124.0(5) [4]
C2–C3–C4	118.5 [d]
C3–C4–C5	118.2(1) [c]
C4–C5–C6	118.8(1) [c]
N1–C6–C5	124.1(5) [4]
C7–C3–C2	120.3(1) [5]
C7–C3–C4	121.3(1) [5]
H–C2–N1	115.6(23) [c]
H–C2–C3	120.4(23) [6]
H–C4–C3	120.3(1) [c]
H–C4–C5	121.5(1) [c]
H–C5–C4	120.9 [d]
H–C5–C6	120.3(1) [c]
H–C6–C5	121.1(23) [6]
H–C6–N1	114.8(24) [c]
C3–C7–N8	180.5(1)

The table data are reprinted from: [68], Copyright 2018, with permission from Elsevier

[a] Parenthesized uncertainties in units of the last significant digit are 3σ and 2σ for the bond lengths and angles, respectively

[b] Parameters with equal superscripts were refined in one group; differences between parameters in each group were assumed at the calculated values of CCSD(T)_ae/cc-pwCVQZ quality

[c] Dependent parameter

[d] Assumed at the value from calculations as indicated above

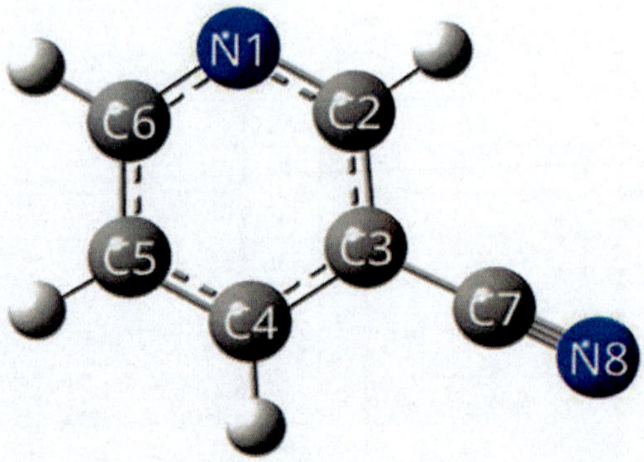

Vibrational corrections to experimental internuclear distances at the temperature of experiment (T_{nozzle} = 336 K) were calculated using scaled quadratic and

cubic force fields (MP2/cc-pVTZ); rovibrational corrections, ΔB_{vib}, to experimental ground-state rotational constants taken from literature were calculated with the same force constants.

No	**60a**
SF	$C_6H_4N_2$
CN	**4-Cyanopyridine**
Symmetry	C_{2v}
Methods	**GED** supplemented by **MW** and **QC**
Ref	[69]

Distances	r_e^{se} [Å] [a,b]
N1–C2	1.335(1)[1]
C2–C3	1.388(1)[1]
C3–C4	1.391(1)[1]
C4–C7	1.438(3)[2]
C7–N8	1.159(3)[2]
C2–H	1.089(12)[3]
C3–H	1.087(12)[3]

Bond angles	θ_e^{se} [deg] [a]
C6–N1–C2	116.8(4)
N1–C2–C3	124.3(4) [c]
C2–C3–C4	117.5(4) [c]
C3–C4–C5	119.5(4)
C7–C4–C3	120.2(3) [c]
H–C2–N1	115.4(4) [c]
H–C2–C3	120.3(5) [c]
H–C3–C2	121.3(5) [c]
H–C3–C4	121.2(5) [c]

[a] Parenthesized uncertainties in units of the last significant digit are 3σ and 2σ for the bond lengths and angles, respectively

[b] Parameters with equal superscripts were refined in one group; differences between parameters in each group were assumed at the calculated values of CCSD(T)_ae/cc-pwCVQZ quality

[c] Dependent parameter

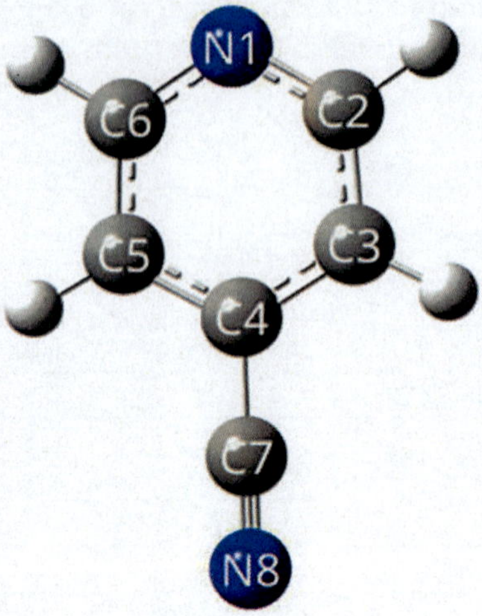

Vibrational corrections to experimental internuclear distances at the temperature of experiment ($T_{nozzle} \approx 335$ K) were calculated with scaled quadratic and cubic force fields from MP2/cc-pVTZ computation; rovibrational corrections, ΔB_{vib}, to experimental ground-state rotational constants taken from literature were calculated with the same force constants.

No	**60b**
Symmetry	C_{2v}
Methods	**MW** supplemented by **QC**
Ref	[70]

Distances	r_e^{se} [Å] [a]
N1...C4	2.7882(6)
N1–C2	1.3351(4)
C2–C3	1.3890(5)
C3–C4	1.3921(5)
C4–C7	1.4368(7)
C7–N8	1.1573(6)
C2–H	1.0815(5)
C3–H	1.0792(5)

Bond angles	θ_e^{se} [deg] [a]
C6–N1–C2	117.30(3)
N1–C2–C3	123.81(5)
C2–C3–C4	118.01(5)
C3–C4–C5	120.47(4)
H–C2–N1	116.16(9)
H–C3–C4	120.93(9)

[a] Parenthesized uncertainties in units of the last significant digit are 1σ

The centimeter-wave rotational spectra of the parent species and all singly substituted ^{13}C and ^{15}N isotopic species were recorded by a pulsed supersonic jet-expansion MB-FTMW spectrometer. As the number of determined ground-state rotational constants was not large enough to determine a complete structure, the predicate-regression mixed-estimation method was used.

Rovibrational corrections, ΔB_{vib}, to experimental ground-state rotational constants were calculated with quadratic and cubic force constants from MP2/cc-pVTZ computations.

The calculated structure of CCSD(T)_ae/cc-pwCVQZ quality agrees well with the determined r_e^{se} structure.

No	**61**
SF	$\mathbf{C_6H_5NO_2}$
CN	**2-Pyridinecarboxylic acid** Picolinic acid
Symmetry	$\mathbf{C_s}$ *(ap-ap)*
Methods	**GED** supplemented by **QC**
Ref	[71]

Distances	r_e^{se} [Å] [a,b]
C2–C3	1.391(3) [1]
C3–C4	1.389(3) [1]
C4–C5	1.389(3) [1]
C5–C6	1.391(3) [1]
N1–C6	1.338(4) [2]
N1–C2	1.340(4) [2]
C2–C7	1.504(4)
C7–O8	1.204(3)

(continued)

(continued)

C7–O9	1.337(4) [2]
C3–H	1.079 [c]
C4–H	1.081 [c]
C5–H	1.080 [c]
C6–H	1.081 [c]
O9–H	0.975(21)
N1…H(–O9)	1.973(27)

Bond angles	θ_e^{se} [deg] [a,b]
C2–C3–C4	117.6(1) [3]
C3–C4–C5	119.2(1) [3]
C4–C5–C6	119.2(1) [3]
C7–C2–C3	120.8(5)
C2–C7–O8	122.4(6)
C2–C7–O9	113.6(6)
C7–O9–H	105.9 [c]
C2–C3–H	119.4 [c]
C3–C4–H	120.5 [c]
C4–C5–H	121.1 [c]
C5–C6–H	121.1 [c]
O9–H…N1	121.7(6) [d]
C3–C2–N1	123.6(6) [d]
C5–C6–N1	122.0(6) [d]
C6–N1–C2	118.4(12) [d]
N1–C2–C7	115.7(9) [d]
O8–C7–O9	124.0(6) [d]
C4–C3–H	123.0(1) [d]
C5–C4–H	120.3(1) [d]
C6–C5–H	119.8(1) [d]
N1–C6–H	116.9(6) [d]

[a] Parenthesized uncertainties in units of the last significant digit are 3σ

[b] Parameters denoted by the same numerical superscripts were refined in one group; differences between parameters in each group were assumed at the calculated values of CCSD(T)_ae/cc-pwCVQZ quality

[c] Assumed at the values calculated at the level of theory as indicated above

[d] Dependent parameter

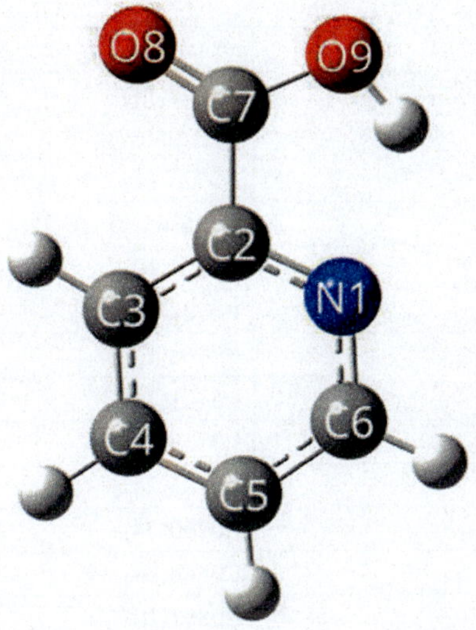

According to predictions of B3LYP/cc-pVTZ, MP2/cc-pVTZ, and CCSD(T)_ae/cc-pwCVTZ calculations, the *ap-ap* conformer of the title molecule with the *antiperiplanar* conformations of the O8–C7–C1–N1 and O8–C7–O9–H fragments is the lowest energy one with respect to the three other possible conformers, *ap-sp, sp-sp,* and *sp-ap*. The energy difference between the lowest energy conformers, *ap-ap* and *ap-sp* (with the *synperiplanar* O8–C7–O9–H fragment) was predicted to be more than 15 kJ mol^{-1}, the corresponding amount of the *ap-sp* conformer is about 2% (at the temperature of experiment, $T_{nozzle} = 375(3)$ K). The *ap-sp* conformer was not detected by GED.

The main conformer was found to be stabilized by the N1…H(–O9) hydrogen bond.

Vibrational corrections to experimental internuclear distances were calculated with quadratic and cubic force constants from MP2/cc-pVTZ computation.

The r_e^{se} and calculated (CCSD(T)_ae/cc-pwCVQZ) structures were found to be in excellent agreement.

No	**62**
SF	**C_6H_6OS**
CN	**2-Acetylthiophene**
Symmetry	**C_s** (*syn*)
Methods	**MW** supplemented by **QC**
Ref	[56]

Distances	r_e^{se} [Å] [a]
C2–S	1.7192(17)
S–C5	1.70705(67)
C2–C3	1.3725(18) [b]
C3–C4	1.41538(87)
C4–C5	1.3654(10)
C2–C6	1.4728(19)
C6–O7	1.2168(17)
C6–C8	1.5101(11)
C3–H	1.0789(24)
C4–H	1.0786(24)
C5–H	1.0760(24)
C–Hs [c]	1.0846(24)
C–Ha [d]	1.0891(20)

Bond angles	θ_e^{se} [deg] [a]
C–S–C	91.404(58)
S–C5–C4	112.592(24)
C3–C4–C5	111.917(21)
S–C2–C6	119.25(14)
C2–C6–O	120.17(13)
C2–C6–C8	117.63(12)
C6–C8–Hs [c]	108.73(23)
C6–C8–Ha [d]	109.90(19)
S–C2–C3	111.50(12) [b]
S–C5–H	120.17(24)
C5–C4–H	124.45(23)
C4–C3–H	124.05(24)

Dihedral angles	τ_e^{se} [deg] [a]
C2–C6–C8–Ha [c]	± 59.35(21)

The table data are reprinted from: [56], Copyright 2022, with permission from the Royal Society of Chemistry

[a] Parenthesized uncertainties in units of the last significant digit are 1σ

[b] Derived parameter

[c] Hs is hydrogen atom in symmetry plane

[d] Ha is hydrogen atom out of symmetry plane

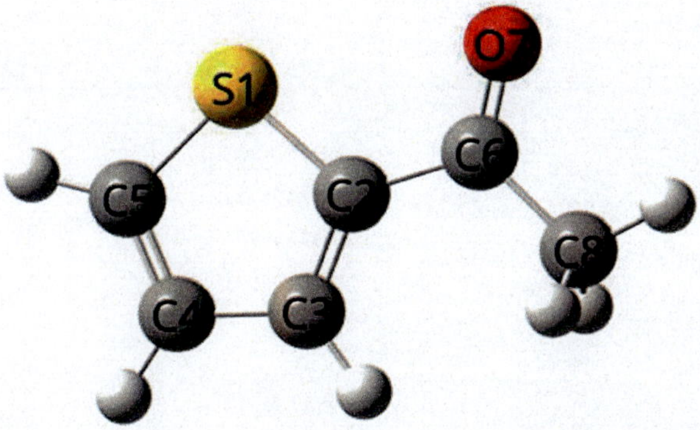

Two conformers, *syn* and *anti*, with the *synperiplanar* and *antiperiplanar* S C–C=O fragments, respectively, were identified in the MW spectra. The spectra of the ^{13}C and ^{34}S isotopologues of the *syn* conformer were also assigned, and the structure of this energetically more favorable conformer was determined.

Rovibrational corrections, ΔB_{vib}, to experimental ground-state rotational constants were calculated with quadratic and cubic force constants from B2PLYP-D3/6–311+G(3df,2pd) computations.

The calculated structure of CCSD(T)_ae/cc-pwCVQZ quality agrees well with the r_e^{se} structure.

The barrier to internal rotation of the methyl group around the C6–C8 bond was determined to be 330.187(35) and 295.957(17) cm^{-1} for the *syn* and *anti* conformers, respectively. Predicted torsional barrier heights (MP2/6–311++G(d,p)) are 320.5 and 284.6 cm^{-1} for the *syn* and *anti* conformers, respectively.

No	**63**
SF	$\mathbf{C_6H_6O_2}$
CN	**2-Acetylfuran**
Symmetry	$\mathbf{C_s}$ (*anti*)
Methods	**MW** supplemented by **QC**
Ref	[72]

Distances	r_e^{se} [Å] [a]
O1–C2	1.3666(5)
C3–C4	1.4238(4)
C4–C5	1.35873(9)
C5–O1	1.3528(3)
C2–C6	1.4685(14)
C6–O7	1.2105(4)

(continued)

(continued)

C6–C8	1.5037(4)
C3–H	1.0754(15)
C4–H	1.0750(16)
C5–H	1.0755(15)
C–Hs [b]	1.0854(16)
C–Ha [c]	1.0897(10)

Bond angles	θ_e^{se} [deg] [a]
C2–O1–C5	106.88(4)
O1–C5–C4	110.925(12)
C3–C4–C5	106.040(14)
O1–C2–C6	118.71(4)
C2–C6–O7	119.56(3)
C2–C6–C8	116.98(3)
C6–C8–Hs [b]	108.85(20)
C6–C8–Ha [c]	110.13(14)
O1–C5–H	115.97(21)
C5–C4–H	126.09(18)
C4–C3–H	128.67(22)

Dihedral angles	τ_e^{se} [deg] [a]
C2–C6–C8–Ha [c]	± 58.99(12)

The table data are reprinted with permission from: [72], Copyright 2021, American Chemical Society

[a] Parenthesized uncertainties in units of the last significant digit are 1σ

[b] Hs is hydrogen atom in symmetry plane

[c] Ha is hydrogen atom out of symmetry plane

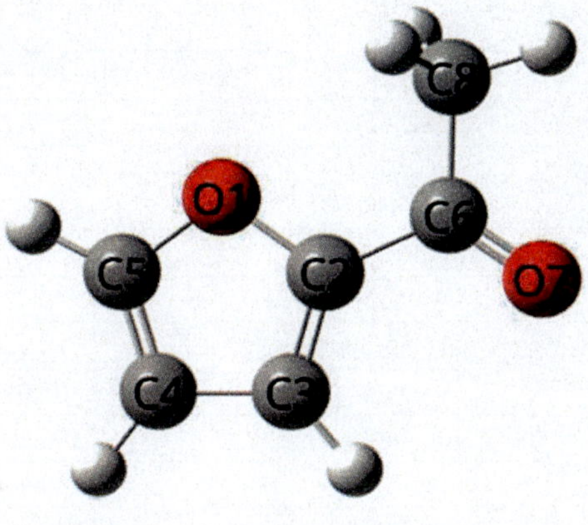

Two conformers, *anti* and *syn*, with the *antiperiplanar* and *synperiplanar* O–C–C=O fragments, respectively, were identified in MW spectra. The spectra of the all ^{13}C and ^{18}O isotopologues of the *anti* conformer were also assigned, and the structure of this energetically more favorable conformer was determined.

Rovibrational corrections, ΔB_{vib}, to experimental ground-state rotational constants were calculated with quadratic and cubic force constants from B2PLYP-D3/6–311+G(3df,2pd) computations.

The calculated structure of CCSD(T)_ae/cc-pwCVQZ quality agrees well with the r_e^{se} structure.

A barrier to internal rotation of the methyl group about the C6–C8 bond was determined to be 319.777(12) and 239.780(13) cm^{-1} for the *anti* and *syn* conformers, respectively.

No	**64**
SF	**C_6H_8O**
CN	**2-Ethylfuran**
Symmetry	**C_1** (I) **C_s** (II)
Methods	**MW** supplemented by **ab initio**
Ref	[73]

Distances	r_e^{se} [Å] [a,b]	
	I	II
O1–C2	1.3072(72)	1.3631(88)
C2–C3	1.3535(98)	1.3534(99)
C3–C4	1.4418(29)	1.4430(50)
C4–C5	1.3175(40)	1.3515(41)
O1–C5	1.3941(54)	1.3651(27)

(continued)

(continued)

C2–C6	1.553(12)	1.4936(76)
C6–C7	1.5282(41)	1.5178(31)

Bond angles	θ_e^{se} [deg] [a,b]	
	I	II
O1–C2–C3	113.86(86)	110.42(54)
C2–C3–C4	103.25(45)	106.08(38)
C3–C4–C5	106.46(15)	106.10(11)
C4–C5–O1	110.43(20)	110.44(15)
O1–C2–C6	114.98(65)	116.46(69)
C3–C2–C6	130.30(54)	133.12(68)
C2–C6–C7	113.91(30)	112.26(41)

Dihedral angles	τ_e^{se} [deg] [a,b]	
	I	II
O1–C2–C6–C7	62.30(59)	180 [c]
C3–C2–C6–C7	−106.25(81)	0 [c]

The table data are reprinted from: [73], Copyright 2020, with permission from Elsevier

[a] Parenthesized uncertainties given in units of the last significant digit were not specified, probably 1σ

[b] The Cartesian coordinates of the atoms were calculated using the Kraitchman's equations with, as input, the semiexperimental rotational constants, the rovibrational corrections, ΔB_{vib}, being calculated with quadratic and cubic force fields from MP2/6-31G(d,p) computations. Then, the internal coordinates were derived from these Cartesian coordinates

[c] Derived from setting all experimental *c*-coordinates to zero

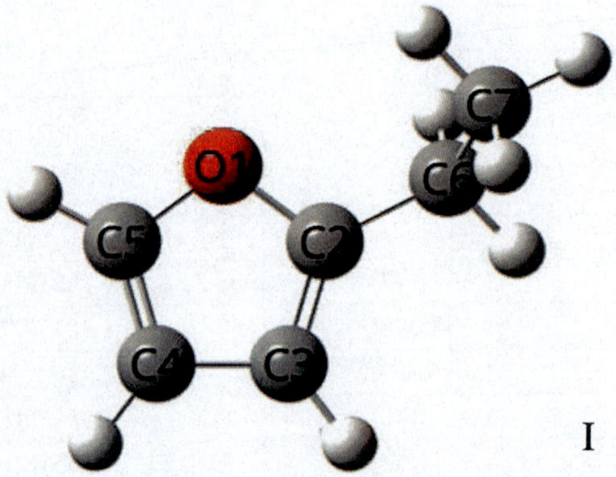

I

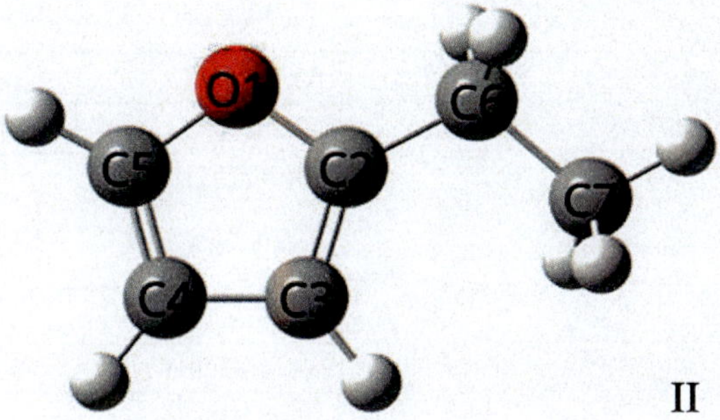

II

The existence of two conformers, I and II, differing in the magnitude of the O–C2–C6–C7 torsional angle (*ca.* 60 and 180°, respectively) was predicted by ab initio. The conformer I with C_1 symmetry is lower in energy in comparison with the conformer II of C_s symmetry by 1.7 kJ mol^{-1} (with ΔE_{ZPE} correction, MP2/6–311++G(d,p)).

The spectra of the parent species and all singly substituted ^{13}C and ^{18}O isotopic species were recorded for both conformers by a molecular jet FTMW spectrometer.

No	**65**
SF	$C_6H_{10}S_2$
CN	**3,3′-Dithiobis-1-propene** Diallyl disulfide
Symmetry	C_1
Methods	**MW** supplemented by **QC**
Ref	[74]

Distances	$r_m^{(1)}$ [Å] [a]
C2–C3	1.3337(15)
C1–C2	1.4870(14)
S–C1	1.8325(10)
S–S	2.0282(13)
S–C1′	1.8332(15)
C1′–C2′	1.4851(16)
C2′–C3′	1.3314(12)

Bond angles	$\theta_m^{(1)}$ [deg] [a]
C3–C2–C1	123.50(16)
C2–C1–S	112.92(10)
C1–S–S	102.284(92)

(continued)

(continued)

S–S–C1′	103.910(35)
S–C1′–C2′	112.24(15)
C1′–C2′–C3′	123.268(51)

Dihedral angles	$\tau_m^{(1)}$ [deg] [a]
S–C1–C2–C3	−114.43(14)
S–S–C1–C2	68.93(27)
C1–S–S–C1′	−93.63(14)
S–S–C1′–C2′	−63.44(16)
S–C1′–C2′–C3′	115.05(24)

[a] Parenthesized uncertainties in units of the last significant digit are 1σ

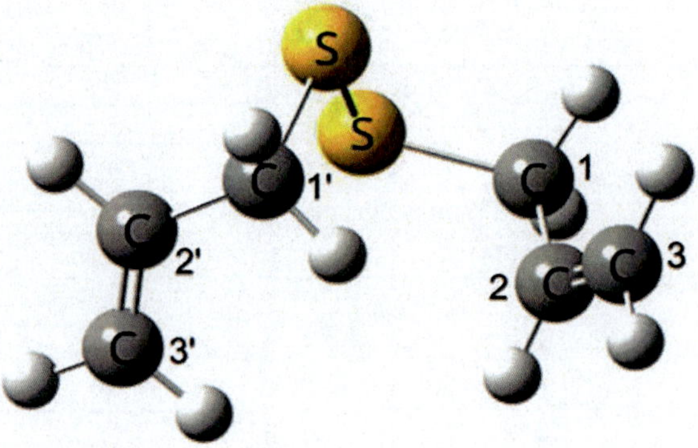

Ground-state rotational constants of the parent species and all heavy-atom isotopic species were determined by chirped-pulse (broadband) high-resolution microwave spectroscopy. Only one conformer (with C_1 point-group symmetry) was detected in the experiment.

The r_e^{se} structure of this floppy molecule with four low-frequency vibrations (from 40 to 98 cm^{-1}, MP2/cc-pV(T+d)) could not been accurately determined for the reason of low accuracy of anharmonic force field.

The determined $r_m^{(1)}$ structure being in good agreement with calculated equilibrium structure of CCSD(T)_ae/cc-pCV(Q+d)Z quality was suggested as a good estimation of r_e^{se} structure.

No	**66**
SF	$C_7HF_{13}O$
CN	**2,2,3,3,4,4,5,5,6,6-Decafluoro-1-(trifluoromethyl)cyclohexan-1-ol**
Symmetry	C_1
Methods	**GED** supplemented by **DFT**
Ref	[65]

Distances	r_e^{se} [Å] [a,b]
C1–C2	1.553(1) [1]
C1–O6	1.413(1) [2]
C1–C8	1.548(1) [1]
C1–C9	1.554(1) [1]
C2–F3	1.323(1) [2]
C2–F4	1.325(1) [2]
C2–F5	1.343(1) [2]
O6–H	0.975(12) [3]
C8–C10	1.544(1) [1]
C8–F13	1.333(1) [2]
C8–F14	1.351(1) [2]
C9–C11	1.542(1) [1]
C9–F15	1.350(1) [2]
C9–F16	1.341(1) [2]
C10–C12	1.534(1) [1]
C10–F17	1.337(1) [2]
C10–F18	1.339(1) [2]
C11–C12	1.533(1) [1]
C11–F19	1.336(1) [2]
C11–F20	1.343(1) [2]
C12–F21	1.339(1) [2]
C12–F22	1.344(1) [2]

Bond angles	θ_e^{se} [deg] [a,b]
C2–C1–O6	110.0(10)
C2–C1–C8	112.3(3) [4]
C2–C1–C9	112.1(3) [4]
C1–C2–F3	111.5(4) [5]
C1–C2–F4	107.9(4) [5]

(continued)

(continued)

C1–C2–F5	105.1(4) [5]
O6–C1–C8	100.7(10)
O6–C1–C9	109.0(6) [6]
C1–O6–H	111.9(10)
C8–C1–C9	112.3(6) [c]
C1–C8–C10	115.6(6) [6]
C1–C8–F13	111.6(2) [7]
C1–C8–F14	109.1(2) [7]
C1–C9–C11	112.4(2) [7]
C1–C9–F15	108.9(2) [7]
C1–C9–F16	112.5(2) [7]
F3–C2–F4	109.3(7) [8]
F3–C2–F5	109.0(7) [8]
C8–C10–F17	109.3(2) [7]
C8–C10–F18	110.3(2) [7]
C9–C11–F19	109.6(2) [7]
C9–C11–F20	109.9(2) [7]
C10–C12–F21	110.6(2) [7]
C10–C12–F22	109.3(2) [7]
C11–C12–F21	110.4(2) [7]
C11–C12–F22	108.9(2) [7]

Dihedral angles	τ_e^{se} [deg] [a,b]
F3–C2–C1–O6	−177.5(11)
C2–C1–O6–H	62.1(11)
C2–C1–C8–C10	−176.0(14)
C2–C1–C8–F13	65.0(6) [9]
C2–C1–C9–C11	172.7(14)
C2–C1–C9–F15	−65.0(6) [9]
O6–C1–C8–F14	−170.0(6) [10]
O6–C1–C9–F16	174.0(6) [10]
H–O6–C1–C9	−61.1(14)
F17–C10–C8–F13	−65.3(6) [9]
F18–C10–C8–F14	169.0(6) [10]
F19–C11–C9–F15	68.8(6) [9]
F20–C11–C9–F16	−163.5(6) [10]

The table data are reprinted from: [65], Copyright 2018, with permission from the Royal Society of Chemistry

[a] Distance and angle values are rounded here to thousandths of Å and tenths of degree, respectively. Parenthesized uncertainties in units of the last significant digit are 3σ
[b] Parameters with equal superscripts were refined in one group. Differences between parameters in each group were assumed at the values derived from the PBE0/cc-pVTZ structure
[c] Dependent parameter

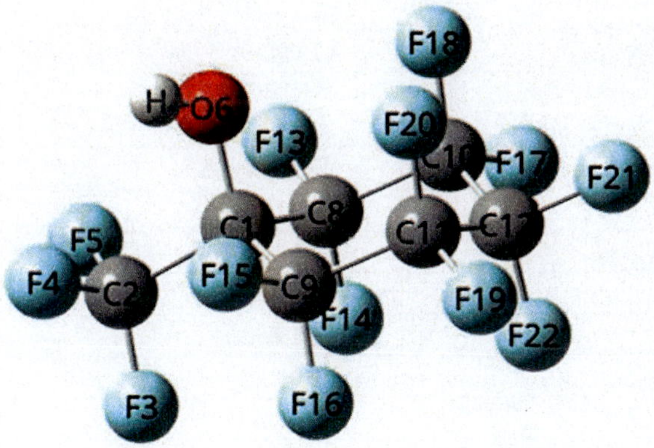

Vibrational corrections to the experimental internuclear distances (T_{nozzle}≈293 K) were calculated with quadratic and cubic force constants from computation at the level of theory as indicated above.

No	**67**
SF	$C_7H_3F_5Te$
CN	**1,2,3,4,5-Pentafluoro-6-(methyltelluro)benzene** Methyl(pentafluorophenyl) telluride
Symmetry	C_1
Methods	**GED** supplemented by **DFT**
Ref	[75]

Distances	r_e^{se} [Å] [a]
C1–C2	1.388(4)
C2–C3	1.387(4)
C3–C4	1.388(4)
C4–C5	1.386(4)
C5–C6	1.386(4)
C1–C6	1.392(4)
C2–F	1.328(5)

(continued)

(continued)

C3–F	1.326(5)
C4–F	1.316(5)
C5–F	1.331(5)
C6–F	1.330(5)
C1–Te	2.077(5)
Te–C7	2.164(10)
C7–H	1.090(6) [b]

Bond angles	θ_e^{se} [deg] [a]
C1–C2–C3	122.0(1)
C2–C3–C4	119.5(1)
C3–C4–C5	119.9(1)
C4–C5–C6	119.5(1)
C1–C6–C5	121.9(1)
C2–C1–C6	117.2(1)
C1–C2–F	120.4(1)
C2–C3–F	120.7(1)
C3–C4–F	120.1(1)
C4–C5–F	119.8(1)
C1–C6–F	120.6(1)
C6–C1–Te	124.9(4)
C2–C1–Te	117.9(4)
C1–Te–C7	93.7(9)
H–C–H	110.4(10) [b]

Dihedral angles	τ_e^{se} [deg] [a]
C2–C1–Te–C7	117.9(14)
C5–C6–C1–Te	−179.5(23)
C6–C1–Te–C7	−63.7(20)

[a] Parenthesized uncertainties in units of the last significant digit are 3σ

[b] Average value

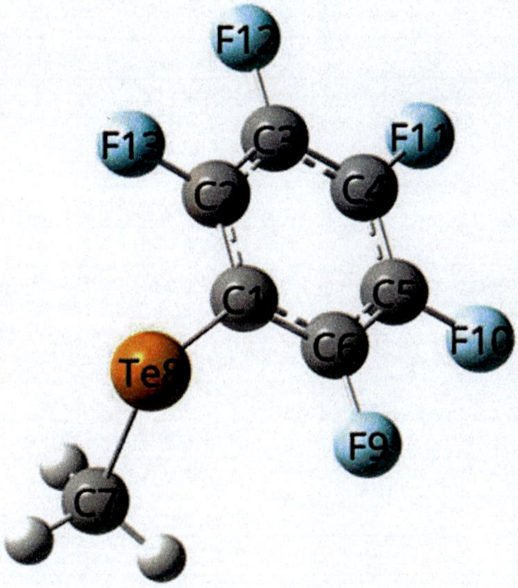

Selected structural parameters are given in these tables.

Vibrational corrections to the experimental internuclear distances at the temperature of experiment (T_{nozzle}≈366 K) were calculated with quadratic and cubic force fields from PBE0-D3/def2-TZVP computation.

No	**68**
SF	$\mathbf{C_7H_6N_2}$
CN	**1*H*-Indazole**
Symmetry	$\mathbf{C_s}$
Methods	**MW** supplemented by **QC**
Ref	[76]

Distances	r_e^{se} [Å] [a]
N1–N2	1.3543(9)
N2–C3	1.3185(9)
C3–C3a	1.4247(14)
C4–C3a	1.4040(16)
C4–C5	1.3791(8)
C5–C6	1.4120(9)
C6–C7	1.3803(7)
C7–C7a	1.4006(16)

(continued)

(continued)

C3a–C7a	1.4064(8)
N1–C7a	1.3610(15)

Bond angles	θ_e^{se} [deg] [a]
N1–N2–C3	105.87(2)
N2–C3–C3a	111.42(5)
C3–C3a–C4	135.94(5)
C3a–C4–C5	118.23(5)
C4–C5–C6	121.15(2)
C5–C6–C7	121.80(2)
C6–C7–C7a	116.72(5)
C7–C7a–N1	131.75(5)
C7a–N1–N2	112.50(6)
C3–C3a–C7a	104.30(15)
N1–C7a–C3a	105.92(16)
C4–C3a–C7a	119.76(15)
C7–C7a–C3a	122.33(14)

The table data are reprinted from: [76], Copyright 2019, with permission from Wiley–VCH Verlag GmbH & Co. KGaA, Weinheim

[a] Parenthesized uncertainties in units of the last significant digit were not specified, probably 1σ

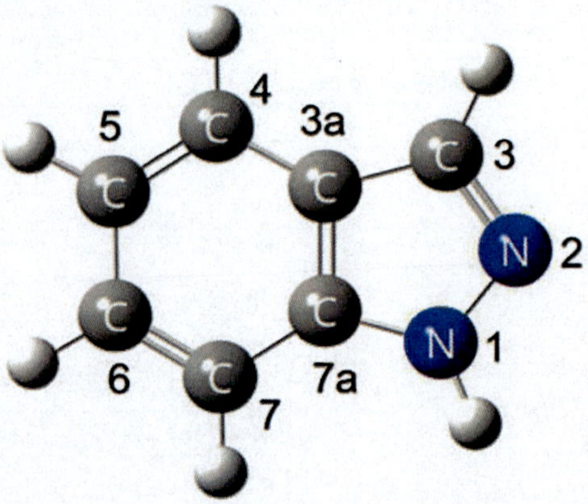

Rotational spectra of the parent species and 17 singly and doubly substituted ^{13}C and ^{15}N isotopic species were measured by a chirped pulse FTMW spectrometer.

Rovibrational corrections, ΔB_{vib}, to experimental ground-state rotational constants were calculated with ab initio quadratic and cubic force fields.

No	**69**
SF	$C_7H_6O_2$
CN	**Benzoic acid**
Symmetry	C_s
Methods	**GED** supplemented by **QC**
Ref	[25]

Distances	r_e^{se} [Å] [a,b]
C1–C7	1.485(4)
C7–O8	1.342(4)
C7–O9	1.202(4)
C1–C2	1.389(5)
C1–C6	1.389(5)
C2–C3	1.383(5)
C3–C4	1.388(5)
C4–C5	1.387(5)
C5–C6	1.385(5)
C–H	1.088(4) [c]
O8–H	0.973(4)

Bond angles	θ_e^{se} [deg] [a,b]
C1–C7–O8	112.2(4)
C1–C7–O9	124.6(4)
C7–C1–C2	116.7(5)
C7–C1–C6	121.8(4)
O8–C7–O9	123.2(5)
C2–C1–C6	121.5(5)
C1–C2–C3	120.4(6)
C1–C6–C5	120.1(4)
C2–C3–C4	115.9(10)
C3–C4–C5	126.0(13)
C4–C5–C6	116.1(10)
C3–C2–H	120.9(6)
C4–C3–H	124.1(10)
C5–C6–H	120.2(4)

(continued)

(continued)

C4–C5–H	124.0(10)
C5–C4–H	114.1(13)

The table data are reprinted from: [25], with permission of AIP Publishing
[a] Parenthesized uncertainties in units of the last significant digit are 1σ
[b] Refined parameters were flexibly restrained to the values of CCSD(T)_ae/cc-pwCVTZ structure
[c] Average value

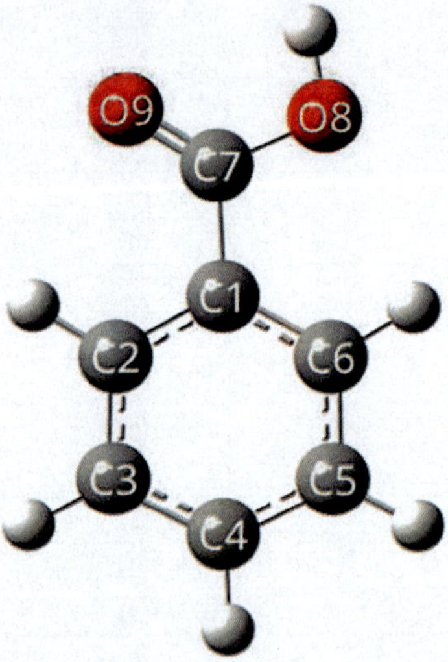

Vibrational corrections to the experimental internuclear distances at the temperature of low-pressure experiment of 288 K were calculated with quadratic and cubic force constants from B3LYP/def2-TZVP computations.

No	**70**
SF	$\mathbf{C_7H_7NOS}$
CN	**2-Hydroxybenzenecarbothioamide** Thiosalicylamide
Symmetry	$\mathbf{C_1}$
Methods	**GED** supplemented by **QC**
Ref	[77]

Distances	r_e^{se} [Å] [a,b]
C1–C2	1.405(2) [1]
C2–C3	1.391(2) [1]

(continued)

(continued)

C3–C4	1.376(2) [1]
C4–C5	1.387(2) [1]
C5–C6	1.376(2) [1]
C6–C1	1.399(2) [1]
C1–C7	1.460(7)
C7–N	1.351(12)
C7–S	1.668(4)
C–H	1.081 [c,d]
N–H	1.006 [c,d]
C–O	1.332(7)
O–H	0.985 [c]

Bond angles	θ_e^{se} [deg] [a]
C1–C2–C3	119.0(2) [2]
C2–C3–C4	121.3(2) [2]
C3–C4–C5	120.4(7) [e]
C4–C5–C6	118.7(9) [e]
C5–C6–C1	122.2(8) [e]
C6–C1–C2	118.3(2) [2]
C2–C1–C7	122.2(2) [2]
C1–C7–S	125.2(5)
C1–C7–N	116.3(8)
C1–C2–O	123.0(5)
S–C7–N	118.4(9) [e]
H–N–H	119.3 [c]

Dihedral angles	τ_e [deg]
C1–C2–C3–C4	2.5 [c]
C2–C3–C4–C5	0.2 [c]
C2–C1–C7–S	28.3 [c]
C6–C1–C7–N	25.9 [c]
C7–S…N–H	177.9 [c]
C7–S…N–H′	14.0 [c]

[a] Parenthesized uncertainties in units of the last significant digit were not specified, probably 3σ

[b] Parameters with the same numerical superscript were refined in one group. Differences between parameters in each group were assumed at the values of MP2/cc-pVTZ structure

[c] Assumed at the value of the structure calculated at the level of theory as indicated above

[d] Average value
[e] Dependent parameter

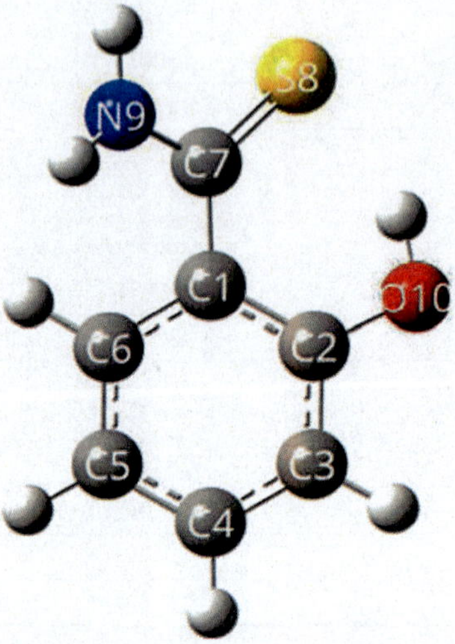

According to predictions of MP2 and B3LYP calculations (with cc-pVTZ basis set), two enantiomers with the C2–C1–C7–S torsional angle of ± 28° (MP2) are separated by a very low barrier (less than 1 kcal mol^{-1}).

In the GED analysis, large-amplitude motion of the thioamide group was described using the model of pseudo-conformers. Differences of structural parameters between the pseudo-conformers were fixed at the values of MP2 structures. Vibrational corrections to the experimental internuclear distances (T_{nozzle} = 401 K) were calculated with MP2 quadratic and cubic force constants.

No	**71**
SF	**C_7H_7NS**
CN	**Thiobenzamide**
Symmetry	**C_1**
Methods	**GED** supplemented by **QC**
Ref	[78]

Distances	r_e^{se} [Å] [a,b]
C1–C2	1.398(2)[1]
C2–C3	1.389(2)[1]
C3–C4	1.395(2)[1]
C4–C5	1.392(2)[1]
C5–C6	1.392(2)[1]
C6–C1	1.401(2)[1]
C1–C7	1.478(9)
C7–N9	1.352(2)[1]
C7–S8	1.641(4)
C–H	1.070(6) [c]
N–H	0.98(1) [c]

Bond angles	θ_e^{se} [deg] [a,b]
C1–C2–C3	120.1(3) [2]
C2–C3–C4	120.3(3) [2]
C3–C4–C5	119.6(3) [2]
C2–C1–C7	119.7(3) [2]
C1–C7–S8	123.4(5)
C1–C7–N9	114.7(5)
C7–N–H [d]	117.7 [e]
C7–N–H [f]	120.3 [e]
C6–C1–C2	119.8(10) [g]
C4–C5–C6	120.6(10) [g]
C5–C6–C1	119.5(15) [g]
S8–C7–N9	121.9(4) [g]

Dihedral angles	τ_e^{se} [deg] [a,b]
C1–C2–C3–C4	0.8 [e]
C2–C3–C4–C5	0.0 [e]
C2–C1–C7–S8	31(4) [3]
C6–C1–C7–N9	29(4) [3]
C1…S8…N–C7	1.1(1) [g]

The table data are reprinted from: [78], Copyright 2018, with permission from Elsevier

[a] Parenthesized uncertainties in units of the last significant digit are 3σ

[b] Parameters with equal superscripts were refined in one group; differences between parameters in each group were assumed at the values from MP2/cc-pVTZ calculations

[c] Average value

[d] N–H is *synperiplanar* to C7–S

[e] Assumed at the value from calculations as above
[f] N–H is *antiperiplanar* to C7–S
[g] Dependent parameter

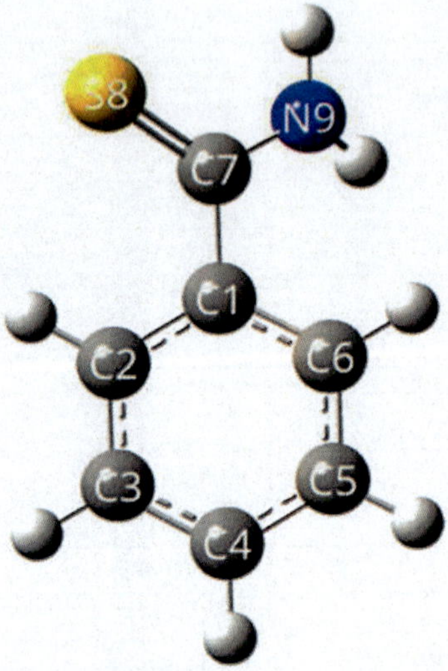

Vibrational corrections to the experimental internuclear distances (T_{nozzle} = 400 K) were calculated with quadratic and cubic force constants from MP2/cc-pVTZ computation.

No	**72**
SF	$C_8H_4O_3$
CN	**1,3-Isobenzofurandione** Phthalic anhydride
Symmetry	C_{2v}
Methods	**MW** supplemented by **QC**
Ref	[79]

Distances	r_e^{se} [Å] [a]
C1–O2	1.3903(4)
C1–C8	1.4798(9)
C8–C9	1.3814(11)
C7–C8	1.3844(8)
C6–C7	1.3921(6)

(continued)

(continued)

C5–C6	1.3984(11)
C1–O10	1.1887(4)
C7–H	1.0803(12)
C6–H	1.0806(12)

Bond angles	θ_e^{se} [deg] [a]
O2–C1–C8	107.21(3)
C1–C8–C9	107.69(10)
C1–O2–C3	110.20(1)
C1–C8–C7	130.40(5)
C6–C7–C8	116.77(4)
C5–C6–C7	121.25(12)
C8–C1–O10	130.40(5)
C8–C7–H	121.24(12)
C7–C6–H	119.56(12)

The table data are reprinted from: [79], Copyright 2022, with permission from Elsevier
[a] Parenthesized uncertainties in units of the last significant digit are 1σ

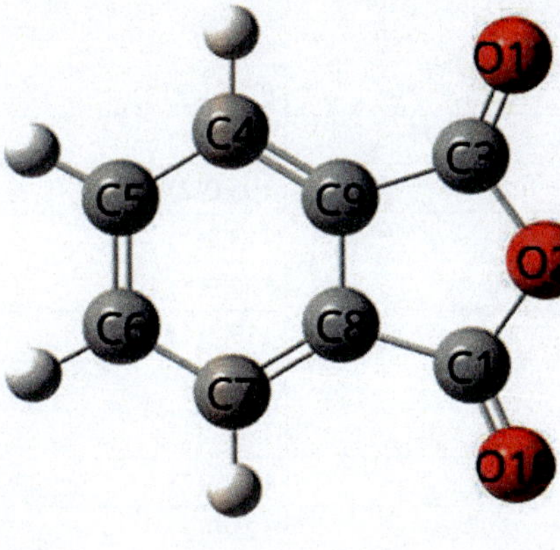

Ground-state rotational constants of the parent and all ^{13}C and ^{18}O isotopic species were taken from literature.

Rovibrational corrections, ΔB_{vib}, to experimental ground-state rotational constants were calculated with quadratic and cubic force constants from B2PLYP/cc-pVTZ computations.

The calculated structure of CCSD(T)_ae/cc-pwCVQZ quality agrees well with the r_e^{se} structure.

No	**73**
SF	**$C_8H_5NO_2$**
CN	**1*H*-Isoindole-1,3(2*H*)-dione** Phthalimide
Symmetry	**C_{2v}**
Methods	**MW** supplemented by **DFT**
Ref	[53]

Distances	r_e^{se} [Å] [a]
C1–C2	1.389(1)
C2–C3	1.490(1)
C3–N4	1.396(3)
C2–C6	1.382(1)
C6–C7	1.395(1)
C7–C8	1.397(1)
C3–O10	1.204(2)
C6–H	1.006(8)
N4–H	1.081(1)
C7–H	1.081(1)

Bond angles	θ_e^{se} [deg] [a]
C1–C2–C3	108.4(1)
N4–C3–C2	105.0(2)
C5–N4–C3	113.2(4)
C6–C2–C3	129.9(1)
C7–C6–C2	117.1(1)
C8–C7–C6	121.2(1)
C2–C3–O10	129.2(1)
C3–N4–H	123.4(2)
C2–C6–H	121.0(1)
C6–C7–H	119.6(1)

[a] Parenthesized uncertainties in units of the last significant digit were not specified, probably 1σ

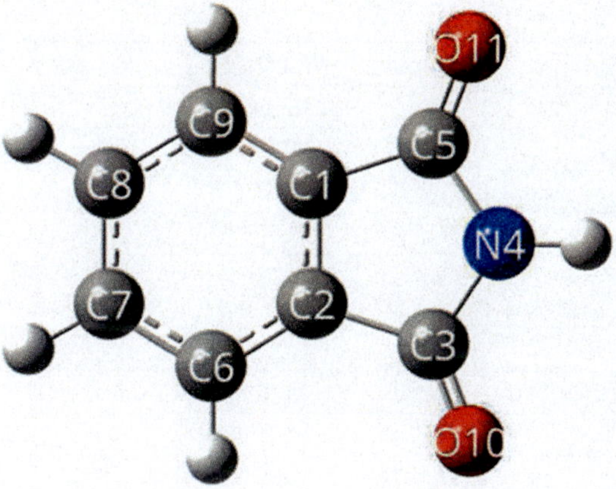

Ground-state rotational constants for the parent and four isotopic (^{13}C1&C2, ^{13}C6&C9, ^{13}C7&C8, and ^{13}C3&C5) species were taken from literature.

Rovibrational corrections, ΔB_{vib}, to experimental ground-state rotational constants were calculated with quadratic and cubic force constants from B2PLYP/cc-pVTZ computations.

No	**74**
SF	$\mathbf{C_8H_6N_2O_2}$
CN	**4-Amino-1*H*-isoindole-1,3(2*H*)-dione** 3-Aminophthalimide
Symmetry	$\mathbf{C_1}$
Methods	**GED** supplemented by **ab initio**
Ref	[80]

Distances	r_e^{se} [Å] [a,b]
C3a–C7a	1.392(3) [1]
C4–C3a	1.391(3) [1]
C5–C4	1.410(3) [1]
C6–C5	1.389(3) [1]
C7–C6	1.403(3) [1]
C7a–C7	1.378(3) [1]
N2–C1	1.397(3) [2]
N2–C3	1.391(3) [2]
N4–C4	1.370(3)
C1–C7a	1.492(5) [3]

(continued)

(continued)

C3–C3a	1.472(5) [3]
C1–O	1.205(3) [4]
C3–O	1.213(3) [4]
N4–H8	1.063(60) [5]
N4–H9	1.059(60) [5]

Bond angles	θ_e^{se} [deg] [b,c]
C3–N2–C1	112.8(2)
C7a–C1–N2	105.4(2)
C3a–C3–N2	105.3(2) [d]
O–C1–N2	125.5(3) [6]
O–C3–N2	125.8(3) [6]
C4–C3a–C3	128.6(5)
C7–C7a–C1	130.2(5)
C4–C3a–C7a	122.1(5) [d]
C5–C4–C3a	115.4(4) [d]
C6–C5–C4	122.3(8) [d]
C7–C6–C5	121.4(7) [d]
C7a–C7–C6	116.3(3)
N4–C4–C5	122.9(3)
C3a–C7a–C7	122.6(6) [d]
C4–N4–H8	116.2(21) [7]
C4–N4–H9	117.8(21) [7]
$\Sigma(-NH_2)$	347.8 [e]

Dihedral angles	τ_e^{se} [deg] [c]
C6–C5–C4–N4	179.5(65)
C5–C4–N4–H8	162.8(106)
C5–C4–N4–H9	22.6(78)

The table data are reprinted from: [80], Copyright 2020, with permission from the Royal Society of Chemistry

[a] Parenthesized uncertainties in units of the last significant digit are estimated total errors including 2.5σ and systematic errors of 0.002*r*

[b] Parameters denoted by the same numerical superscript were refined in one group. Differences between parameter values in each group were assumed at the values of CCSD(T)_ae/cc-pwCVQZ quality

[c] Parenthesized uncertainties in units of the last significant digit are 3σ

[d] Dependent parameter

[e] Sum of the bond angles around the N4 atom

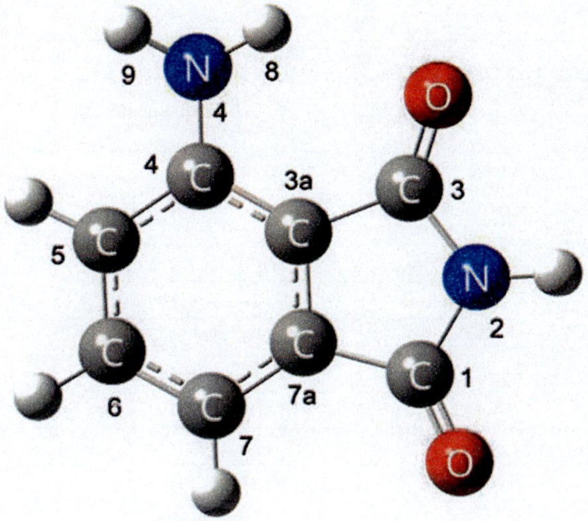

Vibrational corrections to the experimental internuclear distances at the temperature of experiment ($T_{nozzle} = 414$ K) were calculated with quadratic and cubic force constants from MP2/cc-pVTZ computation.

The composite CCSD(T)_ae/cc-pwCVQZ structure agrees well with the r_e^{se} structure.

No	**75**
SF	$\mathbf{C_8H_{12}}$
CN	**Ethynylcyclohexane**
Symmetry	$\mathbf{C_s}$ (equatorial)
Methods	**MW** supplemented by **QC**
Ref	[81]

Distances	r_e^{se} [Å] [a]
C8–H	1.0618(17)
C7–C8	1.2053(11)
C1–C7	1.4633(13)
C1–Hax [b]	1.0954(17)
C1–C2	1.5338(11)
C2–Heq [b]	1.0902(17)
C2–Hax [b]	1.0935(17)
C2–C3	1.5251(11)
C3–Heq [b]	1.0907(17)

(continued)

(continued)

C3–Hax [b]	1.0943(17)
C3–C4	1.5248(10)
C4–Heq [b]	1.0909(17)
C4–Hax [b]	1.0944(17)

Bond angles	θ_e^{se} [deg] [a]
C1–C7–C8	178.39(21)
C7–C1–Hax [b]	107.95(17)
C7–C1–C2	111.060(66)
C1–C2–Heq [b]	109.56(18)
C1–C2–Hax [b]	108.33(20)
Heq–C2–Hax [b]	107.10(26)
C1–C2–C3	110.929(98)
C2–C3–Heq [b]	109.85(17)
C2–C3–Hax [b]	109.13(26)
Heq–C3–Hax [b]	106.88(25)
C2–C3–C4	111.460(89)
C3–C4–Heq [b]	110.332(89)
C3–C4–Hax [b]	109.105(90)
Heq–C4–Hax [b]	106.93(25)
C3–C4–C5	110.940(84)
C2–C1–C6	110.56(10)

Dihedral angles	τ_e^{se} [deg] [a]
C8–C7–C1–Hax [b]	180.0
C3–C2–C1–Hax [b]	−61.91(24)
C8–C7–C1–C2	−61.735(64)
C7–C1–C2–Heq [b]	−57.36(20)
C4–C3–C2–Heq [b]	−178.08(20)
C7–C1–C2–Hax [b]	59.17(29)
C4–C3–C2–Hax [b]	63.69(20)
C7–C1–C2–C3	179.88(17)
C1–C2–C3–Heq [b]	−178.79(20)
Heq–C4–C3–Heq [b]	−59.49(24)
C1–C2–C3–Hax [b]	64.33(20)
Heq–C4–C3–Hax [b]	57.62(33)
C1–C2–C3–C4	−56.09(10)

(continued)

(continued)

C2–C3–C4–Heq [b]	178.13(21)
C2–C3–C4–Hax [b]	−64.68(22)

The table data are reprinted from: [81], with permission of AIP Publishing
[a] Parenthesized uncertainties in units of the last significant digit are 1σ
[b] Hax and Heq are hydrogen atoms in axial and equatorial positions, respectively

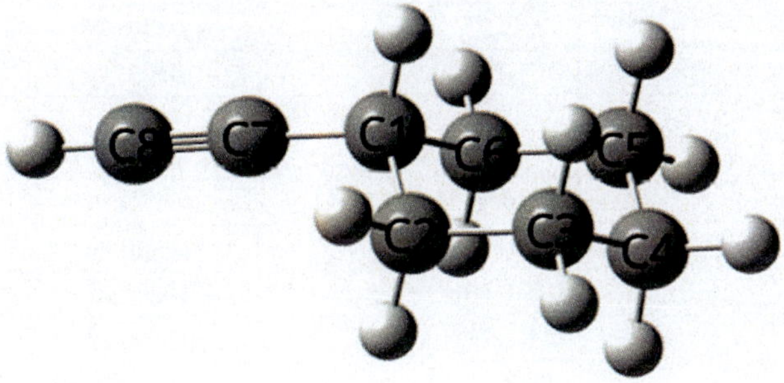

Rotational spectra of the parent species and all ^{13}C isotopic species were recorded by a chirped-pulse FTMW spectrometer.

Rovibrational corrections, ΔB_{vib}, to experimental ground-state rotational constants were calculated with quadratic and cubic force constants from MP2/cc-pVTZ computations.

The composite structure of CCSD(T)_ae/cc-pcVQZ quality agrees well with the r_e^{se} structure.

No	**76**
SF	$C_8H_{18}N_4$
CN	**1,1′-(1,2-Ethanediyl)bis[3,3-dimethyldiaziridine]** Tetramezine
Symmetry	C_i (*meso*) C_2 (enantiomers)
Methods	**GED** supplemented by **QC**
Ref	[82]

Distances	r_e^{se} [Å] [a,b]	
	meso	enantiomers
C1–C2	1.530(4)[1]	1.530(4)[1]
C7–C9	1.518(4)[1]	1.518(4)[1]

(continued)

(continued)

C7–C10	1.519(4)[1]	1.519(4)[1]
N5–N6	1.514(9)[2]	1.514(9)[2]
C1–N5	1.439(3)[3]	1.439(3)[3]
N5–C7	1.426(3)[3]	1.426(3)[3]
N6–C7	1.439(3)[3]	1.439(3)[3]
C1–H	1.085(6)[4,c]	1.086(6)[4,c]
C9–H	1.084(6)[4,c]	1.084(6)[4,c]
C10–H	1.083(6)[4,c]	1.083(6)[4,c]
N6–H	1.0173 [d]	1.0173 [d]

Bond angles	θ_e^{se} [deg] [a,b]	
	meso	enantiomers
C2–C1–N5	107.1(10)[5]	108.4(12)[5]
C9–C7–C10	111.2(13)[6]	111.1(9)[6]
H–C9–H	109.3(18)[7,c]	108.8(18)[7,c]
H–C10–H	109.0(18)[7,c]	109.0(18)[7,c]
H–N6–N5	103.5(1)[8]	103.4(1)[8]
H–N6–C7	107.9(1)[8]	107.9(1)[8]
N5–C7–C9	114.7(9)[9]	114.6(7)[9]
N5–C7–C10	123.2(9)[9]	123.6(7)[9]
N6–C7–C9	119.5(9)[9]	119.5(7)[9]
N6–C7–C10	117.2(9)[9]	117.2(7)[9]
C1–N5–N6	109.8(2)[10]	109.5(2)[10]
C1–N5–C7	118.0(2)[10]	117.9(2)[10]
H–C1–H	109.2 [d]	109.2 [d]
N6–N5–C7	58.5 [e]	58.5 [e]
N5–N6–C7	57.7 [e]	57.7 [e]
N5–C7–N6	63.8 [e]	63.8 [e]

[a] Parenthesized uncertainties in units of the last significant digit are 3σ and 2σ for the bond lengths and angles, respectively

[b] Parameters with the same numerical superscripts were refined in one group; differences between parameters in each group were assumed at the values of MP2/cc-pVTZ structure

[c] Average value

[d] Assumed at the value of calculated structure as indicated above

[e] Dependent parameter

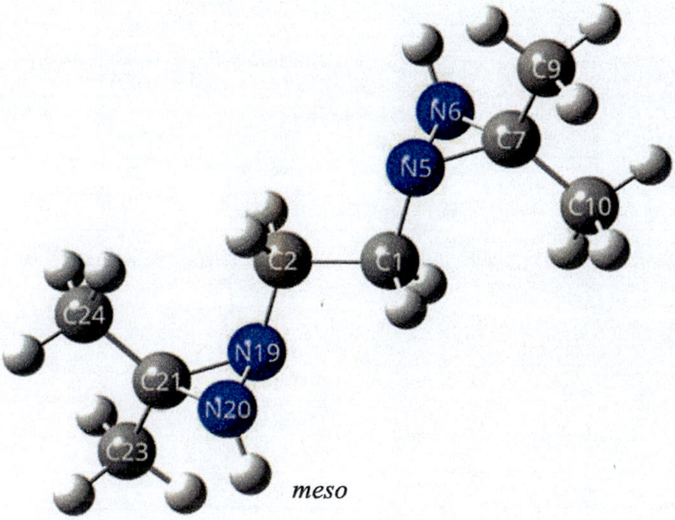

meso

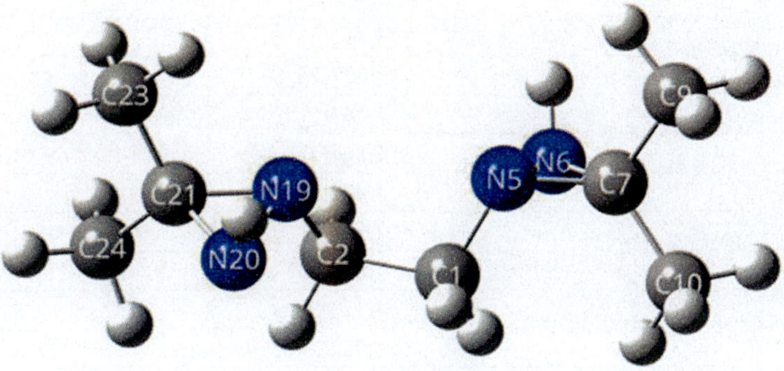

enantiomeric form

According to predictions of MP2 and B3LYP calculations (with cc-pVTZ basis set), the title molecules exist as a mixture of three different diastereomers, achiral *meso* form with C_i point-group symmetry and two enantiomeric forms with C_2 point-group symmetry each. The *meso* form is lower in energy by 4.7 kJ mol^{-1} (MP2) relative to the enantiomers.

According to results of GED analysis, a ratio of *meso*: enantiomers = 70 : 30 (T_{nozzle} = 359 K).

Vibrational corrections to experimental internuclear distances were calculated with scaled quadratic and cubic force fields from B3LYP computations.

No	**77**
SF	$C_9H_6N_2O_3$
CN	**5-Nitroquinolin-8-ol** Nitroxoline
Symmetry	C_1
Methods	**GED** supplemented by **QC**
Ref	[83]

Distances	r_e^{se} [Å] [a,b]
N1–C2	1.311(4)
N11–O13	1.229(1)
N1–C9	1.351(5)
C2–C3	1.395(5)
N11–O12	1.226(1)
C2–H	1.085(27)
C3–C4	1.372(6)
C3–H	1.082(22)
C4–C10	1.413(5)
C4–H	1.077(25)
C5–N11	1.457(5)
C5–C6	1.366(5)
C5–C10	1.422(7)
C6–C7	1.383(5)
C6–H	1.081(23)
C7–C8	1.370(5)
C7–H	1.082(22)
C8–O14	1.325(5)
C8–C9	1.421(5)
O14–H	0.979(18)
C9–C10	1.421(5)

Bond angles	θ_e^{se} [deg] [a,b]
N1–C2–C3	122.7(12)
N1–C2–H	116.6(10)
O13–N11–O12	123.9(4)
O13–N11–C5	118.0(5)

(continued)

(continued)

N1–C9–C8	114.3(4)
N1–C9–C10	123.9(4)
C2–N1–C9	118.3(9)
C2–C3–C4	119.8(10)
C2–C3–H	120.1(20)
O12–N11–C5	118.1(5)
C3–C2–H	120.7(20)
C3–C4–C10	119.8(4)
C3–C4–H	120.3(8)
C4–C3–H	120.2(13)
C4–C10–C5	129.5(6)
C4–C10–C9	115.5(4)
N11–C5–C6	116.3(4)
N11–C5–C10	121.5(6)
C5–C6–C7	121.8(5)
C5–C6–H	118.0(12)
C5–C10–C9	115.0(6)
C6–C5–C10	122.2(6)
C6–C7–C8	119.3(4)
C6–C7–H	120.9(10)
C7–C6–H	120.2(12)
C7–C8–O14	121.3(6)
C7–C8–C9	120.0(4)
C8–C7–H	119.9(11)
O14–C8–C9	118.7(5)
C8–O14–H	104.3(40)
C8–C9–C10	121.7(5)
C10–C4–H	119.9(8)

Dihedral angle	τ_e^{se} [deg] [a]
τ [c]	8(3)

The table data are reprinted with permission from: [83], Copyright 2018, American Chemical Society

[a] Parenthesized uncertainties in units of the last significant digit are 1σ

[b] All bond lengths and angles were refined with restrictions based on the corresponding parameters from computation at the MP2_ae/cc-pwCVTZ level of theory

[c] Twist angle of the nitro group with respect to the hydroxyquinoline plane

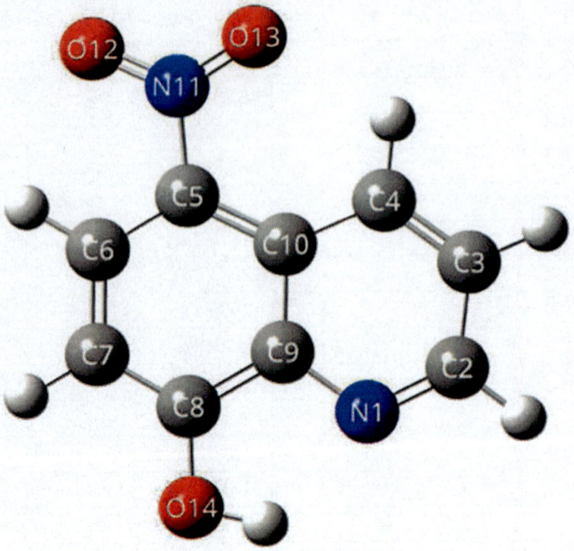

Vibrational corrections to the experimental internuclear distances (T_{nozzle}≈431 K) were calculated with quadratic and cubic force constants from B3LYP/cc-pVTZ computation.

The barrier height to planarity was estimated to be 80 cm^{-1} (MP2_ae/cc-pwCVTZ).

No	**78**
SF	**$C_9H_7NO_2$**
CN	**1-Methyl-1*H*-indole-2,3-dione** 1-Methylisatin
Symmetry	**C_s**
Methods	**GED** supplemented by **QC**
Ref	[84]

Distances	r_e^{se} [Å] [a,b]
C8–C9	1.403(7) [1]
C8–N1	1.401(7) [1]
C2–N1	1.379(7) [1]
C2–C3	1.560(7) [1]
C3–C9	1.474(7) [1]
C7–C8	1.385(7) [1]

(continued)

(continued)

C6–C7	1.401(7) [1]
C5–C6	1.395(7) [1]
C4–C5	1.396(7) [1]
C4–C9	1.384(7) [1]
C2–O11	1.206(5) [2]
C3–O12	1.204(5) [2]
N1–C10	1.443(7) [1]
C–H (ring)	1.088(13) [c]
C–H (methyl)	1.089 [c,d]

Bond angles	θ_e^{se} [deg] [a,b]
C9–C8–N1	111.5(7) [3]
C2–N1–C8	111.8(7) [3]
C3–C2–N1	104.4(6) [e]
C9–C3–C2	106.3(10) [e]
C8–C9–C3	106.0(11) [e]
C7–C8–C9	121.6(6) [4]
C6–C7–C8	117.6(6) [4]
C5–C6–C7	122.4(6) [4]
C4–C5–C6	118.2(6) [4]
C9–C4–C5	121.0(13) [e]
C8–C9–C4	119.3(14) [e]
O11–C2–N1	126.3(16)
O12–C3–C2	125.3(15)
C10–N1–C2	123.9 [d]

The table data are reprinted from: [84], Copyright 2018, with permission of Taylor & Francis Ltd, http://tandfoline.com

[a] Parenthesized uncertainties in units of the last significant digit are 3σ and include a systematic error of 0.001r

[b] Parameters with the same numerical superscripts were refined in one group; differences between parameters in each group were assumed at the values of the composite CCSD(T)_ae/cc-pwCVQZ structure

[c] Average value

[d] Assumed at the value of the structure calculated at the level of theory as indicated above

[e] Dependent parameter

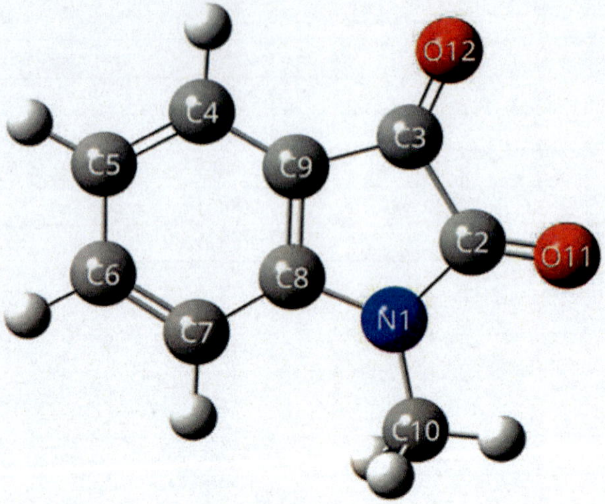

Vibrational corrections to experimental internuclear distances at the temperature of experiment (T_{nozzle} = 405 K) were calculated using quadratic and cubic force fields from B2PLYP/cc-pVTZ computation.

No	**79**
SF	**$C_9H_{20}F_3PSi$**
CN	**Bis(1,1-dimethylethyl)[(trifluorosilyl)methyl]phosphine**
Symmetry	nearly $\mathbf{C_s}$ (*anti*) $\mathbf{C_1}$ (*gauche*)
Methods	**GED** supplemented by **QC**
Ref	[85]

Distances	r_e^{se} [Å] [a,b]	
	anti	*gauche*
P…Si	3.011(23)	3.133(23)
P–C(Si)	1.856(2)	1.845(2)
P–C(*tert*-butyl)	1.872(2)	1.876(2)
Si–C	1.810(2)	1.811(2)
Si–F	1.570(1) [c]	1.571(1) [c]
C–C	1.516(1) [c]	1.516(1) [c]
C–H	1.082(3) [c]	1.082(3) [c]

Bond angle	θ_e^{se} [deg] [a]	
	anti	*gauche*
P–C–Si	110.4(12)	118.0(12)

Dihedral angles	τ_e^{se} [deg] [a]	
	anti	*gauche*
Si–C–P–C	243(4)	271(5)
Si–C–P–C	129(4)	159(5)
F–Si–C–P	165(18)	202(14)

[a] Parenthesized uncertainties in units of the last significant digit are 3σ

[b] Differences in the bond lengths between the conformers were assumed at the values of the structure calculated at the TPSS-D3BJ/def2-TZVP level of theory

[c] Average value

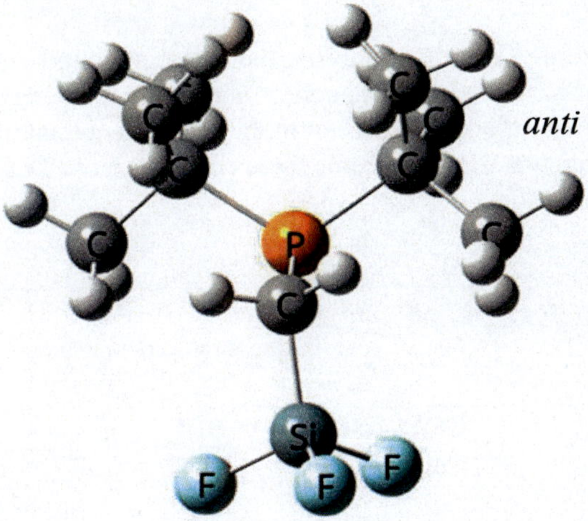

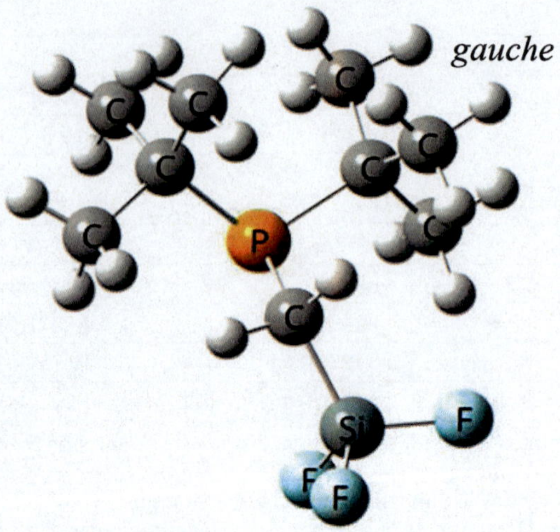

Two conformers, *anti* and *gauche*, differing in the magnitude of the Si–C–P–C torsional angle were predicted to be close in energy by ab initio and DFT calculations. The *anti* conformer is about 1.3 kJ mol^{-1} lower in energy than the *gauche* conformer (MP2/cc-pVTZ).

According to results of GED analysis, the title compound exists as a mixture of 61(11)% *anti* and 39(11)% *gauche* conformers at the temperature of experiment ($T_{nozzle} = 398$ K). Vibrational corrections to the experimental internuclear distances were calculated with quadratic and cubic force constants from TPSS computation as indicated above.

No	**80**
SF	**$C_{10}H_{14}MnO_4$**
CN	**Bis(2,4-pentanedionato-κ*O*,κ*O′*)manganese**
Symmetry	**D_{2d}**
Methods	**GED** supplemented by **DFT**
Ref	[86]

Distances	r_e^{se} [Å] [a]
Mn–O	2.035(4)
O–C	1.263(3)
C(O)–C(H)	1.398(3)
C(O)–C(methyl)	1.503(3)
O…O	2.848(15)
C(H)–H	1.083(68)

(continued)

(continued)

C(methyl)–H	1.071(13) [b]
Mn…C(O)	2.976(8)
Mn…C(H)	3.342(13)
Mn…C(methyl)	4.330(10)

Bond angles	θ_e^{se} [deg] [c]
O–Mn–O	88.8(7)
Mn–O–C	127.4(7)
C–C(H)–C	125.7(6)
O–C–C(methyl)	115.7(6)
H–C(methyl)–C	111.9 [b,d]

The table data are reprinted with permission from: [86], Copyright 2019, American Chemical Society

[a] Parenthesized uncertainties in units of the last significant digit include 2.5σ and systematic error of 0.002*r*

[b] Mean value

[c] Parenthesized uncertainties in units of the last significant digit are 3σ

[d] Assumed at the value from UB3LYP/cc-pVTZ calculation

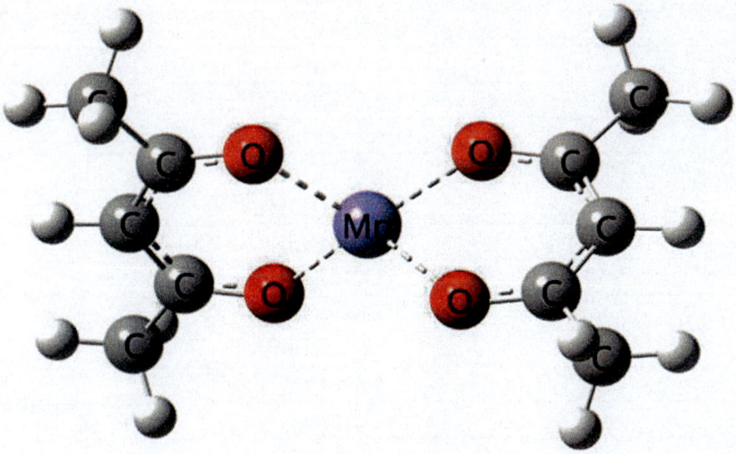

Vibrational corrections to the experimental internuclear distances at the temperature of experiment ($T_{effusion\ cell}$≈470 K) were calculated with quadratic and cubic force constants from computation at the level of theory as indicated above.

No	**81**
SF	$\mathbf{C_{11}H_{12}F_5NTe}$
CN	***N,N*-Dimethyl-3-[(2,3,4,5,6-pentafluorophenyl)telluro]-1-propanamine** (*N,N*-Dimethylaminopropyl)(pentafluorophenyl) telluride
Symmetry	$\mathbf{C_1}$
Methods	**GED** supplemented by **DFT**
Ref	[75]

Distances	r_e^{se} [Å] [a]
C1–C6	1.401(9)
C5–C6	1.374(8)
C4–C5	1.384(10)
C3–C4	1.397(9)
C2–C3	1.367(9)
C1–C2	1.387(9)
C6–F	1.332(11)
C5–F	1.334(11)
C4–F	1.303(8)
C3–F	1.320(11)
C2–F	1.326(11)
C1–Te	2.144(7)
Te–C7	2.151(9)
C7–C8	1.523(10)
C7–H	1.095(13) [b]
C8–C9	1.533(10)
C8–H	1.101(13) [b]
C9–N	1.464(10)
C9–H	1.104(13) [b]
N–C11	1.452(10)
N–C10	1.465(10)
C11–H	1.101(13) [b]
C10–H	1.102(13) [b]
Te…N	2.918(31)

Bond angles	θ_e^{se} [deg] [a]
C1–C6–C5	122.3(2)
C6–C5–C4	119.2(2)
C5–C4–C3	119.9(2)
C4–C3–C2	119.5(2)
C3–C2–C1	122.4(2)
C2–C1–C6	116.7(2)
C6–C1–Te	122.6(2)
C1–C6–F	120.6(2)
C6–C5–F	121.0(2)
C5–C4–F	120.1(2)
C4–C3–F	119.6(2)
C3–C2–F	117.1(2)
C2–C1–Te	120.7(2)
C1–Te–C7	88.6(7)
Te–C7–C8	112.7(4)
C7–C8–C9	113.4(4)
C8–C9–N	112.8(5)
C9–N–C11	112.4(2)
C9–N–C10	112.5(2)
C10–N–C11	111.0(2)

Dihedral angles	τ_e^{se} [deg] [a]
C6–C1–Te–C7	−38.9(21)
C1–Te–C7–C8	−152.1(11)
Te–C7–C8–C9	−59.1(8)
C7–C8–C9–N	63.9(8)
C8–C9–N–C11	78.7(8)
C8–C9–N–C10	−155.1(8)

[a] Parenthesized uncertainties in units of the last significant digit are 3σ

[b] Average value

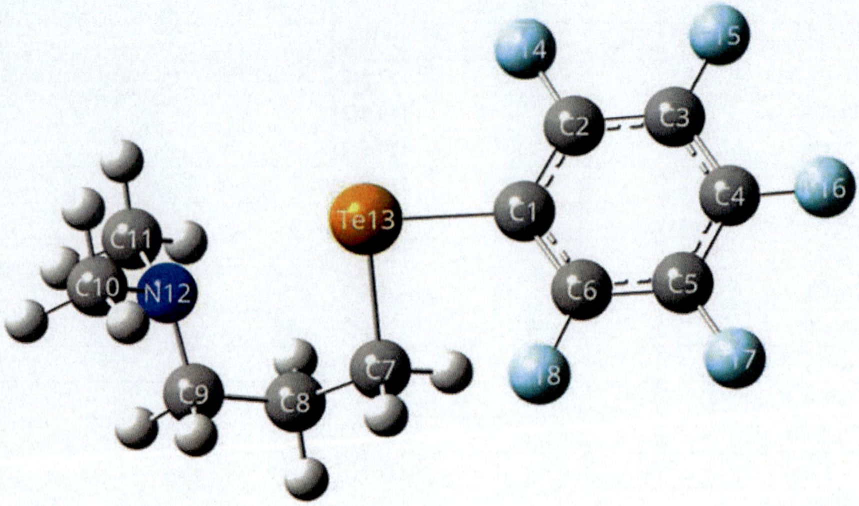

Selected structural parameters are given in these tables.

Vibrational corrections to the experimental internuclear distances at the temperature of experiment (T_{nozzle} = 444 K) were calculated with quadratic and cubic force constants from PBE0-D3/def2-TZVP computation.

No	**82**
SF	$C_{12}H_8O_4Si$
CN	**2,2′-Spirobis[1,3,2-benzodioxasilole]**
Symmetry	D_{2d}
Methods	**GED** supplemented by **DFT**
Ref	[87]

Distances	r_e^{se} [Å] [a]
Si–O5	1.635(2)
O–C	1.377(4)
C9–C10	1.403(11)
C9–C11	1.373(4)
C11–C13	1.394(4)
C11–H	1.102(11)
C13–C14	1.390(17)
C13–H	1.103(11)

Bond angles	θ_e^{se} [deg] [a]
O5–Si–O6	97.7(4)
O5–Si–O7	115.7(2)
Si–O5–C9	108.5(5)
O5–C9–C10	112.6(3)
O5–C9–C11	126.7(6)
C10–C9–C11	120.7(5)
C9–C11–C13	118.8(11)
C9–C11–H	120.4(24)
C11–C13–C14	120.5(6)
C11–C13–II	119.3(24)

Dihedral angle	τ_e^{se} [deg] [a]
C9–O5–Si–O7	123.3(1)

The table data are reprinted with permission from: [87], Copyright 2021, American Chemical Society

[a] Distance and angle values are rounded here to thousandths of Å and tenths of degree, respectively. Parenthesized uncertainties in units of the last significant digit are 3σ

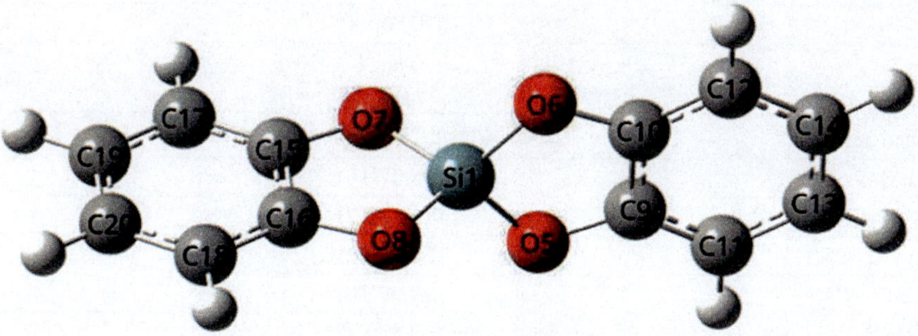

Vibrational corrections to the experimental internuclear distances at the temperature of experiment ($T_{nozzle} = 289$ K) were calculated using quadratic and cubic force fields from TPSSh/def2-TZVPP computation.

Differences between values of similar parameters were fixed at the values of the structure calculated at the same level of theory as indicated above.

No	**83**
SF	$C_{12}H_{10}S_2$
CN	**1,1′-Dithiobisbenzene** Diphenyl disulfide
Symmetry	C_2
Methods	**MW** supplemented by **QC**
Ref	[88]

Distances	$r_m^{(2a)}$ [Å] [a]
S–S	2.0204(12)
C1–S	1.7770(12)
C1–C2	1.3976(12)
C2–C3	1.3888(12)
C3–C4	1.3913(12)
C4–C5	1.3904(11)
C5–C6	1.3915(12)
C2–H	1.08290(98)
C3–H	1.08150(98)
C4–H	1.08100(98)
C5–H	1.08160(98)
C6–H	1.08150(98)

Bond angles	$\theta_m^{(2a)}$ [deg] [a]
C1–S–S	105.07(16)
S–C1–C2	115.82(17)
C1–C2–C3	120.39(12)
C2–C3–C4	119.99(10)
C3–C4–C5	119.570(72)
C4–C5–C6	120.67(10)
C1–C2–H	119.96(19)
C2–C3–H	119.49(19)
C3–C4–H	120.20(45)
C4–C5–H	120.10(19)
C5–C6–H	120.57(19)

Dihedral angles	$\tau_m^{(2a)}$ [deg] [a]

(continued)

(continued)

C–S–S–C	−87.26(20)
S–S–C1–C2	−167.59(34)
S–C1–C2–C3	−177.55(22)

The table data are reprinted from: [88], Copyright 2018, with permission from Wiley–VCH Verlag GmbH & Co. KGaA, Weinheim

[a] Parenthesized uncertainties in units of the last significant digit are 1σ

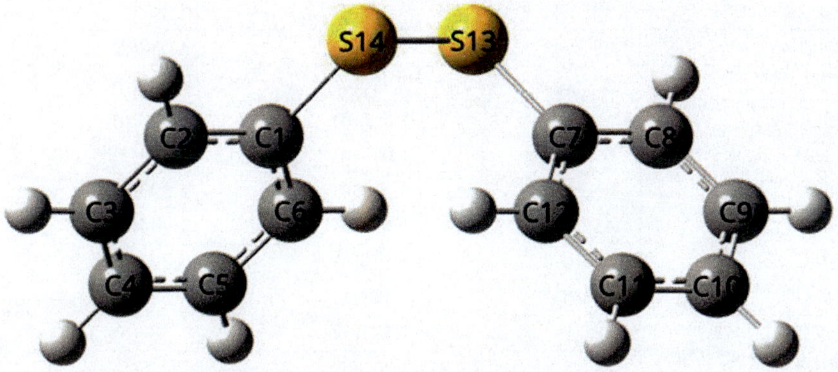

According to predictions of B3LYP and MP2 calculations (with def2-TZVP basis set), both lowest energy conformers of the title molecule have C_2 point-group symmetry. They are differing by orientations of the phenyl rings with respect to the S–C bonds (τ(S–S–C1–C2) of ca. 165° versus −83°).

Only one conformer was detected in the supersonic-jet microwave spectroscopy experiment.

The r_e^{se} structure of this floppy molecule with three low-frequency vibrations (<50 cm^{-1}) associated to phenyl torsions could not be accurately determined for the reason of low accuracy of anharmonic force field.

The determined $r_m^{(2a)}$ structure was found to be a good approximation of equilibrium structure.

No	**84**
SF	$C_{14}H_9F_5O_2$
CN	**1,2,3,4,5-Pentafluoro-6-(2-phenoxyethoxy)benzene** 1-(Pentafluorophenoxy)-2-(phenoxy)ethane
Symmetry	C_1 (I, III, IV)
Methods	**GED** supplemented by **DFT**
Ref	[89]

Distances	r_e^{se} [Å] [a]		
	I	III	IV
C1–C2	1.399(5)	1.399(5)	1.397(5)
C1–C6	1.394(5)	1.399(5)	1.401(5)
C1–O12	1.367(6)	1.368(6)	1.370(6)
C2–C3	1.386(5)	1.390(5)	1.388(5)
C3–C4	1.393(5)	1.397(5)	1.396(5)
C4–C5	1.387(5)	1.390(5)	1.392(5)
C5–C6	1.395(5)	1.398(5)	1.392(5)
C–H (ring)	1.082(6) [b]	1.082(6) [b]	1.082(7) [b]
O12–C13	1.420(6)	1.416(6)	1.423(6)
C13–C14	1.526(6)	1.508(6)	1.528(6)
C13–H	1.095(6) [b]	1.097(6) [b]	1.094(7) [b]
C14–H15	1.093(6) [b]	1.092(6) [b]	1.093(7) [b]
C14–O17	1.432(6)	1.433(6)	1.434(6)
O17–C18	1.347(6)	1.355(6)	1.335(6)
C18–C19	1.384(5)	1.382(5)	1.389(5)
C18–C23	1.385(5)	1.381(5)	1.383(5)
C19–C20	1.381(5)	1.376(5)	1.375(5)
C19–F28	1.325(6)	1.323(6)	1.320(6)
C20–C21	1.376(5)	1.378(5)	1.380(5)
C20–F27	1.323(6)	1.323(6)	1.320(6)
C21–C22	1.379(5)	1.375(5)	1.375(5)
C21–F26	1.323(6)	1.326(6)	1.321(6)
C22–C23	1.378(5)	1.379(5)	1.381(5)
C22–F25	1.325(6)	1.324(6)	1.324(6)
C23–F24	1.321(6)	1.326(6)	1.328(6)

Bond angles	θ_e^{se} [deg] [a]		
	I	III	IV
C2–C1–C6	119.5(1)	119.8(1)	119.6(1)
C2–C1–O12	115.6(1)	115.9(1)	116.5(1)
C1–O12–C13	122.4(10)	123.6(10)	123.1(10)
O12–C13–C14	114.2(3)	107.7(3)	112.1(3)
C13–C14–O17	112.9(3)	112.9(3)	116.2(3)
C14–O17–C18	117.3(9)	116.5(8)	117.9(9)
O17–C18–C19	122.3(1)	120.2(1)	117.4(1)

(continued)

(continued)

O17–C18–C23	119.5(1)	121.6(1)	125.0(1)
C19–C18–C23	118.1(1)	118.2(1)	117.6(1)
C18–C19–F28	119.8(1)	119.5(1)	119.3(1)
C18–C23–F24	119.6(1)	119.6(1)	120.0(1)

Dihedral angles	τ_e^{se} [deg] [a]		
	I	III	IV
C6–C1–O–C	−17.8(32)	14.6(59)	−47.1(45)
C1–O12–C–C	−65.9(34)	166.7(51)	−104.8(28)
O–C–C–O	−59.8(21)	−71.0(36)	58.9(19)
C18–O–C–C	75.9(14)	71.6(36)	36.7(24)
C23–C18–O–C	72.8(22)	−105.0(18)	71.4(17)

The table data are reprinted from: [89], Copyright 2020, with permission from Wiley–VCH GmbH

[a] Parenthesized uncertainties in units of the last significant digit are 1σ

[b] Average value

I

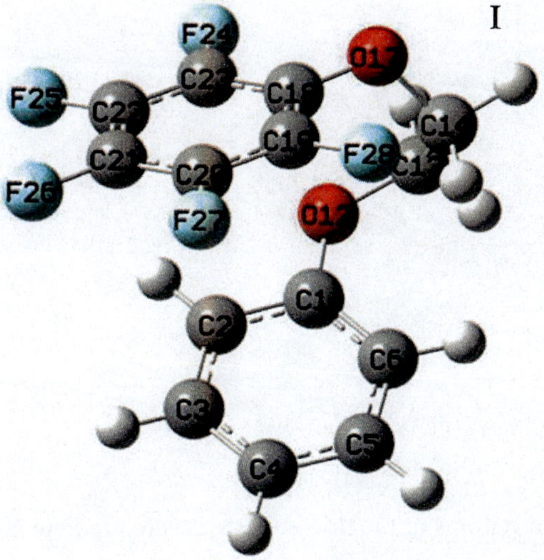

III

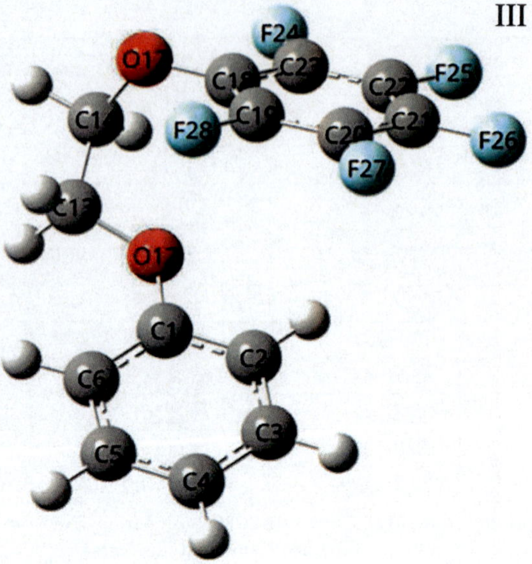

IV

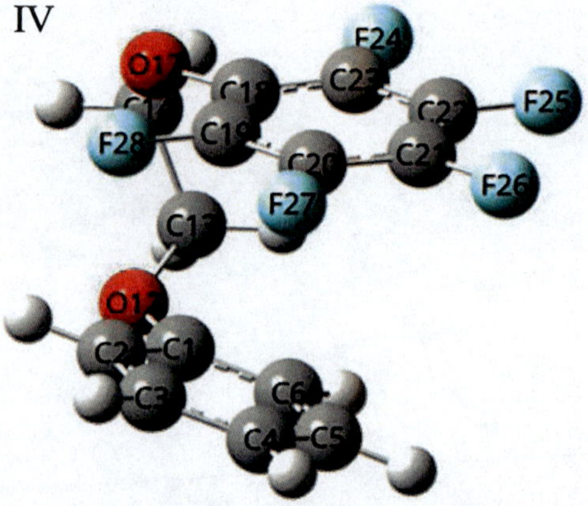

Ten stable conformations of the title molecule were predicted by PBE0-D3/def2-TZVP calculations.

The best agreement between experimental electron scattering intensities and their theoretical counterparts was reached for each of three models of single conformers, I, III, and IV, with folded structures (see tables for selected parameters). It was impossible to prove whether each of these conformers exist at the experimental conditions ($T_{nozzle} \approx 440$ K).

Refined structural parameters were constrained to the values calculated at the PBEh-3c level of theory.

Vibrational corrections to the experimental internuclear distances were estimated by molecular dynamics method.

No	**85**
SF	$C_{26}H_{16}N_8$
CN	**7,11:20,24-Diimino-5,26:18,13-dinitrilodibenzo[c,n][1,6,12,17]tetraazacyclodocosine** Hemiporphyrazine
Symmetry	C_s
Methods	**GED** supplemented by **DFT**
Ref	[90]

Distances	r_e^{se} [Å] [a]
C2–N*m*	1.383(3) [b]
Cα–N*m*	1.274(6)
N*p*–C2	1.344(3)
C2–C3	1.399(3) [b]
C3–C4	1.378(3) [b]
N*i*–Cα	1.387(3) [b]
Cα–Cβ	1.469(3) [b]
Cβ–Cγ	1.382(3) [b]
Cγ–Cδ	1.387(3) [b]
Cβ–Cβ	1.385(10)
Cδ–Cδ	1.407(12)

[a] Parenthesized uncertainties in units of the last significant digit include 2.5σ and systematic error of 0.002*r*

[b] All internuclear distances denoted by this superscript were refined in one group. Differences between parameters in this group were fixed at the values of B3LYP/cc-pVTZ structure

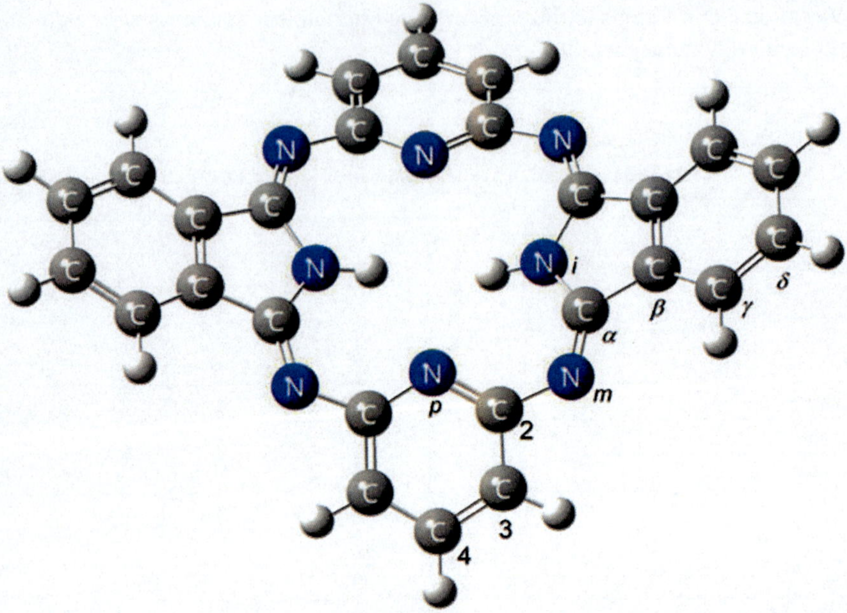

Vibrational corrections to the experimental internuclear distances at the temperature of experiment ($T_{effusion\ cell} \approx 577$ K) were taken into account by method of molecular dynamics simulations.

No	**86**
SF	**$C_{28}H_{18}N_6$**
CN	**Dicarbahemiporphyrazine**
Symmetry	**C_{2v}**
Methods	**GED** supplemented by **DFT**
Ref	[90]

Distances	r_e^{se} [Å] [a]
C2–N*m*	1.408(3) [b]
Cα–N*m*	1.271(4)
C1–C2	1.405(3)
C2–C3	1.402(3) [b]
C3–C4	1.392(3) [b]
N*i*–Cα	1.400(3) [b]
Cα–Cβ	1.478(3) [b]
Cβ–Cγ	1.389(3) [b]

(continued)

(continued)

Cγ–Cδ	1.395(3) [b]
Cβ–Cβ	1.390(5)
Cδ–Cδ	1.399(8)

The table data are reprinted from: [90], Copyright 2019, with permission from Elsevier

[a] Parenthesized uncertainties in units of the last significant digit include 2.5σ and systematic error of 0.002*r*

[b] All internuclear distances denoted by this superscript were refined in one group. Differences between parameters in this group were fixed at the values of B3LYP/cc-pVTZ structure

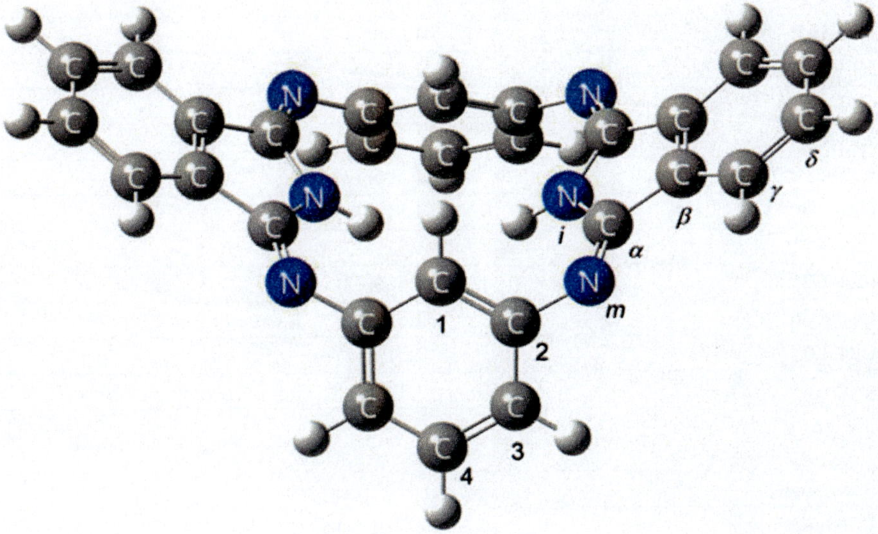

Vibrational corrections to the experimental internuclear distances at the temperature of experiment ($T_{effusion\ cell}$≈574 K) were calculated with quadratic and cubic force constants from computation at the level of theory as indicated above.

No	**87**
SF	$C_{28}H_{40}O_4Si$
CN	**4,4′,6,6′-Tetra-*tert*-butyl-2,2′-spirobis[[1,3,2]benzodioxasilole]**
Symmetry	C_2
Methods	**GED** supplemented by **DFT**
Ref	[87]

Distances	r_e^{se} [Å] [a]

(continued)

(continued)

Si–O5	1.647(4)
Si–O6	1.646(4)
O5–C9	1.393(12)
O6–C10	1.388(12)
C9–C10	1.542(5)
C9–C11	1.364(12)
C10–C12	1.351(12)
C11–C13	1.381(12)
C11–C21	1.503(12)
C12–C14	1.377(12)
C12–H	1.096(7)
C13–C14	1.458(60)
C13–H	1.093(7)
C14–C24 [b]	1.506(12)
C21–C29	1.506(12)
C21–C30	1.513(12)
C21–C31	1.513(12)
C24–C35 [b]	1.506(12)
C24–C36 [b]	1.512(12)
C24–C37 [b]	1.512(12)
C–H	1.107(7) [c]

Bond angles	θ_e^{se} [deg] [a]
O5–Si–O6	101.3(16)
O5–Si–O7	113.7(9)
O5–Si–O8	113.7(9)
Si–O5–C9	108.3(14)
Si–O6–C10	108.0(14)
O5–C9–C10	110.7(11)
O5–C9–C11	131.3(21)
O6–C10–C9	111.6(11)
O6–C10–C12	130.0(21)
C10–C9–C11	118.0(17)
C9–C10–C12	118.4(17)
C9–C11–C13	120.6(31)
C9–C11–C21	120.0(22)
C10–C12–C14	123.6(31)

(continued)

(continued)

C10–C12–C22	117.5(22)
C13–C11–C21	119.4(26)
C11–C13–C14	122.3(17)
C11–C13–C23	115.8(22)
C11–C21–C29	113.5(6)
C11–C21–C30	111.1(6)
C11–C21–C31	111.0(6)
C14–C12–C22	118.8(26)
C12–C14–C13	117.1(15)
C12–C14–C24 [b]	117.2(22)
C14–C13–C23	122.0(33)
C13–C14–C24 [b]	125.6(32)
C14–C24–C35 [b]	114.0(6)
C14–C24–C36 [b]	110.9(6)
C14–C24–C37	110.9(6)
C29–C21–C30	114.0(6)
C29–C21–C31	110.9(6)
C30–C21–C31	94.8(36)
C35–C24–C36 [b]	109.9(6)
C35–C24–C37 [b]	109.9(6)
C36–C24–C37 [b]	100.3(34)

Dihedral angles	τ_e^{se} [deg] [a]
C9–O5–Si–O7	122.4(4)
C9–O5–Si–O8	−122.3(4)
O5–Si–O8–C16	−122.4(4)
C10–O6–Si–O8	122.4(4)

[a] Distance and angle values are rounded here to thousandths of Å and tenths of degree, respectively. Parenthesized uncertainties in units of the last significant digit are 3σ

[b] C24 is the central carbon atom in *tert*-butyl group

[c] Average value for methyl groups

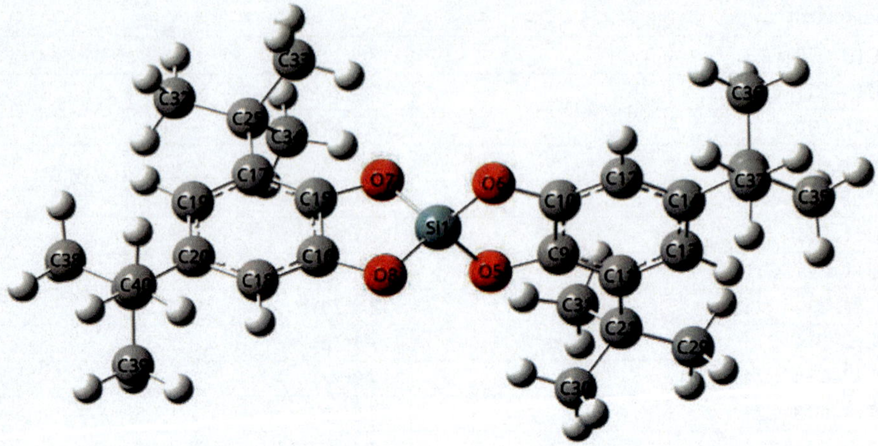

Values of angles including hydrogen atom and differences between values of similar parameters were taken from TPSSh/def2-TZVPP calculations.

Vibrational corrections to the experimental internuclear distances at the temperature of experiment (T_{nozzle}≈291 K) were calculated with quadratic and cubic force fields from computation at the level of theory as indicated above.

No	**88**
SF	$C_{60}F_{36}$
CN	**Hexatriacontafluorofullerene-C_{60}**
Symmetry	C_1
Methods	**GED** supplemented by **QC**
Ref	[91]

Distances	r_e^{se} [Å] [a,b]
C–C	1.374(15)
C–C [c]	1.338(15)
C–C(F)	1.481(15)
(F)C–C(F)	1.589(15)
C–F	1.360(39)

[a] Parenthesized uncertainties in units of the last significant digit are 1σ

[b] Average values

[c] Double bond

The existence of three isomers of fluorofullerene (with T, C_3, and C_1 point-group symmetry) was predicted by RI-MP2/def2-TZVPP computations. The T isomer is lower in energy by 6.2 and 8.0 kcal mol^{-1} in comparison with the C_3 and C_1 isomers, respectively. The C_1 and C_3 isomers were found to be present by GED (T_{nozzle} = 563 K) in amounts of 81(4) and 16(4)%, respectively. The presence of small amount of the T isomer (up to 5%) could not be excluded. For each isomer, bond angles and dihedral angles were assumed at the values of MP2 structure. Vibrational corrections to the experimental internuclear distances were calculated with quadratic and cubic force constants from BP86/SV computations. The structural data are given here for the C_1 isomer.

No	**89**
SF	**ClSi**
CN	**Chlorosilyliumylidene**
Symmetry	$C_{\infty v}$
Methods	**MW**
Ref	[92]

Distance	r_e [Å] [a]
Si–Cl	1.9439729(10)

[a] Parenthesized uncertainty in units of the last significant digit is 1σ

FTMW spectra for the $J = 1 \leftarrow 0$ and $2 \leftarrow 1$ rotational transitions were measured for the parent species and the ^{37}Cl isotopic species in the ground and the first excited vibrational states of the ground $^1\Sigma^+$ electronic state.

No	**90**
SF	$Cl_2O_5S_2$
CN	**Disulfuryl dichloride**
Symmetry	C_2
Methods	**GED** supplemented by **QC**
Ref	[93]

Distances	r_e^{se} [Å] [a]
S–Cl	1.985(1)
S–O5	1.408(1) [b]
S–O6	1.412(1) [b]
S–O1	1.617(1)

Bond angles	θ_e^{se} [deg] [a]
O1–S–Cl	98.7(3)
O5–S–O6	125.1(2)
S–O1–S	127.2(2)

Dihedral angles	τ_e^{se} [deg] [a]
S–O1–S–Cl	80.5(4)
Cl–S…S–Cl	155.2(9) [c]

[a] Parenthesized uncertainties in units of the last significant digit are probably 3σ

[b] Difference between the S=O bond lengths was assumed at the value of MP2/aug-cc-pVTZ structure

[c] Dependent parameter

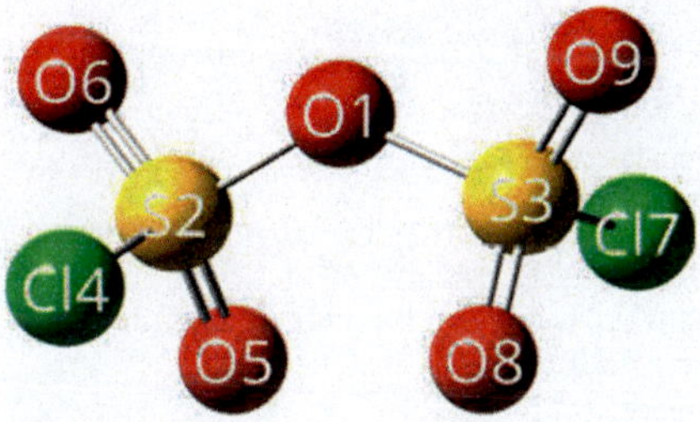

According to predictions of MP2/cc-pVTZ calculations, the title molecules exist as a mixture of three conformers, *anti*, *syn,* and *gauche* (in a ratio of 70.4 : 16.3 : 13.3) differing in the magnitude of the Cl–S…S–Cl torsional angle (78.8°, 38.4°, and 78.7°, respectively).

Structural parameters were determined only for the *anti* conformer. Vibrational corrections to the experimental internuclear distances (T_{nozzle} = 293 K) were calculated with quadratic and cubic force constants from B3PW91/aug-cc-pVTZ computation.

The presence of other conformers could neither be confirmed nor excluded by GED.

No	**91**
SF	**F_4Si**
CN	**Tetrafluorosilane**
Symmetry	**T_d**
Methods	**IR**
Ref	[94]

Distance	r_e [Å] [a]
Si–F	1.5516985(30)

[a] Parenthesized uncertainty given in units of the last significant digit was not specified, probably 1σ

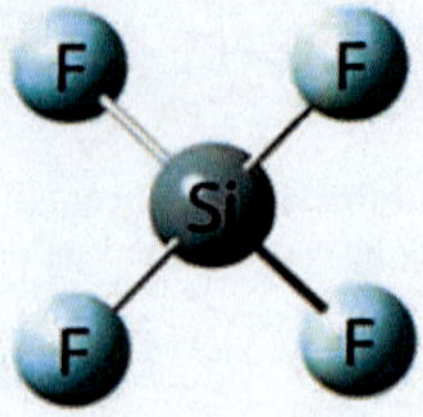

High-resolution infrared spectrum of $^{28}SiF_4$ isotopologue was recorded in the ν_2, $\nu_1 - \nu_4$, and $\nu_4 - \nu_2$ regions.

No	**92**
SF	**F_5HOTe**
CN	**Tellurium fluoride hydroxide** Teflic acid
Symmetry	**C_{4v}** (skeleton)
Methods	**MW** supplemented by **QC**
Ref	[95]

Distances	r_e^{se} [Å] [a]
H–O	0.970(282)
O–Te	1.8686(116)
Te–Fax [b]	1.819(21)
Te–Feq [c]	1.827(20)

Bond angle	θ_e^{se} [deg] [a]
O–Te–Feq [c]	91.06(104)

The table data are reprinted from: [95], with permission of AIP Publishing

[a] Parenthesized uncertainties in units of the last significant digit are estimated upper limits for possible deviations

[b] Fax is F atom in axial position

[c] Feq is F atom in equatorial position

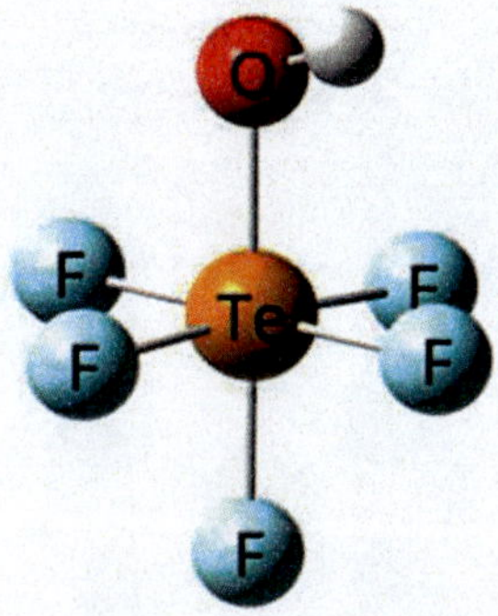

The rotational spectra of the parent species and the singly (^{122}Te, ^{124}Te, ^{126}Te, ^{128}Te, ^{18}O, and D) and doubly (^{126}Te^{18}O, ^{128}Te^{18}O, and ^{18}OD) substituted isotopic species were recorded by a pulsed supersonic jet-expansion FTMW spectrometer.

Rovibrational corrections, ΔB_{vib}, to experimental ground-state rotational constants were calculated with quadratic and cubic force constants from MP2 computations (with SDD effective core potential for Te and aug-cc-pVTZ basis set for H, O, and F).

No	**93**
SF	**F_6KrS**
CN	**Sulfur hexafluoride – krypton (1/1)**
Symmetry	C_{3v} (effective symmetry)
Method	**IR**
Ref	[21]

Distance	r_e [Å] [a]
S…Kr	4.3938(28)

[a] Parenthesized uncertainty given in units of the last significant digit was not specified, probably 1σ

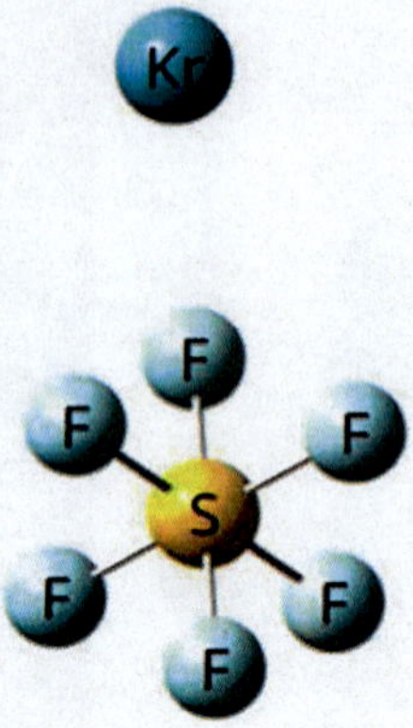

The vibration–rotation spectrum of the title van der Waals heterocluster was recorded in the ν_3 region of SF_6 using a high-resolution infrared spectrometer in combination with a cavity quantum cascade laser coupled to a pulsed supersonic jet.

No	**94**
SF	**F_6SXe**
CN	**Sulfur hexafluoride – xenon (1/1)**
Symmetry	C_{3v} (effective symmetry)
Method	**IR**
Ref	[21]

Distance	r_e [Å] [a]
S…Xe	4.5925(13)

[a] Parenthesized uncertainty given in units of the last significant digit was not specified, probably 1σ

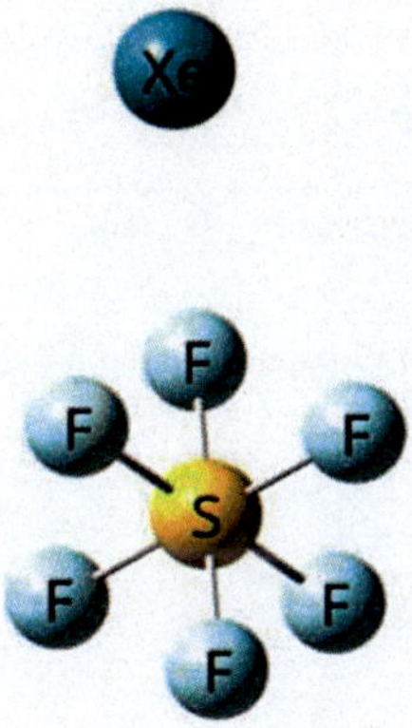

The vibration–rotation spectrum of the title van der Waals heterocluster was recorded in the ν_3 region of SF_6 using high-resolution infrared spectrometer in combination with a cavity quantum cascade laser coupled to a pulsed supersonic jet.

No	**95**
SF	**GeH_4**
CN	**Germane**
Symmetry	**T_d**
Methods	**IR**
Ref	[96]

Distance	r_e [Å] [a]
Ge–H	1.51714(25)

[a] Parenthesized uncertainty in units of the last significant digit are 1σ

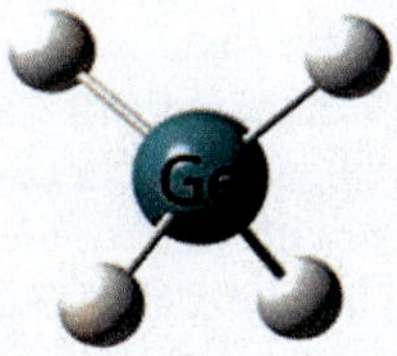

The absorption spectra of the title molecule were recorded in the bending region using a high-resolution FT spectrometer. The ν_2 and ν_4 bands of five isotopologues (^{70}Ge, ^{72}Ge, ^{73}Ge, ^{74}Ge, and ^{76}Ge) were analyzed.

No	**96**
SF	H_2O_2
CN	**Hydrogen peroxide**
Symmetry	C_2
Methods	**MW** supplemented by **ab initio**
Ref	[97]

Distances	r_e^{se} [Å] [a]
O–H	0.9617(2)
O–O	1.4524(3)

Bond angle	θ_e^{se} [deg] [a]
O–O–H	99.76(6)

Dihedral angle	τ_e^{se} [deg] [a]
O–O–H	113.6(3)

The table data are reprinted from: [97], Copyright 2017, with permission from Elsevier Inc
[a] Parenthesized uncertainties in units of the last significant digit are probably 1σ

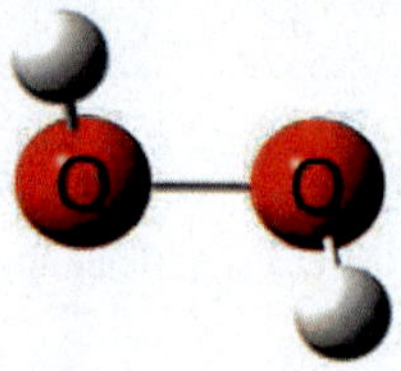

Experimental values of rotational constants were taken from literature.

Rovibrational corrections were obtained from a variational calculation with an ab initio potential energy surface taken from literature (computed at the CCSD(T)-F12b/aug-cc-pV5Z level of theory, with valence electrons correlated).

The r_e^{se} structure was found to be in excellent agreement with ab initio structure extrapolated from the CCSD(T)_ae/cc-pCVTZ-5Z geometry.

No	97
SF	H_2S_2
CN	**Disulfane** Hydrogen disulfide
Symmetry	C_2
Methods	**MW** supplemented by **QC**
Ref	[98]

Distances	r_e^{se} [Å] [a,b]	r_e^{se} [Å] [a,c]	r_e^{se} [Å] [a,d]
S–S	2.0513(3)	2.0516(7)	2.0514(6)
S–H	1.3403(14)	1.3400(42)	1.3404(26)

Bond angle	θ_e^{se} [deg] [a,b]	θ_e^{se} [deg] [a,c]	θ_e^{se} [deg] [a,d]
S–S–H	98.13(2)	98.06(4)	98.09(3)

Dihedral angle	τ_e^{se} [deg] [a,b]	τ_e^{se} [deg] [a,c]	τ_e^{se} [deg] [a,d]
H–S–S–H	90.72(2)	90.75(5)	90.72(4)

[a] Parenthesized uncertainties in units of the last significant digit are 1σ
[b] With B2PLYP-D3BJ/aug-cc-pVTZ force field
[c] With B3LYP-D3BJ/aug-cc-pVTZ force field
[d] With MP2/aug-cc-pVTZ force field

The experimental ground-state rotational constants for the parent species, two singly substituted (^{34}S and D), and doubly substituted (D_2) isotopic species were taken from literature.

Rovibrational corrections, ΔB_{vib}, to experimental ground-state rotational constants were calculated with quadratic and cubic force constants from QC computations (see footnotes [b–d]).

The structure r_e^{se} structure was found to be more accurate than the CCSD(T)/aug-cc-pVTZ optimized geometry.

No	**98**
SF	SO_2
CN	**Sulfur dioxide**
Symmetry	C_{2v}
Methods	**MW** supplemented by **ab initio**
Ref	[99]

Distance	r_e [Å] [a]	r_e^{se} [Å] [a,b]
S–O	1.4307897(99)	1.43079396(93)

Bond angle	θ_e [Å] [a]	θ_e^{se} [Å] [a,b]
O–S–O	119.32893(27)	119.328783(55)

[a] Parenthesized uncertainties given in units of the last significant digit are 1σ

[b] Rovibrational corrections ΔB_{vib} to experimental ground-state rotational constants were calculated with anharmonic force field refined using experimental data (α constants, centrifugal distortion constants, etc.)

Rotational spectra taken from literature were refitted, and ground-state rotational constants were determined.

Experimental equilibrium structure was determined taken into account the γ-terms. The result is the median of the values for four isotopic species.

9.4 Compound Index

All reviewed compounds are listed here in alphabetical order of their sum formulae in Hill system, *i.e.,* in the same order as in Sect. 9.3. Besides the chemical names and the sum formulae, the structural formulae are also given for the unambiguous identification of the molecules.

Each structural formula shows connectivity of atoms in the compound molecule and (plausible) bond orders, but it is not responsible for presentation of the three-dimensional molecular structure, *i.e.,* it is uniformed for all possible stable molecular conformations (conformers).

Document no.	Molecular formula	Structural formula	Chemical name
1	ArF_6S	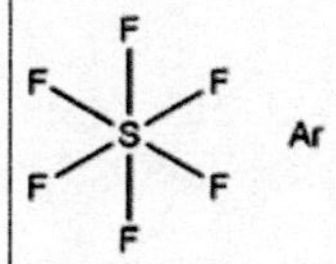	Sulfur hexafluoride – argon (1/1)
2	CCl_2NO_2P	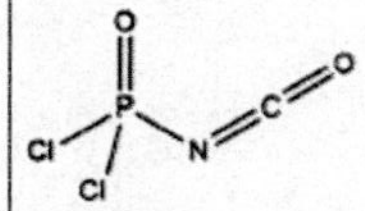	Phosphorisocyanatidic dichloride
3	CCl_2O_4S	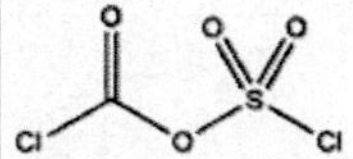	Chlorosulfuric acid chlorocarbonyl ester
4	CF_2NOPS	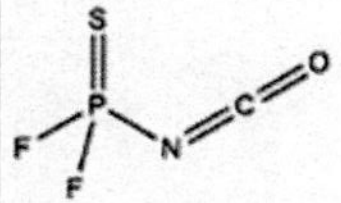	Phosphorisocyanatidothioic difluoride
5	CHI_3	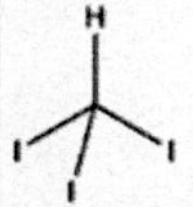	Triiodomethane
6	CHO	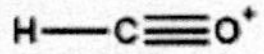	Methylidyneoxonium

(continued)

(continued)

Document no.	Molecular formula	Structural formula	Chemical name
7	CHO	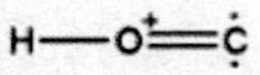	Oxoniomethylene
8	CH_2BrNO_2	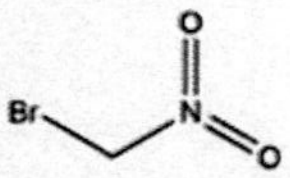	Bromonitromethane
9	CH_2FNO_3	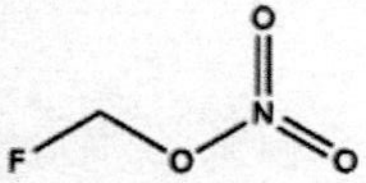	Nitric acid fluoromethyl ester
10	CH_2OS	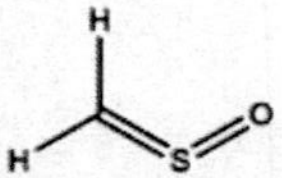	Methanethione *S*-oxide
11	CH_2S	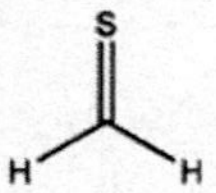	Methanethione
12	CH_3NO_3	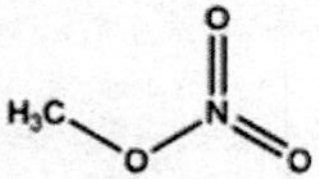	Nitric acid methyl ester

(continued)

(continued)

Document no.	Molecular formula	Structural formula	Chemical name
13	CI_4	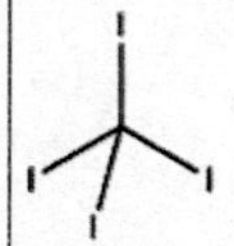	Tetraiodomethane
14	CN_2O_2	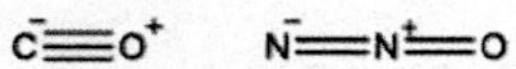	Carbon monoxide – dinitrogen monoxide (1/1)
15	$C_2Cl_3F_3$	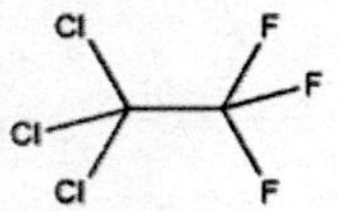	1,1,1-Trichloro-2,2,2-trifluoroethane
16	C_2H_2ClF	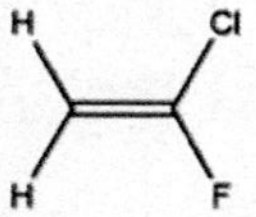	1-Chloro-1-fluoroethene
17	C_2H_2S	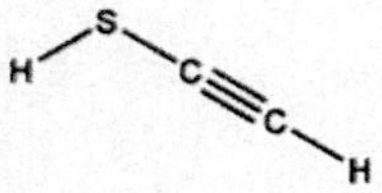	Ethynethiol
18	C_2H_3P	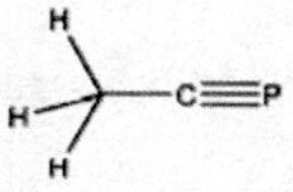	Ethylidynephosphine

(continued)

(continued)

Document no.	Molecular formula	Structural formula	Chemical name
19	$C_2H_6O_2S_2$	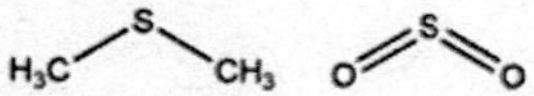	1,1′-Thiobismethane – sulfur dioxide (1/1)
20	$C_2H_{11}B_{10}Br$	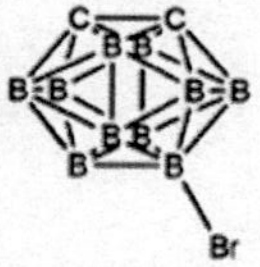	9-Bromo-1,2-dicarbadodecaborane(12)
21	$C_2H_{12}B_{10}S_2$	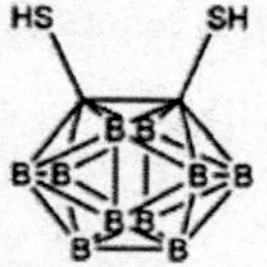	1,2-Dicarbadodecaborane(12)-1,2-dithiol
22	$C_2H_{12}B_{10}Se_2$	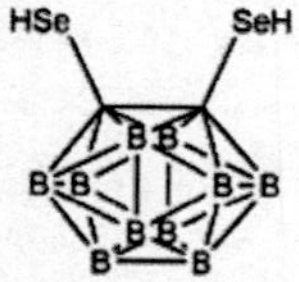	1,2-Dicarbadodecaborane(12)-1,2-diselenol
23	$C_2N_4O_5$	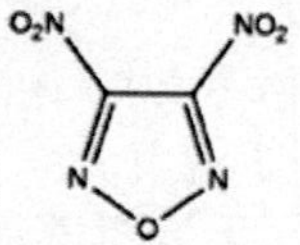	3,4-Dinitrofurazan

(continued)

(continued)

Document no.	Molecular formula	Structural formula	Chemical name
24	C_3	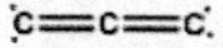	1,2-Propadiene-1,3-diylidene
25	C_3HNS		Isothiocyanatoethyne
26	$C_3H_2F_4$	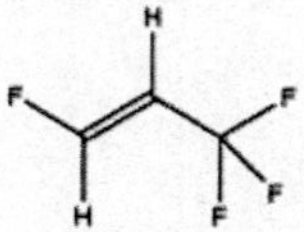	(*E*)-1,3,3,3-Tetrafluoro-1-propene
27	$C_3H_2F_4$	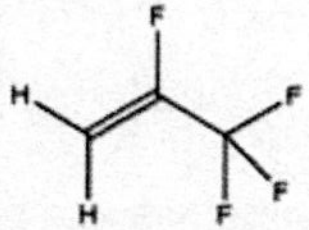	2,3,3,3-Tetrafluoro-1-propene
28	C_3H_2O	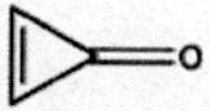	2-Cyclopropen-1-one
29	C_3H_3N	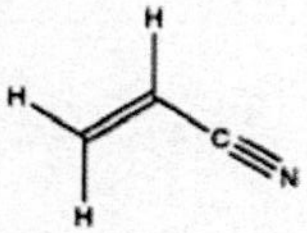	2-Propenenitrile

(continued)

(continued)

Document no.	Molecular formula	Structural formula	Chemical name
30	C_3H_3N	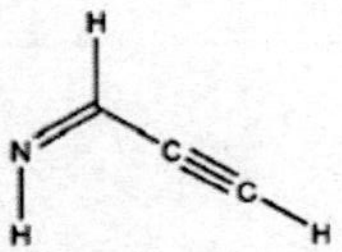	(1Z)-2-Propyn-1-imine
31	C_3H_3NS	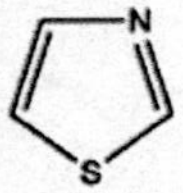	Thiazole
32	$C_3H_3N_3O_4$	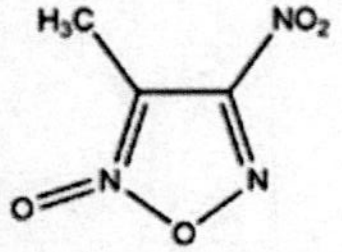	3-Methyl-4-nitro-1,2,5-oxadiazole 2-oxide
33	$C_3H_3N_3O_4$	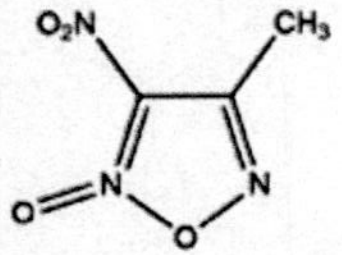	4-Methyl-3-nitro-1,2,5-oxadiazole 2-oxide
34	C_3H_4	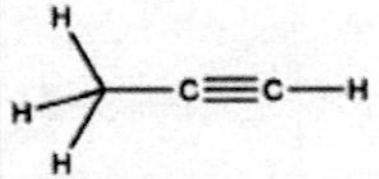	1-Propyne

(continued)

(continued)

Document no.	Molecular formula	Structural formula	Chemical name
35	$C_3H_6Cl_3N$	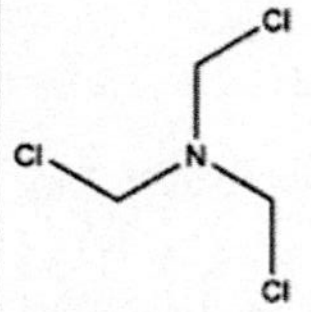	1-Chloro-*N*,*N*-bis(chloromethyl)methanamine
36	C_3O_2	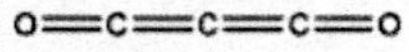	1,2-Propadiene-1,3-dione
37	$C_4H_3FN_2O_2$	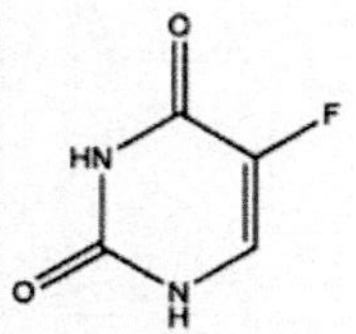	5-Fluoro-2,4(1*H*,3*H*)-pyrimidinedione
38	$C_4H_3NO_2$	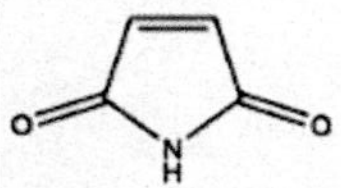	1*H*-Pyrrol-2,5-dione
39	C_4H_4	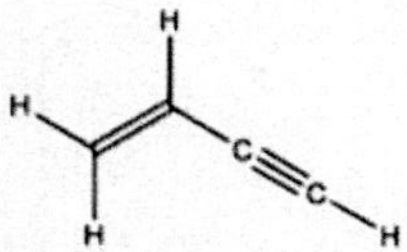	1-Buten-3-yne

(continued)

(continued)

Document no.	Molecular formula	Structural formula	Chemical name
40	$C_4H_4N_2$	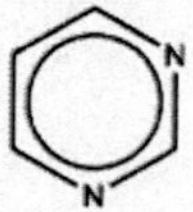	Pyrimidine
41	$C_4H_4O_3$	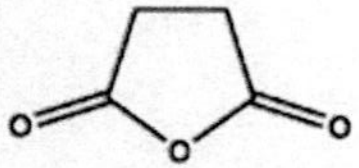	Dihydro-2,5-furandione
42	C_4H_4S		Thiophene
43	C_4H_6		1,3-Butadiene
44	$C_4H_6O_2$	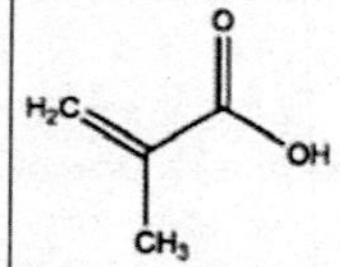	2-Methyl-2-propenoic acid

(continued)

(continued)

Document no.	Molecular formula	Structural formula	Chemical name
45	$C_4H_8O_4$	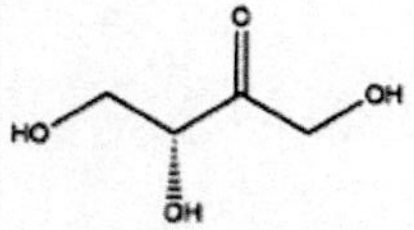	(*R*)-1,3,4-Trihydroxybutan-2-one
46	C_5BrF_4N	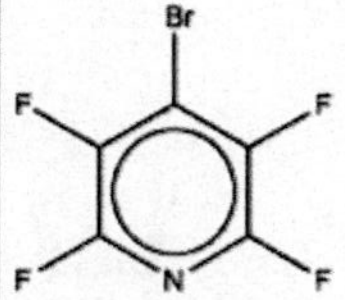	4-Bromo-2,3,5,6-tetrafluoropyridine
47	C_5ClF_4N	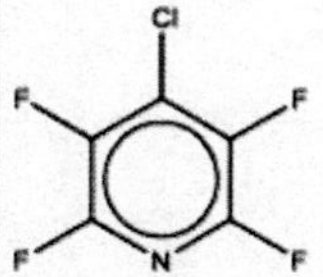	4-Chloro-2,3,5,6-tetrafluoropyridine
48	C_5F_4IN	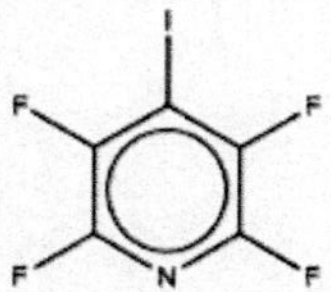	4-Iodo-2,3,5,6-tetrafluoropyridine

(continued)

(continued)

Document no.	Molecular formula	Structural formula	Chemical name
49	C_5F_5N	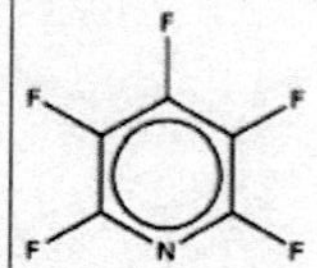	Pentafluoropyridine
50	$C_5H_4N_2O_4$	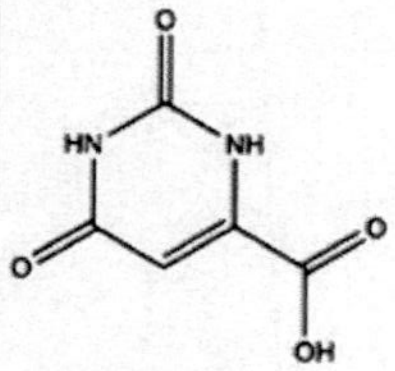	1,2,3,6-Tetrahydro-2,6-dioxo-4-pyrimidinecarboxylic acid
51	C_5H_4OS	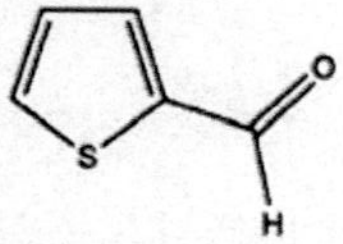	2-Thiophenecarboxaldehyde
52	$C_5H_5N_3O$	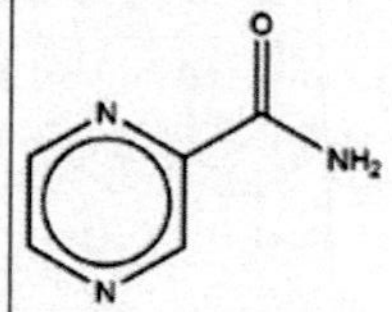	2-Pyrazinecarboxamide

(continued)

(continued)

Document no.	Molecular formula	Structural formula	Chemical name
53	$C_5H_8O_2$	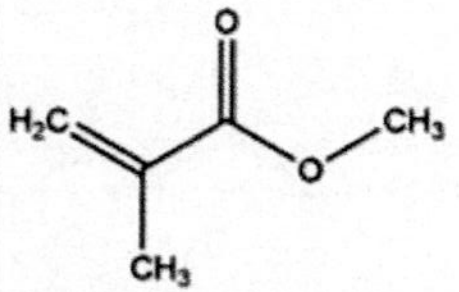	2-Methyl-2-propenoic acid methyl ester
54	$C_5H_{10}N_2$	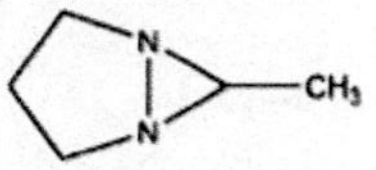	6-Methyl-1,5-diazabicyclo[3.1.0]hexane
55	$C_6H_2F_{10}O_2$	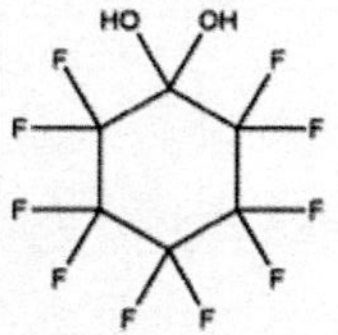	2,2,3,3,4,4,5,5,6,6-Decafluorocyclohexane-1,1-diol
56	C_6H_4	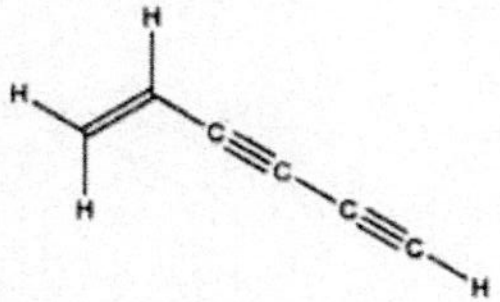	1-Hexene-3,5-diyne

(continued)

(continued)

Document no.	Molecular formula	Structural formula	Chemical name
57	$C_6H_4Cl_2$	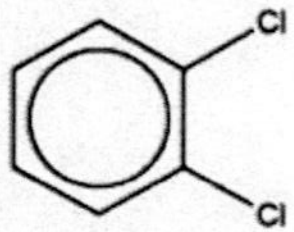	1,2-Dichlorobenzene
58	$C_6H_4N_2$	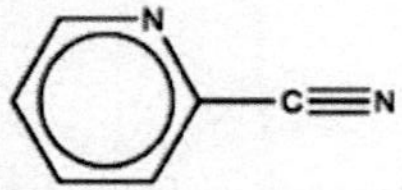	2-Cyanopyridine
59	$C_6H_4N_2$	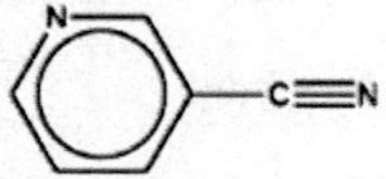	3-Cyanopyridine
60	$C_6H_4N_2$	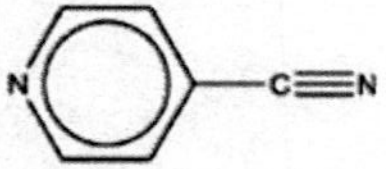	4-Cyanopyridine
61	$C_6H_5NO_2$	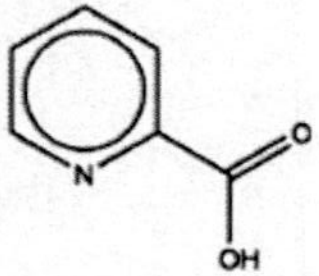	2-Pyridinecarboxylic acid

(continued)

(continued)

Document no.	Molecular formula	Structural formula	Chemical name
62	C_6H_6OS	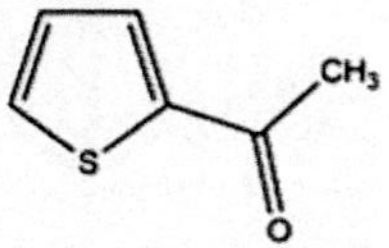	2-Acetylthiophene
63	$C_6H_6O_2$	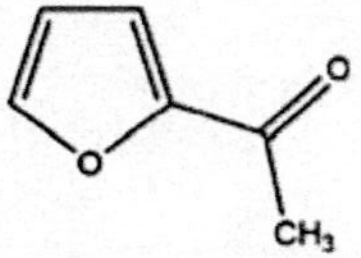	2-Acetylfuran
64	C_6H_8O	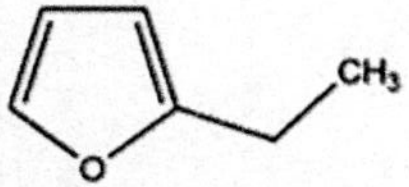	2-Ethylfuran
65	$C_6H_{10}S_2$	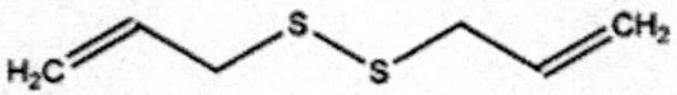	3,3′-Dithiobis-1-propene
66	$C_7HF_{13}O$	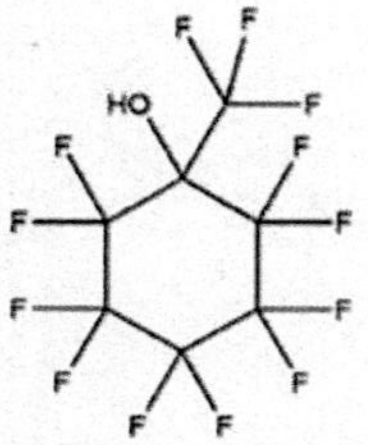	2,2,3,3,4,4,5,5,6,6-Decafluoro-1-(trifluoromethyl)cyclohexan-1-ol

(continued)

(continued)

Document no.	Molecular formula	Structural formula	Chemical name
67	$C_7H_3F_5Te$	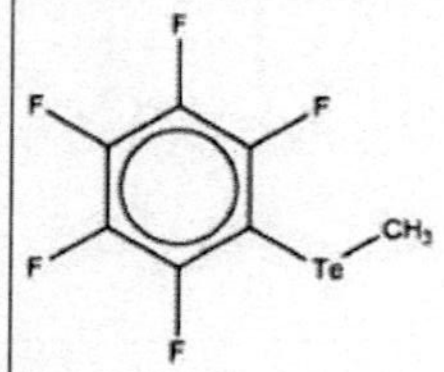	1,2,3,4,5-Pentafluoro-6-(methyltelluro)benzene
68	$C_7H_6N_2$	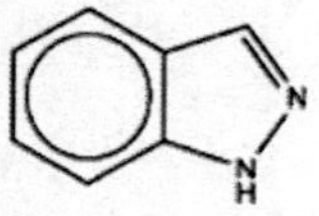	1*H*-Indazole
69	$C_7H_6O_2$	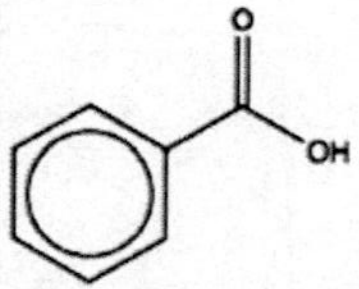	Benzoic acid
70	C_7H_7NOS	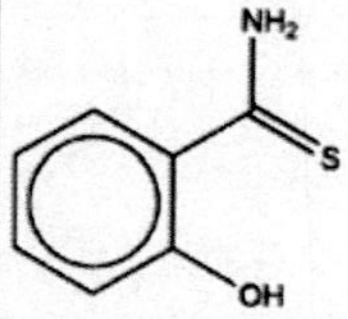	2-Hydroxybenzenecarbothioamide

(continued)

(continued)

Document no.	Molecular formula	Structural formula	Chemical name
71	C_7H_7NS	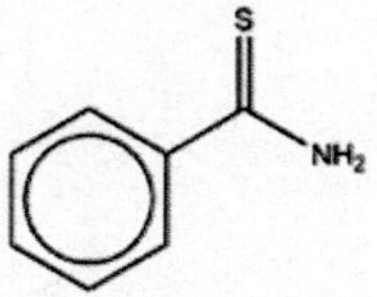	Thiobenzamide
72	$C_8H_4O_3$	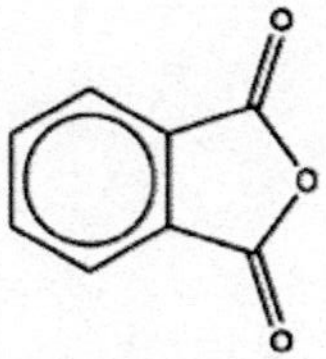	1,3-Isobenzofurandione
73	$C_8H_5NO_2$	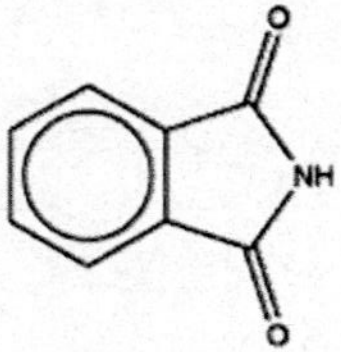	1*H*-Isoindole-1,3(2*H*)-dione

(continued)

(continued)

Document no.	Molecular formula	Structural formula	Chemical name
74	$C_8H_6N_2O_2$	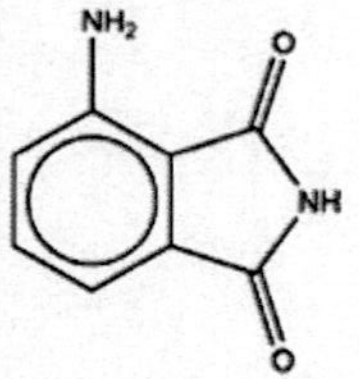	4-Amino-1*H*-isoindole-1,3(2*H*)-dione
75	C_8H_{12}	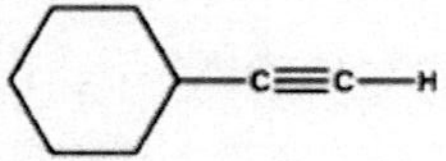	Ethynylcyclohexane
76	$C_8H_{18}N_4$	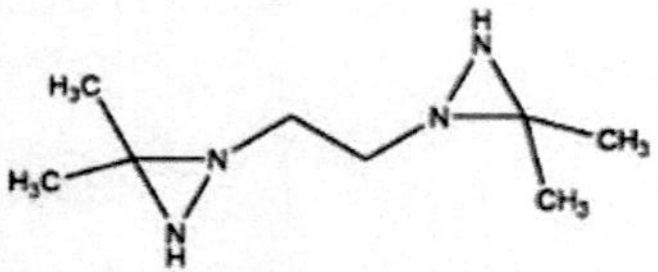	1,1′-(1,2-Ethanediyl)bis[3,3-dimethyl-diaziridine]
77	$C_9H_6N_2O_3$	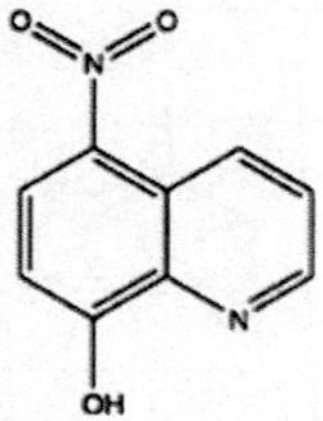	5-Nitroquinolin-8-ol

(continued)

(continued)

Document no.	Molecular formula	Structural formula	Chemical name
78	$C_9H_7NO_2$	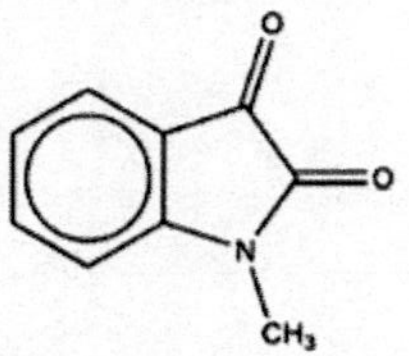	1-Methyl-1*H*-indole-2,3-dione
79	$C_9H_{20}F_3PSi$	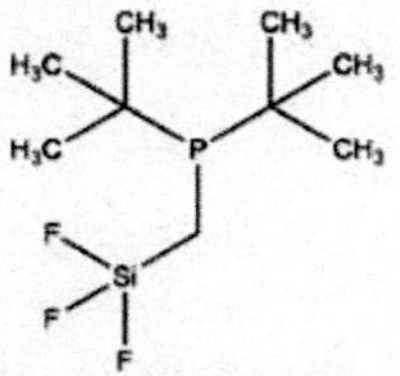	Bis(1,1-dimethylethyl)[(trifluorosilyl)methyl]-phosphine
80	$C_{10}H_{14}MnO_4$	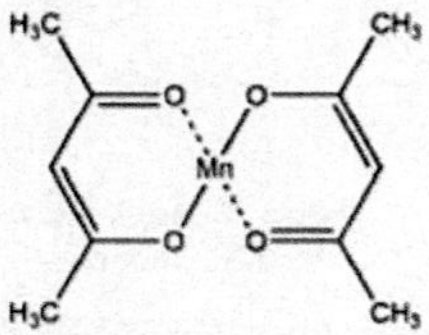	Bis(2,4-pentanedionato-κ*O*,κ*O*′)manganese

(continued)

(continued)

Document no.	Molecular formula	Structural formula	Chemical name
81	$C_{11}H_{12}F_5NTe$	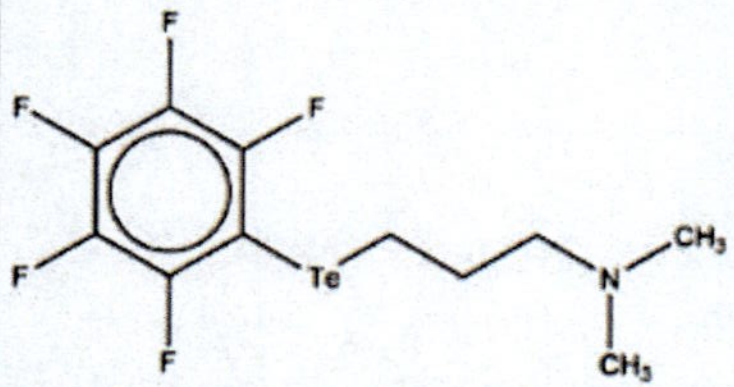	*N*,*N*-Dimethyl-3-[(2,3,4,5,6-pentafluorophenyl)telluro]-1-propanamine
82	$C_{12}H_8O_4Si$	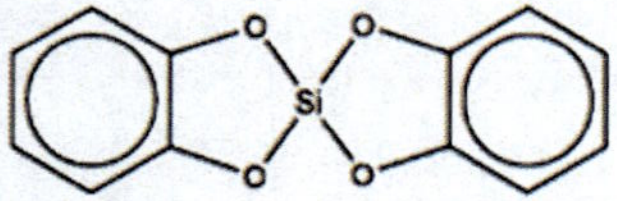	2,2′-Spirobis[1,3,2-benzodioxasilole]
83	$C_{12}H_{10}S_2$	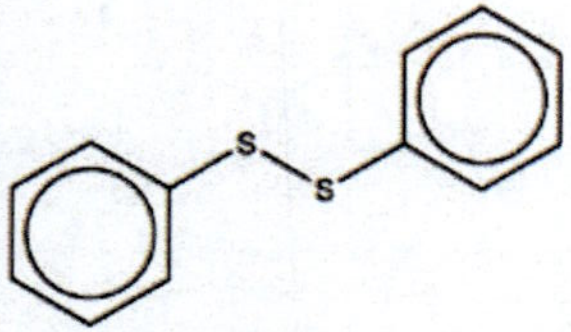	1,1′-Dithiobisbenzene

(continued)

(continued)

Document no.	Molecular formula	Structural formula	Chemical name
84	$C_{14}H_9F_5O_2$	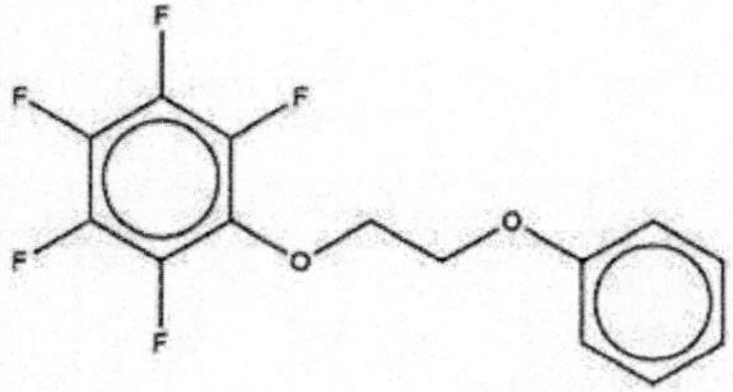	1,2,3,4,5-Pentafluoro-6-(2-phenoxyethoxy)benzene
85	$C_{26}H_{16}N_8$	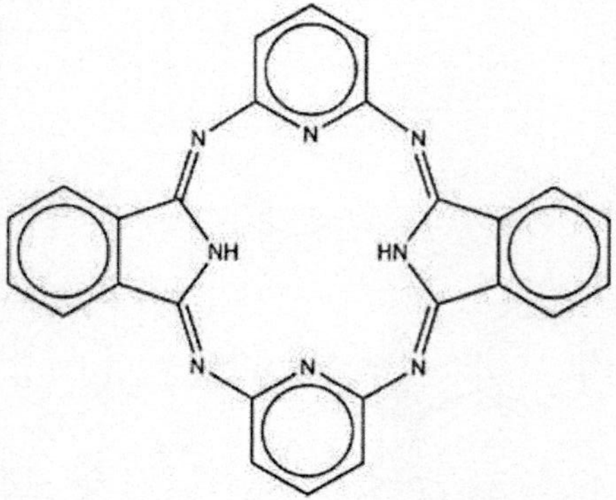	7,11:20,24-Diimino-5,26:18,13-dinitrilodi-benzo[c,n][1,6,12,17]tetraazacyclodocosine

(continued)

(continued)

Document no.	Molecular formula	Structural formula	Chemical name
86	$C_{28}H_{18}N_6$	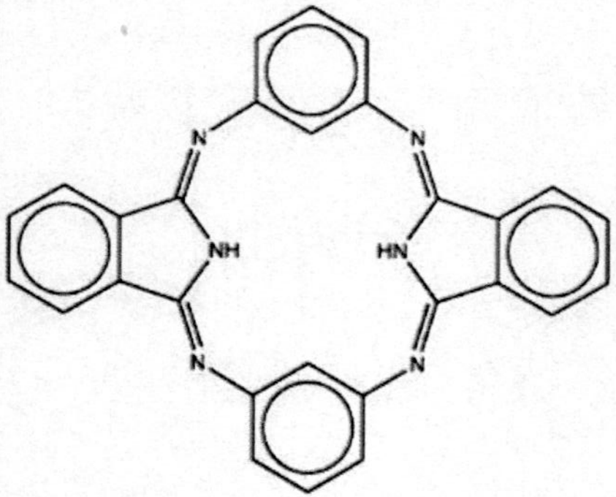	Dicarbahemiporphyrazine
87	$C_{28}H_{40}O_4Si$	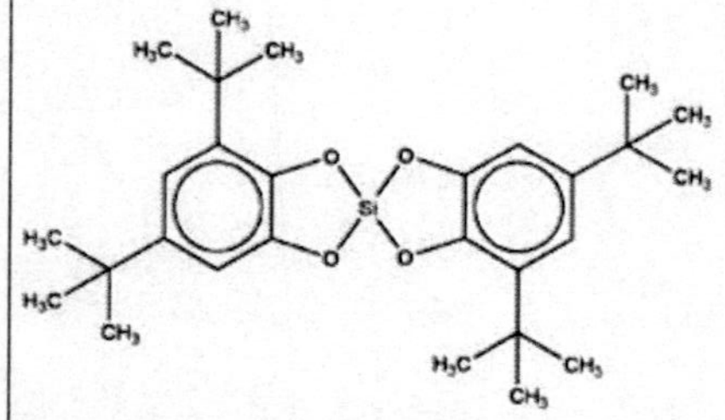	4,4′,6,6′-Tetra-*tert*-butyl-2,2′-spirobis[[1,3,2]benzodioxasilole]

(continued)

(continued)

Document no.	Molecular formula	Structural formula	Chemical name
88	$C_{60}F_{36}$	 (F atoms are omitted)	Hexatriacontafluorofullerene-C_{60}
89	ClSi		Chlorosilyliumylidene
90	$Cl_2O_5S_2$	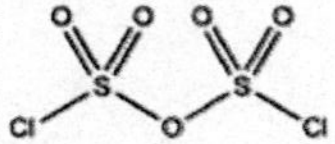	Disulfuryl dichloride
91	F_4Si	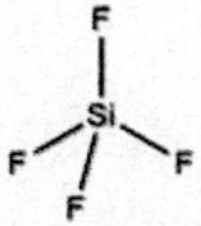	Tetrafluorosilane
92	F_5HOTe	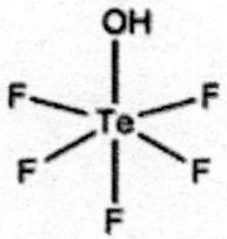	Tellurium fluoride hydroxide

(continued)

(continued)

Document no.	Molecular formula	Structural formula	Chemical name
93	F_6KrS	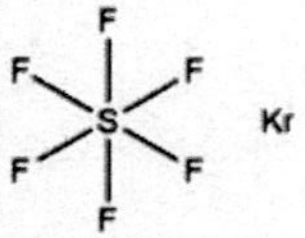	Sulfur hexafluoride – krypton (1/1)
94	F_6SXe	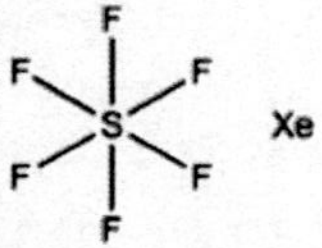	Sulfur hexafluoride – xenon (1/1)
95	GeH_4	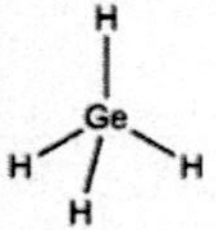	Germane
96	H_2O_2	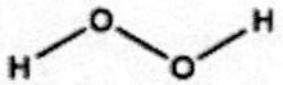	Hydrogen peroxide

(continued)

(continued)

Document no.	Molecular formula	Structural formula	Chemical name
97	H_2S_2	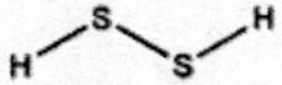	Disulfane
98	O_2S		Sulfur dioxide

References

1. Graner G, Hirota E, Iijima T, Kuchitsu K, Ramsay DA, Vogt J, Vogt N (1998) Structure data of free polyatomic molecules. In: Kuchitsu K (ed) Landolt-Börnstein, new series: molecules and radicals, vol II/25A. Springer-Verlag, Berlin, p 359
2. Graner G, Hirota E, Iijima T, Kuchitsu K, Ramsay DA, Vogt J, Vogt N (1999) Structure data of free polyatomic molecules. In: Kuchitsu K (ed) Landolt-Börnstein, new series: molecules and radicals, vol II/25B. Springer-Verlag, Berlin, p 512
3. Graner G, Hirota E, Iijima T, Kuchitsu K, Ramsay DA, Vogt J, Vogt N (2000) Structure data of free polyatomic molecules. In: Kuchitsu K (ed) Landolt-Börnstein, new series: molecules and radicals, vol II/25C. Springer-Verlag, Berlin, p 481
4. Graner G, Hirota E, Iijima T, Kuchitsu K, Ramsay DA, Vogt J, Vogt N (2003) Structure data of free polyatomic molecules. In: Kuchitsu K (ed) Landolt-Börnstein, new series: molecules and radicals, vol II/25D. Springer-Verlag, Berlin, p 569
5. Hirota E, Iijima T, Kuchitsu K, Ramsay DA, Vogt J, Vogt N (2006) Structure data of free polyatomic molecules. In: Kuchitsu K, Vogt N, Tanimoto M (eds) Landolt-Börnstein, new series: molecules and radicals, vol II/28A. Springer, Berlin, p 169
6. Hirota E, Iijima T, Kuchitsu K, Ramsay DA, Vogt J, Vogt N (2006) Structure data of free polyatomic molecules. In: Kuchitsu K, Vogt N, Tanimoto M (eds) Landolt-börnstein, new series: molecules and radicals, vol II/28B. Springer, Berlin, p 192
7. Hirota E, Iijima T, Kuchitsu K, Ramsay DA, Vogt J, Vogt N (2007) Structure data of free polyatomic molecules. In: Kuchitsu K, Vogt N, Tanimoto M (eds) Landolt-Börnstein, new series: molecules and radicals, vol II/28C. Springer, Berlin, p 187
8. Hirota E, Iijima T, Kuchitsu K, Ramsay DA, Vogt J, Vogt N (2007) Structure data of free polyatomic molecules. In: Kuchitsu K, Vogt N, Tanimoto M (eds) Landolt-Börnstein, new series: molecules and radicals, vol II/28D. Springer, Berlin, p 229
9. Hirota E, Kuchitsu K, Steimle T, Vogt J, Vogt N (2014) Structure data of free polyatomic molecules. In: Kuchitsu K, Vogt N, Tanimoto M (eds) Landolt-Börnstein, new series: molecules and radicals, vol II/30A. Springer, Berlin, p 414
10. Hirota E, Kuchitsu K, Steimle T, Vogt J, Vogt N (2014) Structure data of free polyatomic molecules. In: Kuchitsu K, Vogt N, Tanimoto M (eds) Landolt-Börnstein, new series: molecules and radicals, vol II/30B. Springer, Berlin, p 447
11. Vogt N, Vogt J (2019) Structure data of free polyatomic molecules. Springer Nature, Switzerland, p 926
12. Vogt J, Mez-Starck B, Vogt N, Hutter W (1999) MOGADOC—a database for gas-phase molecular spectroscopy and structure. J Mol Struct 485:249–254
13. Vogt J, Vogt N, Kramer R (2003) Visualization and substructure retrieval tools in the MOGADOC database (Molecular Gas-phase Documentation). J Chem Inf Comput Sci 43:357–361
14. Vogt J, Vogt N (2003) Statistical tools of the MOGADOC database. Struct Chem 14:137–141
15. Vogt N, Popov E, Rudert R, Kramer R, Vogt J (2010) 3D visualization of molecular structures in the MOGADOC database. J Mol Struct 978:201–204
16. Vogt J, Vogt N, Rudert R, Popov E, Schlagenhauf S, Deutzmann K, Kramer R (2015) New features in the MOGADOC database. Struct Chem 26:1725–1727
17. Demaison J, Vogt N (2020) Accurate structure determination of free molecules. Springer Nature, Switzerland, p 277
18. Hill EA (1900) On a system of indexing chemical literature, adopted by the classification division of the U.S. patent office. J Am Chem Soc 22:478–494
19. GausView, graphical user interface for GAUSSIAN program package. Retrieved from https:/ gaussian.com/gaussview6/
20. Mohr PJ, Taylor BN, Newell DB (2008) CODATA recommended values of the fundamental physical constants. J Phys Chem Ref Data 37:1187–1284

21. Asselin P, Turner AC, Bruel L, Brenner V, Gaveau M-A, Mons M (2018) Rovibrational laser jet-cooled spectroscopy of SF_6–rare gas complexes in the ν_3 region of SF_6. Phys Chem Chem Phys 20:28105–28113
22. Schwabedissen J, Li D, Reuter CG, Stammler H-G, Mitzel NW, Bernhardt E, Zeng X (2018) Conformation and structure of dichlorophosphoryl isocyanate in the gaseous and solid phases. Z Anorg Allg Chem 644:1415–1422
23. Romano RM, Betancourt AM, Della Védova CO, Zeng X, Beckers H, Willner H, Schwabedissen J, Mitzel NW (2018) Preparation and properties of chlorosulfuryl chloroformate, $ClC(O)OSO_2Cl$. Inorg Chem 57:14834–14842
24. Schwabedissen J, Trapp PC, Stammler H-G, Mitzel NW, Wu Z, Chu X, Zeng X (2020) Spectroscopic properties, conformation and structure of difluorothiophosphoryl isocyanate in the gaseous and solid phase. ChemistryOpen 9:913–920
25. Vishnevskiy YV, Blomeyer S, Reuter CG (2020) Low pressure gas electron diffraction: an experimental setup and case studies. Rev Sci Instrum 91:074104-1–074104-9
26. Mladenović M, Lewerenz M (2018) Comparison of spectroscopic strategies to determine molecular geometries and the impact of nuclear versus atomic masses: the example of HCO^+ and HOC^+. Mol Phys 116(23–24):3607–3620
27. Winterhoff G, Galleguillos Kempf SC, Jensen P, Bunker PR (2018) Empirical potential energy surface and bending angle probability densities for the electronic ground state of HCO^+. J Mol Spectrosc 354:71–82
28. Tarasov YI, Kochikov IV, Bazhanova ZG (2018) Internal rotation and equilibrium structure of the bromonitromethane molecule according to gas electron diffraction data and quantum chemical calculations. J Struct Chem 59(3):512–519
29. Reichel M, Krumm B, Vishnevskiy YV, Blomeyer S, Schwabedissen J, Stammler H-G, Karaghiosoff K, Mitzel NW (2019) Solid-state and gas-phase structures and energetic properties of the dangerous methyl and fluoromethyl nitrates. Angew Chem Int Ed 58:18557–18561
30. Demaison J, Vogt N, Ksenafontov DN (2020) Accuracy of semiexperimental equilibrium structures: sulfine as an example. J Mol Spectrosc 1206:127676-1–127676-7
31. Müller HSP, Maeda A, Thorwirth S, Lewen F, Schlemmer S, Medvedev IR, Winnewisser M, De Lucia FC, Herbst E (2019) Laboratory spectroscopic study of isotopic thioformaldehyde, H2CS, and determination of its equilibrium structure. Astron Astrophys 621:A143-1–A143-11
32. Demaison J, Vogt N, Jin Y, Saragi RT, Juanes M, Lesarri A (2021) How accurate is the determination of equilibrium structures for van der Waals complexes? The dimer $N_2O\ldots CO$ as an example. J Chem Phys 154:194302-1–194302-12
33. Kisiel Z, Pszczółkowski L, Białkowska-Jaworska E, Jaworski M, Uriarte I, Basterretxea FJ, Cocinero EJ (2018) Rotational spectroscopy update for the newly identified atmospheric ozone depleter CF_3CCl_3. J Mol Spectrosc 352:1–9
34. Gambi A, Charmet AP, Stoppa P, Tasinato N, Ceselin G, Barone V (2019) Molecular synthons for accurate structural determinations: the equilibrium geometry of 1-chloro-1-fluoroethene. Phys Chem Chem Phys 21:3615–3625
35. Lee KLK, Martin-Drumel M-A, Lattanzi V, McGuire BA, Caselli P, McCarthy MC (2019) Gas phase detection and rotational spectroscopy of ethynethiol, HCCSH. Mol Phys 117:1381–1391
36. Degli Esposti C, Melosso M, Bizzocchi L, Tamassia F, Dore L (2020) Determination of a semi-experimental equilibrium structure of 1-phosphapropyne from millimeter-wave spectroscopy of CH_3CP and CD_3CP. J Mol Struct 1203:127429-1–127429-9
37. Obenchain DA, Spada L, Alessandrini S, Rampino S, Herbers S, Tasinato N, Mendolicchio M, Kraus P, Gauss J, Puzzarini C, Grabow J-U, Barone V (2018) Unveiling the sulfur–sulfur bridge: accurate structural and energetic characterization of a homochalcogen intermolecular bond. Angew Chem Int Ed 57:15822–15826
38. Holub J, Vishnevskiy YV, Fanfrlík J, Mitzel NW, Tikhonov D, Schwabedissen J, McKee ML, Hnyk D (2020) Bromination mechanism of *closo*-1,2–$C_2B_{10}H_{12}$ and the structure of the resulting 9-Br-*closo*-1,2–$C_2B_{10}H_{11}$ determined by gas electron diffraction. ChemPlusChem 85:2606–2610

39. Baše T, Holub J, Fanfrlík J, Hnyk D, Lane PD, Wann DA, Vishnevskiy YV, Tikhonov D, Reuter CG, Mitzel NW (2019) Icosahedral carbaboranes with peripheral hydrogen–chalcogenide groups: structures from gas electron diffraction and chemical shielding in solution. Chem Eur J 25:2313–2321
40. Belyakov AV, Losev VA, Rykov AN, Shishkov IF, Kuznetsov VV, Khakhalev AV, Sheremetev AB (2022) Combined gas-phase electron diffraction and coupled cluster determination of the molecular structure of 3,4-dinitrofurazan—a propellant ingredient. J Mol Struct 1250:131669-1–131669-8
41. Breier AA, Giesen TF, Ross SC, Yamada KMT (2020) Improved bond length determination technique for C_3 and other linear molecules with a large amplitude bending vibration. J Mol Struct 1219:128329-1–128329-11
42. Sun W, Davis RL, Thorwirth S, Harding ME, van Wijngaarden J (2018) A highly flexible molecule: the peculiar case of ethynyl isothiocyanate HCCNCS. J Chem Phys 149:104304-1–104304-7
43. Schwabedissen J, Glodde T, Vishnevskiy YV, Stammler H-G, Flierl L, Kornath AJ, Mitzel NW (2020) Structures and properties of *trans*-1,3,3,3-tetrafluoropropene (HFO-1234ze) and 2,3,3,3-tetrafluoropropene (HFO-1234yf) refrigerants. ChemistryOpen 9:921–928
44. Müller HSP, Brahmi AM, Guillemin J-C, Lewen F, Schlemmer S (2021) Rotational spectroscopy of isotopic cyclopropenone, *c*-H_2C_3O, and determination of its equilibrium structure Astron Astrophys 647:A179-1–A179-10
45. Thorwirth S, Harding ME, Dudek JB, McCarthy MC (2018) Equilibrium molecular structures of vinyl carbon chains: vinyl acetylene, vinyl diacetylene, and vinyl cyanide. J Mol Struct 350:10–17
46. Melli A, Tonolo F, Barone V, Puzzarini C (2021) Extending the applicability of the semi-experimental approach by means of "Template Molecule" and "Linear Regression" models on top of DFT computations J Phys Chem A 125:9904–9916
47. Esselman BJ, Zdanovskaia MA, Owen AN, Stanton JF, Woods RC, McMahon RJ (2021) Precise equilibrium structure of thiazole (*c*-C_3H_3NS) from twenty-four isotopologues. J Chem Phys 155:054302-1–054302-15
48. Belyakov AV, Oskorbin AA, Losev VA, Rykov AN, Shishkov IF, Fershtat LL, Larin AA, Kuznetsov VV , Makhova NN (2020) The equilibrium molecular structure of 3-methyl-4-nitro- and 4-methyl-3-nitrofuroxans by gas-phase electron diffraction and coupled cluster calculations. J Mol Struct 1222:128856-1–128856-9
49. Müller HSP, Thorwirth S, Lewen F (2020) Rotational spectroscopy of singly ^{13}C substituted isotopomers of propyne and determination of a semi-empirical equilibrium structure. J Mol Struct 1207:127769-1–127769-8
50. Rankine CD, Atkinson SJ, Waterland MR, Masters SL, Wann DA (2018) The structure of tris(chloromethyl)amine in the gas phase using quantum chemical calculations and gas electron diffraction and as a solid and melt using Raman spectroscopy. Struct Chem 29:803–813
51. Tarasov YI, Kochikov IV (2018) Pseudo-conformer models for linear molecules: joint treatment of spectroscopic, electron diffraction and *ab initio* data for the C_3O_2 molecule. J Mol Struct 1162:117–124
52. Vogt N, Ksenafontov DN, Savelev D, Rykov AN (2020) Molecular structure of 5-fluorouracil from gas-phase electron diffraction data and quantum-chemical calculations. Mendeleev Comm 30:660–662
53. Belyakov AV, Kulishenko RY, Oskorbin AA (2019) Semiexperimental equilibrium molecular structures of the maleimide and phthalimide. Mol Phys 117:1097–1103
54. Heim ZN, Amberger BK, Esselman BJ, Stanton JF, Woods RC, McMahon RJ (2020) Molecular structure determination: equilibrium structure of pyrimidine (*m*-$C_4H_4N_2$) from rotational spectroscopy (r_e^{SE}) and high-level ab initio calculation (r_e) agree within the uncertainty of experimental measurement. J Chem Phys 152:104303-1–104303-12
55. Jahn MK, Obenchain DA, Nair KPR, Grabow J-U, Vogt N, Demaison J, Godfrey PD, McNaughton D (2020) The puzzling hyper-fine structure and an accurate equilibrium geometry of succinic anhydride. Phys Chem Chem Phys 22:5170–5177

56. Dindić C, Ludovicy J, Terzi V, Lüchow A, Vogt N, Demaison J, Nguyen HVL (2022) Determination of the semiexperimental equilibrium structure of 2-acetylthiophene in the presence of methyl internal rotation and substituent effects compared to thiophene. Phys Chem Chem Phys 24:3804–3815
57. Baraban JH, Martin-Drumel M-A, Changala PB, Eibenberger S, Nava M, Patterson D, Stanton JF, Ellison GB, McCarthy MC (2018) The molecular structure of *gauche*-1,3-butadiene: experimental establishment of non-planarity. Angew Chem Int Ed 57:1821–1825
58. Herbers S, Kraus P, Grabow J-U (2019) Accurate equilibrium structures of methyl methacrylate and methacrylic acid by microwave spectroscopy and dispersion corrected calculations. J Chem Phys 150:144308-1–144308-10
59. Insausti A, Alonso ER, Tercero B, Santos JI, Calabrese C, Vogt N, Corzana F, Demaison J, Cernicharo J, Cocinero EJ (2021) Laboratory observation of, astrochemical search for, and structure of elusive erythrulose in the interstellar medium. J Phys Chem Lett 12:1352–1359
60. Schwabedissen J, Trapp PC, Stammler H-G, Neumann B, Lamm J-H, Vishnevskiy YV, Körte LA, Mitzel NW (2019) Halogen bonds of halotetrafluoropyridines in crystals and co-crystals with benzene and pyridine. Chem Eur J 25:7339–7350
61. Altova EP, Rykov AN, Vogt N, Shishkov IF (2021) Equilibrium molecular structure of orotic acid from gas-phase electron diffraction data and quantum-chemical calculations. Mendeleev Comm 31:81–83
62. Hakiri R, Derbel N, Bailey WC, Nguyen HVL, Mouhib H (2020) The heavy atom structures and ^{33}S quadrupole coupling constants of 2-thiophenecarboxaldehyde: insights from microwave spectroscopy. Mol Phys 118:1728406-1–1728406-13
63. Otlyotov AA, Girichev GV, Rykov AN, Glodde T, Vishnevskiy YV (2020) Molecular structure of pyrazinamide: a critical assessment of modern gas electron diffraction data from three laboratories. J Phys Chem A 124:5204–5211
64. Khaikin LS, Ageev GG, Rykov AN, Grikina OE, Shishkov IF, Kochikov IV, Kuznetsov VV, Makhova NN, Bukalov SS, Leites LA (2020) Equilibrium molecular structure and spectra of 6-methyl-1,5-diazabicyclo[3.1.0]hexane: joint analysis of gas phase electron diffraction, quantum chemistry, and spectroscopic data. Phys Chem Chem Phys 22:22477–22492
65. Schaab J, Schwab M, Kratzert D, Schwabedissen J, Stammler H-G, Mitzel NW, Krossing I (2018) The perfluorinated alcohols c-$C_6F_{11}OH$, c-C_6F_{10}-1,1-$(OH)_2$ and c-C_6F_{10}-1-$(CF_3)OH$. Chem Commun 54:9294–9297
66. Kisiel Z, Białkowska-Jaworska E, Chen J, Pszczółkowski L, Gawryś P, Kosarzewski J (2020) Rotational spectroscopy and precise molecular structure of 1,2-dichlorobenzene. J Mol Spectrosc 374:111380-1–111380-10
67. Vogt N, Khaikin LS, Rykov AN, Grikina OE, Batiukov AA, Vogt J, Kochikov IV, Shishkov IF (2019) The equilibrium molecular structure of 2-cyanopyridine from combined analysis of gas-phase electron diffraction and microwave data and results of ab initio calculations. Struct Chem 30:1699–1706
68. Khaikin LS, Vogt N, Rykov AN, Grikina OE, Vogt J, Kochikov IV, Ageeva ES, Shishkov IF (2018) The equilibrium molecular structure of 3-cyanopyridine according to gas-phase electron diffraction and microwave data and the results of quantum-chemical calculations. Mendeleev Comm 28:236–238
69. Khaikin LS, Vogt N, Rykov AN, Grikina OE, Demaison J, Vogt J, Kochikov IV, Shishova YD, Ageeva ES, Shishkov IF (2018) The equilibrium molecular structure of 4-cyanopyridine according to a combined analysis of gas-phase electron diffraction and microwave data and coupled-cluster computations. Russ J Phys Chem A 92:1970–1974
70. Vogt N, Nair KPR, Grabow J-U, Demaison J (2018) Microwave rotational spectrum and ab initio computations on 4-cyanopyridine: molecular structure and hyperfine interactions. Mol Phys 116:3530–3537
71. Vogt N, Marochkin II, Rykov AN (2018) Experiment and theory at the convergence limit: accurate equilibrium structure of picolinic acid by gas-phase electron diffraction and coupled-cluster computations. Phys Chem Chem Phys 20:9787–9795

72. Dindić C, Lüchow A, Vogt N, Demaison J, Nguyen HVL (2021) Equilibrium structure in the presence of methyl internal rotation: microwave spectroscopy and quantum chemistry study of the two conformers of 2-acetylfuran. J Phys Chem A 125:4986–4997
73. Nguyen HVL (2020) The heavy atom substitution and semi-experimental equilibrium structures of 2-ethylfuran obtained by microwave spectroscopy. J Mol Struct 1208:127909-1–127909-8
74. Demaison J, Vogt N, Saragi RT, Juanes M, Rudolph HD, Lesarri A (2019) How flexible is the disulfide linker? A combined rotational–computational investigation of diallyl disulfide. Phys Chem Chem Phys 21:19732–19736
75. Glodde T, Vishnevskiy YV, Zimmermann L, Stammler H-G, Neumann B, Mitzel NW (2021) The nature of chalcogen-bonding-type tellurium–nitrogen interactions: a first experimental structure from the gas phase. Angew Chem Int Ed 60:1519–1523
76. Uriarte I, Reviriego F, Calabrese C, Elguero J, Kisiel Z, Alkorta I, Cocinero EJ (2019) Bond length alternation observed experimentally: the case of 1*H*-indazole. Chem Eur J 25(43):10172–10178
77. Kolesnikova IN, Rykov AN, Shuvalov MV, Shishkov IF (2019) Internal rotation and intramolecular hydrogen bonding in thiosalicylamide: gas phase electron diffraction study supported by quantum chemical calculations. Struct Chem 30:1993–2001
78. Kolesnikova IN, Putkov AE, Rykov AN, Shishkov IF (2018) Thiobenzamide: structure of a free molecule as studied by gas electron diffraction and quantum chemical calculations. J Mol Struct 1161:76–82
79. Belyakov AV, Vogt N, Demaison J, Kulishenko RY, Oskorbin AA (2022) Semiexperimental equilibrium molecular structure of phthalic anhydride. Chem Phys Lett 795:139540-1–139540-5
80. Vogt N, Savelev D, Giricheva NI, Girichev GV (2020) Accurate equilibrium structure of 3-aminophthalimide from gas electron diffraction and coupled-cluster computations and diverse structural effects due to electron density transfer. Phys Chem Chem Phys 22:27539–27546
81. Vogt N, Demaison J, Rudolph H-D, Juanes M, Fernández J, Lesarri A (2018) Semiexperimental and mass-dependent structures by the mixed regression method: accurate equilibrium structure and failure of the Kraitchman method for ethynylcyclohexane. J Chem Phys 148:064306-1–064306-11
82. Khaikin LS, Kochikov IV, Rykov AN, Grikina OE, Ageev GG, Shishkov IF, Kuznetsov VV, Makhova NN (2019) Equilibrium structures of the tetramezine diastereomers and their ratio: joint analysis of gas phase electron diffraction, quantum chemistry, and spectroscopic data. Phys Chem Chem Phys 21:5598–5613
83. Tikhonov DS, Sharapa DI, Otlyotov AA, Solyankin PM, Rykov AN, Shkurinov AP, Grikina OE, Khaikin LS (2018) Nitroxoline molecule: planar or not? A story of battle between π–π conjugation and interatomic repulsion. J Phys Chem A 122(6):1691–1701
84. Belyakov AV, Nikolaenko KO, Oskorbin AA, Vogt N, Rykov AN, Shishkov IF (2019) Semiexperimental equilibrium structure of 1-methylisatin from gas-phase electron diffraction data and structural changes in isatin due to 1-methyl and 5-fluoro substituents as predicted by coupled cluster computations. Mol Phys 117(14):1850–1857
85. Kinder TA, Blomeyer S, Franke M, Depenbrock F, Neumann B, Stammler H-G, Mitzel NW (2019) Small neutral geminal silicon/phosphorus frustrated Lewis pairs. Eur J Inorg Chem 2019(35):3933–3939
86. Berger RJF, Girichev GV, Giricheva NI, Otlyotov AA, Petrova AA (2019) Ligand coordination in bis(β-diketonato) d metals: the Mn(II) case of D_{2h} versus D_{2d} symmetry. Inorg Chem 58(7):4344–4349
87. Hartmann D, Thorwart T, Müller R, Thusek J, Schwabedissen J, Mix A, Lamm J-H, Neumann B, Mitzel NW, Greb L (2021) The structure of bis(catecholato)silanes: phase adaptation by dynamic covalent chemistry of the Si–O bond. J Am Chem Soc 143(44):18784–18793
88. Demaison J, Vogt N, Saragi RT, Juanes M, Rudolph HD, Lesarri A (2019) The S-S bridge: a mixed experimental-computational estimation of the equilibrium structure of diphenyl disulfide. ChemPhysChem 20(3):366–373

89. Weddeling J-H, Vishnevskiy YV, Neumann B, Stammler H-G, Mitzel NW (2020) Inter- and intramolecular aryl–aryl interactions in partially fluorinated ethylenedioxy-bridged bisarenes. Chem Eur J 26(68):16111–16121
90. Otlyotov AA, Zhabanov YA, Pogonin AE, Kuznetsova AS, Islyaikin MK, Girichev GV (2019) Gas-phase structures of hemiporphyrazine and dicarbahemiporphyrazine: key role of interactions inside coordination cavity. J Mol Struct 1184:576–582
91. Belyakov AV, Kulishenko RY, Johnson RD, Shishkov IF, Rykov AN, Markov VY, Khinevich VE, Goryunkov AA (2020) Structure of $C_{60}F_{36}$: a gas-phase electron diffraction and quantum chemical computational study of a remarkably distorted fluorofullerene. J Phys Chem A 124(49):10216–10224
92. Tanaka K, Harada K, Cabezas C, Endo Y (2018) Fourier transform microwave spectroscopy of the $SiCl^+$ ion. J Mol Struct 345:39–45
93. Betancourt AM, Schwabedissen J, Romano RM, Della Védova CO, Beckers H, Willner H, Stammler H-G, Mitzel NW (2018) Disulfuryl dichloride $ClSO_2OSO_2Cl$: a conformation and polymorphism chameleon. Chem Eur J 24(41):10409–10421
94. Boudon V, Richard C, Manceron L (2022) High-resolution spectroscopy and analysis of the fundamental modes of $^{28}SiF_4$. Accurate experimental determination of the Si–F bond length. J Mol Spectrosc 383:111549-1–111549-14
95. Herbers S, Obenchain DA, Kraus P, Wachsmuth D, Grabow J-U (2018) Blurring out hydrogen: the dynamical structure of teflic acid. J Chem Phys 148:194307-1–194307-5
96. Richard C, Boudon V, Rizopoulos A, Auwera JV, Tchana FK (2020) Line positions and intensities for the ν_2/ν_4 bands of 5 isotopologues of germane near 11.5 μm. J Quant Spectr Rad Transf 260:107474-1–107474-14
97. Baraban JH, Changala PB, Stanton JF (2018) The equilibrium structure of hydrogen peroxide. J Mol Spectrosc 343:92–95
98. Ye H, Mendolicchio M, Kruse H, Puzzarini C, Biczysko M, Barone V (2020) The challenging equilibrium structure of HSSH: another success of the rotational spectroscopy/quantum chemistry synergism. J Mol Struct 1211:127933-1–127933-10
99. Demaison J, Liévin J (2021) Accuracy of the equilibrium structure of sulphur dioxide. Mol Phys https://doi.org/10.1080/00268976.2021.1950857